# Differential Equations and Their Applications

## Third Edition

**ZAFAR AHSAN**

*Department of Mathematics*
*Aligarh Muslim University*
*Aligarh (UP)*

**PHI Learning Private Limited**
Delhi-110092
2017

₹ 525.00

**DIFFERENTIAL EQUATIONS AND THEIR APPLICATIONS, Third Edition**
Zafar Ahsan



**ISBN-978-81-203-5269-8**



**Sixteenth Printing (Third Edition) ... ... May, 2017**

Published by Asoke K. Ghosh, PHI Learning Private Limited, Rimjhim House, 111, Patparganj Industrial Estate, Delhi-110092 and Printed by Syndicate Binders, A-20, Hosiery Complex, Noida, Phase-II Extension, Noida-201305 (N.C.R. Delhi).

*To*

***My Parents***

# Contents

## 10. Calculus of Variations and Its Applications 543–602

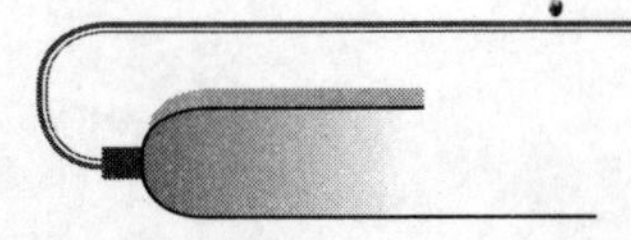

# Preface

I very much appreciate the favourable responses to the second edition, and hope that the third edition may prove more useful. This new edition is the result of thorough revision of the second edition, keeping the spirit of the first edition. Errors (typographical or others) that I have found, or pointed out to me by colleagues and students, have of course been corrected and it is hoped that this new edition may be free from errors. A few additions have been made here, which are in accordance with the new choice-based credit system of University Grants Commission for all Indian universities. New sections have been added in different chapters. In Chapters 5, 7, 8, 9 and 10 a separate section, consisting of number of solved examples, has been added. In Section 9.7, some more methods for finding the complimentary function and particular integrals have been included, and two more sections, namely Section 9.8.1, consisting of equations reducible to linear partial differential equations, and Section 9.9, consisting of general method for solving the second order non-linear partial differential equations (Monge's method) have been added, while the section on Fourier series is almost rewritten. In Chapter 10, a section on Lagrange's equations of motion has been included as an application of calculus of variations.

I am thankful to my publisher, PHI Learning, in particular, the editorial and production teams for their excellent efforts in bringing out the new edition in a nice way. Finally, I wish to thank my family members for encouragement and patience they have rendered to me during the preparation of this new edition. Any suggestion for further improvement shall be welcomed.

**Zafar Ahsan**

# Preface to the Second Edition

The first edition of *Differential Equations and Their Applications* has been favourabley received by a large number of users. The second edition reflects the suggestions and experiences of these users to whom I am extremely thankful.

The spirit of this new edition is same as that of the previous one. As far as possible, the efforts have been made to keep this new edition free from typographic and other errors. Few additions have been made, which are in accordance wih the UGC curriculum (2001) in mathematics for all Indian universities. In Chapter 5, the series solutions of a differential equation has been included and the method of Frobenius for solving a differential equation has been discussed in detail with the help of a number of examples. The series solutions for some of the differential equations of special functions, namely, Bessel, Legendre and hypergeometric equations have also been derived in this chapter. The major change of this edition is the inclusion of calculus of variations as Chapter 10. The calculus of variations has a number of applications in any field of study, where optimization is needed, e.g., the path of a guided missile, economic growth, pest control, spread of a contagious disease, cancer chemotherapy and immune system, etc. This chapter deals with the methods of finding the extremals of a given functional and thus leads to the solution of differential equations. Some of the applications of the calculus of variations have also been given.

For a successful publication of the book many other persons are involved, I wish to thank all of them personally: my friend Professor J.L. Lopez Bonilla, Instituto Politecnico Nacional, Mexico, for fruitful suggestions and pointing out a few errors in the first edition; my friend and colleague Professor Mursaleen, for critically reading the contents of Chapter 10 and correcting a number of errors (these fellows are, of course, not responsible for any error that remains) and finally my publisher, PHI Learning, in particular, the editorial and production teams for their nice efforts to bring out the new edition so quickly. Lastly, I wish to thank my family members for lots of encouragement and patience they have shown during the preparation of the material for this new edition.

Any suggestions for further improvement shall be welcomed.

**Zafar Ahsan**

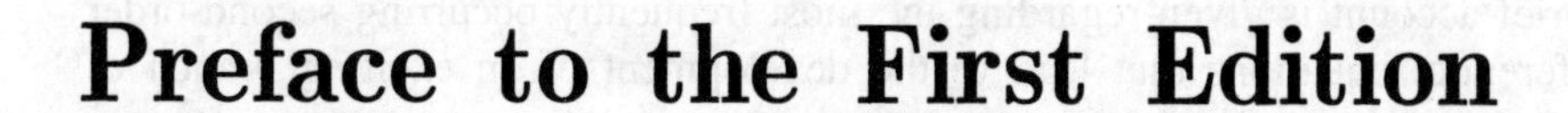

# Preface to the First Edition

It is an incontrovertible fact that differential equations form the most important branch of modern mathematics and, in fact, occupy the position at the centre stage of both pure and applied mathematics. This is obvious because the mathematical understanding of any physical situation usually consists of the following steps:

1. Understanding the various parameters of the situation and then making a rough mathematical model;
2. Posing a corresponding precise mathematical problem, and analyzing it, trying to find an exact or approximate solution;
3. Comparing the result with the experimental data to check the validity of the model.

Step 2 is nothing but forming the corresponding differential equations and then evolving techniques of pure mathematics to arrive at the solution.

It is also obvious that differential equations form the basis of applied mathematics. As far as their role in pure mathematics is concerned, attempts to get their exact solutions lead us to the Existence Theorems and the Theory of Functions, Differential Geometry, and deep results in Functional Analysis.

It is impossible to describe all the roles played by differential equations in pure and applied mathematics in a single book. However, the author's objective in writing this book is two-fold:

1. To provide the reader with an easier and systematic way of solving ordinary and partial differential equations;
2. To find the possible applications of differential equation in such diverse areas as biology, physiology medicine and economics, along with the applications in physical and engineering sciences.

In Chapter 1, a brief introduction to the definitions and terminology is presented. This is followed by a discussion about how differential equations arise naturally from geometrical and physical points of view.

Chapter 2 discusses about the different methods for solving differential equations of first order and first degree. Chapter 3 covers differential equations of first order but not of first degree.

Chapter 4 includes a wide variety of problems which are chosen from different disciplines. The applications of first-order differential equations to carbon dating and radioactivity, mixture problem, estimation of time of death, absorption of drugs in cell, the problem of epidemiology, the motion of a rocket, the path of a guided missile, electric circuits, chemical reactions, are just a few of the attractions of this chapter.

In Chapter 5, the higher-order linear differential equations, the concepts of solutions, and the methods of obtaining these solutions are analyzed. Also, a brief account is given regarding the most frequently occurring second-order differential equations that lead to the development of an exciting branch of mathematics, viz. Special Functions.

Chapter 6 covers the applications of higher-order differential equations in rectilinear motion, resonance, the hanging cable, civil and electrical engineering, economics, cardiology and detection of diabetes.

Chapter 7 deals with simultaneous differential equations and their applications. It first focusses on the basic definition and the methods of solution and then on their applications.

The Laplace transforms and their applications constitute Chapter 8. The chapter gives the solution of the tautochrone problem and analyzes the theory of automatic control and servomechanics, along with some other applications of transform methods for solving differential equations.

Chapter 9 provides an elementary treatment of partial differential equations and their applications. The applications covered include vibration of string and membranes, heat conduction equation, transmission lines and nuclear fission.

The emphasis throughout the text is on solving problems. At the end of each chapter, a carefully selected set of problems is given. The answers to each problem set are provided at the end of the book. More than 330 problems have been solved completely while the number of unsolved problems is 480.

This book can be used as a textbook for the undergraduate and postgraduate students of science and is also suitable for engineering students of various universities.

In preparing the book, I have consulted many standard works. I am indebted to the authors of those works. I wish to express my gratitude to (Late) Prof. S. Izhar Husain whose excellent training has stimulated and enlivened my interest in mathematics and inspired me to write this book.

I also wish to thank my colleague and friend, Dr. M.M.R. Khan, whose suggestions and comments proved fruitful and clarified many points. My thanks are also due to Mr. S. Fazal Hasnain Naqvi for the excellent typing of the manuscript. The partial financial assistance provided by Aligarh Muslim University for the preparation of the manuscript is gratefully acknowledged. Finally, I wish to express my sincere thanks to the publisher, PHI Learning, in particular, the editorial and production teams, for their meticulous processing of the manuscript.

Any suggestions or comments for improving the contents will be warmly appreciated.

**Zafar Ahsan**

CHAPTER 1

# Basic Concepts

## 1.1 INTRODUCTION

In this text, we deal with differential equations and their applications. It is well known that differential equations are very useful to students of applied sciences. It may be worthwhile to list the various differential equations which have arisen in the different fields of engineering and the sciences. Such listing is intended to indicate to students that differential equations can be applied to many practical fields, even though it must be emphasized that the subject is of great interest in itself. The differential equations have been compiled from those occurring in advanced textbooks and research journals. Some of these equations are as follows:

$$xy'' + y' + xy = 0 \tag{1}$$

$$\frac{d^2x}{dt^2} = -kx \tag{2}$$

$$\frac{d^2I}{dt^2} + 5\frac{dI}{dt} + 8I = 100 \sin 20t \tag{3}$$

$$EIy^{\text{iv}} = w(x) \tag{4}$$

$$y'' = \frac{W}{H}\sqrt{1 + y'^2} \tag{5}$$

$$v + m\frac{dv}{dm} = v^2 \tag{6}$$

$$\frac{\partial^2 V}{\partial x^2} + \frac{\partial^2 V}{\partial y^2} + \frac{\partial^2 V}{\partial z^2} = 0 \tag{7}$$

$$\frac{\partial V}{\partial t} = k\left(\frac{\partial^2 V}{\partial x^2} + \frac{\partial^2 V}{\partial y^2}\right) \tag{8}$$

$$\frac{\partial^2 V}{\partial t^2} = a^2 \frac{\partial^2 V}{\partial x^2} \tag{9}$$

$$\frac{\partial^4 \phi}{\partial x^4} + 2\frac{\partial^4 \phi}{\partial x^2 \partial y^2} + \frac{\partial^4 \phi}{\partial y^4} = F(x, y) \tag{10}$$

Equation (1) arises in the field of mechanics, heat, electricity, aerodynamics, stress analysis, and so on.

Equation (2) has a wide range of application in the field of mechanics, in relation to simple harmonic motion, as in small oscillations of a simple pendulum. It could, however, arise in many other allied areas.

Equation (3) determines the current $I$ as a function of time $t$ in an alternating current circuit.

Equation (4) is an important equation in civil engineering in the theory of bending and deflection of beams.

Equation (5) arises in problems concerning suspension cables.

Equation (6) occurs in problems on rocket flight.

Equation (7) is the famous Laplace's equation which occurs in heat, electricity, aerodynamics, potential theory, gravitation and many other fields.

Equation (8) is found in the theory of heat conduction as well as in the diffusion of neutrons in an atomic pile for the production of nuclear energy. It also occurs in the study of Brownian motion.

Equation (9) is used in connection with the vibration of strings, bars, membranes, as well as in the propagation of electric signals and nuclear reactors.

Equation (10) is widely used in the theory of stress analysis (in the theory of slow motion of viscous fluid and the theory of an elastic body).

These are a few of the many equations which could occur and a few of the fields from which they are taken. Studies of differential equations such as these by pure mathematicians, applied mathematicians, theoretical and applied physicists, chemists, engineers and other scientists throughout the years have led to the conclusion that there are certain definite methods by which many of these equations can be solved. The history of these discoveries is, in itself, extremely interesting. However, there are many unsolved equations; some of them are of great importance. The use of modern giant calculating machines, in determining the solution of such equations which are vital for research involving national defence as well as many other endeavours, is still in progress.

It is one of the aims of this book to provide an introduction to some of the important real life problems appearing in many areas of science with which most of the researchers should be acquainted.

## 1.2 DEFINITION AND TERMINOLOGY

### 1.2.1 Differential Equations

An equation involving independent and dependent variables and the derivatives or differentials of one or more dependent variables with respect to one or more independent variables is called a differential equation.

Apart from Eqs. (1)–(10), the following relations are also some of the examples of differential equations:

$$\frac{dy}{dx} = \sin x + \cos x \tag{11}$$

$$y = \sqrt{x}\,\frac{dy}{dx} + \frac{k}{\frac{dy}{dx}} \tag{12}$$

$$k\frac{d^2 y}{dx^2} = \left[1 + \left(\frac{dy}{dx}\right)^2\right]^{3/2} \tag{13}$$

$$y = x\frac{dy}{dx} + k\left[\sqrt{1 + \left(\frac{dy}{dx}\right)^2}\right] \tag{14}$$

$$\frac{d^3 x}{dt^3} + \frac{d^2 x}{dt^2} + \left(\frac{dx}{dt}\right)^4 = e^t \tag{15}$$

$$\frac{\partial^3 v}{\partial t^3} = k\left(\frac{\partial^2 v}{\partial x^2}\right)^2 \tag{16}$$

There are two main classes of differential equations:

(i) Ordinary differential equations.
(ii) Partial differential equations.

**Ordinary differential equations.** A differential equation which involves derivatives with respect to a single independent variable is known as an *ordinary differential equation.*

Equations (1)–(6) and (11)–(15) are examples of ordinary differential equations.

**Partial differential equations.** A differential equation which contains two or more independent variables and partial derivatives with respect to them is called a *partial differential equation.* Equations (7)–(10) and (16) are examples of partial differential equations.

### 1.2.2 Order of Differential Equations

The order of the highest order derivative involved in a differential equation is called the *order* of a differential equation.

Equations (6), (11), (12) and (14) are of the first order; Eqs. (1)–(3), (5), (7)–(9) and (13) are of the second order; while Eqs. (15) and (16) and Eqs. (4) and (10) are equations of the third and fourth orders, respectively.

### 1.2.3 Degree of a Differential Equation

The *degree* of a differential equation is the degree of the highest order derivative present in the equation, after the differential equation has been made free from the radicals and fractions as far as the derivatives are concerned.

Equations (1)–(11), except Eq. (5), are of degree one. Equations (5) and (12)–(14) can respectively be written as

$$(y'')^2 = \frac{W^2}{H^2}(1+y'^2)$$

$$y\frac{dy}{dx} = \sqrt{x}\left(\frac{dy}{dx}\right)^2 + k$$

$$k^2\left(\frac{d^2y}{dx^2}\right)^2 = \left[1+\left(\frac{dy}{dx}\right)^2\right]^3$$

$$\left(y - x\frac{dy}{dx}\right)^2 = k^2\left[1+\left(\frac{dy}{dx}\right)^2\right]$$

The above equations are of the second degree.

## 1.3 LINEAR AND NONLINEAR DIFFERENTIAL EQUATIONS

A differential equation in which the dependent variables and all its derivatives present occur in the first degree only and no products of dependent variables and/or derivatives occur is known as a *linear differential equation.* A differential equation which is not linear is called a *nonlinear differential equation.* Thus, Eq. (11) is a linear equation of order one and Eqs. (2) and (7) are linear equations of order two. Equations (12)–(16) are nonlinear equations.

## 1.4 SOLUTION OF A DIFFERENTIAL EQUATION

A *solution* of a differential equation is a relation between the dependent and independent variables, not involving the derivatives such that this relation and the derivatives obtained from it satisfies the given differential equation. For example, $y = ce^{2x}$ is a solution of the differential equation $dy/dx - 2y = 0$, because $dy/dx = 2ce^{2x}$ and $y = ce^{2x}$ satisfy the given differential equation.

## 1.5 ORIGINS AND FORMATION OF DIFFERENTIAL EQUATIONS

In the discussion that follows we shall see how specific differential equations arise not only out of consideration of families of geometric curves, but also how differential equations result from an attempt to describe, in mathematical terms, physical problems in science and engineering. It would not be too presumptive to state that differential equations form the basis of subjects such as physics and electrical engineering, and even provide an important working tool in such diverse fields as biology, physiology, medicine, statistics, sociology, psychology and economics. Both theoretical and applied differential equations are active fields of current research. Several of the examples and problems in this section will serve as previews of topics discussed in Chapters 4 and 7–9.

### 1.5.1 Differential Equation of a Family of Curves

Suppose we are given an equation containing $n$ arbitrary constants. Then by differentiating it successively $n$ times we get $n$ equations more containing $n$ arbitrary constants and derivatives. Now by eliminating $n$ arbitrary constants from the above $(n + 1)$ equations and obtaining an equation which involves derivatives upto the $n$th order, we get a differential equation of order $n$. We now work out in detail some examples to illustrate the method of forming differential equations.

**EXAMPLE 1.1** Find the differential equation of the family of curves $y = c_1e^{2x} + c_2e^{-2x}$, where $c_1$ and $c_2$ are arbitrary constants.

***Solution*** Given

$$y = c_1e^{2x} + c_2e^{-2x} \tag{17}$$

Differentiating Eq. (17) twice with respect to $x$, we get

$$\frac{dy}{dx} = 2c_1e^{2x} - 2c_2e^{-2x} \tag{18}$$

$$\frac{d^2y}{dx^2} = 4c_1e^{2x} + 4c_2e^{-2x} = 4(c_1e^{2x} + c_2e^{-2x}) \tag{19}$$

From Eqs. (17) and (19), we obtain

$$\frac{d^2y}{dx^2} - 4y = 0 \tag{20}$$

Thus, the two arbitrary constants $c_1$ and $c_2$ have been eliminated from Eqs. (17)–(19). Hence Eq. (20) is the required differential equation of the family of curves given by Eq. (17).

**EXAMPLE 1.2** Find the differential equation corresponding to the family of curves $y = c(x - c)^2$, where $c$ is an arbitrary constant.

***Solution*** Given

$$y = c(x - c)^2 \tag{21}$$

Differentiating both sides with respect to $x$, we get

$$\frac{dy}{dx} = 2c(x - c) \tag{22}$$

or

$$\left(\frac{dy}{dx}\right)^2 = 4c^2(x - c)^2 \tag{23}$$

Dividing Eq. (23) by (21), we obtain

$$\frac{1}{y}\left(\frac{dy}{dx}\right)^2 = 4c \quad \text{or} \quad c = \frac{1}{4y}\left(\frac{dy}{dx}\right)^2$$

Substituting this value of $c$ in Eq. (22), we get

$$\frac{dy}{dx} = 2\cdot\frac{1}{4y}\left(\frac{dy}{dx}\right)^2\left[x - \frac{1}{4y}\left(\frac{dy}{dx}\right)^2\right]$$

or

$$2y = \frac{dy}{dx}\left[x - \frac{1}{4y}\left(\frac{dy}{dx}\right)^2\right]$$

or

$$8y^2 = 4xy\frac{dy}{dx} - \left(\frac{dy}{dx}\right)^3$$

which is the required differential equation for the family of curves (21).

**EXAMPLE 1.3** Find the differential equation that describes the family of circles passing through the origin.

***Solution*** The general form of the circles passing through the origin is

$$(x - h)^2 + (y - k)^2 = \left(\sqrt{h^2 + k^2}\right)^2$$

or

$$x^2 - 2xh + y^2 - 2ky = 0 \tag{24}$$

Using implicit differentiation twice, we find

$$x - h + yy' - ky' = 0 \tag{25}$$

and

$$1 + yy'' + (y')^2 - ky'' = 0 \tag{26}$$

Now, Eq. (24) yields

$$h = \frac{x^2 + y^2 - 2ky}{2x}$$

Putting this in Eq. (25), we get

$$x - \frac{x^2 + y^2 - 2ky}{2x} + yy' - ky' = 0 \tag{27}$$

Now, solving Eq. (27) for $k$, we obtain

$$k = \frac{x^2 - y^2 + 2xyy'}{2(xy' - y)} \tag{28}$$

Substituting this value in Eq. (26) and simplifying, we get the following nonlinear differential equation:

$$1 + yy'' + (y')^2 - \frac{x^2 - y^2 + 2xyy'}{2(xy' - y)} y'' = 0$$

or

$$(x^2 + y^2)y'' + 2[(y')^2 + 1]\,(y - xy') = 0$$

which is the required differential equation that describes the family of circles passing through the origin.

**EXAMPLE 1.4** Find the differential equation of all circles of radius $r$.

***Solution*** The equation of all circles of radius $r$ is given by

$$(x - h)^2 + (y - k)^2 = r^2 \tag{29}$$

where $h$ and $k$ are the coordinates of the centre and are taken as arbitrary constants.

Differentiating Eq. (29) with respect to $x$, we get

$$(x - h) + (y - k)\frac{dy}{dx} = 0 \tag{30}$$

Again, differentiating Eq. (30) with respect to $x$, we obtain

$$1 + (y - k)\frac{d^2y}{dx^2} + \left(\frac{dy}{dx}\right)^2 = 0 \tag{31}$$

Equation (31) yields

$$(y - k) = -\frac{\left[1 + \left(\frac{dy}{dx}\right)^2\right]}{\frac{d^2y}{dx^2}} \tag{32}$$

Substituting this value of $(y - k)$ in Eq. (30), we get

$$(x - h) = \frac{\left[1 + \left(\frac{dy}{dx}\right)^2\right]\frac{dy}{dx}}{\frac{d^2y}{dx^2}} \tag{33}$$

Now, substituting the values of $(x - h)$ and $(y - k)$ from Eqs. (33) and (32), respectively, in Eq. (29), we obtain

$$\frac{\left[1+\left(\frac{dy}{dx}\right)^2\right]\left(\frac{dy}{dx}\right)^2}{\left(\frac{d^2y}{dx^2}\right)^2}+\frac{\left[1+\left(\frac{dy}{dx}\right)^2\right]^2}{\left(\frac{d^2y}{dx^2}\right)^2}=r^2$$

or

$$\left[1+\left(\frac{dy}{dx}\right)^2\right]^3=r^2\left(\frac{d^2y}{dx^2}\right)^2$$

which is the differential equation of all circles of radius $r$.

**EXAMPLE 1.5** Show that the differential equation of a general parabola is

$$\frac{d^2}{dx^2}\left[\left(\frac{d^2y}{dx^2}\right)^{-2/3}\right]=0$$

***Solution*** The equation of a general parabola is

$$a^2x^2 + 2abxy + b^2y^2 + 2gx + 2fy + c = 0$$

or

$$(ax + by)^2 + 2gx + 2fy + c = 0 \tag{34}$$

Differentiating Eq. (34) with respect to $x$, we get

$$\frac{dy}{dx}=-\frac{a^2x+aby+g}{abx+b^2y+f} \tag{35}$$

Differentiating Eq. (35) with respect to $x$, we obtain

$$\frac{d^2y}{dx^2}=-\frac{(af-bg)^2}{(abx+b^2y+f)^3}$$

Thus

$$\left(\frac{d^2y}{dx^2}\right)^{-1/3}=\left[-\frac{(abx+b^2y+f)^3}{(af-bg)^2}\right]^{1/3}=-\frac{abx+b^2y+f}{(af-bg)^{2/3}}$$

or

$$\left(\frac{d^2y}{dx^2}\right)^{-2/3}=\frac{(abx+b^2y+f)^2}{(af-bg)^{4/3}}=A(abx+b^2y+f)^2$$

where $A = 1/(af - bg)^{4/3}$. Therefore,

$$\frac{d}{dx}\left(\frac{d^2y}{dx^2}\right)^{-2/3}=2A(abf-b^2g) \qquad \left[\text{Using Eq. (35) for } \frac{dy}{dx}\right]$$

or $$\frac{d}{dx}\left(\frac{d^2y}{dx^2}\right)^{-2/3} = \frac{2b}{(af-bg)^{1/3}} = \text{constant} = k \text{ (say)}$$

Differentiating with respect to $x$, yields

$$\frac{d^2}{dx^2}\left(\frac{d^2y}{dx^2}\right)^{-2/3} = 0$$

which proves the result.

**EXAMPLE 1.6** The equation to a system of confocal ellipse is

$$\frac{x^2}{a^2+k} + \frac{y^2}{b^2+k} = 1$$

where $k$ is an arbitrary constant. Find the corresponding differential equation.

***Solution*** Given

$$\frac{x^2}{a^2+k} + \frac{y^2}{b^2+k} = 1 \tag{36}$$

Differentiating Eq. (36) with respect to $x$, we get

$$\frac{2x}{a^2+k} + \frac{2y}{b^2+k}\frac{dy}{dx} = 0 \tag{37}$$

Denote $dy/dx = p$; then from Eq. (37)

$$\frac{2x}{a^2+k} + \frac{2y}{b^2+k}p = 0$$

Thus $$k = \frac{-pya^2 - b^2x}{x+py}$$

Also $$a^2+k = \frac{(a^2-b^2)x}{x+py}, \qquad b^2+k = \frac{-(a^2-b^2)py}{x+py}$$

Substituting these values of $(a^2 + k)$ and $(b^2 + k)$ in Eq. (36), after simplification, we have

$$(x^2-y^2) + xy\left(p - \frac{1}{p}\right) = a^2 - b^2$$

as the required differential equation, where $p = dy/dx$.

***Example* 1.7** Find the differential equation corresponding to the family of curves $x^2 + y^2 + 2c_1x + 2c_2y + c_3 = 0$, where $c_1$, $c_2$ and $c_3$ are arbitrary constants.

***Solution*** The given equation of the curve is

$$x^2 + y^2 + 2c_1x + 2c_2y + c_3 = 0 \tag{38}$$

Differentiating Eq. (38) with respect to $x$, we get

$$2x + 2y\frac{dy}{dx} + 2c_1 + 2c_2\frac{dy}{dx} = 0 \tag{39}$$

Differentiating again, we obtain

$$2 + 2\left(y\frac{d^2y}{dx^2} + \frac{dy}{dx}\frac{dy}{dx}\right) + 2c_2\frac{d^2y}{dx^2} = 0$$

or

$$1 + y\frac{d^2y}{dx^2} + \left(\frac{dy}{dx}\right)^2 + c_2\frac{d^2y}{dx^2} = 0 \tag{40}$$

Differentiating Eq. (40) with respect to $x$ yields

$$(y + c_2)\frac{d^3y}{dx^3} + 3\frac{dy}{dx}\frac{d^2y}{dx^2} = 0 \tag{41}$$

From Eq. (40)

$$c_2\frac{d^2y}{dx^2} = -\left(\frac{dy}{dx}\right)^2 - y\frac{d^2y}{dx^2} - 1$$

and Eq. (41) gives

$$c_2\frac{d^3y}{dx^3} = -y\frac{d^3y}{dx^3} - 3\frac{dy}{dx}\frac{d^2y}{dx^2}$$

Dividing these two equations, we get

$$\frac{d^3y}{dx^3}\left[1 + \left(\frac{dy}{dx}\right)^2\right] = 3\frac{dy}{dx}\left(\frac{d^2y}{dx^2}\right)^2$$

which is the required differential equation.

**Remark.** From the above examples we observe that Examples 1.2 and 1.6 have only one arbitrary constant and the differential equations thus formed are of the first order. On the other hand, Examples 1.1, 1.3 and 1.4 contain two arbitrary constants and the resulting differential equations are of the second order, while Example 1.7 contains three arbitrary constants and we get the third order differential equation. Thus we see that the number of arbitrary constants in a solution of a differential equation depends upon the order of the differential equation and is the same as its order. Hence, a differential equation of $n$ order will contain $n$ arbitrary constants.

## 1.5.2 Physical Origins of Differential Equations

In the above discussions we have seen how differential equations arise through geometrical consideration and, in this section, with the help of some examples, we shall look for the physical origins of differential equations.

**EXAMPLE 1.8** We know that freely falling objects, close to the surface of the earth, accelerate at a constant rate $g$. Acceleration is the derivative of velocity, which in turn, is the derivative of the distance $s$. Thus, if we assume that the upward direction is positive, the equation

$$\frac{d^2 s}{dt^2} = -g$$

is the differential equation governing the vertical distance that the falling body travels. The negative sign is used since the weight of the body is a force directed opposite to the positive direction.

Now, if we assume that a stone is tossed off the roof of a building of height $s_0$ (Fig. 1.1) with an initial upward velocity $v_0$, then we have to solve the differential equation

$$\frac{d^2 s}{dt^2} = -g, \qquad 0 < t < t_1$$

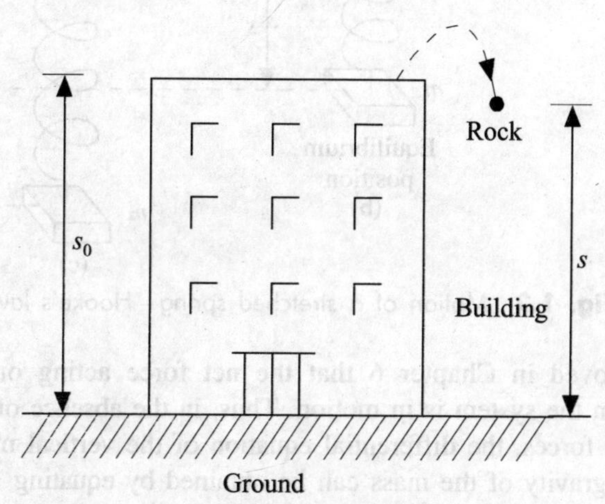


**Fig. 1.1** A stone is tossed off from the roof of a building.

with initial conditions $s(0) = s_0$, $s'(0) = v_0$. Here $t = 0$ is taken to be the initial time when the stone leaves the roof of the building and $t_1$ is the time required to hit the ground. As the stone is thrown upward, it would be assumed that $v_0 > 0$. The formulation of the problem does not include the force of air resistance acting on the body.

**EXAMPLE 1.9** To find the vertical displacement $x(t)$ of a mass attached to a spring (see Fig. 1.2) we use two different laws: Newton's second law of motion and Hooke's law. The former states that the net force acting on the system in motion is $F = ma$, where $m$ is the mass and $a$ is the acceleration, while the latter states that the restoring force of a stretched spring is proportional to the elongation $s + x$, i.e. the restoring force is $k(s + x)$, where $k > 0$ is a constant. In Fig. 1.2b, $s$ is the elongation of the spring after the mass $m$ has been attached and the system hangs at rest in the equilibrium position. When the system is in motion, the variable $x$ represents a directed distance of the mass beyond the equilibrium position (see Fig. 1.2c).

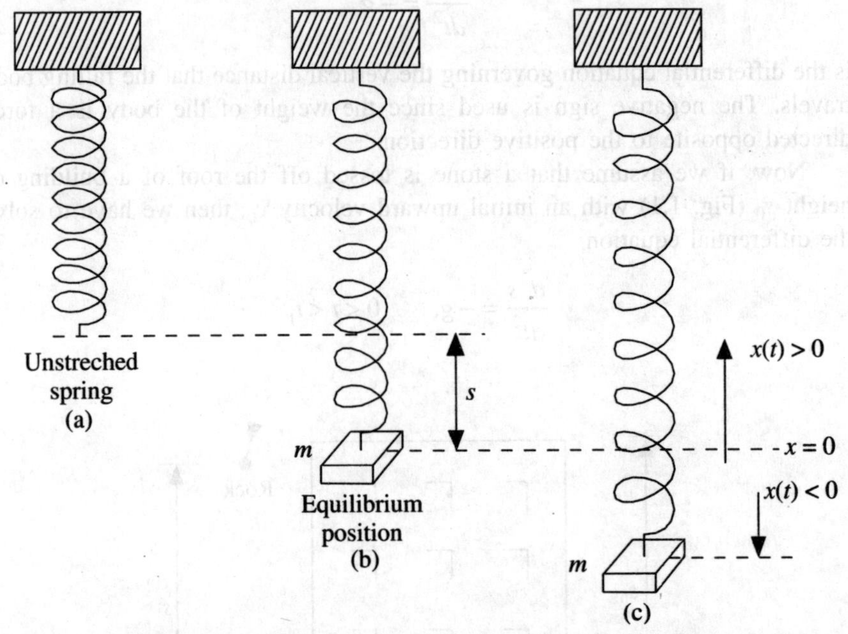


**Fig. 1.2** Motion of a stretched spring—Hooke's law.

It will be proved in Chapter 6 that the net force acting on the mass is $F = -kx$, when the system is in motion. Thus, in the absence of damping and other external forces, the differential equation of the vertical motion through the centre of gravity of the mass can be obtained by equating

$$m\frac{d^2x}{dt^2} = -kx$$

where the negative sign denotes the restoring force of the spring acting opposite to the direction of motion, i.e. towards the equilibrium position. Often, we write the differential equation as

$$\frac{d^2x}{dt^2} + w^2x = 0 \tag{42}$$

where $w^2 = k/m$.

**Units.** Three commonly used system of units are summarized in Table 1.1. In each unit the unit for time is second(s).

**Table 1.1** Systems of Units

| Quantity | FPS | MKS | CGS |
|---|---|---|---|
| Force | pound (lb) | newton (N) | dyne |
| Mass | slug | kilogramme (kg) | gram (g) |
| Distance | foot (ft) | metre (m) | centimetre (cm) |
| Acceleration due to gravity $g$ | 32 ft/s$^2$ | 9.8 m/s$^2$ | 980 cm/s$^2$ |

The gravitational force exerted by the earth on a body of mass $m$ is called its weight $W$. In the absence of air resistance, the only force acting on a freely falling body is its weight. Hence, from Newton's second law, the mass $m$ and weight $W$ are related by $W = mg$. For example, in engineering system a mass of 1/4 slug corresponds to an 8 lb weight. Since $m = W/g$, a 64 lb weight corresponds to a mass of 64/32 = 2 slugs. In the CGS system, a weight of 2450 dynes has a mass of 2450/980 = 2.5 g. In the MKS system, a weight of 50 N has a mass of 50/9.8 = 5.1 kg. We note 1 N = $10^5$ dynes = 0.2247 lb.

In the following example, we shall derive the differential equation which governs the motion of a simple pendulum.

**EXAMPLE 1.10** A mass $m$ havting weight $W$ is suspended from the end of a rod of length $l$. For vertical motion (see Fig. 1.3), we want to determine the displacement angle $\theta$, measured from the vertical, as a function of time (we consider $\theta > 0$ to the right of $OP$ and $\theta < 0$ to the left of $OP$). We know that an arc $s$ of a circle of radius $l$ is related to the central angle $\theta$ through the formula $s = l\theta$. Hence, the angular acceleration is $a = d^2s/dt^2 = l(d^2\theta/dt^2)$. From Newton's second law we have

$$F = ma = ml\frac{d^2\theta}{dt^2}$$

From Fig. 1.3, we see that the tangential component of the force due to the weight $W$ is $mg \sin \theta$. Neglecting the mass of the rod, we have

$$ml\frac{d^2\theta}{dt^2} = -mg \sin \theta$$

or

$$\frac{d^2\theta}{dt^2} + \frac{g}{l} \sin \theta = 0 \tag{43}$$

The nonlinear differential Eq. (43) cannot be solved in terms of the elementary functions; so we make further simplifying assumptions. If the angular displacements are not too large, then $\sin \theta \approx \theta$ and Eq. (43) can be replaced by the second order linear differential equation

$$\frac{d^2\theta}{dt^2} + \frac{g}{l}\,\theta = 0 \tag{44}$$

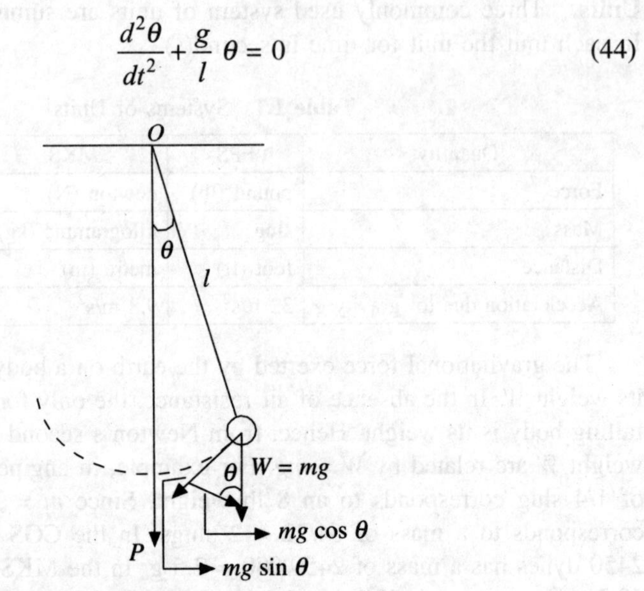


**Fig. 1.3** Motion of a simple pendulum.

If we put $w^2 = g/l$, we see that Eq. (42) has exactly the same structure as the differential Eq. (44) describing the motion of a spring. The fact that one basic differential equation can describe many diverse physical or even economic pheomena is a common occurrence in the study of applicable mathematics.

**EXAMPLE 1.11** Consider the single loop series circuit containing an inductor, resistor and capacitor (Fig. 1.4). Kirchhoff's second law states that the sum of the voltage drops across each part of the circuit is same as the impressed voltage $E(t)$. If $q(t)$ denotes the charge of the capacitor at any time $t$, then $i(t)$ is given by $i = dq/dt$.

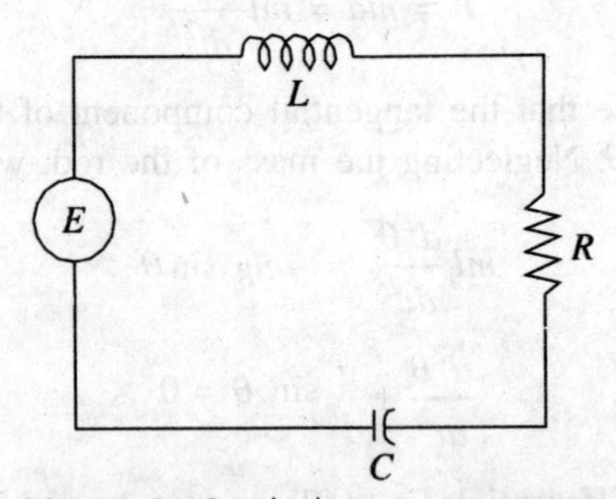


**Fig. 1.4** Single loop series circuit.

We also know that the voltage drops across:

$$\text{an inductor} = L\frac{di}{dt} = L\frac{d^2q}{dt^2}$$

# Differential Equations and Their Applications

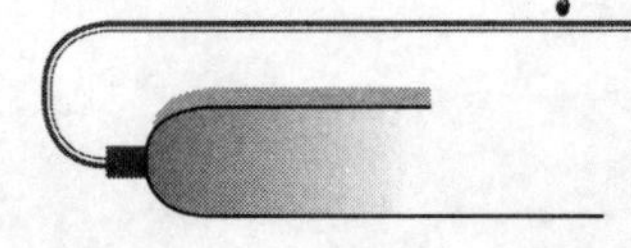

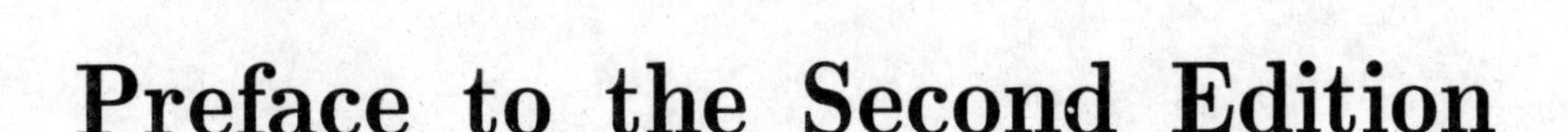

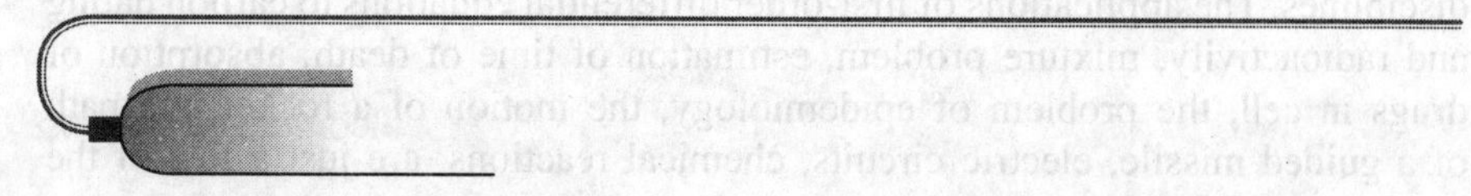

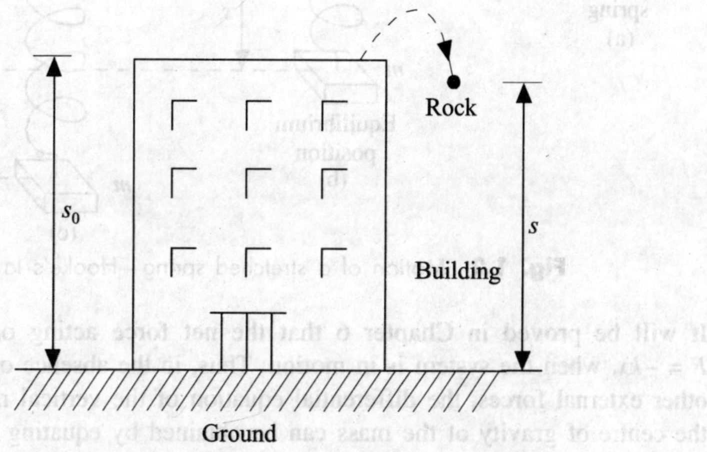

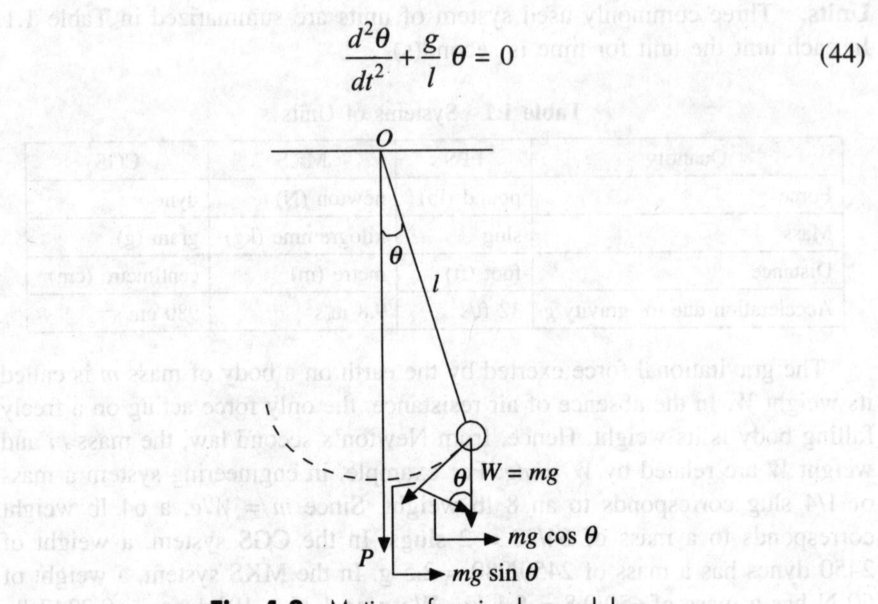



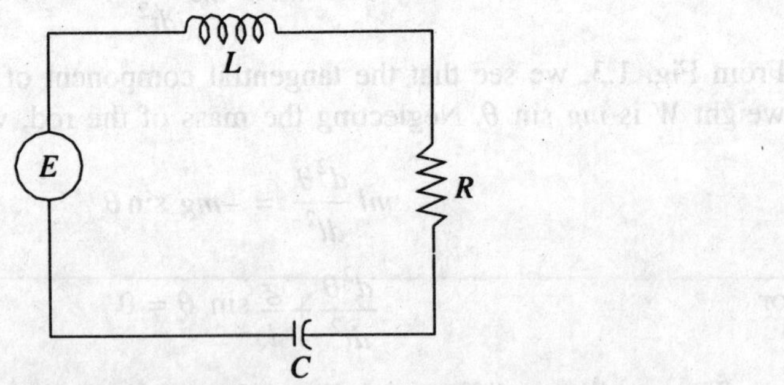


**Fig. 1.4** Single loop series circuit.

We also know that the voltage drops across:

$$\text{an inductor} = L\frac{di}{dt} = L\frac{d^2q}{dt^2}$$

Thus, the given equation is exact. Now

$$\underset{(y=\text{constant})}{\int Mdx} = \int (1 + e^{x/y})dx = x + ye^{x/y}, \qquad \underset{(\text{terms not having } x)}{\int Ndy} = 0$$

Therefore, the required solution is

$$\int Mdx + \int Ndy = c$$

or

$$x + ye^{x/y} = c$$

**EXAMPLE 2.31** Solve $(\sin x \cos y + e^{2x})\, dx + (\cos x \sin y + \tan y)\, dy = 0$.

***Solution*** Here

$$M = \sin x \cos y + e^{2x}, \qquad N = \cos x \sin y + \tan y, \qquad \frac{\partial M}{\partial y} = \frac{\partial N}{\partial x}$$

which show that the given equation is exact. Now

$$\underset{(y=\text{constant})}{\int Mdx} = \int (\sin x \cos y + e^{2x})\, dx = -\cos x \cos y + \frac{1}{2}e^{2x}$$

and

$$\underset{(\text{terms not having } x)}{Ndy} = \int \tan y\, dy = \log \sec y$$

Therefore, the solution of the given equation is

$$\frac{1}{2}e^{2x} - \cos x \cos y + \log \sec y = c$$

## 2.8 INTEGRATING FACTORS

A non-exact differential equation can always be made exact by multiplying it by some functions of $x$ and $y$. Such a function is called an *integrating factor.* Although a differential equation of the type $Mdx + Ndy = 0$ always has an integrating factor, there is no general method of finding them. Here we shall explain some of the methods for finding the integrating factors.

**Method I.** In some cases the integrating factor is found by inspection. Using the following few exact differentials, it is easy to find the integrating factors:

(a) $d\left(\frac{x}{y}\right) = \frac{ydx - xdy}{y^2}$ (b) $d\left(\frac{y}{x}\right) = \frac{xdy - ydx}{x^2}$

(c) $d(xy) = xdy + ydx$ (d) $d\left(\frac{x^2}{y}\right) = \frac{2yxdx - x^2dy}{y^2}$

(e) $d\left(\frac{y^2}{x}\right) = \frac{2xydy - y^2dx}{x^2}$ (f) $d\left(\frac{x^2}{y^2}\right) = \frac{2xy^2dx - 2x^2ydy}{y^4}$

(g) $d\left(\dfrac{y^2}{x^2}\right) = \dfrac{2x^2 y dy - 2xy^2 dx}{x^4}$ (h) $d\left(\dfrac{1}{xy}\right) = -\dfrac{xdy + ydx}{x^2 y^2}$

(i) $d\left(\log \dfrac{y}{x}\right) = \dfrac{xdy - ydx}{xy}$ (j) $d\left(\log \dfrac{x}{y}\right) = \dfrac{ydx - xdy}{xy}$

(k) $d\left(\tan^{-1} \dfrac{x}{y}\right) = \dfrac{ydx - xdy}{x^2 + y^2}$ (l) $d\left(\tan^{-1} \dfrac{y}{x}\right) = \dfrac{xdy - ydx}{x^2 + y^2}$

(m) $d\left(\dfrac{e^x}{y}\right) = \dfrac{ye^x dx - e^x dy}{y^2}$ (n) $d\left(\dfrac{e^y}{x}\right) = \dfrac{xe^y dy - e^y dx}{x^2}$

(o) $d\left(-\dfrac{1}{xy}\right) = \dfrac{xdy + ydx}{x^2 y^2}$ (p) $d\left[\dfrac{1}{2}\log (x^2 + y^2)\right] = \dfrac{xdx + ydy}{x^2 + y^2}$

**EXAMPLE 2.32** Solve $(1 + xy)y\,dx + (1 - xy)x\,dy = 0$.

***Solution*** The given equation can be written as

$$(ydx + xdy) + (xy^2dx - x^2ydy) = 0$$

or $\quad d(yx) + xy^2dx - x^2ydy = 0 \qquad$ [see equation in (c) above]

Dividing by $x^2y^2$, we get

$$-d\left(\frac{1}{xy}\right) + \frac{1}{x}\,dx - \frac{1}{y}\,dy = 0$$

Integrating, we get

$$-\frac{1}{xy} + \log x - \log y = c$$

or $\quad -\dfrac{1}{xy} + \log \dfrac{x}{y} = c \quad$ or $\quad \log \dfrac{x}{y} = c + \dfrac{1}{xy}$

which is the required solution.

**EXAMPLE 2.33** Solve $(x^3e^x - my^2)\,dx + mxy\,dy = 0$.

***Solution*** The given equation can be written as

$$x^3e^xdx + m(xy\,dy - y^2dx) = 0$$

Dividing by $x^3$, we get

$$e^x\,dx + m\frac{xy\,dy - y^2\,dx}{x^3} = 0$$

or $$e^x\,dx + \frac{m}{2}\,\frac{x^2\,2y\,dy - y^2\,2x\,dx}{x^4} = 0$$

or $$e^x\,dx + \frac{1}{2}m\,d\left(\frac{y^2}{x^2}\right) = 0 \qquad \text{[from exact differential (g)]}$$

Integrating, we get

$$e^x + \frac{1}{2} m \frac{y^2}{x^2} = c$$

or
$$2x^2 e^x + my^2 = 2cx^2$$

which is the required solution.

**EXAMPLE 2.34** Solve $x\,dy - y\,dx = a(x^2 + y^2)\,dy$.

***Solution*** The given equation can be written as

$$\frac{x\,dy - y\,dx}{x^2 + y^2} = a\,dy$$

or
$$d\left(\tan^{-1}\frac{y}{x}\right) = a\,dy \quad \text{[see exact differential (l)]}$$

Integrating, we get the required solution as

$$\tan^{-1}\frac{y}{x} = ay + c$$

**Method II.** If the differential equation $Mdx + Ndy = 0$ is homogeneous and $Mx + Ny \neq 0$, then $1/(Mx + Ny)$ is the integrating factor.

***Proof*** We have

$$Mdx + Ndy = \frac{1}{2}\left[(Mx + Ny)\left(\frac{dx}{x} + \frac{dy}{y}\right) + (Mx - Ny)\left(\frac{dx}{x} - \frac{dy}{y}\right)\right] \quad (29)$$

Multiplying by $1/(Mx + Ny)$, we get

$$\frac{Mdx + Ndy}{Mx + Ny} = \frac{1}{2}\left(\frac{y\,dx + x\,dy}{xy} + \frac{Mx - Ny}{Mx + Ny}\,\frac{ydx - xdy}{xy}\right)$$

$$= \frac{1}{2}\left[d(\log xy) + f\left(\frac{y}{x}\right) d\left(\log\frac{x}{y}\right)\right]$$

Since $Mdx + Ndy = 0$ is homogeneous, we obtain

$$\frac{Mdx + Ndy}{Mx + Ny} = \frac{1}{2}\left[d(\log xy) + f(e^{\log(y/x)})\,d\left(\log\frac{x}{y}\right)\right]$$

$$= \frac{1}{2}\left[d(\log xy) + F\left(\log\frac{x}{y}\right) d\left(\log\frac{x}{y}\right)\right]$$

which is an exact differential. Therefore, if $1/(Mx + Ny)$ is an integrating factor, then $Mdx + Ndy = 0$ is an exact differential equation.

**EXAMPLE 2.35** Solve $x^2ydx - (x^3 + y^3)dy = 0$.

***Solution*** Here, $Mx + Ny \neq 0 = -y^4$, and the given equation is homogeneous. Thus, the integrating factor is $-1/y^4$. Multiplying the given equation by this factor, we get

$$\frac{-x^2}{y^3}dx + \left(\frac{x^3}{y^4} + \frac{1}{y}\right)dy = 0$$

Now $$M = \frac{-x^2}{y^3}, \quad N = \frac{x^3}{y^4} + \frac{1}{y}; \quad \frac{\partial M}{\partial y} = \frac{\partial N}{\partial x}$$

This shows that the resulting differential equation is exact. Thus

$$\underset{(y=\text{constant})}{\int M dx} = -\int \frac{x^2}{y^3}dx = -\frac{x^3}{3y^3}$$

$$\underset{(\text{terms not having } x)}{\int N dy} = \int \frac{dy}{y} = \log y$$

Therefore, the required solution is

$$\int M\,dx + \int N\,dy = c$$

or $$-\frac{x^3}{3y^3} + \log y = c \quad \text{or} \quad x^3 = 3y^3(\log y - c)$$

**EXAMPLE 2.36** Solve $(x^2y - 2xy^2)dx - (x^3 - 3x^2y)dy = 0$.

***Solution*** The given equation is homogeneous and $Mx + Ny = x^2y^2 \neq 0$. The integrating factor is $1/x^2y^2$. Multiplying the given equation by this factor, we get

$$\left(\frac{1}{y}dx - \frac{x}{y^2}dy\right) - \frac{2}{x}dx + \frac{3}{y}dy = 0$$

or $$d\left(\frac{x}{y}\right) - \frac{2}{x}dx + \frac{3}{y}dy = 0$$

Integrating term by term, we obtain

$$\frac{x}{y} - 2\log x + 3\log y = c$$

which is the required solution.

**Method III.** If in the differential equation $Mdx + Ndy = 0$, $M = yf_1(xy)$ and $N = xf_2(xy)$. Then $1/(Mx - Ny)$ is an integrating factor.

***Proof*** Multiply Eq. (29) by $1/(Mx - Ny)$ and then proceed in the same way as in method II, the proof follows.

**EXAMPLE 2.37** Solve $(xy + 2x^2y^2)ydx + (xy - x^2y^2)xdy = 0$.

***Solution*** Here, $M = yf_1(xy)$ and $N = xf_2(xy)$. Thus, the integrating factor $= 1/(Mx - Ny) = 1/(3x^3y^3)$. Multiplying the given equation by this factor, we get

$$\frac{1}{3}\left(\frac{1}{x^2y}+\frac{2}{x}\right)dx+\frac{1}{3}\left(\frac{1}{xy^2}-\frac{1}{y}\right)dy=0$$

which is an exact differential equation, whose solution is

$$\frac{1}{3}\left(-\frac{1}{xy}+2\log x\right)+\frac{1}{3}(-\log y)=C$$

or
$$2\log x-\log y=\frac{1}{xy}+C_1 \qquad (C_1=3C)$$

**EXAMPLE 2.38** Solve $(x^2y^2 + xy + 1)ydx + (x^2y^2 - xy + 1)xdy = 0$.

***Solution*** Here, $M = yf_1(xy)$, $N = xf_2(xy)$, and thus the integrating factor is $1/(2x^2y^2)$. Multiplying the given equation by this factor, we get

$$(ydx+xdy)+\left(\frac{1}{x}dx-\frac{1}{y}dy\right)+\left(\frac{1}{x^2y}dx+\frac{1}{xy^2}dy\right)=0$$

or
$$d(xy)+\left(\frac{dx}{x}-\frac{dy}{y}\right)+\frac{ydx+xdy}{x^2y^2}=0$$

or
$$d(xy)+\frac{dx}{x}-\frac{dy}{y}+\frac{d(xy)}{x^2y^2}=0$$

Integrating term by term, we get

$$xy+\log x-\log y-(1/xy)=c$$

which is the required solution.

**Method IV.** The equation, $Mdx + Ndy = 0$ has $e^{\int f(x)\,dx}$ as the integrating factor if $1/N(\partial M/\partial y - \partial N/\partial x)$ is a function of $x$ [say $f(x)$].

***Proof*** Multiplying the given equation by $e^{\int f(x)\,dx}$, we get

$$Me^{\int f(x)\,dx}\,dx+Ne^{\int f(x)\,dx}\,dy=0$$

which is of the form $M_1dx + N_1dy = 0$. Now

$$\frac{\partial M_1}{\partial y}=\frac{\partial}{\partial y}[Me^{\int f(x)\,dx}]=\frac{\partial M}{\partial y}e^{\int f(x)dx}$$

and
$$\frac{\partial N_1}{\partial x}=\frac{\partial}{\partial x}[Ne^{\int f(x)dx}]$$

$$=e^{\int f(x)\,dx}\left[Nf(x)+\frac{\partial N}{\partial x}\right]$$

$$= e^{\int f(x)\,dx}\left(\frac{\partial M}{\partial y} - \frac{\partial N}{\partial x} + \frac{\partial N}{\partial x}\right)$$

$$= \frac{\partial M}{\partial y}\, e^{\int f(x)\,dx}$$

Thus $$\frac{\partial M_1}{\partial y} = \frac{\partial N_1}{\partial x}$$

which shows that the given equation is exact.

**Remark.** If $\frac{1}{N}\left(\frac{\partial M}{\partial y} - \frac{\partial N}{\partial x}\right) = k$ (a const.), then I.F. $= e^{\int k dx}$..

**EXAMPLE 2.39** Solve $(x^2 + y^2)dx - 2xydy = 0$.

***Solution*** Here

$$M = x^2 + y^2, \qquad N = -2xy, \qquad \frac{1}{N}\left(\frac{\partial M}{\partial y} - \frac{\partial N}{\partial x}\right) = -\frac{2}{x}$$

Thus $$\text{I.F.} = e^{\int f(x)\,dx} = \exp\left(\int -\frac{2}{x}\,dx\right) = e^{-2\log x} = \frac{1}{x^2}$$

Multiplying the given equation by $1/x^2$, we get

$$\left(1 + \frac{y^2}{x^2}\right)dx - \frac{2y}{x}\,dy = 0$$

or $$dx + d\left(-\frac{y^2}{x}\right) = 0$$

Integrating term by term, we obtain

$$x - \frac{y^2}{x} = C$$

which is the required solution.

**EXAMPLE 2.40** Solve $\left(y + \frac{1}{3}y^3 + \frac{1}{2}x^2\right)dx + \frac{1}{4}(x + xy^2)dy = 0$.

***Solution*** Here

$$\frac{1}{N}\left(\frac{\partial M}{\partial y} - \frac{\partial N}{\partial x}\right) = \frac{3}{x}$$

Thus $$\text{I.F.} = e^{\int f(x)\,dx} = x^3$$

Multiply the given equation by this factor. Then we get

$$2x^5\,dx + (x^4dy + 4x^3y\,dx) + \frac{1}{3}(x^4 3y^2 dy + 4x^3y^3dx) = 0$$

or $$2x^5dx + d(x^4y) + \frac{1}{3}d(x^4y^3) = 0$$

Integrating term by term, we have

$$x^6 + 3x^4y + x^4y^3 = 3C$$

which is the required solution.

**Method V.** If $1/M(\partial N/\partial x - \partial M/\partial y)$ is a function of $y$ [say $f(y)$], then $e^{\int f(y)\,dy}$ is the integrating factor of $Mdx + Ndy = 0$.

***Proof*** Proceed in a similar way as that of Method IV.

**Remark.** If $1/M(\partial N/\partial x - \partial M/\partial y) = k$ (a constant), then $e^{\int k\,dy}$ is the integrating factor.

**EXAMPLE 2.41** Solve $(xy^3 + y)\,dx + 2(x^2y^2 + x + y^4)\,dy = 0$.

***Solution*** Here

$$\frac{1}{M}\left(\frac{\partial N}{\partial x} - \frac{\partial M}{\partial y}\right) = \frac{1}{y}$$

Thus $$\text{I.F.} = e^{\int f(y)dy} = y$$

Multiplying the given equation by $y$, we get

$$(xy^4 + y^2)dx + 2(x^2y^3 + xy + y^5)dy = 0$$

Here $$M_1 = xy^4 + y^2, \quad N_1 = 2x^2y^3 + 2xy + 2y^5,$$

Then $$\underset{(y\,=\,\text{constant})}{\int M_1 dx} = \frac{1}{2}x^2y^4 + xy^2 \quad \text{and} \quad \underset{(\text{terms not having } x)}{\int N_1 dy} = \frac{2}{6}y^6$$

Therefore, the required solution is

$$3x^2y^4 + 6xy^2 + 2y^6 = 6C$$

**EXAMPLE 2.42** Solve $(xy^2 - x^2)dx + (3x^2y^2 + x^2y - 2x^3 + y^2)\,dy = 0$.

***Solution*** Here

$$\frac{1}{M}\left(\frac{\partial N}{\partial x} - \frac{\partial M}{\partial y}\right) = 6$$

which is a constant, thus I.F. $= e^{\int 6dy} = e^{6y}$. Multiplying the given equation by this factor, we get

$$(xy^2 - x^2)e^{6y}dx + (3x^2y^2 + x^2y - 2x^3 + y^2)e^{6y}dy = 0$$

Here

$$M_1 = xy^2e^{6y} - x^2e^{6y} \quad \text{and} \quad N_1 = 3x^2y^2e^{6y} + x^2ye^{6y} - 2x^3e^{6y} + y^2e^{6y}$$

Thus
$$\int\limits_{(y=\text{constant})} M_1 dx = e^{6y}\left(\frac{1}{2}x^2y^2 - \frac{1}{3}x^3\right)$$

and
$$\int\limits_{(\text{terms not having } x)} N_1 dy = \int y^2 e^{6y} dy = e^{6y}\left(\frac{1}{6}y^2 - \frac{1}{18}y + \frac{1}{108}\right)$$

by integrating by parts. Therefore, the required solution is

$$e^{6y}\left(\frac{1}{2}x^2y^2 - \frac{1}{3}x^3 + \frac{1}{6}y^2 - \frac{1}{18}y + \frac{1}{108}\right) = C$$

**Method VI.** If the equation $Mdx + Ndy = 0$ is of the form

$$x^a y^b(mydx + nxdy) + x^c y^d(pydx + qxdy) = 0$$

where $a$, $b$, $c$, $d$, $m$, $n$, $p$ and $q$ are constants, then $x^h y^k$ is the integrating factor, where $h$, $k$ are constants and can be obtained by applying the condition that after multiplication by $x^h y^k$ the given equation is exact.

**EXAMPLE 2.43** Solve $(y^2 + 2x^2y)dx + (2x^3 - xy)dy = 0$.

***Solution*** The given equation can be written as

$$y(y + 2x^2)dx + x(2x^2 - y)dy = 0$$

Let $x^h y^k$ be the integrating factor. Multiplying the given equation by this factor, we have

$$(x^h y^{k+2} + 2x^{h+2}y^{k+1})dx + (2x^{h+3}y^k - x^{h+1}y^{k+1})dy = 0 \qquad (30)$$

Here,
$$M = x^h y^{k+2} + 2x^{h+2}y^{k+1}$$
$$N = 2x^{h+3}y^k - x^{h+1}y^{k+1}$$

If Eq. (30) is exact, then $\partial M/\partial y = \partial N/\partial x$, or

$$(k + 2)\, x^h y^{k+1} + 2(k + 1)x^{h+2}y^k = -(h + 1)x^h y^{k+1} + 2(h + 3)\, x^{h+2}y^k$$

Equating the coefficients of $x^h y^{k+1}$ and $x^{h+2}y^k$ on both sides and solving, we get

$$h = -\frac{5}{2}, \qquad k = -\frac{1}{2}$$

Therefore, the integrating factor is

$$x^h y^k = x^{-5/2}y^{-1/2}$$

Multiplying the given equation by this factor, we get

$$(x^{-5/2}y^{3/2} + 2x^{-1/2}y^{1/2})dx + (2x^{1/2}y^{-1/2} - x^{-3/2}y^{1/2})dy = 0$$

In this equation, we have

$$M_1 = x^{-5/2}y^{3/2} + 2x^{-1/2}y^{1/2}, \qquad N_1 = 2x^{1/2}y^{-1/2} - x^{-3/2}y^{1/2}$$

and the equation is exact. Also

$$\underset{(y=\text{constant})}{\int M_1 dx} = -\frac{2}{3}x^{-3/2}y^{3/2} + 4x^{1/2}y^{1/2}$$

and

$$\underset{(\text{terms not having } x)}{\int N_1 dy} = 0$$

Hence, the required solution of the given equation is

$$\int M_1 dx + \int N_1 dy = C$$

or

$$-\frac{2}{3}x^{-3/2}y^{3/2} + 4x^{1/2}y^{1/2} = C$$

**EXAMPLE 2.44** Solve $(2ydx + 3xdy) + 2xy(3ydx + 4xdy) = 0$.

***Solution*** We can write the given equation as

$$(2y + 6xy^2)dx + (3x + 8x^2y)dy = 0$$

Let $x^h y^k$ be the integrating factor. Multiplying the given equation by this factor, we get

$$(2x^h y^{k+1} + 6x^{h+1}y^{k+2})dx + (3x^{h+1}y^k + 8x^{h+2}y^{k+1})dy = 0$$

If this equation is exact, we must then have $\partial M/\partial y = \partial N/\partial x$. Thus

$$2(k + 1)y^k x^h + 6(k + 2)y^{k+1}x^{h+1} = 3(h + 1)x^h y^k + 8(h + 2)x^{h+1}y^{k+1}$$

Equating the coefficients of $x^h y^k$ and $x^{h+1}y^{k+1}$ on both sides and solving, we get $h = 1$, $k = 2$. Thus, the integrating factor is $x^h y^k = xy^2$. Now, the multiplication of the given equation by $xy^2$ yields

$$(2xy^3 + 6x^2y^4)dx + (3x^2y^2 + 8x^3y^3)dy = 0$$

Here, $M_1 = 2xy^3 + 6x^2y^4$, $N_1 = 3x^2y^2 + 8x^3y^3$, and the equation is exact. Therefore, the solution is

$$\underset{(y=\text{constant})}{\int M_1 dx} + \underset{(\text{terms not having } x)}{\int N_1 dy} = C$$

or

$$x^2y^3 + 2x^3y^4 = C$$

## 2.9 CHANGE OF VARIABLES

By suitable substitution we can reduce a given differential equation which does not directly come under any of the forms discussed so far to one of these forms. This procedure of reducing the given differential equation by substitution is called the change of dependent (or independent) variable.

**EXAMPLE 2.45** Solve $xdx + ydy = \dfrac{a^2(xdy - ydx)}{x^2 + y^2}$.

***Solution*** Let $x = r\cos\theta$, and $y = r\sin\theta$. Then $r^2 = x^2 + y^2$ and $y/x = \tan\theta$. Differentiating these relations and substituting the respective value in the given equation, we get after simplification, the relation

$$2rdr = 2a^2 d\theta$$

which on integration yields

$$r^2 = 2a^2\theta + C \quad \text{or} \quad x^2 + y^2 = 2a^2 \tan^{-1}\frac{y}{x} + C$$

which is the required solution.

**EXAMPLE 2.46** Solve $\sec^2 y\left(\dfrac{dy}{dx}\right) + 2x\tan y = x^3$.

***Solution*** Putting $\tan y = v$, the given equation reduces to

$$\frac{dv}{dx} + 2xv = x^3$$

which is a linear differential equation whose solution is

$$ve^{x^2} = \int x^3 e^{x^2}\,dx + c \quad \text{or} \quad \tan y = \frac{1}{2}(x^2 - 1) + Ce^{-x^2}$$

## 2.10 TOTAL DIFFERENTIAL EQUATIONS

An ordinary differential equation of the first order and first degree involving three variables is of the form

$$P + Q\frac{dy}{dx} + R\frac{dz}{dx} = 0 \tag{31}$$

where $P$, $Q$, $R$ are functions of $x$ and $x$ is the independent variable.

In terms of differentials, Eq. (31) has the form

$$Pdx + Qdy + Rdz = 0 \tag{32}$$

Equation (32) is integrable only when

$$P\left(\frac{\partial Q}{\partial z} - \frac{\partial R}{\partial y}\right) + Q\left(\frac{\partial R}{\partial x} - \frac{\partial P}{\partial z}\right) + R\left(\frac{\partial P}{\partial y} - \frac{\partial Q}{\partial x}\right) = 0 \tag{33}$$

To get a solution of Eq. (32), we have the following rule.

*Rule:* If Eq. (33) is satisfied, take one of the variables, say $z$, as constant so that $dz = 0$. Then integrate the equation $Pdx + Qdy = 0$. Replace the arbitrary constant appearing in its integral by $\phi(z)$. Now differentiate the integral just obtained with respect to $x$, $y$, $z$. Finally, compare this result with the given differential equation to determine $\phi(z)$.

**EXAMPLE 2.47** Solve $(y^2 + yz)\,dx + (z^2 + zx)\,dy + (y^2 - xy)\,dz = 0$.

***Solution*** Here, $P = y^2 + yz$, $Q = z^2 + zx$, $R = y^2 - xy$. It can be shown that Eq. (33) is satisfied here. Consider $z$ as constant so that the given equation takes the form

$$(y^2 + yz)\,dx + (z^2 + zx)\,dy = 0$$

or
$$\frac{dx}{z(z+x)} + \frac{dy}{y(y+z)} = 0$$

Integrate it to get

$$\log(z + x) + \log y - \log(y + z) = \text{constant}$$

or
$$\frac{y(z+x)}{y+z} = \phi(z) = \text{constant} \tag{34}$$

or
$$y(z + x) - \phi(z)(y + z) = 0$$

Differentiate with respect to $x$, $y$, $z$, to obtain

$$y\,dx + [z + x - \phi(z)]dy + [y - (y + z)\phi'(z) - \phi(z)]dz = 0 \tag{35}$$

Comparing Eq. (35) with the given equation, we get

$$\frac{y^2 + yz}{y} = \frac{z^2 + zx}{z + x - \phi(z)} = \frac{y^2 - xy}{y - (y+z)\,\phi'(z) - \phi(z)}$$

The relation

$$\frac{y^2 + yz}{y} = \frac{z^2 + zx}{z + x - \phi(z)}$$

reduces to Eq. (34) and thus gives no information about $\phi(z)$. Therefore, take

$$\frac{y^2 + yz}{y} = \frac{y^2 - xy}{y - (y+z)\phi'(z) - \phi(z)}$$

and simplify to get

$$y^2 - xy = y^2 - xy - (y + z)^2\,\phi'(z) \qquad \text{[use Eq. (34)]}$$

or
$$(y + z)^2\,\phi'(z) = 0$$

which gives $\phi'(z) = 0$ and $\phi(z) = C$.

Hence, the required solution, from Eq. (34), is

$$y(z + x) = (y + z)C$$

**REMARK.** Sometimes the integral is obtained simply by regrouping the terms in the given equation as illustrated in the following example.

**EXAMPLE 2.48** Solve $x\,dx + z\,dy + (y + 2z)\,dz = 0$.

***Solution*** After regrouping the terms, the given equation can be written as

$$x\,dx + (y\,dz + z\,dy) + 2z\,dz = 0$$

By integrating this equation, we obtain

$$\frac{x^2}{2} + yz + z^2 = C$$

## 2.11 SIMULTANEOUS TOTAL DIFFERENTIAL EQUATIONS

The equation

$$\begin{aligned} Pdx + Qdy + Rdz &= 0 \\ P'dx + Q'dy + R'dz &= 0 \end{aligned} \tag{36}$$

where $P$, $Q$, $R$ and $P'$, $Q'$, $R'$ are any functions of $x$, are called *simultaneous total differential equations*.

(a) If each of the above equation is integrable and has solution $\phi(x, y, z) = C$ and $\psi(x, y, z) = C'$, respectively, then taken these equations together form the solution of (36).

(b) If one or both of the Eq. (36) is not integrable, then we write them as

$$\frac{dx}{QR' - RQ'} = \frac{dy}{RP' - PR'} = \frac{dz}{PQ' - QP'}$$

and solve these by the methods given below.

## 2.12 EQUATIONS OF THE FORM *dx/P* = *dy/Q* = *dz/R*

### 2.12.1 Method of Grouping

Note that if it is possible to take two fractions $\frac{dx}{P} = \frac{dz}{R}$, from which $y$ can be cancelled or is absent, leaving the equation in $x$ and $z$ only. Then integrate it giving

$$\phi(x, z) = C \tag{37}$$

Again, note that if one variable, say $x$, is absent or can be cancelled (may be with the help of Eq. 37) from the equation $\frac{dy}{Q} = \frac{dz}{R}$, then integrate it to get

$$\psi(y, z) = C' \tag{38}$$

The two independent solutions, Eqs. (37) and 38, taken together form the complete solution of the given equation (see also Section 9.4).

**EXAMPLE 2.49** Solve $\frac{dx}{z^2 y} = \frac{dy}{z^2 x} = \frac{dz}{y^2 x}$.

***Solution*** Taking $dx/(z^2y) = dy/(z^2x)$, we get $xdx - ydy = 0$, which on integration, yields

$$x^2 - y^2 = C \tag{39}$$

Now take $dy/(z^2x) = dz/(y^2x)$, to get

$$y^2dy - z^2dz = 0$$

which on integration gives

$$y^3 - z^3 = C' \tag{40}$$

Equations (39) and (40) taken together constitute the required solution of the given equation.

## 2.12.2 Method of Multipliers

By a proper choice of multipliers $l$, $m$, $n$ which are not necessarily constants, we write

$$\frac{dx}{P} = \frac{dy}{Q} = \frac{dz}{R} = \frac{ldx + mdy + ndz}{lP + mQ + nR}$$

such that $lP + mQ + nR = 0$. Then $ldx + mdy + ndz = 0$ can be solved giving the solution as

$$\phi(x, y, z) = C \tag{41}$$

Again look for another set of multipliers $\lambda$, $\mu$, $\gamma$ such that $\lambda P + \mu Q + \gamma R = 0$, giving $\lambda dx + \mu dy + \gamma dz = 0$, which on integration gives the solution as

$$\psi(x, y, z) = C' \tag{42}$$

Equations (41) and (42) taken together form the required solution.

**EXAMPLE 2.50** Solve $\dfrac{dx}{x(y^2 - z^2)} = \dfrac{dy}{-y(z^2 + x^2)} = \dfrac{dz}{z(x^2 + y^2)}$.

***Solution*** Using the multipliers $x$, $y$, $z$,

$$\text{Each fraction} = \frac{xdx + ydy + zdz}{x^2(y^2 - z^2) - y^2(z^2 + x^2) + z^2(x^2 + y^2)}$$

$$= \frac{xdx + ydy + zdz}{0}$$

Thus, $xdx + ydy + zdz = 0$, which on integration gives

$$x^2 + y^2 + z^2 = C \tag{43}$$

Again, using the multipliers $\dfrac{1}{x}, -\dfrac{1}{y}, -\dfrac{1}{z}$,

$$\text{Each fraction} = \frac{\frac{1}{x}dx - \frac{1}{y}dy - \frac{1}{z}dz}{(y^2 - z^2) + (z^2 + x^2) - (x^2 + y^2)}$$

$$= \frac{\frac{1}{x}dx - \frac{1}{y}dy - \frac{1}{z}dz}{0}$$

Thus, $\frac{1}{x}dx - \frac{1}{y}dy - \frac{1}{z}dz = 0$ which becomes, on integration,

$$\log x - \log y - \log z = \text{constant}$$

or
$$yz = C'x \tag{44}$$

Hence, the solution of the given equation is

$$x^2 + y^2 + z^2 = C, \quad yz = C'x$$

## EXERCISES

Solve the following differential equations:

1. $(xy^2 + x)dx + (yx^2 + y)dy = 0.$
2. $(e^y + 1)\cos x dx + e^y \sin x dy = 0.$
3. $\frac{dy}{dx} = xy + x + y + 1.$
4. $\frac{dy}{dx} = (4x + y + 1)^2.$
5. $\sqrt{(1+x^2+y^2+x^2y^2)} + xy\frac{dy}{dx} = 0.$
6. $(2ax + x^2)\frac{dy}{dx} = a^2 + 2ax.$
7. $\frac{dy}{dx} + \frac{x^2+3y^2}{3x^2+y^2} = 0$
8. $x^2y dx - (x^3 + y^3)dy = 0.$
9. $x(x - y)\frac{dy}{dx} = y(x + y).$
10. $(1 + e^{x/y})dx + e^{x/y}\left(1 - \frac{x}{y}\right)dy = 0.$
11. $x \sin\left(\frac{y}{x}\right)\frac{dy}{dx} = y \sin\left(\frac{y}{x}\right) - x.$
12. $\frac{dy}{dx} = \frac{y}{x} + \tan\frac{y}{x}.$
13. $\frac{dy}{dx} = \frac{x+y+1}{x-y}.$
14. $(x - y - 2)dx = (2x - 2y - 3)dy.$
15. $\frac{dy}{dx} = \frac{1-3x-3y}{2(x+y)}.$
16. $(6x + 2y - 10)\frac{dy}{dx} = 2x + 9y - 20.$
17. $\frac{dy}{dx} = \frac{x+y+1}{2x+2y+3}.$
18. $(7y - 3x + 3)dy + (3y - 7x + 7)dx = 0.$
19. $x(x-1)\frac{dy}{dx} - (x-2)y = x^3(2x-1).$
20. $\cos^3 x\frac{dy}{dx} + y\cos x = \sin x.$
21. $x\frac{dy}{dx} + 2y = x^2 \log x.$
22. $\frac{dy}{dx} = \frac{\sin^2 x}{1+x^3} - \frac{3x^2}{1+x^3}y$
23. $\frac{dy}{dx} + \frac{y}{(1-x)\sqrt{x}} = 1 - \sqrt{x}.$
24. $(1 + y^2) + (x - e^{\tan^{-1}y})\frac{dy}{dx} = 0.$
25. $(x + \tan y)dy = \sin 2y\, dx.$
26. $xy - \frac{dy}{dx} = y^3 e^{-x^2}$

**27.** $(x+y+1)\dfrac{dy}{dx}=1$.

**28.** $(2x-10y^3)\dfrac{dy}{dx}+y=0$.

**29.** $x\dfrac{dy}{dx}+y=xy^3$.

**30.** $\dfrac{dy}{dx}+\dfrac{y}{x}=y^2\sin x$.

**31.** $3\dfrac{dy}{dx}+\dfrac{2y}{x+1}=\dfrac{x^3}{y^2}$.

**32.** $\dfrac{dy}{dx}=\dfrac{x^2+y^2+1}{2xy}$.

**33.** $\dfrac{dy}{dx}=2y\tan x+y^2\tan^2 x$.

**34.** $\cos x\,dy=y(\sin x-y)dx$.

**35.** $x\dfrac{dy}{dx}+y\log y=xye^x$.

**36.** $\dfrac{dy}{dx}=x^3y^3-xy$.

**37.** $3e^x\tan y+(1-e^x)\sec^2 y\dfrac{dy}{dx}=0$.

**38.** $(x^2y^3+xy)dy-dx=0$.

**39.** $(x^2-2xy-y^2)dx-(x+y)^2dy=0$.

**40.** $y\sin 2x\,dx-(1+y^2+\cos^2 x)dy=0$.

**41.** $\left[y\left(1+\dfrac{1}{x}\right)+\cos y\right]dx+(x+\log x-x\sin y)\,dy=0$.

**42.** $(e^y+1)\cos x\,dx+e^y\sin x\,dy=0$.

**43.** $ydx-xdy+(1+x^2)dx+x^2\sin y\,dy=0$.

**44.** $2xy^2dx=e^x(dy-ydx)$.

**45.** $ydx-xdy+\log x\,dx=0$.

**46.** $y(2x^2y+e^x)dx-(e^x+y^3)dy=0$.

**47.** $y^2+x^2\dfrac{dy}{dx}=xy\dfrac{dy}{dx}$.

**48.** $(x^3y^3+x^2y^2+xy+1)ydx+(x^3y^3-x^2y^2-xy+1)xdy=0$.

**49.** $y(1-xy)dx-x(1+xy)dy=0$.

**50.** $(x^4y^4+x^2y^2+xy)ydx+(x^4y^4-x^2y^2+xy)xdy=0$.

**51.** $(xy\sin xy+\cos xy)ydx+(xy\sin xy-\cos xy)xdy=0$.

**52.** $(x^3-2y^2)dx+2xydy=0$.

**53.** $(x^2+y^2+2x)dx+2ydy=0$.

**54.** $(2x^2y-3y^2)dx+(2x^3-12xy+\log y)dy=0$.

**55.** $(3x^2y^4+2xy)dx+(2x^3y^3-x^2)dy=0$.

**56.** $(20x^2+8xy+4y^2+3y^3)ydx+4(x^2+xy+y^2+y^3)xdy=0$.

**57.** $(3x+2y^2)ydx+2x(2x+3y^2)dy=0$.

**58.** $x(4ydx+2xdy)+y^3(3ydx+5xdy)=0$.

**59.** $x(3ydx+2xdy)+8y^4(ydx+3xdy)=0$.

**60.** $(2x^2y-3y^4)dx+(3x^3+2xy^3)dy=0$.

**61.** $\dfrac{xdx + ydy}{xdy - ydx} = \sqrt{\dfrac{a^2 - x^2 - y^2}{x^2 + y^2}}$.

**62.** $(y + z)dx + (z + x)dy + (x + y)dz = 0$.

**63.** $yzdx - 2xzdy + (xy - zy^3)dz = 0$.

**64.** $(x + z)^2dy + y^2(dx + dz) = 0$.

**65.** $\dfrac{dx}{x^2} = \dfrac{dy}{y^2} = \dfrac{dz}{nxy}$.

**66.** $\dfrac{dx}{mz - ny} = \dfrac{dy}{nx - lz} = \dfrac{dz}{ly - mx}$.

**67.** $\dfrac{dx}{y - zx} = \dfrac{dy}{yz + x} = \dfrac{dz}{x^2 + y^2}$.

**68.** $\dfrac{dx}{x(y^2 - z^2)} = \dfrac{dy}{y(z^2 - x^2)} = \dfrac{dz}{z(x^2 - y^2)}$.

**69.** $\dfrac{dx}{x^2 - y^2 - z^2} = \dfrac{dy}{2xy} = \dfrac{dz}{2xz}$.

**70.** Using the substitution $y^2 = v - x$, reduce the equation

$$y^3 \frac{dy}{dx} + x + y^2 = 0$$

to the homogeneous form and hence solve the equation.

CHAPTER 3

# Equations of the First Order but not of the First Degree

The most general form of a differential equation of the first order but not of the first degree (say *n*th degree) is

$$\left(\frac{dy}{dx}\right)^n + P_1\left(\frac{dy}{dx}\right)^{n-1} + P_2\left(\frac{dy}{dx}\right)^{n-2} + \cdots + P_{n-1}\left(\frac{dy}{dx}\right) + P_n = 0 \qquad (1)$$

or $$p^n + P_1 p^{n-1} + P_2 p^{n-2} + \cdots + P_{n-1}\, p + P_n = 0$$

where $p = dy/dx$ and $P_1, P_2, \ldots, P_n$ are functions of $x$ and $y$. This equation can also be written as

$$F(x, y, p) = 0 \qquad (2)$$

The above equation however cannot be solved in this general form. We will discuss here the situations where a solution of this equation exists. Let us consider two cases:

CASE I. In this case, the first member of Eq. (1) can be resolved into rational factors of the first degree.

CASE II. Here the member cannot be thus factored.

## 3.1 CASE I

### 3.1.1 Equations Solvable for *p*

Suppose a differential equation can be solved for $p$ and is of the form

$$[p - f_1(x, y)]\,[p - f_2(x, y)] \ldots [p - f_n(x, y)] = 0$$

Equating each factor to zero we get equations of the first order and the first degree. Let their solutions be

$$\phi_1(x, y, c_1) = 0, \quad \phi_2(x, y, c_2) = 0, \ldots, \quad \phi_n(x, y, c_n) = 0$$

Without any loss of generality, we can write

$$c_1 = c_2 = \cdots = c_n = c$$

as they are arbitrary constants. Therefore, the solution of Eq. (1) can be put in the form

$$\phi_1(x, y, c)\ \phi_2(x, y, c) \cdots \phi_n(x, y, c) = 0$$

**EXAMPLE 3.1** Solve $(p - xy)(p - x^2)(p - y^2) = 0$.

***Solution*** Here, $p = xy, x^2, y^2$. If $p = xy$, then $dy/y = xdx$, which on integration gives $\log y = (1/2)x^2 + c_1$. If $p = x^2$, then integration yields $y = (1/3)x^3 + c_2$. If $p = y^2$, then the solution is $-1/y = x + c_3$. Therefore, the required solution of the given equation is

$$\left(\log y - \frac{1}{2}x^2 - c\right)\left(y - \frac{1}{3}x^3 - c\right)\left(x + \frac{1}{y} + c\right) = 0$$

**EXAMPLE 3.2** Solve $(p + y + x)(xp + y + x)(p + 2x) = 0$.

***Solution*** Here, we have

$$p + y + x = 0, \qquad xp + y + x = 0, \qquad p + 2x = 0$$

If $p + y + x = 0$, then $dy/dx + y + x = 0$. Put $x + y = v$; then this equation becomes

$$\frac{dv}{1 - v} = dx \quad \text{or} \quad -\log(1 - v) = x + c_1$$

or

$$(1 - v) = e^{-x-c1} = ce^{-x}$$

or

$$1 - x - y - ce^{-x} = 0 \tag{3}$$

If $xp + y + x = 0$, then $dy/dx + (1/x)y = 1$, whose solution is

$$yx = \frac{1}{2}c_2 - \frac{1}{2}x^2 \quad \text{or} \quad 2xy + x^2 - c_2 = 0 \tag{4}$$

Finally, if $p + 2x = 0$, then the solution is

$$y + x^2 - c_3 = 0 \tag{5}$$

From Eqs. (3)–(5), the solution of the given equation is

$$(1 - x - y - ce^{-x})(2xy + x^2 - c)(y + x^2 - c) = 0$$

**EXAMPLE 3.3** Solve $p^2 + 2py \cot x = y^2$.

***Solution*** Solving the given equation for $p$, we get

$$p = \frac{1}{2}\left[-2y \cot x \pm \sqrt{(4y^2 \cot^2 x + 4y^2)}\right] = -y \cot \frac{x}{2},\ y \tan \frac{x}{2}$$

If $p = -y \cot(x/2)$, then by integrating, we have

$$\log y = -2 \log \sin \frac{x}{2} + \log A$$

Solving, we get

$$y = \frac{A}{\sin^2(x/2)} \quad \text{or} \quad y(1 - \cos x) = 2A = c_1$$

If $p = y \tan (x/2)$, then by integrating, we get

$$\log y = 2 \log \sec \frac{x}{2} + \log B$$

or

$$y(1 + \cos x) = 2B = c_2$$

Therefore, the required solution is

$$[y(1 - \cos x) - c]\,[y(1 + \cos x) - c] = 0$$

**EXAMPLE 3.4** Solve $x^2p^2 + xyp - 6y^2 = 0$.

***Solution*** Solving the given equation for $p$, we have

$$p = \frac{2y}{x}, \quad -\frac{3y}{x}$$

If $p = 2y/x$, then $\log y = 2 \log x + \log c_1$, or $y = c_1x^2$; and if $p = -3y/x$, then $yx^3 = c_2$. Therefore, the required solution is

$$(y - cx^2)\,(yx^3 - c) = 0$$

**EXAMPLE 3.5** Solve $xy^2(p^2 + 2) = 2py^3 + x^3$.

***Solution*** The given equation can be written as

$$(yp - x)\,[xyp + (x^2 - 2y^2)] = 0$$

If $yp - x = 0$, then integration yields

$$y^2 - x^2 = c_1 \tag{6}$$

If $xyp + x^2 - 2y^2 = 0$, then

$$2y\frac{dy}{dx} - \frac{4y^2}{x} = -2x$$

or

$$\frac{dv}{dx} - \frac{4}{x}v = -2x$$

where $v = y^2$. This is a linear differential equation in $v$ and its solution is

$$\frac{v}{x^4} = c_2 + \frac{1}{x^2}$$

or

$$y^2 = c_2x^4 + x^2 \tag{7}$$

From Eqs. (6) and (7), the required solution is

$$(y^2 - x^2 - c)(y^2 - cx^4 - x^2) = 0$$

**EXAMPLE 3.6** Solve $x^2p^3 + y(1 + x^2y)p^2 + y^3p = 0$.

***Solution*** The given equation can be written as

$$p(x^2p + y)(p + y^2) = 0$$

If $p = 0$, then

$$dy = 0 \quad \text{or} \quad y = c_1 \tag{8}$$

If $x^2p + y = 0$, then

$$\frac{dy}{dx} + \frac{y}{x^2} = 0$$

which is a linear equation and the solution is

$$ye^{-1/x} = c_2 \tag{9}$$

If $p + y^2 = 0$, then integration yields

$$xy + c_3y - 1 = 0 \tag{10}$$

Therefore, the required solution is

$$(y - c)(ye^{-1/x} - c)(xy + cy - 1) = 0$$

## 3.2 CASE II

Equation (2) may have one or more of the following properties:

(a) It may be solvable for $y$.
(b) It may be solvable for $x$.
(c) It may not contain either $x$ or $y$.
(d) It may be homogeneous in $x$ and $y$.
(e) It may be of the first degree in $x$ and $y$.

### 3.2.1 Equations Solvable for *y*

If the differential equation $f(x, y, p) = 0$ is solvable for $y$, then

$$y = f(x, p) \tag{11}$$

Differentiating with respect to $x$, gives

$$p = \frac{dy}{dx} = \phi\left(x, p, \frac{dp}{dx}\right) \tag{12}$$

which is an equation in two variables $x$ and $p$, and it will give rise to a solution of the form

$$F(x, p, c) = 0 \tag{13}$$

The elimination of $p$ between Eqs. (11) and (13) gives a relation between $x$, $y$ and $c$, which is the required solution. When the elimination of $p$ between these equations is not easily done, the values of $x$ and $y$ in terms of $p$ can be found, and these together will constitute the required solution.

***Example 3.7*** Solve $y + px = x^4p^2$.

***Solution*** The given equation is

$$y = x^4p^2 - xp \tag{14}$$

Differentiating with respect to $x$ yields, after simplification,

$$\frac{dp}{p} + 2\frac{dx}{x} = 0$$

which on integration gives

$$p = \frac{c}{x^2}$$

Substitution of this value of $p$ in Eq. (14) gives the required solution as

$$xy + c = c^2x$$

**EXAMPLE 3.8** Solve $y = \sin p - p \cos p$.

***Solution*** Differentiating the given equation with respect to $x$, we get

$$\sin p \, dp = dx$$

Integrating, we have

$$\cos p = c - x \tag{15}$$

From the given equation, we also have

$$p \cos p = \sin p - y \quad \text{or} \quad p = \frac{\sqrt{1-\cos^2 p} - y}{\cos p}$$

or

$$c - x = \cos\left(\frac{\sqrt{1-c^2-x^2+2cx} - y}{c-x}\right)$$

which is the required solution.

**EXAMPLE 3.9** Solve $y = yp^2 + 2px$.

***Solution*** The given equation can be written as

$$y = \frac{2px}{1-p^2} \tag{16}$$

Differentiating Eq. (16) with respect to $x$, we get

$$\frac{2dp}{p(p^2-1)} = \frac{dx}{x}$$

or

$$\left(\frac{1}{p-1} + \frac{1}{p+1} - \frac{2}{p}\right)dp = \frac{dx}{x}$$

which on integration gives

$$\log(p-1)+\log(p+1)-2\log p=\log x+\log c$$

or
$$\frac{p^2-1}{p^2}=cx \quad \text{or} \quad p=\frac{1}{\sqrt{1-cx}}$$

Substituting this value of $p$ in Eq. (16), we get

$$2x\sqrt{1-cx}+cxy=0$$

which is the required solution.

### 3.2.2 Equations Solvable for *x*

When the differential Eq. (2) is solvable for $x$, then we have

$$x=f(y,p)$$

Differentiating with respect to $y$, gives

$$\frac{1}{p}=\phi\left(y,p,\frac{dp}{dy}\right)$$

from which a relation between $p$ and $y$ may be obtained

$$F(y,p,c)=0 \text{ (say)}$$

Between this and the given equation, $p$ may be eliminated, or $x$ and $y$ expressed in terms of $p$ as in Section 3.2.1.

**EXAMPLE 3.10** Solve $y^2 \log y = xyp + p^2$.

***Solution*** The given equation can be written as

$$x=\frac{1}{p}y\log y-\frac{p}{y}$$

Differentiating with respect to $y$, we get, after simplification,

$$\frac{dp}{p}=\frac{dy}{y}$$

which on integration gives

$$\log p=\log y+\log c=\log cy \qquad \text{or} \qquad p=cy$$

Eliminating $p$ from this and the given equation, we obtain

$$\log y=cx+c^2$$

which is the required solution.

**EXAMPLE 3.11** Solve $xp^3 = a + bp$.

***Solution*** Solving for $x$, we get

$$x = \frac{a}{p^3} + \frac{b}{p^2} \tag{17}$$

Differentiating Eq. (17) with respect to $y$, we get

$$dy + \left(\frac{3a}{p^3} + \frac{2b}{p^2}\right) dp = 0$$

Integrating it, we have

$$y = \frac{3a}{2p^2} + \frac{2b}{p} + c \tag{18}$$

Equations (17) and (18) constitute the required solution.

**EXAMPLE 3.12** Solve $x = y + a \log p$.

***Solution*** Differentiating the given equation with respect to $y$, we obtain

$$dy = \frac{a}{1 - p} dp$$

which on integration gives

$$y = c - a \log (1 - p) \tag{19}$$

From the given equation, we have

$$x = c - a \log (1 - p) + a \log p \tag{20}$$

Equations (19) and (20) give the required solution.

### 3.2.3 Equations That do not Contain *x* (or *y*)

If the equation has the form $f(y, p) = 0$ and is solvable for $p$, it will then give $dy/dx = \phi(y)$, which is integrable.

If it is solvable for $y$, then

$$y = F(p),$$

which is the case of Section 3.2.1. When the equation is of the form $f(x, p) = 0$, it will give $dy/dx = \phi(x)$, which is also integrable. But if it is solvable for $x$, then $x = F(p)$, which is the case of Section 3.2.2.

It may be mentioned that in equations having either of the properties (*c*) and not solvable for $p$, on solving for $x$ or $y$, the differentiation is made w.r.t. absent variable.

By differentiating the equation given in Sections 3.2.1–3.2.3, we have a chance of obtaining a differential equation, by means of which another relation may be found between $p$ and $x$ or $y$ in addition to the original relation. These two relations will then be used either for the elimination of $p$ or for the expression of $x$ and $y$ in terms of $p$.

### 3.2.4 Equations Homogeneous in *x* and *y*

When the equation is homogeneous in $x$ and $y$, it can be written as

$$F\left(\frac{dy}{dx}, \frac{y}{x}\right) = 0$$

It is then possible to solve it for $dy/dx$ and proceed as in Section 2.3, or solve it for $y/x$, and obtain $y = xf(p)$ which is given in Section 3.2.1.

Proceeding as in Section 3.2.1, and differentiating with respect to $x$, we get

$$p = f(p) + xf'(p)\frac{dp}{dx}$$

from which

$$\frac{dx}{x} = \frac{f'(p)\,dp}{p - f(p)}$$

where the variables are separated.

### 3.2.5 Equations of the First Degree in *x* and *y*—Clairaut's Equation

When the given Eq. (2) is of the first degree in $x$ and $y$, then

$$y = xf_1(p) + f_2(p) \tag{21}$$

Equation (21) is known as *Lagrange's equation.* To solve it, we differentiate with respect to $x$ to obtain

$$p = \frac{dy}{dx} = f_1(p) + xf_1'(p) + f_2'(p)\frac{dp}{dx}$$

or

$$\frac{dp}{dx} - \frac{f_1'(p)}{p - f_1(p)}x = \frac{f_2'(p)}{p - f_1(p)} \tag{22}$$

which is a linear equation in $x$ and $p$, and hence can be solved in the form

$$x = \phi(p, c) \tag{23}$$

Eliminating $p$ from Eqs. (21) and (23), we get the required solution. If it is not possible to eliminate $p$, then the values of $x$ and $y$ in terms of $p$ can be found from Eqs. (21) and (23), and these will constitute the required solution.

If $f_1(p) = p$ and $f_2(p) = f(p)$, then Eq. (21) reduces to

$$y = px + f(p) \tag{24}$$

Equation (24) is known as *Clairaut's equation*. To solve it, we differentiate with respect to $x$ to obtain

$$p = \frac{dy}{dx} = [x + f'(p)]p' + p$$

or

$$[x + f'(p)]\frac{dp}{dx} = 0 \tag{25}$$

If $dp/dx = 0$, then $p = c$ = constant. Eliminating $p$ between this and Eq. (24), we get

$$y = cx + f(c) \tag{26}$$

which is the required solution of Clairaut's equation.

REMARK. If we eliminate $p$ between

$$x + f'(p) = 0$$

and Eq. (24), we get a solution which does not contain any arbitrary constant, and hence, is not a particular case of solution (26). Such a solution is called *singular solution.*

Sometimes by a suitable substitution, an equation can be reduced to Clairaut's form.

**EXAMPLE 3.13** Solve $(y - px)(p - 1) = p$.

***Solution*** The given equation can be written as

$$y = xp + \frac{p}{p-1}$$

Differentiation yields after simplification

$$\frac{dp}{dx}\left[x - \frac{1}{(p-1)^2}\right] = 0$$

Therefore, $\frac{dp}{dx} = 0$ or $p = c$

From this and the given equation, the elimination of $p$ gives

$$(y - cx)(c - 1) = c$$

which is the required solution.

***Example 3.14*** Solve $p = \log(px - y)$.

***Solution*** The given equation is

$$y = px - e^p.$$

Differentiating with respect to $x$, we get

$$(x - e^p)\frac{dp}{dx} = 0$$

or

$$\frac{dp}{dx} = 0 \quad \text{or} \quad p = c$$

Eliminating $p$ from this and the given equation, we get the required solution as

$$c = \log (cx - y)$$

**EXAMPLE 3.15** Find the general and singular solution of the differential equation $y = px + \sqrt{(a^2 p^2 + b^2)}$.

***Solution*** Differentiating the given equation with respect to $x$, we have

$$[x + a^2p(a^2p^2 + b^2)^{-1/2}]\frac{dp}{dx} = 0 \tag{27}$$

If $dp/dx = 0$, then $p = c$. With this and Eq. (27), the general solution is

$$y = cx + \sqrt{a^2c^2 + b^2}$$

Also, from Eq. (27)

$$x + a^2p(a^2p^2 + b^2)^{-1/2} = 0$$

or

$$p = \frac{bx}{a\sqrt{a^2 - x^2}}$$

Using this value of $p$ and Eq. (27), the singular solution is obtained as

$$y^2a^2(a^2 - x^2) = b^2(x^2 + a^2)$$

**EXAMPLE 3.16** Reduce $xyp^2 - (x^2 + y^2 + 1)\,p + xy = 0$ to Clairaut's form and find its singular solution.

***Solution*** Let $x^2 = u$ and $y^2 = v$, then the given equation becomes

$$u\left(\frac{dv}{du}\right)^2 - (u + v - 1)\frac{dv}{du} + v = 0$$

or

$$uP^2 - (u + v + 1)\,P + v = 0, \qquad \text{where } P = \frac{dv}{du}$$

or

$$v = uP + \frac{P}{P-1} \tag{28}$$

which is of Clairaut's form. Differentiating it with respect to $u$, we get

$$\left[u - \frac{1}{(P-1)^2}\right]\frac{dP}{du} = 0 \tag{29}$$

To get the singular solution, we consider

$$u - \frac{1}{(P-1)^2} = 0$$

which gives $P = 1 + (1/\sqrt{u})$. Putting this in Eq. (28), we get the required solution as

$$y^2 = (x + 1)^2$$

**EXAMPLE 3.17** Solve $(px - y)(py + x) = h^2 p$.

***Solution*** Putting $x^2 = u$ and $y^2 = v$, the given equation takes the form

$$u\left(\frac{dv}{du}\right)^2 + (u - v - h^2)\frac{dv}{du} - v = 0$$

or $$uP^2 + (u - v - h^2)P - v = 0, \quad \text{where } P = \frac{dv}{du}$$

or $$v = uP - \frac{h^2 P}{P+1}$$

which is of Clairaut's form and has the solution as

$$v = uc - \frac{h^2 c}{c+1}$$

where $u = x^2$ and $v = y^2$.

**EXAMPLE 3.18** Solve $y = 2px + y^2 p^3$.

***Solution*** Multiplying the given equation by $y$ and then putting $y^2 = v$, we obtain

$$v = x\frac{dv}{dx} + \frac{1}{8}\left(\frac{dv}{dx}\right)^3 = xP + \frac{1}{8}P^3$$

where $P = dv/dx$. This equation is of Clairaut's form and the solution is

$$v = cx + \frac{1}{8}c^3 \quad \text{or} \quad y^2 = cx + \frac{1}{8}c^3$$

## EXERCISES

Solve the following differential equations:

**1.** $p^2 - 7p + 12 = 0$. **2.** $yp^2 + (x - y)p - x = 0$.

**3.** $p^3 + 2xp^2 - y^2p^2 - 2xy^2p = 0$. **4.** $4y^2p^2 + 2xyp(3x + 1) + 3x^3 = 0$.

**5.** $p^3(x + 2y) + 3p^2(x + y) + (y + 2x)p = 0$.

**6.** $x^2p^2 - 2xyp + (2y^2 - x^2) = 0$. **7.** $xp^2 - 2yp + ax = 0$.

**8.** $y = 2px + \tan^{-1}(xp^2)$.

**9.** $y = 3x + \log p$.

**10.** $x^2 + p^2x = yp$.

**11.** $p = \tan\left(x - \dfrac{p}{1+p^2}\right)$.

**12.** $x^2 = a^2(1 + p^2)$.

**13.** $x = y - p^2$.

**14.** $p^3 - p(y + 3) + x = 0$.

**15.** $y = 2px + y^{n-1}p^n$.

**16.** $y = 2p + 3p^2$.

**17.** $x(1 + p^2) = 1$.

**18.** $y^2 + xyp - x^2p^2 = 0$.

**19.** $y = yp^2 + 2px$.

**20.** $(x - a)p^2 + (x - y)p - y = 0$.

**21.** $\sin px \cos y = \cos px \sin y + p$.

**22.** $xy(y - px) = x + py$.

**23.** $(x^2 + y^2)(1 + p)^2 - 2(x + y)(1 + p)(x + py) + (x + yp)^2 = 0$.
[*Hint*: Put $x^2 + y^2 = u$ and $x + y = v$.]

**24.** Solve $x^2p^2 + yp(2x + y) + y^2 = 0$ by reducing it to Clairaut's form by using the substitution $y = u$ and $xy = v$.

**25.** Use the transformation $x^2 = u$ and $y^2 = v$ to solve the equation

$$axyp^2 + (x^2 - ay^2 - b)p - xy = 0.$$

CHAPTER 4

# Applications of First Order Differential Equations

In this chapter, we shall give some applications of the differential equations which appeared in Chapters 2 and 3. The examples worked out here have been chosen from the fields of engineering, physics, chemistry, geology, biological and social sciences, national defence, and so on.

## 4.1 GROWTH AND DECAY

The initial value problem

$$\frac{dx}{dt} = kx, \qquad x(t_0) = x_0 \tag{1}$$

where $k$ is a constant, occurs in many physical theories involving either growth or decay. For example, in biology it is often observed that the rate at which certain bacteria grow is proportional to the number of bacteria present at any time. Over short intervals of time, the population of small animals, such as rodents, can be predicted quite accurately by the solution of Eq. (1). The constant $k$ can be obtained from the solution of the differential equation by using a subsequent measurement of the population at a time $t_1 > t_0$.

In physics, an initial value problem such as the one shown in Eq. (1) provides a model for approximating the remaining amount of a substance which is disintegrating through radioactivity. Equation (1) can also be used to determine the temperature of a cooling body. In chemistry, the amount of a substance remaining during a reaction is also described by Eq. (1).

We shall now illustrate, with the help of the following solved examples, how Eq. (1) works.

**EXAMPLE 4.1** A culture initially has $N_0$ number of bacteria. At $t = 1$ hr. the number of bacteria is measured to be $(3/2)N_0$. If the rate of growth is proportional to the number of bacteria present, determine the time necessary for the number of bacteria to triple.

***Solution*** The present problem is governed by the differential equation

$$\frac{dN}{dt} = kN \tag{2}$$

subject to $N(0) = N_0$.
Separating the variables in (2) and solving, we have

$$N = N(t) = ce^{kt}$$

At $t = 0$, we have $N_0 = ce^0 = c$ and so $N(t) = N_0e^{kt}$. At $t = 1$, we have (3/2) $N_0 = N_0e^k$ or $e^k = 3/2$, which gives $k = \log(3/2) = 0.4055$. Thus

$$N(t) = N_0e^{0.4055t}$$

To find the time at which the bacteria have tripled, we solve

$$3N_0 = N_0e^{0.4055t}$$

for $t$, to get

$$0.4055t = \log 3$$

or $$t = \frac{\log 3}{0.4055} = 2.71 \text{ hr (approx.)}$$

*Note:* The function $N(t)$, using the laws of exponents, can also be written as

$$N(t) = N_0(e^k)^t = N_0\left(\frac{3}{2}\right)^t$$

since, $e^k = 3/2$. This latter solution provides a convenient method for computing $N(t)$ for small positive integral values of $t$; it also shows the influence of the subsequent experimental observation at $t = 1$ on the solution for all time. Also, it may be noticed that the actual number of bacteria present at time $t = 0$ is quite irrelevant in finding the time required to triple the number in the culture. The necessary time to triple, say, 100 or 10,000 bacteria is still approximately 2.71 hours.

As shown in the Fig. 4.1, exponential function $e^{kt}$ increases as $t$ increases for $k > 0$, and decreases as $t$ decreases. Thus, problems describing growth, such as population, bacteria, or even capital, are characterized by a positive value of $k$, whereas problems involving decay, as in radioactive disintegration, will yield a negative value.

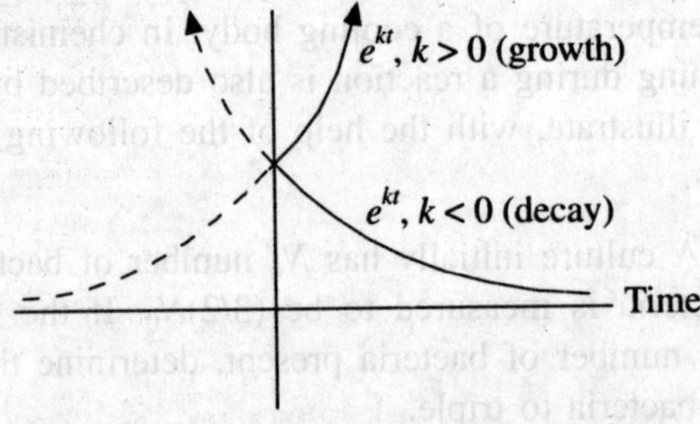


**Fig. 4.1** Exponential growth and decay.

**EXAMPLE 4.2** Bacteria in a certain culture increase at a rate proportional to the number present. If the number $N$ increases from 1000 to 2000 in 1 hour, how many are present at the end of 1.5 hours?

***Solution*** The differential equation $dN/dt = kN$ has the solution $N = N_0e^{kt}$. Since, $N = 1000$ when $t = 0$, $N_0 = 1000$. Setting $N = 2000$, and $t = 1$ in $N = 1000\,e^{kt}$, we obtain, $e^k = 2$. Thus, $N(t) = 1000(e^k)^t = 1000(2)^t$, and $N(1.5) = 1000(2)^{1.5} = 2828.4271$.

**EXAMPLE 4.3** In a culture of yeast, the amount $A$ of active yeast grows at a rate proportional to the amount present. If the original amount $A_0$ doubles in 2 hours, how long does it take for the original amount to triple?

***Solution*** The amount $A_0$ grows exponentially according to $A = A_0e^{kt}$. Since, $A = 2A_0$ when $t = 2$, $2A_0 = A_0e^{2k}$ and $e^{2k} = 2$. Thus, at time $t$

$$A = A_0(e^{2k})^{t/2} = A_0(2)^{t/2}$$

Setting $A = 3A_0$ and solving for $t$, we get

$$3A_0 = A_0(2)^{t/2}$$

or $$\log 3 = \frac{t}{2}\log 2, \qquad t = \frac{2\log 3}{\log 2} = 3.1699 \text{ hr}$$

**EXAMPLE 4.4** Bacteria in a certain culture increase at a rate proportional to the number present. If the number doubles in one hour, how long does it take for the number to triple?

***Solution*** Let $y$ denote the numbers present at time $t$. Then the function denoted by $y = f(t)$, satisfies the differential equation

$$\frac{dy}{dt} = ky$$

and the conditions $t = 0$, $y = y_0$; $t = 1$, $y = 2y_0$. Writing the differential equation in linear form, i.e.

$$\frac{dy}{dt} - ky = 0$$

we see that $e^{\int -k\,dt} = e^{-kt}$ is an integrating factor and hence, the equation has the solution

$$ye^{-kt} = \int 0\,dt + c \qquad \text{or} \qquad y = ce^{kt}$$

From the condition $t = 0$, $y = y_0$, we find that $y_0 = c$. The additional condition $t = 1$, $y = 2y_0$ is now used to find the constant of proportionality $k$. From $2y_0 = y_0e^{k(1)}$, we obtain $k = \log 2$. Hence, $y = y_0e^{t\log 2}$. Substituting $y = 3y_0$ and solving for $t$ yields

$$3y_0 = y_0e^{t\log 2} \quad \text{or} \quad \log 3 = t\log 2$$

Hence, the number will be triple in

$$t = \frac{\log 3}{\log 2} = 1.5849 \text{ hr}$$

## 4.2 DYNAMICS OF TUMOUR GROWTH

It has been observed experimentally that free-living dividing cells, such as bacteria cells, grow at a rate proportional to the volume of dividing cells at that moment. Let $V(t)$ denote the volume of dividing cells at time $t$. Then

$$\frac{dV}{dt} = kV \tag{3}$$

for some positive constant $k$. The solution of Eq. (3) is

$$V(t) = V_0 e^{k(t - t_0)} \tag{4}$$

where $V_0$ is the volume of dividing cells at time $t_0$ (initial time). Thus, free-living dividing cells grow exponentially with time, whereas solid tumours do not grow exponentially with time. As the tumour becomes larger, the doubling time of the total tumour volume continuously increases. A number of researchers have shown that the data for many solid tumours is fitted remarkably well, over almost a thousand-fold increase in tumour volume, by the equation

$$V(t) = V_0 \exp\left[\frac{k}{a}(1 - e^{-at})\right] \tag{5}$$

where $k$ and $a$ are positive constants.

Equation (5) is usually known as a *Gompertzian relation*. It states that tumour grows more and more slowly with the passage of time, and that it ultimately approaches the limiting volume $V_0 e^{k/a}$. Medical scientists have long been concerned with explaining this deviation from simple exponential growth. An insight into this problem can be gained by finding a differential equation satisfied by $V(t)$.

Differentiation of Eq. (5) yields

$$\frac{dV}{dt} = V_0 k e^{-at} \exp\left[\frac{k}{a}(1 - e^{-at})\right] = k e^{-at} V \tag{6}$$

Equation (6) can also be arranged as

$$\frac{dV}{dt} = (k e^{-at})V \tag{6a}$$

$$\frac{dV}{dt} = k(e^{-at} V) \tag{6b}$$

With these arrangements of Eq. (6), two theories have been evolved for the dynamics of tumour growth. According to the first theory, the retarding effect of tumour growth is due to an increase in the mean generation time of the cells, without a change in the proportion of reproducing cells. As time goes on, the reproducing cells mature or age, and thus divide more slowly. This theory corresponds to Eq. (6a). On the other hand, the second theory corresponding to Eq. (6b), suggests that the mean generation time of the dividing cells remains constant, and the retardation of growth is due to a loss in reproductive cells in the tumour. One possible explanation for this is that a necrotic region develops in the centre of the tumour. This necrosis appears at a critical size for a particular type of tumour, and thereafter the necrotic "core" increases rapidly as the total tumour mass increases. According to this theory a necrotic core develops because in many tumours the supply of blood, and thus of oxygen and nutrients, is almost completely confined to the surface of the tumour and a short distance beneath it. As the tumour grows, the supply of oxygen to the central core by diffusion becomes more and more difficult resulting in the formation of a necrotic core.

## 4.3 RADIOACTIVITY AND CARBON DATING

The science of radiogeology applies our knowledge of radioactivity to geology. It is known that uranium 238 undergoes radioactive decay with half-life $T = 4.55$ billion years. During decay, it becomes radium 226 and eventually ends as non-radioactive lead 206. Radioactive dating uses this knowledge to estimate the date of events that took place long ago. One technique uses the ratio of lead to uranium in a rock formation to estimate the time that has elapsed since the lava solidified and formed the mass of the rock. The age of the solar system and hence, of the earth has been estimated by radioactive dating as 4.5 billion years. Other elements such as potassium (half-life 13.9 billion years) and rubidium (half-life 50 billion years) are also used in radioactive dating.

An interesting description of use of Pb 210 or white lead (half-life 22 years) is in determining whether a given oil painting is authentic or a forgery. For instance, it can be proved that the beautiful painting *Disciples at Emmaus* which was bought by the Rembrandt Society of Belgium for \$170,000 was a modern forgery (for details see [11]).

An important breakthrough in radiography occurred in 1947 when an American chemist Willard Frank Libby discovered radiocarbon (a radioactive isotope of carbon), designated as carbon-14 ($^{14}C$). For this discovery and its application to radiogeology and radiochronology, Libby received the Nobel Prize in Chemistry in 1960. The basis of this method is as follows: The atmosphere of the earth is continuously bombarded by cosmic rays. These cosmic rays produce neutrons in the earth's atmosphere, and these neutrons combine with nitrogen to produce $^{14}C$. This radiocarbon ($^{14}C$) is incorporated in carbon dioxide and thus moves through the atmosphere to be absorbed by plants. In turn, radiocarbon is built in animal tissues by eating the plants. In living tissues, the rate of ingestion of $^{14}C$ exactly balances the rate of disintegration of $^{14}C$. When an organism dies, though, it ceases to ingest $^{14}C$, its $^{14}$C concentration begins to decrease through disintegration of the $^{14}$C present. Now, it is a physically accepted fact that the rate of bombardment of the earth's atmosphere by cosmic rays has always been constant. This implies that the original rate of disintegration of $^{14}$C in a sample such as charcoal is the same as the rate measured today.* This assumption enables us to find the age of a sample of charcoal.

Let $N(t)$ denote the amount of carbon-14 present in a sample at time $t$, and $N_0$ denote the amount present at time $t = 0$ when the sample was formed. If $k$ denotes the decay constant of $^{14}$C (half-life 5568 years), then $dN(t)/dt = -kN(t)$, $N(0) = N_0$ and, consequently, $N(t) = N_0 e^{-kt}$. Now, the present rate $R(t)$ of disintegration of $^{14}$C in the sample is given by $R(t) = kN(t) = kN_0 e^{-kt}$, and the original rate of disintegration is $R(0) = kN_0$. Thus, $R(t)/R(0) = e^{-kt}$ so that $t = (1/k) \log [R(0)/R(t)]$. Hence, if we measure $R(t)$, the present rate of disintegration of $^{14}$C in the charcoal, and observe that $R(0)$ must equal the rate of disintegration of $^{14}$C in a comparable amount of living wood, then we can compute the age $t$ of the charcoal.

We shall work out now some examples about radioactive decay and carbon dating.

**EXAMPLE 4.5** It is found that 22 per cent of the original radiocarbon in a wooden archaeological specimen has decomposed. Use the half-life $T = 5568$ yr. of $^{14}$C to compute the number of years since the specimen was a part of a living tree. (This should yield a good estimate of the time elapsed since the specimen, a wooden bowl, was used in an ancient civilization.)

***Solution*** Let $C$ be the amount of $^{14}$C present at time $t$ and $C_0$ the amount present at $t = 0$. Then $C = C_0 e^{kt}$ gives $0.78C_0 = C_0 e^{kt}$, and finally, $t = \log 0.78/k$. From $T = -\log 2/k = 5568$, we obtain $k = -\log 2/5568$ and, therefore,

$$t = \frac{5568 \log 0.78}{-\log 2} = 1996 \text{ yr. (approx.)}$$

At arbitrary time $t$,

$$C = C_0 \exp\left(\frac{-\log 2}{5568}\right) t = C_0 (2)^{-t/5568}$$

**EXAMPLE 4.6** It is found that 0.5 per cent of radium disappears in 12 years. (a) What percentage will disappear in 1000 years? (b) What is the half-life of radium?

---

* Since the mid-1950's, the testing of nuclear weapons has significantly increased the amount of radiocarbon in our atmosphere. Ironically, this unfortunate state of affairs provides yet another extremely powerful method of detecting art forgeries. Most of the materials of artists, such as linseed oil and canvas paper, come from plants and animals, and so will contain the same concentration of carbon-14 as the atmosphere at the time the plant or animal dies. Thus linseed oil (which is derived from the flax plant) that was produced during the last few years will contain a much greater concentration of $^{14}$C than linseed oil produced before 1950.

***Solution*** Let $A$ be the quantity of radium in grammes, present after $t$ years. Then $dA/dt$ represents the rate of disintegration of radium. According to the law of radioactive decay, we have

$$\frac{dA}{dt} \propto A \qquad \text{or} \qquad \frac{dA}{dt} = aA$$

Since $A$ is positive and is decreasing, then $dA/dt < 0$ and we see that the constant of proportionality $a$ must be negative. Writing $a = -k$, we get

$$\frac{dA}{dt} = -kA$$

Let $A_0$ be the amount, in grammes of radium present initially. Then $0.005A_0$ g disappears in 12 years, so that $0.995A_0$ g remains. We thus, have $A = A_0$ at $t = 0$, and $A = 0.995A_0$ g at $t = 12$ (years). The solution of the above equation is $A = Ce^{-kt}$. Since, $A = A_0$ at $t = 0$ and $C = A_0$, hence

$$A = A_0e^{-kt}$$

Also, at $t = 12$ and $A = 0.995A_0$, then

$$0.995A_0 = A_0e^{-12k} \quad \text{or} \quad e^{-12k} = 0.995 \quad \text{or} \quad e^{-k} = (0.995)^{1/12} \tag{7}$$

Hence

$$A = A_0e^{-kt} = A_0(e^{-k})^t = A_0(0.995)^{t/12} \tag{8}$$

or, if we solve for $k$ in Eq. (7), we find $k = 0.000418$, so that

$$A = A_0e^{-0.000418t} \tag{9}$$

Thus, we have

(a) When $t = 1000$, from Eqs. (8) and (9), $A = 0.658A_0$, so that 34.2 percent will disappear in 1000 years.

(b) The half-life of a radioactive substance is defined as the time it takes for 50 per cent of the substance to disappear. In our case, we have $A = 1/2A_0$ and using Eq. (9), we find $e^{-0.000418t} = 1/2$ or $t = 1672.1770$ years.

**EXAMPLE 4.7** A breeder reactor converts the relatively stable uranium 238 into the isotope plutonium 239. After 15 years it is found that 0.043 per cent of the initial amount $A_0$ of the plutonium has disintegrated. Find the half-life of this isotope, if the rate of disintegration is proportional to the remaining amount.

***Solution*** Let $A(t)$ denote the amount of the plutonium remaining at any time. Then, the solution of the initial value problem

$$\frac{dA}{dt} = kA, \qquad A(0) = A_0$$

is $A(t) = A_0e^{kt}$.

If 0.043 per cent of the atoms of $A_0$ have disintegrated then 99.957 percent of the substance remains. To find $k$, we solve $0.99957A_0 = A_0e^{15k}$ to get $e^{15k} = 0.99957$ or, $k = \log(0.99957)/15 = -0.00002867$. Hence, $A(t) = A_0 e^{-0.00002867t}$.

Now, the half-life is the corresponding value of time for which $A(t) = A_0/2$. Solving for $t$, we get

$$\frac{A_0}{2} = A_0 e^{-0.00002867t}$$

or
$$t = \frac{\log 2}{0.00002867} = 24176.74156 \text{ yr.}$$

**EXAMPLE 4.8** A fossilized bone is found to contain 1/1000 the original amount of $^{14}C$. Determine the age of the fossil.

***Solution*** We have

$$A(t) = A_0 e^{kt}$$

When $t = 5568$ yr., $A(t) = A_0/2$, from which we can find the value of $k$ as

$$\frac{A_0}{2} = A_0 e^{5568k}$$

or
$$k = -\frac{\log 2}{5568} = -0.000124488$$

Therefore, $A(t) = A_0 e^{-0.000124488t}$, where $A(t) = 1/1000$, gives

$$t = \frac{\log 2}{0.000124488} = 55489.32651 \text{ yr}$$

## 4.4 COMPOUND INTEREST

Interest is defined as a charge for the borrowed money. It is difficult to give a precise definition of interest since there exist many different methods for computing interest. If a principal of $P$ rupees, invested at an interest rate $r$ per annum grows to $P(1 + r)$ rupees in 1 year, $P(1 + r)^2$ rupees in 2 years and $P(1 + r)^t$ rupees in $t$ years, $r$ is called the rate of interest per annum compounded annually. If the interest rate per annum is $r$ and interest is compounded twice a year, $P$ rupees grows to $P(1 + \frac{r}{2})$ rupees in 6 months, $P(1 + \frac{r}{2})^2$ in a year, $P(1 + \frac{r}{2})^3$ rupees in 1.5 years, and $P(1 + \frac{r}{2})^t$ in $t$ years. If $P$ rupees invested at interest rate $r$ per annum with interest compounded $k$ times per year, then the amount $a$ of the original investment at the end of $t$ year is

$$a = P\left(1 + \frac{r}{k}\right)^{kt} = f(t) \tag{10}$$

The quantity $r/k$ is the interest rate applied at each compounding, and $e^{kt}$ is the total number of compoundings in $t$ years. For example, 1 rupee invested at 10 per cent per annum compounded annually ($r = 10$) grows to $1(1 + 0.1)^t = 1.10$ rupee in 1 year. One rupee invested at 10 per cent per annum compounded twice a year ($k = 2$) grows to $1(1 + 0.05)^2 = 1.1025$ rupees in 1 year. In the second instance,

10 per cent is called the nominal interest rate per annum and 10.25 per cent the effective interest rate per annum. Similarly, if the interest is compounded quarterly ($k = 4$), the same rupee becomes $1(1 + 0.025)^4 = 1.1038$ rupee in 1 year and the effective interest rate per annum is 10.38 per cent.

It can be proved easily that for fixed values of $P$, $r$ and $t$, the value of $a$ in Eq. (10) increases as $k$ increases. As $k \to \infty$, the value of $a$ does not increase without limit, but rather as

$$\lim_{k\to+\infty} a = \lim_{k\to+\infty} P\left(1+\frac{r}{k}\right)^{kt} = P \lim_{k\to+\infty}\left[\left(1+\frac{r}{k}\right)^{k/r}\right]^{rt}$$

$$= P\left[\lim_{k\to+\infty}\left(1+\frac{r}{k}\right)^{k/r}\right]^{rt} = Pe^{rt}$$

where $e \simeq 2.71828$.

It is reasonable to argue that money should earn interest continuously. But even if the interest is compounded every second, then the function $f(t)$ in Eq. (10) is a step function and is not continuous. In order to have a continuous model for continuous compounding, define $A$ as

$$A = Pe^{rt} = F(t) \tag{11}$$

This definition looks more reasonable as $da/dt$ in Eq. (10) is zero or undefined, while $dA/dt$ is more useful and meaningful. Also, the graph of $f$ in Eq. (10) and $F$ in Eq. (11) are virtually indistinguishable when $k$ is very large. Equation (11), yields on differentiation

$$\frac{dA}{dt} = Pre^{rt} = rA \tag{12}$$

Thus, we see that money invested at compound interest compounded continuously grows according to the differential equation of the organic growth. Conversely, a quantity, such as the number of bacteria in a culture grows according to the compound interest law.

**EXAMPLE 4.9** If ₹ 10,000 is invested at 6 per cent per annum, find what amount has accumulated after 6 years if interest is compounded: (a) annually, (b) quarterly, and (c) continuously.

***Solution***

(a) $a = 10{,}000(1.06)^6 =$ ₹ 14,185.19

(b) $a = 10{,}000\left(1+\dfrac{0.06}{4}\right)^{4(6)} = 10{,}000(1.015)^{24} =$ ₹ 14,295.03

(c) $A = 10{,}000e^{0.36} =$ ₹ 14,333.29

**EXAMPLE 4.10** How long does it take for a given amount of money to double at 6 per cent per annum compounded: (a) annually, and (b) continuously?

***Solution*** (a) We have $2P = P(1.06)^t$. Therefore,

$$\log 2 = t \log 1.06 \quad \text{or} \quad t = \frac{\log 2}{\log 1.06} = 11.89566 \text{ yr.}$$

(b) From question $2P = Pe^{0.06t}$. Then

$$\log 2 = 0.06t \quad \text{or} \quad t = \left(\frac{\log 2}{0.06}\right) = 11.55245 \text{ yr.}$$

**EXAMPLE 4.11** A savings and loan company advertises an interest rate per annum of 7.5 per cent compounded continuously. Find the effective interest rate per annum.

***Solution*** The interest for 1 year will be

$$Pe^1 - P = \frac{P}{e^1 - 1}$$

Thus $$e^{0.075} - 1 = 0.0779$$

and the effective rate per annum is 7.79 per cent. Many companies advertise an effective rate of 7.90 per cent. This is obtained by using a 360-day year and multiply 7.79 by 365/360. A company can thus make a more attractive offer without exceeding the nominal rate 7.5 per cent set by law. This illustrates our statement that there are many ways of calculating the interest.

REMARK. The nominal rate $j$, which by continuous compounding is equivalent to an effective rate $i$, is called the *force of interest.* It is useful in comparing various business propositions.

## 4.5 BELT OR CABLE FRICTION

Consider a belt (Fig. 4.2) wrapped around a drum of radius $r$. Suppose that tensions $T_1$ and $T_0$ at $B$ and $A$, respectively, are such that the belt is on the point of slipping and that $T_1 > T_0$. The tension in the belt varies from $T = T_0$, corresponding to $\theta = 0$ to $T = T_1$, corresponding to $\theta = \angle AOB$, known as the *angle of wrap.* Let $T$ denote the tension at the point $P$ corresponding to a fixed but arbitrary value of $\theta$ and $T + \Delta T$ be the tension at a nearby point corresponding to $\theta + \Delta\theta$. The arc of the belt of length $\Delta s = r\Delta\theta$ is in equilibrium under the action of the forces having the magnitudes $T$, $T + \Delta T$, $\Delta N$, $\Delta F$, where $\Delta N$ is the magnitude of the normal force the drum exerts on the section of the belt of length $\Delta s$, and $\Delta F$ is the magnitude of the force of friction opposing slipping. The force having magnitude $T + \Delta T$ acts at an angle $\Delta\theta$ with the tangential direction $t$ along which the force of magnitude $T$ acts. As the section of the belt of length $\Delta s$ is in equilibrium, the sum of the components of the four external forces acting on it in any direction must be zero. Choosing the tangential and normal directions $t$ and $n$, we get

$$T + \Delta F - (T + \Delta T) \cos \Delta\theta = 0$$

$$\Delta N - (T + \Delta T) \sin \Delta\theta = 0$$

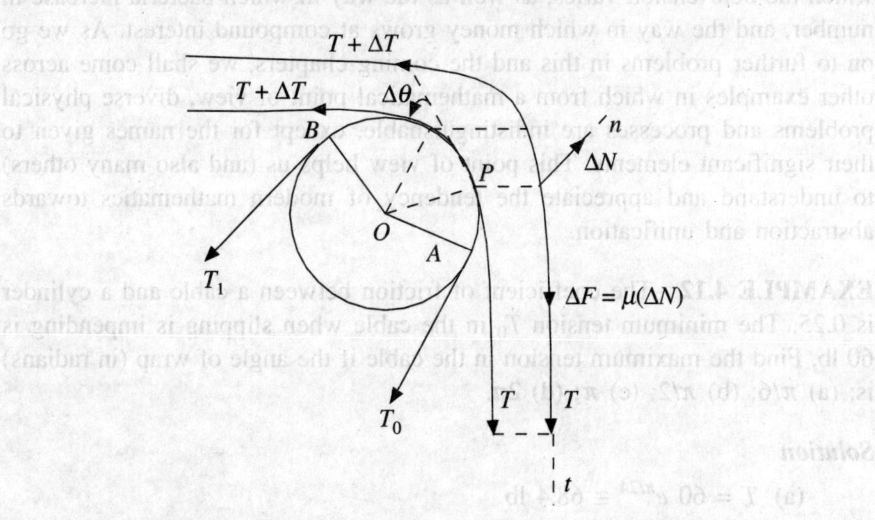


**Fig. 4.2** A belt wrapped around a drum of radius $r$.

Since, the belt is on the point of slipping, we can replace $\Delta F$ by $\mu(\Delta N)$, where $\mu$ is the coefficient of friction. Now eliminating $\Delta N$, we obtain

$$T + \mu(T + \Delta T) \sin \theta - (T + \Delta T) \cos \theta = 0$$

which, after division by $\Delta\theta$ and rearrangement, becomes

$$T\frac{1 - \cos \Delta\theta}{\Delta\theta} + \mu(T + \Delta T)\frac{\sin \Delta\theta}{\Delta\theta} - \frac{\Delta T}{\Delta\theta}\cos \Delta\theta = 0$$

Now taking limit as $\Delta\theta \to 0$, we get

$$\frac{dT}{d\theta} = \mu T \tag{13}$$

which has the solution

$$T = T_0 e^{\mu\theta} \tag{14}$$

(The angle $\theta$ must be measured in radians because

$$\lim_{x \to 0} \frac{\sin x}{x} = 1$$

only when $x$ is measured in radians.)

This application has been introduced here with some objectives. First, we want to show how we can derive a differential equation by making assumptions about a complicated physical situation. (For more details about the friction of the belt, see [21]). Second, we want to show that the same mathematical model often governs many apparently unrelated physical situations. We have seen that the differential equation of the organic growth governs the way in which the belt tension varies, as well as the way in which bacteria increase in number, and the way in which money grows at compound interest. As we go on to further problems in this and the coming chapters, we shall come across other examples in which from a mathematical point of view, diverse physical problems and processes are indistinguishable, except for the names given to their significant elements. This point of view helps us (and also many others) to understand and appreciate the tendency of modern mathematics towards abstraction and unification.

**EXAMPLE 4.12** The coefficient of friction between a cable and a cylinder is 0.25. The minimum tension $T_0$ in the cable when slipping is impending is 60 lb. Find the maximum tension in the cable if the angle of wrap (in radians) is: (a) $\pi/6$; (b) $\pi/2$; (c) $\pi$; (d) $2\pi$.

***Solution***

(a) $T = 60\ e^{\pi/24} \cong 68.4$ lb

(b) $T = 60\ e^{\pi/8} \cong 88.9$ lb

(c) $T = 60\ e^{\pi/4} \cong 131.6$ lb

(d) $T = 60\ e^{\pi/2} \cong 288.6$ lb.

**EXAMPLE 4.13** The coefficient of friction between the fixed horizontal cylinder and the cable is 0.3 (Fig. 4.3). Find what magnitude of the force $F$ is just sufficient: (a) to start the 300 lb body moving upward; (b) to prevent the 300 lb body from moving downward.

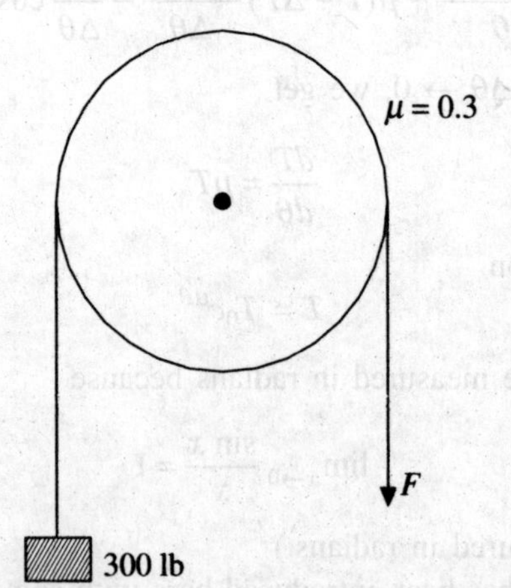


**Fig. 4.3** Movement of a cable on a fixed horizontal cylinder.

***Solution***

(a) $F = 300\, e^{0.3\pi} \cong 769.9$ lb

(b) $300 = F\, e^{0.3\pi}$ or $F = 300\, e^{-0.3\pi} \cong 116.9$ 1b.

## 4.6 TEMPERATURE RATE OF CHANGE (NEWTON'S LAW OF COOLING)

Under certain conditions, the temperature rate of change of a body is proportional to the difference between the temperature $T$ of the body and the temperature $T_0$ of the surrounding medium. This is known as *Newton's law of cooling*. Here, we shall consider the case in which $T_0$ remains constant and also suppose that heat flows rapidly enough that the temperature $T$ of the body is the same at all points of the body at a given time $t$.

If $T = f(t)$ denotes the temperature of the body at time $t$, then $f$ satisfies the differential equation

$$\frac{dT}{dt} = k(T - T_0) \tag{15}$$

where $k < 0$ (Fig. 4.4). This equation can be solved by separating the variables.

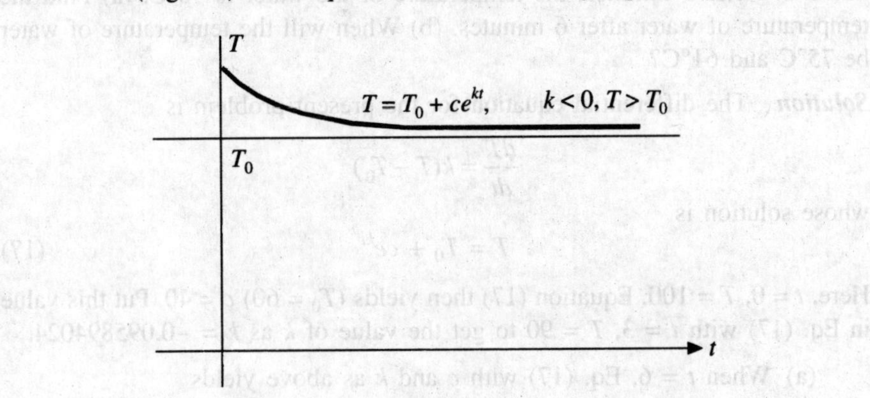


**Fig. 4.4** Newton's law of cooling.

**EXAMPLE 4.14** A body whose temperature $T$ is initially 200°C is immersed in a liquid when temperature $T_0$ is constantly 100°C. If the temperature of the body is 150°C at $t = 1$ minute, what is its temperature at $t = 2$ minutes?

***Solution*** Separating the variables in Eq. (15), we get

$$\frac{dT}{T - 100} = kdt$$

and the solution is

$$\log (T - 100) = kt + C \tag{16}$$

When $t = 0$, $T = 200$, we find that $C = \log 100$. Also, at $t = 1$, $T = 150$ and Eq. (16) gives

$$\log 50 = k(1) + \log 100$$

or
$$k = -\log 2$$

Now, substituting $C = \log 100$ and $k = -\log 2$ in Eq. (16), we obtain

$$\log (T - 100) = -t \log 2 + \log 100$$

or
$$T = 100(1 + 2^{-t})$$

Thus, at $t = 2$ min, $T = 125°C$.

**Remark.** Consider Eq. (15), which has a solution of the form

$$\log (T - T_0) = kt + \log c \quad \text{or} \quad T = T_0 + ce^{kt}$$

Now, as $t \to \infty$, then $T \to T_0$ and, consequently, $dT/dt = k(T - T_0) \to 0$. That is, as $t$ becomes large, the difference between the temperature of the body and the temperature of the surrounding medium approaches zero, and the rate at which the body cools also approaches zero (see Fig. 4.4).

**EXAMPLE 4.15** Water is heated to the boiling point temperature 100°C. It is then removed from heat and kept in a room which is at a constant temperature of 60°C. After 3 minutes, the temperature of the water is 90°C. (a) Find the temperature of water after 6 minutes. (b) When will the temperature of water be 75°C and 61°C?

***Solution*** The differential equation for the present problem is

$$\frac{dT}{dt} = k(T - T_0)$$

whose solution is

$$T = T_0 + ce^{kt} \tag{17}$$

Here, $t = 0$, $T = 100$. Equation (17) then yields ($T_0 = 60$) $c = 40$. Put this value in Eq. (17) with $t = 3$, $T = 90$ to get the value of $k$ as $k = -0.095894024$.

(a) When $t = 6$, Eq. (17) with $c$ and $k$ as above yields

$$T = 60 + 40\, e^{-0.095894024t} = 82.5°C$$

(b) If $T = 75°C$ and $T = 61°C$, then from Eq. (17), we get

$$75° = 60° + 40\, e^{-0.095894024t} \quad \text{or} \quad t = 10.2 \text{ min.}$$

and

$$61° = 60° + 40\, e^{-0.095894024t} \quad \text{or} \quad t = 38.5 \text{ min.}$$

**EXAMPLE 4.16** When a cake is removed from an oven, its temperature is measured at 300°F. Three minutes later its temperature is 200°F. How long will it take to cool off to a room temperature of 70°F?

***Solution*** Here, we have to solve the initial value problem

$$\frac{dT}{dt} = k(T - T_0), \quad T(0) = 300 \tag{18}$$

and find the value of $k$ so that $T(3) = 200$. Equation (18), after separating the variables, has a solution of the form

$$T = T_0 + ce^{kt}$$

to give the value of $c = 230$, and thus $k = -0.19018$, so that

$$T(t) = 70 + 230e^{-0.19018t} \tag{19}$$

However, Eq. (19) gives no finite solution to $T(t) = 70$ since $\lim_{t \to \infty} T(t) = 70$. Yet intuitively, we expect the cake will be at room temperature after a long period of time. This period is given in the tabular form in Fig. 4.5(b).

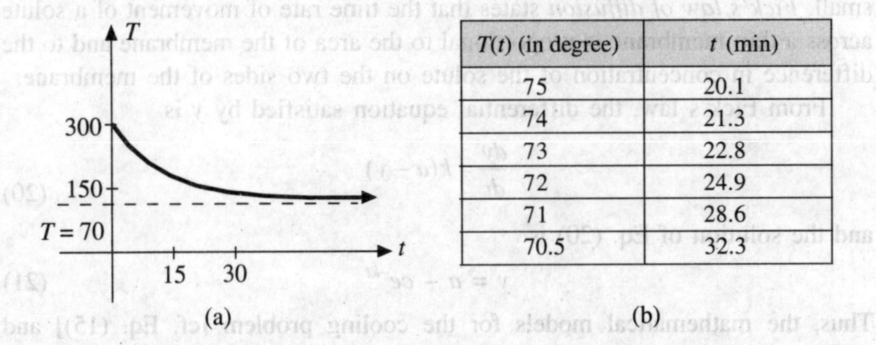


| $T(t)$ (in degree) | $t$ (min) |
|---|---|
| 75 | 20.1 |
| 74 | 21.3 |
| 73 | 22.8 |
| 72 | 24.9 |
| 71 | 28.6 |
| 70.5 | 32.3 |

**Fig. 4.5** Cooling time for a cake at room temperature.

Figures 4.5(a) and (b) show that the cake will approximately be at room temperature in about half an hour.

**EXAMPLE 4.17** *Estimation of time of murder.* The body of a murder victim was discovered at 11.00 p.m. The doctor took the temperature of the body at 11.30 p.m., which was 94.6°F. He again took the temperature after one hour when it showed 93.4°F, and noticed that the temperature of the room was 70°F. Estimate the time of death. (Normal temperature of human body = 98.6°F.)

***Solution*** The differential equation governing the present situation is

$$\frac{dT}{dt} = k(T - T_0)$$

whose solution is $T = T_0 + ce^{kt}$.

At $t = 0$, $T = 94.6$ and this gives $c = 24.6$. When $t = 60$ min., $T = 93.4$, and this gives

$$k = \frac{1}{60}\log\left(\frac{23.4}{24.6}\right) = -0.000361988$$

Now, using the values of $c$, $k$, $T = 98.6$ and $T_0 = 70$ in the solution of the given differential equation, after simplification we get

$$t = -3.012573443$$

Therefore, the estimated time of death is

$$11.30 - 3.0 = 8.30 \text{ p.m. (approx.)}$$

## 4.7 DIFFUSION

Let $y$ denote the concentration in mg/cm$^3$ of a drug or chemical in a small body, and let $y_0$ be the concentration at time $t = 0$. Suppose, the body is placed in a container or vat in which the concentration of the drug or chemical is $a$, where $a > y_0$. The concentration in the small body will increase but $a$ remains constant. This is a reasonable assumption if the vat is large and the body is small. *Fick's law of diffusion* states that the time rate of movement of a solute across a thin membrane is proportional to the area of the membrane and to the difference in concentration of the solute on the two sides of the membrane.

From Fick's law, the differential equation satisfied by $y$ is

$$\frac{dy}{dt} = k(a - y) \tag{20}$$

and the solution of Eq. (20) is

$$y = a - ce^{-kt} \tag{21}$$

Thus, the mathematical models for the cooling problem [cf. Eq. (15)] and diffusion problems are essentially the same, except that, here $a > y_0$ and $k > 0$. Consequently, $dy/dt$ is positive for $t \geq 0$ and $y \to a$ as $t \to \infty$ (see Fig. 4.6).

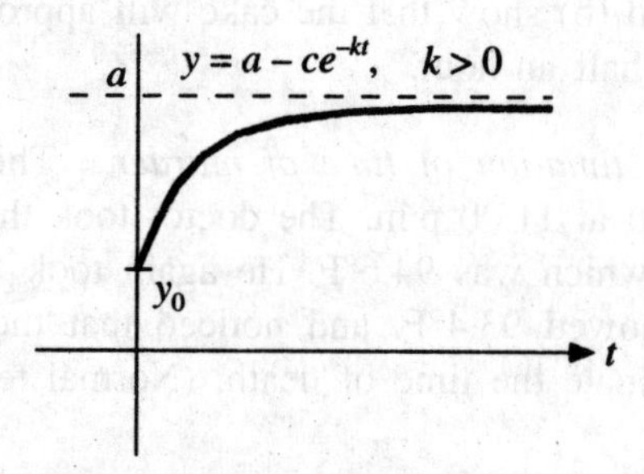


**Fig. 4.6** Fick's law of diffusion.

This model is adequate for describing many important phenomena, although it must be modified in a number of physical situations. The body might be a human organ, and often the concentration $y_0$ is zero. The diffusion model is essentially the model for cooling with the decreasing temperature replaced by an increasing temperature.

**EXAMPLE 4.18** The concentration of the potassium in a kidney is 0.0025 mg/cm$^3$. The kidney is placed in a vat in which the potassium concentration is 0.0040 mg/cm$^3$. In 2 hours, the potassium concentration in the kidney is found to be 0.0030 mg/cm$^3$. What would be the concentration of potassium in the kidney 4 hours after it was placed in the vat? How long does it take for the concentration to reach 0.0035 mg/cm$^3$? Assume that the vat is sufficiently large and that the vat concentration $a$ = 0.0040 mg/cm$^3$ remains constant.

***Solution*** This problem can be solved using Eq. (21). Here, $t = 0$, $y = 0.0025$ and $a = 0.0040$. Put these in Eq. (21) to get $c = 0.0015$. Use this value of $c$ with $t = 2$ and $y = 0.0030$ in Eq. (21) to obtain $k = 0.088$. Now the concentration after 4 hours is

$$y = a - ce^{-kt} = 0.004 - (0.0015)e^{(-0.088)(4)}$$

or

$$y = 0.0033 \text{ mg/cm}^3$$

Also, the time required to reach the concentration level $y = 0.0035$ is

$$0.0035 = 0.0040 - (0.0015)e^{-(0.088)t} \quad \text{or} \quad t = 5.42 \text{ hr. (approx.)}$$

**EXAMPLE 4.19** *Intravenous feeding of glucose.* Infusion of glucose into the bloodstream is an important medical technique. To study this process, we define $G(t)$ as the amount of the glucose in the bloodstream of a patient at time $t$. Assume that the glucose is infused into the bloodstream at a constant rate of $k$ g/min. At the same time, the glucose is converted and removed from the bloodstream at a rate proportional to the amount of the glucose present. Then, the function $G(t)$ satisfies the differential equation

$$\frac{dG}{dt} = k - aG$$

which is a linear equation in $G$ with the integrating factor as $e^{at}$ and, therefore, the solution is

$$G(t) = \frac{k}{a} + ce^{-at}$$

At $t = 0$, $c = G(0) - k/a$, the solution can be written as

$$G(t) = \frac{k}{a} + \left[G(0) - \frac{k}{a}\right]e^{-at}$$

As $t \to \infty$, the concentration of glucose approaches an equilibrium value $k/a$.

More details about the infusion of the glucose into the veins are given in [23].

**EXAMPLE 4.20** *Nerve excitation.* The cells of a nerve fibre may be conceived as an electric system. The protoplasm contains a large number of different ions, both cations (positive electric charge) and anions (negative electric charge). When an electric current is applied to a nerve fibre, the cations move to the cathode, the anions to the anode, and the electric equilibrium is disturbed. This phenomenon leads to the excitation of the nerve.

Based on the observation that the excitation originates at the cathode, N. Rashevsky, developed a theory which postulates that two different kinds of cations are responsible for the process. One is exciting and the other kind is inhibiting. These two kinds are said to be *antagonistic factors.*

Let $E = E(t)$ be the concentration of the exciting cations and $F = F(t)$ be the concentration of the inhibiting cations near the cathode at any time $t$. The theory then states that excitation occurs whenever the ratio $E/F$ exceeds a certain value. Denoting this value by $C$, we have excitation when $E/F \geq C$ and there will be no excitation if $E/F < C$. Let $E_0$ and $F_0$ be the concentrations at rest of exciting and inhibiting cations, respectively. When $E$ increases and $F$ remains limited, there is excitation. When $E$ does not increase as fast as $F$, then there is no excitation.

Let $I$ be the intensity of the stimulant current. For convenience sake, assume that $I$ is constant during a certain time interval. Rashevsky showed that the excitation of nerves can be described by the differential equations

$$\frac{dE}{dt} = JI - K(E - E_0) \quad \text{and} \quad \frac{dF}{dt} = LI - M(F - F_0)$$

where $J$, $K$, $L$, $M$ are positive constants.

The above equations can be easily solved for $E$ and $F$, and finally the ratio $E/F$ determines whether excitation occurs and when. For more details see [9].

## 4.8 BIOLOGICAL GROWTH

A fundamental problem in biology is that of growth, whether it is the growth of a cell, an organ, a human, a plant or population. We have already dealt with the problem of growth (see Section 4.1), where we saw that the fundamental differential equation was Eq. (2). Now

$$\frac{dN}{dt} = kN \tag{22}$$

has the solution $N = ce^{kt}$, which at $t = 0$ gives $c = N_0$, and

$$N = N_0 e^{kt} \tag{23}$$

From this we see that growth occurs if $k > 0$, while decay (or shrinkage) occurs if $k < 0$. The model given by Eqs. (22) and (23) is called the *Malthusian law of growth,* named after T.R. Malthus (1766–1834). One obvious drawback of Eq. (22) and the corresponding solution [Eq. (23)] is that if $k > 0$ then we have $N \to \infty$ as $t \to \infty$, so that as time passes, growth is unlimited. This conflicts with reality, for after a certain period of time, we know that a cell or individual stops growing having attained a maximum size. We shall now modify Eq. (22) to include these biological facts as follows:

Suppose $N$ denotes the height of a human being (we can also consider the size of a cell) and assume that the rate of change of height depends on the height in a more complicated manner than simple proportionality as shown in Eq. (22). Thus, we have

$$\frac{dN}{dt} = F(N), \qquad N = N_0 \quad \text{for} \quad t = 0 \tag{24}$$

where $N_0$ represents the height at some specified time, $t = 0$, and $F$ is some suitable function but as yet unknown. Since the linear function $F(N) = kN$ is not suitable, we consider a next order of approximation given by a quadratic function $F(N) = aN - bN^2$, where we choose constant $b > 0$ in order to inhibit the growth of $N$ as demanded by reality. Thus, Eq. (24) becomes

$$\frac{dN}{dt} = aN - bN^2, \qquad N = N_0 \quad \text{for } t = 0 \tag{25}$$

Equation (25) is termed as a *logistic equation* and the growth governed by it is called *logistic growth*. The model represented by this equation is referred to as the *Verhulst-Pearl model*.

Separating the variables in Eq. (25), we obtain

$$\frac{dN}{aN - bN^2} = dt \quad \text{or} \quad \int \frac{dN}{N(a - bN)} = t + c$$

i.e.

$$\int \frac{1}{a}\left(\frac{1}{N} + \frac{b}{a - bN}\right) dN = t + c$$

or

$$\frac{1}{a}[\log N - \log(a - bN)] = t + c \tag{26}$$

Using $t = 0$ and $N = N_0$, we see that

$$c = \frac{1}{a}[\log N_0 - \log(a - bN_0)]$$

and Eq. (26) thus becomes

$$\frac{1}{a}[\log N - \log(a - bN)] = t + \frac{1}{a}[\log N_0 - \log(a - bN_0)]$$

Solving for $N$, yields

$$N = \frac{a}{b + \left(\dfrac{a}{N_0} - b\right)e^{-at}} = \frac{a/b}{1 + \left(\dfrac{a/b}{N_0} - 1\right)e^{-at}} \tag{27}$$

This is known as the *Verhulst formula.*

If we take the limit of Eq. (27) as $t \to \infty$, we get (since $a > 0$)

$$N_{\max} = \lim_{t \to \infty} N = \frac{a}{b} \tag{28}$$

which shows that there is a limit to the growth of $N$ as desired by the biological facts and tends to indicate the correctness of our mathematical model.

To apply Eq. (27), suppose that at times $t = 1$ and $t = 2$, the values of $N$ are $N_1$ and $N_2$, respectively, then from Eq. (27), we obtain

$$N_1 = \frac{a/b}{1 + \left(\frac{a/b}{N_0} - 1\right) e^{-a}}, \quad N_2 = \frac{a/b}{1 + \left(\frac{a/b}{N_0} - 1\right) e^{-2a}}$$

or

$$\frac{b}{a}(1 - e^{-a}) = \frac{1}{N_1} - \frac{e^{-a}}{N_0}, \quad \frac{b}{a}(1 - e^{-2a}) = \frac{1}{N_2} - \frac{e^{-2a}}{N_0} \tag{29}$$

To find $b/a$ and $a$ in terms of $N_0$, $N_1$, $N_2$, we have to proceed as follows: Divide the members of the second equation in relation (29) by the corresponding members of the first equation so that $b/a$ is eliminated. This yields

$$1 + e^{-a} = \frac{\dfrac{1}{N_2} - \dfrac{e^{-2a}}{N_0}}{\dfrac{1}{N_1} - \dfrac{e^{-a}}{N_0}} \tag{30}$$

or

$$e^{-a} = \frac{N_0(N_2 - N_1)}{N_2(N_1 - N_0)} \tag{31}$$

Putting this in the first equation of (29), we get

$$\frac{b}{a} = \frac{N_1^2 - N_0 N_2}{N_1(N_0 N_1 - 2N_0 N_2 + N_1 N_2)} \tag{32}$$

The value of Eqs. (31) and (32) can be used to write Eq. (27) in terms of suitably choosen values of $N_0$, $N_1$ and $N_2$. The limiting value of $N$ is

$$N_{\max} = \lim_{t \to \infty} N = \frac{N_1 (N_0 N_1 - 2N_0 N_2 + N_1 N_2)}{N_1^2 - N_0 N_2} \tag{33}$$

We shall now apply these results in the following examples.

**EXAMPLE 4.21** The mean height in inches of male children at various stages is shown in Table 4.1. Use the data to predict the mean height of adult males at full growth.

**TABLE 4.1** Mean Height of Male Children at Various Ages

| Age (years) | Height (inches) |
|---|---|
| Birth | 19.4 |
| 1 | 31.3 |
| 2 | 34.5 |
| 3 | 37.2 |
| 4 | 40.3 |
| 5 | 43.9 |
| 6 | 48.1 |
| 7 | 52.5 |
| 8 | 56.8 |

***Solution*** To cover the full set of data given in the table, let $t$ = 0, 1, 2 correspond to the ages at birth, 4 and 8 years, respectively. Thus, we have

$$N_0 = 19.4, \qquad N_1 = 40.3, \qquad N_2 = 56.8$$

Substituting these values in Eq. (33), we get

$$N_{\max} = 66.9$$

which is the required mean height.

**EXAMPLE 4.22** The population of a country for the years 1900–1960 is given in Table 4.2. Using this data find: (a) the theoretically maximum population, (b) the population in 1990; and (c) the population of 1870.

**Table 4.2** Population of a Country

| Year | Population (in millions) |
|---|---|
| 1900 | 76.0 |
| 1910 | 92.0 |
| 1920 | 105.7 |
| 1930 | 122.8 |
| 1940 | 131.7 |
| 1950 | 151.1 |
| 1960 | 179.3 |

***Solution*** Let $t$ = 0, 1, 2 correspond to the years 1900, 1930, and 1960, respectively. Then

$$N_0 = 76.0, \qquad N_1 = 122.8, \qquad N_2 = 179.3 \tag{34}$$

(a) From Eq. (33), we have

$$N_{\max} = 346.3$$

which is the maximum population of the country.

(b) From Eq. (34), equation (31) yields $e^{-a} = 0.5117$. Using this and $a/b = 346.3$ from (a), we obtain from Eq. (27)

$$N = \frac{346.3}{1 + 3.557\,(0.5117)^t} = \frac{346.3}{1 + 3.557e^{-0.6678t}} \tag{35}$$

Since, the year 1990 corresponds to $t$ = 3, thus from Eq. (35) by putting $t$ = 3, we get $N$ = 234.5, which is the population in 1990.

(c) The year 1870 corresponds to $t$ = –1. Again, from Eq. (35) we get $N$ = 43.6, which is the population in 1870.

The graph of Eq. (27) has the general appearance shown in Fig. 4.7. This graph reveals that as $t$ increases from 0, $N$ increases from $N_0$ and gets closer and closer to $N_{\max}$. The curve has an increasing slope from $t$ = 0 to a time corresponding to point $P$, and thereafter has a decreasing slope. Thus, point $P$ is a point of inflexion and is obtained by the usual methods of calculus.

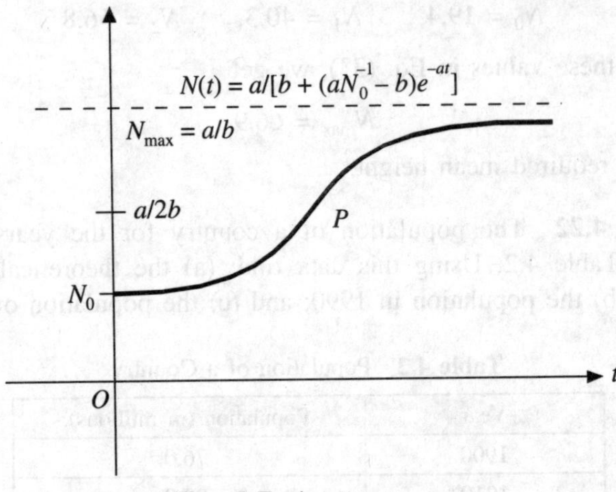


**Fig. 4.7** Logistic curve.

The S-shaped curve in Fig. 4.7 is called *the logistic curve* (from the Greek word "logistikos", meaning rational) and has proved to be of great accuracy in predicting the growth patterns, in a limited space, of certain types of bacteria, protozoa, water fleas (Daphnia) and fruit-flies (Drosophila).

## 4.9 A PROBLEM IN EPIDEMIOLOGY

An important problem in biology and medicine deals with the occurrence, spreading and control of a contagious disease, i.e. one which can be transmitted from one individual to another. The science that deals with this study is called *epidemiology*, and if a large number of population gets the disease, we say that there is an *epidemic*.

Problems involving the spread of disease can be very complicated. For example, it is known that some individuals may not actually get a disease even when exposed for long periods of time to others having the disease. In such case, we say that the individual has an immunity to the disease either because having had the disease before he has build up resistance to recurrence or by having initial resistance (natural immunity) he is not able to contract the disease. In some cases, individuals are immune to a disease but are capable of transmitting it to others; such individuals are called carriers. An example to this effect is the case of typhoid fever.

To have a simple mathematical description for the spread of a disease suppose that there is a large but finite population. Let us restrict ourselves to the students in some large college or university, who remain on campus for a relatively long period and do not have access to other communities. We presuppose that there are only two types of students, those who have a contagious disease (called infected), and those who do not have the disease, (i.e. unaffected) but are capable of contracting it on exposure to an infected student. If there are some infected students initially, then we want to find a formula for the number of infected students at any time.

Let $N_i$ denote the number of infected students at any time $t$ and $N_u$ the uninfected students. Then, if $N$ is the total number of students (assumed to be constant), we have

$$N = N_i + N_u \tag{36}$$

Here, $dN_i/dt$ is the time rate of change in the number of infected students and should depend in some way on $N_i$, and thus $N_u$. Assuming that $dN_i/dt$ is a quadratic function of $N_i$ as an approximation, we get

$$\frac{dN_i}{dt} = a_0 + a_1 N_i + a_2 N_i^2 \tag{37}$$

where $a_0$, $a_1$, $a_2$ are constants. Now we would expect $dN_i/dt = 0$, where $N_i = 0$, i.e. there are no infected students and where $N_i = N$, i.e. all students are infected. Then from Eq. (37), we have

$$a_0 = 0 \quad \text{and} \quad a_1 N + a_2 N^2 = 0 \quad \text{or} \quad a_2 = -a_1/N$$

so that Eq. (37) becomes

$$\frac{dN_i}{dt} = a_1 N_i - \frac{a_1 N_i^2}{N} = \frac{a_1}{N} N_i (N - N_i) = kN_i(N - N_i) \tag{38}$$

where $k = a_1/N$ and the initial conditions are

$$N_i = N_0 \quad \text{at} \quad t = 0 \tag{39}$$

The Eq. (38) with Eq. (39) has the solution as

$$N_i = \frac{N}{1 + \left(\frac{N}{N_0} - 1\right) e^{-kNt}} \tag{40}$$

The graph of Eq. (40) is the logistic curve (Fig. 4.7). It follows that results already obtained in connection with 'growth equation' apply as well as to the "disease equations" (38) or (40). From the shape of the logistic curve we see that initially there is a gradual increase in the number of infected students, followed by a rather sharp rise in their number near the inflection point, and finally a tapering off. The limiting case occurs where all students become infected, as seen by Eq. (40) that $N_i \to N$ as $t \to \infty$. But from a realistic point of view, this would not happen because the infected students once they are discovered would be isolated or quarantined so as to prevent others from contacting it. The problem of epidemics where quarantine is taken into consideration is more complicated and is considered in Chapter 7.

In case where large populations are involved, such as in a city or a country, it is easy to see why disaster could prevail even with attempts to quarantine. This was the case in the disease 'black plague', which ravaged much of Europe for several years in the middle of the 14th century.

For additional information on population growth see [14].

**EXAMPLE 4.23** *Spread of flu virus.* A student carrying a flu virus returns to an isolated college hostel of 1000 students. If it is assumed that the rate at which the virus spreads is proportional not only to the number $N_i$ of infected students but also to the students not infected. Find the number of infected students after 6 days when it is further observed that after 4 days $N_i(4) = 50$.

***Solution*** Assuming that no one leaves the hostel throughout the duration of the disease, we must then solve the initial value problem

$$\frac{dN_i}{dt} = kN_i(N - N_i), \quad N(0) = 1 = kN_i(1000 - N_i)$$

We have, from Eq. (40)

$$N_i = N(t) = \frac{1000}{1 + 999\,e^{-1000\,kt}} \tag{41}$$

Now, using $N_i = N(4) = 50$, we can determine $k$.

$$50 = \frac{1000}{1 + 999\,e^{-1000\,k \times 4}} \quad \text{or} \quad k = 0.0009906$$

Thus, Eq. (41) becomes

$$N_i = N(t) = \frac{1000}{1 + 999\,e^{-0.9906t}}$$

or

$$N_i = N(6) = \frac{1000}{1 + 999\,e^{-5.9436}} = 276 \text{ students}$$

Additional calculated values of $N(t)$ are given in Fig. 4.8.

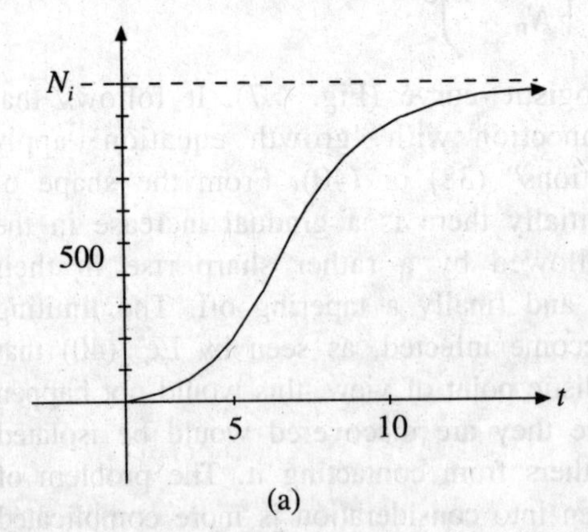


| $t$ (days) | $N$ (No. of infected students) |
|---|---|
| 4 | 50 (observed) |
| 5 | 124 |
| 6 | 276 |
| 7 | 507 |
| 8 | 735 |
| 9 | 882 |
| 10 | 953 |

(b)

**Fig. 4.8** (a) Logistic curve for infected students; (b) Number of infected students during a time interval.

## 4.10 THE SPREAD OF TECHNOLOGICAL INNOVATIONS

Once an innovation has been introduced by one firm, how soon do others adopt it? What are the factors which determine how rapidly the innovation spreads? These questions are of vital importance to economists, sociologists and ultimately governments. Here we shall develop a model for the spread of an innovation.

Assume that a region contains $R$ companies, all of which can derive possible advantage by adopting a new product. Let $N(t)$ represent the number of these companies that have adopted the new product at time $t$, and $N(0) = N_0$ denote the number of $R$ companies that have already adopted the innovation at time $t = 0$. Also, assume that the rate at which $N(t)$ increases is proportional to the number $N(t)$ that have already adopted the innovation, and also to the number $R - N(t)$ of the remaining adopters. We thus have the differential equation

$$\frac{dN}{dt} = kN(R - N) \tag{42}$$

where the constant of proportionality $k$ is positive. Equation (42) can be written as

$$\frac{dN}{dt} = (kR)N - kN^2 \tag{43}$$

Comparing it with Eq. (25), we see that $kR$ and $k$ correspond to $a$ and $b$, respectively. Thus, $N$ grows logistically and approaches the limiting value $(kR)/k = R$ as $t \to \infty$. The solution of Eq. (42) can be obtained by replacing $a$ by $kR$ and $b$ by $k$ in Eq. (27).

## 4.11 MIXTURE PROBLEM

Consider a tank containing $G_0$ gallons of a solution in which $P_0$ lb of a substance $S$ is dissolved. A second solution flows into the tank at a given rate $r_1$ gal/min., this solution containing $P_1$ lb/gal of $S$. Finally, the mixture in the tank flows out at a given rate $r_2$ gal/min. Find out the number of pounds $P$ of $S$ in the tank at time $t > 0$.

Assume that the mixture in the tank is well-stirred, so that at any given time $t_k$, $P(t) = P(t_k)$ has the same value at each point in the tank. The rate at which $P$ changes with time $t$ is

$$\frac{dP}{dt} = \text{(rate } S \text{ flowin)} - \text{(rate } S \text{ flowout)} \tag{44}$$

Equation (44) is called the *equation of continuity*. It states that the mass of the quantity $S$ is conserved, i.e. no amount of $S$ is created or destroyed in the process.

The rate at which $S$ flows into the tank is $P_1 r_1$ lb/min. The rate at which $S$ flows out in the tank is $[P/G(t)]r_2$ lb/min., where $[P/G(t)]$ is the concentration of $S$ in the tank at the time $t$; $G(t)$ is the number of gallons in the tank at time $t$. If $r_1 = r_2$, $G(t)$ will have the constant value $G_0(t)$. Many interesting cases arise, each leading to a different differential equation, e.g. either $r_1$ or $r_2$ may be zero, $P_0$ or $P_1$ may be zero and so on.

**EXAMPLE 4.24** A tank contains 100 gallon brine in which 10 lb of salt are dissolved. Brine containing 2 lb salt per gallon flows into the tank at 5 gal/min. If the well-stirred mixture is drawn off at 4 gal/min., find: (a) the amount of the salt in the tank at time $t$, and (b) the amount of the salt in the tank at $t = 10$ min.

***Solution*** Let $P(t)$ denote the number of pounds of salt in the tank and $G(t)$, the number of gallons of brine at time $t$. Then $G(t) = 100 + t$. Also, $P(0) = 10$ and $G(0) = 100$. Since $5(2) = 10$ lb salt is added to the tank per minute and $[P/(100 + t)](4)$ lb salt per minute is extracted from the tank, $P$ then satisfies the differential equation

$$\frac{dP}{dt} = 10 - \frac{4P}{100 + t} \tag{45}$$

Equation (45) can be written as

$$\frac{dP}{dt} + \frac{4P}{100 + t} = 10$$

which is a linear differential equation and the solution is

$$P(100 + t)^4 = \int 10(100 + t)^4 dt = 2(100 + t)^5 + c$$

Putting $t = 0$ and $P = 10$, we get $c = -190(100)^4$. Thus

(a) $P(t) = 2(100 + t) - 190(100)^4 (100 + t)^{-4}$

(b) $P(10) = 2(100 + 10) - 190(100)^4 + (100 + 10)^{-4} = 90.2$ lb

**EXAMPLE 4.25** Initially 50 pounds of salt is dissolved in a large tank having 300 gallons of water. A brine solution is pumped into the tank at a rate of 3 gal/min., and a well-stirred solution is then pumped out at the same rate. If the concentration of the solution entering is 2 lb/gal, find the amount of salt in the tank at any time. How much salt is present after 50 min. and after a long time?

***Solution*** Let $P(t)$ be the amount of salt in the tank at any time $t$. The rate at which $P(t)$ changes is

$$\begin{aligned} \frac{dP}{dt} &= \text{(rate of substance entering)} - \text{(rate of substance leaving)} \\ &= R_1 - R_2 \end{aligned} \tag{46}$$

Now $\qquad R_1 = (3 \text{ gal/min.}) (2 \text{ lb/gal}) = 6 \text{ lb/min.}$

$$R_2 = (3 \text{ gal/min.}) \left( \frac{P}{300} \text{ lb/gal} \right) = \frac{P}{100} \text{ lb/min} .$$

Thus, Eq. (46) takes the form

$$\frac{dP}{dt} = 6 - \frac{P}{100} \tag{47}$$

with the initial condition as $P(0) = 50$. The solution of the linear differential Eq. (47) is

$$P = 600 + ce^{-t/100} \tag{48}$$

When $t = 0$, $P = 50$, we have $c = -550$. Thus, Eq. (48) becomes

$$P(t) = 600 - 550e^{-t/100} \tag{49}$$

At $t = 0$, we find that $P(50) = 266.41$ lb. Also as $t \to \infty$, it is seen from Eq. (49) and Fig. 4.9, that $P \to 600$ which is what we would expect; over a long period of time the number of pounds of salt in the solution must be (300 gal) (2 lb/gal) = 600 lb.

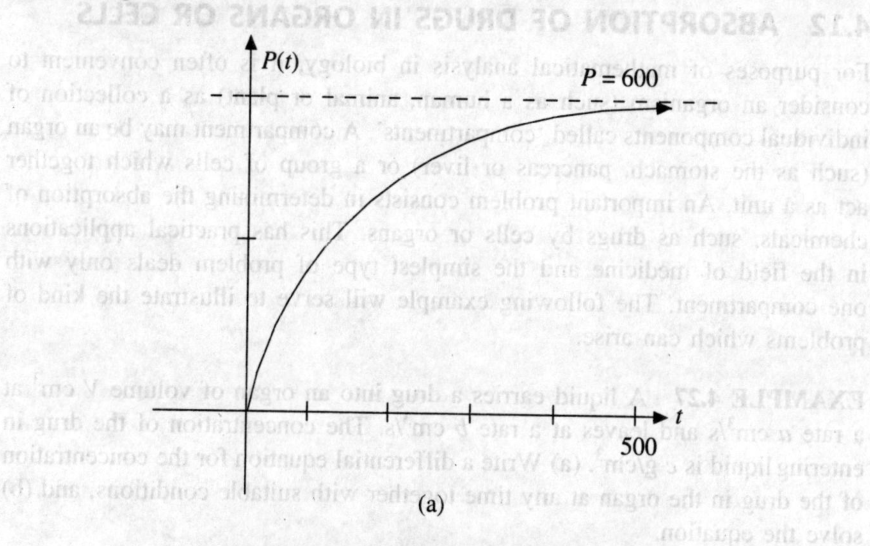


| $t$ (min.) | $P$ (lb) |
|---|---|
| 50 | 266.41 |
| 100 | 397.67 |
| 150 | 477.27 |
| 200 | 525.57 |
| 300 | 572.62 |
| 400 | 589.93 |

(b)

**Fig. 4.9** (a) Schematic diagram for the amount of salt in the tank; (b) Amount of salt present during a time interval.

**EXAMPLE 4.26** A tank contains 40 gallons of brine for which the concentration is initially 3 lb salt/gal. A salt solution of 2 lb salt/gal enters the tank at the rate of 5 gal/min and the well-stirred mixture is drawn off at the same rate. Find the time required for the amount of salt in the tank to be reduced to 100 lb.

***Solution*** Let $P$ denote the number of pounds of salt in the tank at time $t$. When $t = 0$, $P(0) = 120$ gallon. Salt flows into the tank at 2(5) = 10 lb/min. Since the concentration of the salt in the tank at time $t$ is $P/40$ lb/gal, salt flows out of the tank at (P/40) (5) lb/min. From Eq. (44), we get

$$\frac{dP}{dt} = 2(5) - \frac{P}{40}(5) = \frac{1}{8}(80 - P) \tag{50}$$

This differential equation is a special case of the diffusion Eq. (20). Thus, we have another example of two different physical situations served by the same mathematical model. We leave the solution of Eq. (50) as an exercise.

## 4.12 ABSORPTION OF DRUGS IN ORGANS OR CELLS

For purposes of mathematical analysis in biology, it is often convenient to consider an organism (such as a human, animal or plant) as a collection of individual components called 'compartments'. A compartment may be an organ (such as the stomach, pancreas or liver) or a group of cells which together act as a unit. An important problem consists in determining the absorption of chemicals, such as drugs by cells or organs. This has practical applications in the field of medicine and the simplest type of problem deals only with one compartment. The following example will serve to illustrate the kind of problems which can arise.

**EXAMPLE 4.27** A liquid carries a drug into an organ of volume $V$ cm$^3$ at a rate $a$ cm$^3$/s and leaves at a rate $b$ cm$^3$/s. The concentration of the drug in entering liquid is $c$ g/cm$^3$. (a) Write a differential equation for the concentration of the drug in the organ at any time together with suitable conditions, and (b) solve the equation.

***Solution*** (a) Figure 4.10 shows a single compartment of volume $V$ together with an inlet and an outlet. Let $x$ be the concentration of drugs in the organ, then the amount of the drug in the organ at any time $t$ is

$$(V \text{ cm}^3)(x \text{ g/cm}^3) = xV \text{ g} \tag{51}$$

The amount entering the organ at any time $t$ is

$$(a \text{ cm}^3/\text{s})\ (c \text{ g/cm}^3) = ac \text{ g/s} \tag{52}$$

and the amount leaving the organ is

$$(b \text{ cm}^3/\text{s})\ (x \text{ g/cm}^3) = bx \text{ g/s} \tag{53}$$

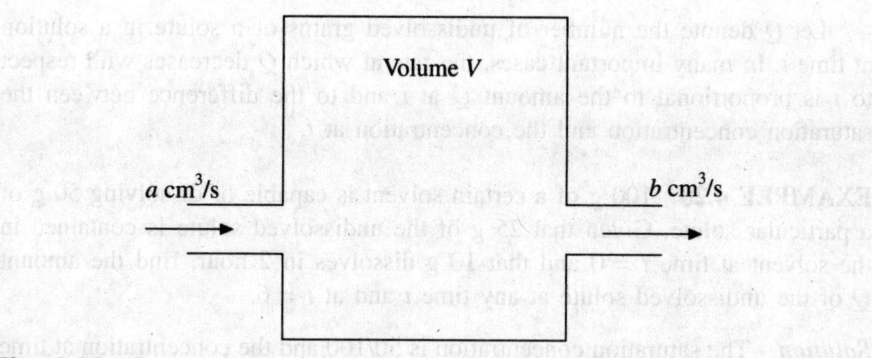


**Fig. 4.10** Concentration of the drug in an organ—single compartment analysis.

Now, the differential equation describing the present situation is given by [from Eqs. (51) to (53)]

$$\frac{d}{dt}(xV) = ac - bx \tag{54}$$

and the suitable conditions are

$$x = x_0 \quad \text{at} \quad t = 0 \tag{55}$$

(b) Equation (54) can be written as

$$V\frac{dx}{dt} = ac - bx$$

Separating the variables and using Eq. (55), the solution of this equation is

$$x = \frac{ac}{b} + \left(x_0 - \frac{ac}{b}\right)e^{-bt/V} \tag{56}$$

We now have

CASE I. When $a = b$. Here the rate at which the drug enters is same as the rate at which the drug leaves and Eq. (56) becomes

$$x = c + (x_0 - c)e^{-bt/V}$$

CASE II. When $a = b$ and $x_0 = 0$. That is, the inflow and outflow rates are equal and the initial concentration of the drug in the organ is zero. In this case, we have

$$x = c(1 - e^{-bt/V})$$

## 4.13 RATE OF DISSOLUTION

A liquid capable of dissolving another substance is known as a *solvent*, and the dissolved substance is called the *solute*. The concentration of the solution at any time $t$ is the ratio of the solute to the solvent at time $t$; and when the concentration attains its maximum possible value, the solution is said to be *saturated*.

Let $Q$ denote the number of undissolved grams of a solute in a solution at time $t$. In many important cases, the rate at which $Q$ decreases with respect to $t$ is proportional to the amount $Q$ at $t$ and to the difference between the saturation concentration and the concentration at $t$.

**EXAMPLE 4.28** 100 g of a certain solvent is capable of dissolving 50 g of a particular solute. Given that 25 g of the undissolved solute is contained in the solvent at time $t = 0$ and that 10 g dissolves in 2 hour, find the amount $Q$ of the undissolved solute at any time $t$ and at $t = 6$.

***Solution*** The saturation concentration is 50/100 and the concentration at time $t$ is $(25 - Q)/100$. Thus, the differential equation of the present problem is

$$\frac{dQ}{dt} = kQ\left(\frac{50}{100} - \frac{25 - Q}{100}\right) = kQ\,\frac{25 + Q}{100}$$

where the constant of proportionality $k$ is negative.

Separating the variables, we get

$$\frac{100\,dQ}{Q(25 + Q)} = k\,dt$$

Resolving the left-hand side into a partial fraction, we get

$$4\left(\frac{1}{Q} - \frac{1}{25 + Q}\right)dQ = k\,dt$$

Integrating it, we obtain

$$\log Q - \log(25 + Q) = \frac{kt}{4} - \log c$$

or

$$\log\frac{25 + Q}{cQ} = -\frac{kt}{4}$$

Put $t = 0$ and $Q = 25$ to obtain $c = 2$. Again put $t = 2$ and $Q = 25 - 10 = 15$ in $(25 + Q)/2Q = e^{-kt/4}$ to get $e^{-k/4} = (4/3)^{1/2}$. Hence

$$\frac{25 + Q}{2Q} = e^{-kt/4} \quad \text{becomes} \quad \frac{25 + Q}{Q} = 2\left(\frac{4}{3}\right)^{t/2}$$

Therefore,
$$Q = Q(t) = \frac{25}{2(4/3)^{t/2} - 1}$$

$$Q(6) = \frac{25}{2(4/3)^{6/2} - 1} = 6.68 \text{ g (approx.)}$$

If this is the number of grams of undissolved solute at $t = 6$, $25 - 6.68$ or 18.32 g dissolved in the first 6 hours. Also as $t \to \infty$, $Q \to 0$.

## 4.14 CHEMICAL REACTIONS—LAW OF MASS ACTION

In some types of chemical reactions, the rate at which a chemical $A$ transforms into a second chemical $B$ is proportional to the amount $Q$ of $A$ remaining untransformed at time $t$. The quantity $Q$ satisfies the differential equation

$$\frac{dQ}{dt} = kQ, \qquad k < 0$$

A chemical reaction of this type is called a *first order reaction*. An example of this kind of reaction is the conversion of *t*-butyl chloride into *t*-butyl alcohol:

$$(CH_3)_3CCl + NaOH \rightarrow (CH_3)_3COH + NaCl$$

Only the concentration of the *t*-butyl chloride controls the rate of reaction.

Consider now two substances X and Y which combine to form a third substance Z. Assume that $x$ grams of X reacts with $y$ grams of Y and that the molecular structures of X and Y are such that $a$ grams of X combines with $b$ grams of Y to form $(a + b)$ grams of Z. The constants $x$ and $y$ represent initial amounts of X and Y. Let $Q(t)$ represent the number of grams of Z present at time $t$, with $Q(0) = 0$.

The rate at which $Q$ changes is governed by the *law of mass action*, according to which if no temperature change is involved, the rate at which two substances X and Y react to form a third substance is proportional to the product of the amounts of X and Y untransformed at time $t$.

Since the molecules of X and Y combine in the ratio of $a$ to $b$, Q($t$) consists of $[a/(a + b)]Q$ g X and $[b/(a + b)]$ g Y. Hence $[x - aQ/(a + b)]$ g X and $[y - bQ/(a + b)]$ g Y remain untransformed at time $t$. By the law of mass action, $Q$ satisfies the differential equation

$$\frac{dQ}{dt} = k\left(x - \frac{aQ}{a+b}\right)\left(y - \frac{bQ}{a+b}\right) \tag{57}$$

where the constant $k$ (of proportionality) is positive.

This kind of reaction is called a *second order reaction*, since $dQ/dt$ is a quadratic function of $Q$. For example, consider the reaction

$$CH_3Cl + NaOH \rightarrow CH_3OH + NaCl$$

In this reaction for every molecule of methyl chloride, one molecule of sodium hydroxide is consumed, thus forming one molecule of methyl alcohol and one molecule of sodium chloride. In this case, the rate at which the reaction proceeds is proportional to the product of the remaining concentrations of $CH_3Cl$ and of NaOH.

Equation (57) can be written as

$$\frac{dQ}{dt} = \frac{kab}{(a+b)^2}\left(\frac{a+b}{a}x - Q\right)\left(\frac{a+b}{b}y - Q\right) = K(A - Q)\,(B - Q) \tag{58}$$

where

$$K = \frac{kab}{(a+b)^2}, \qquad A = \frac{a+b}{a}x, \qquad B = \frac{a+b}{b}y$$

If $A = B$, i.e. if $x/a = y/b$, then

$$(A - Q)^{-2}\, dQ = K dt \quad \text{and} \quad (A - Q)^{-1} = Kt + c$$

Putting $t = 0$ and $Q = 0$, we get $c = A^{-1}$. Thus

$$Q = A - \frac{A}{KAt + 1} \quad \text{or} \quad Q = \frac{a+b}{a}x - \frac{a^{-1}(a+b)^2}{kbxt + (a+b)}$$

As $t \to \infty$, $Q \to \frac{a+b}{a}x = A = B$.

If $A \neq B$, $A$ and $B$ can be labelled so that $A > B$. We write Eq. (58) as

$$\frac{dQ}{(A-Q)(B-Q)} = K dt$$

or

$$\frac{1}{A-B}\left(\frac{-1}{A-Q} + \frac{1}{B-Q}\right) dQ = K dt$$

Integrating, we get

$$\frac{1}{A-B}[\log (A - Q) - \log (B - Q)] = Kt + c_1$$

or

$$\frac{A-Q}{B-Q} = ce^{[K(A - B)t]}$$

where $c = e^{[c_1(A - B)]}$. Solving for $Q$, we obtain

$$Q = \frac{A - Bce^{K(A-B)t}}{1 - ce^{K(A-B)t}} \tag{59}$$

Put $t = 0$ and $Q = 0$ to obtain $c = A/B$ and Eq. (59) reduces to

$$Q(t) = \frac{AB[1 - e^{-K(A-B)t}]}{A - B\, e^{-K(A-B)t}} \tag{60}$$

As $t \to \infty$, $Q \to B = [(a + b)/b]y$ (as happened in the case $A = B$).

Thus, the limiting value that $Q$ approaches depends on the amount $y$ of the substance Y present initially, and on the ratio $b/(a + b)$.

The dissolution equation $dQ/dt = kQ(25 + Q)/100$ given in Example 4.28 has the same form as that form as that of Eq. (58) when substitution $u = 25 - Q$ is made. The quantity '$u$' representing the amount of solute dissolved at time $t$, corresponds to $Q$, the amount of compound Z formed at time $t$.

The law of mass action can also be applicable to more complicated reactions involving more than two compounds. An $n$th order process involving $n$ compounds is described by the differential equation

$$\frac{dQ}{dt} = k(x_1 - r_1Q)(x_2 - r_2Q) \ldots (x_n - r_nQ)$$

where $x_i$ denotes the initial amount of the $i$th compound, $r_i$ the fraction of $Q(t)$ contributed by the $i$th compound and $r_1 + r_2 + \cdots + r_n = 1$.

**EXAMPLE 4.29** 2 g of substance Y combines with 1 g substance X to form 3 g of substance Z. When 100 g Y is thoroughly mixed with 50 g X, it is found that in 10 min. 50 g Z has been formed. How many grams of Z can be formed in 20 min.? How long does it take to form 60 g Z?

***Solution*** From Eq. (57), we have

$$\frac{dQ}{dt} = k\left(50 - \frac{Q}{3}\right)\left(100 - \frac{2Q}{3}\right) = R(150 - Q)^2$$

where $R$ is a constant. Separating the variables and integrating, yields

$$(150 - Q)^{-1} = Rt + c$$

Putting $t = 0$ and $Q = 0$ to obtain $c = 1/150$, and $t = 10$, $Q = 50$ yields $R = 1/3000$. Now, putting $t = 20$ in

$$\frac{1}{150 - Q} = \frac{t}{3000} + \frac{1}{150} \tag{61}$$

and solving for $Q$, we get $Q(20) = 75$ g. Putting $Q = 60$ in Eq. (61) and solve for $t$, we get $t = 40/3$ min.

**EXAMPLE 4.30** Two chemicals X and Y react to form another chemical Z. It is found that the rate at which Z is formed varies as the product of the instantaneous amount of chemicals X and Y present. The formation requires 2 lb of X for each pound of Y. If 10 lb of Y are present initially, and if 6 lb of Z are formed in 20 minutes, find the amount of chemical Z at any time.

***Solution*** Let $Q$ lb be the amount of Z formed in time $t$ hours; then $dQ/dt$ is the rate of its formation. To form $Q$ lb of Z, we need $2Q/3$ lb of X and $Q/3$ lb of Y, since twice as much of chemical X as Y is needed. Thus, the amount of X present at time $t$ when $Q$ lb of Z formed is $10 - 2Q/3$, and the amount of Y at this time is $20 - Q/3$. Hence

$$\frac{dQ}{dt} = k\left(10 - \frac{2Q}{3}\right)\left(20 - \frac{Q}{3}\right) = k(15 - Q)(60 - Q)$$

with the conditions $Q = 0$ at $t = 0$; $Q = 6$ at $t = 1/3$.
Separating the variables, we get

$$\int \frac{dQ}{(15 - Q)(60 - Q)} = k\int dt = kt + c_1$$

or $$\int \frac{1}{45}\left(\frac{1}{15 - Q} - \frac{1}{60 - Q}\right) dQ = kt + c_1 \quad \text{or} \quad \frac{60 - Q}{15 - Q} = ce^{45kt}$$

From the given conditions, we find that $c = 4$ and consequently, $e^{15k} = 3/2$. Thus

$$\frac{60-Q}{15-Q} = 4(e^{15k})^{3t} = 4\left(\frac{3}{2}\right)^{3t}$$

which when solved for $Q$, yields

$$Q = Q(t) = \frac{15\left[1-(2/3)^{3t}\right]}{1-(1/4)(2/3)^{3t}}$$

As $t \to \infty$, $Q \to 15$ lb.

**EXAMPLE 4.31** The compound Z is formed when two chemicals X and Y are combined. The resulting reaction between the two chemicals is such that for each gram of X, 4 g of Y is used. It is observed that 30 g of the compound Z is formed in 10 minutes. Determine the amount of Z at any time if the rate of reaction is proportional to the amounts of X and Y remaining when initially there are 50 g of X and 32 g of Y. How much of the compound Z is present after 15 minutes? Interpret the solution as $t \to \infty$.

***Solution*** Let $Q(t)$ denote the number of grams of the compound Z present at any time, then the present problem is described by the differential equation

$$\frac{dQ}{dt} = k\left(50 - \frac{Q}{5}\right)\left(32 - \frac{4}{5}Q\right) = k(250-Q)(40-Q)$$

Separating the variables, we get

$$\frac{dQ}{(250-Q)(40-Q)} = kdt$$

or

$$\frac{-1/210}{250-Q}\,dQ + \frac{1/210}{40-Q}\,dQ = kdt$$

Integrating, we get

$$\log\frac{250-Q}{40-Q} = 210kt + c_1$$

or

$$\frac{250-Q}{40-Q} = c_2\,e^{210kt} \tag{62}$$

When $t = 0$, $Q = 0$, then $c_2 = 25/4$. Also when $t = 10$ and $Q = 30$, we have $210k = 0.1258$. Using these in Eq. (62) and solving for $Q$, we get

$$Q = Q(t) = 1000\,\frac{1-e^{-0.1258t}}{25-4\,e^{-0.1258t}} \tag{63}$$

The behaviour of $Q$ as a function of time is shown in Fig. 4.11(a). From Fig. 4.11(b) and Eq. (63) it is clear that $Q \to 40$ as $t \to \infty$. This means there are 40 g of compound Z formed, leaving behind

$$50 - \frac{1}{5}(40) = 42 \text{ g of chemical X}$$

$$32 - \frac{4}{5}(40) = 0 \text{ g of chemical Y}$$

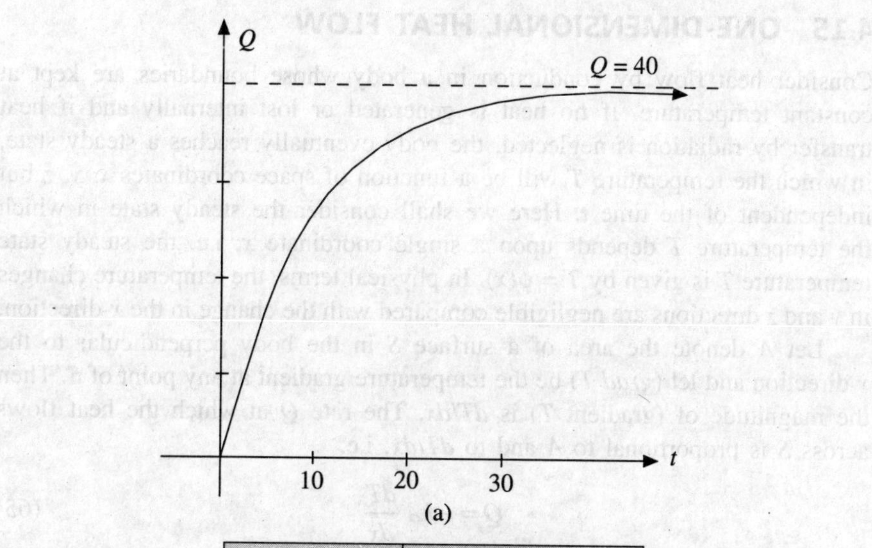


(a)

| $t$ (min.) | $Q$ (grams) |
|---|---|
| 10 | 30 (measured) |
| 15 | 34.78 |
| 20 | 37.25 |
| 25 | 38.54 |
| 30 | 39.22 |
| 35 | 39.59 |

(b)

**Fig. 4.11** Law of mass action.

**Remark.** Another type of reaction is one in which an amount $a$ of one substance $A$ is transformed into a second substance $B$ in such a way that the rate at which the amount $x$ of $B$ present at time $t$ increases is proportional to $x$ and to the amount $(a - x)$ of $A$ remaining untransformed at time $t$. This kind of reaction is called an *autocatalytic reaction* and is described by the differential equation*

---

*The solution of Eq. (64) is

$$x = ax_0[(a - x_0)e^{-akt} + x_0]^{-1} \quad \text{as} \quad t \to \infty,\ x \to a$$

Also, for $x = a/2$,

$$\left(\frac{dx}{dt}\right)_{\max} = \frac{ka^2}{4}, \qquad t = \frac{1}{ak}\log\frac{a - x_0}{x_0}$$

$$\frac{dx}{dt} = kx(a - x), \quad k > 0 \tag{64}$$

(For various other chemical and biochemical reactions see [15]).

## 4.15 ONE-DIMENSIONAL HEAT FLOW

Consider heat flow by conduction in a body whose boundaries are kept at constant temperature. If no heat is generated or lost internally and if heat transfer by radiation is neglected, the body eventually reaches a steady state, in which the temperature $T$ will be a function of space coordinates $x$, $y$, $z$ but independent of the time $t$. Here we shall consider the steady state in which the temperature $T$ depends upon a single coordinate $x$; i.e. the steady state temperature $T$ is given by $T = \phi(x)$. In physical terms, the temperature changes in $y$ and $z$ directions are negligible compared with the change in the $x$-direction.

Let $A$ denote the area of a surface $S$ in the body perpendicular to the $x$-direction and let (*grad T*) be the temperature gradient at any point of $S$. Then the magnitude of (gradient $T$) is $dT/dx$. The rate $Q$ at which the heat flows across $S$ is proportional to $A$ and to $dT/dx$; i.e.

$$Q = -kA\frac{dT}{dx} \tag{65}$$

where $k$, the constant of proportionality, is called the *thermal conductivity* of the medium, and the minus sign indicates that heat flows in the direction of decreasing temperature. If $x$ is measured in cm, $A$ in cm$^2$, $T$ in 0°C (absolute temperature), and $t$ in seconds, then $Q$ is measured in cal/s.

If $S'$ is any other surface such that the heat flowing across $S'$ is the same as the heat flowing across $S$, then $Q$ is constant for all such surfaces. This is a form of the equation of continuity and indicates that heat is neither created nor being lost inside a closed surface.

**EXAMPLE 4.32** The iron bar in Fig. 4.12 is 100 cm long has a constant cross-sectional area of 4 cm$^2$ and is perfectly insulated laterally so that heat flow takes place only in the $x$-direction. If the left end of the bar is kept at 0°C and the right end at 60°C, what is $T$ in terms of $x$ (the value of $k$ for iron is 0.15)?

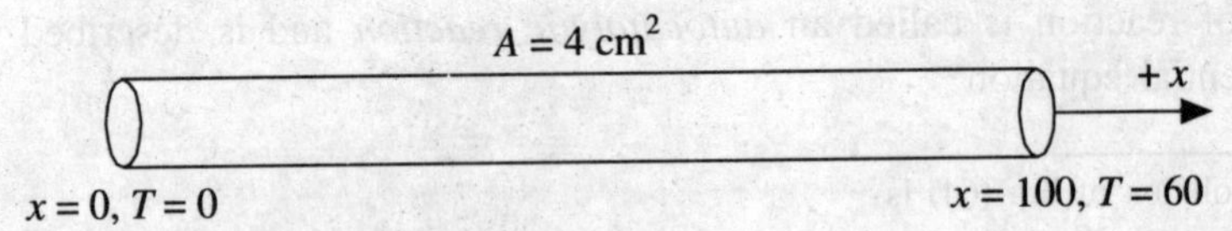


**Fig. 4.12** One-dimensional heat flow.

***Solution*** First by solving the two-point boundary value problem, we have

$$Q = -0.15(4)\,\frac{dT}{dx}, \quad x = 0, \quad T = 0; \quad x = 100,\ T = 60°$$

From $dT/dx = -\,5Q/3$ we obtain $T = -(5/3)Qx + c$.

From $x = 0$, $T = 0$, we find $c = 0$ and from $x = 100$, $T = 60$, we get

$$Q = \frac{-3(60)}{5(100)} = -0.36$$

Hence
$$T = -\frac{5}{3}(-0.36)x = 0.6\,x$$

The heat flow is in the direction of the negative $x$-axis at the rate 0.36 cal/s across every section perpendicular to the $x$-axis.

**EXAMPLE 4.33** A hollow spherical brass ($k = 0.26$) shell has an inner radius 4 cm and an outer radius 10 cm. If the inner surface temperature is kept at 100°C and the outer surface temperature at 20°C, what is the temperature $T$ in terms of $r$, the radial distance from the centre of the shell? What is the temperature on the sphere where $r = 7$ cm and for what value of $r$ is $T = 60$°C?

***Solution*** The same amount of heat per second flows across every spherical surface having its centre at the centre of the shell, radius $r$, and the surface area $A = 4\pi r^2$. The flow is in the radial direction and hence, Eq. (65) with $x$ replaced by $r$ becomes

$$Q = -(0.26)\,(4\pi r^2)\,\frac{dT}{dr}$$

Separation of the variables yields

$$dT = Br^{-2}\,dr, \qquad B = -\frac{Q}{(0.26)(4\pi r^2)}$$

Integration yields $T = -(B/r) + C$. Setting $r = 4$, $T = 100$ and $r = 10$, $T = 20$, we obtain $100 = -0.25\,B + C$ and $20 = -0.1\,B + C$ which gives $B = -1600/3$ and $C = -100/3$. Thus

$$T = T(r) = \frac{1600}{3r} - \frac{100}{3}$$

when $r = 7$, $\;T \cong 42.9$°C; $\;T = 60$, $\;r \cong 5.7$ cm

The value of $Q$ can be obtained from $Q = (60.26)(4\pi)B$.

**EXAMPLE 4.34** A long steel pipe, of thermal conductivity $k = 0.15$ cgs units, has an inner radius 10 cm and an outer radius of 20 cm. The inner surface is kept at 200°C and the outer surface is kept at 50°C.

(a) Find the temperature as a function of distance from the common axis of concentric cylinders.

(b) Find the temperature when $r = 15$ cm.

(c) How much heat is lost per minute in a portion of the pipe which is 20 m long?

***Solution*** It is obvious that the isothermal surfaces are the concentric cylinders. The area of such a surface having a radius $r$ and length $l$ is $2\pi rl$. Equation (65) in this case can be written as

$$Q = -kA\frac{dT}{dr} = -k(2\pi rl)\frac{dT}{dr}$$

Since $k = 0.15$, $l = 20$ m = 2000 cm, we have

$$Q = -600\pi r\frac{dT}{dr}$$

Separating the variables, we have

$$-600\pi T = Q \log r + c$$

which, on using $T = 200°C$ at $r = 10$, and $T = 50°C$ at $r = 20$, gives

$$-600\pi(200) = Q \log 10 + c$$

and
$$-600\pi(50) = Q \log 20 + c$$

from which, we obtain $Q = 408{,}000$ and $c = -1{,}317{,}000$. Thus, (a) $Q = 699 - 216 \log r$. (b) If $r = 15$, we find that $T = 114°C$. For the (c) part, from this equation, we get $Q = 408{,}000 \times 60$ cal/min. = 24,480,000 cal/min.

## 4.16 ELECTRIC CIRCUIT

In a series circuit containing only a resistor and an inductor, Kirchhoff's second law states that the sum of the voltage drop across the inductor $L(di/dt)$ and the voltage drop across the resistor ($iR$) is same as the impressed voltage [$E(t)$] on the circuit (see Fig. 4.13). Therefore, for the current $i(t)$, the differential equation is

$$L\frac{di}{dt} + Ri = E(t) \tag{66}$$

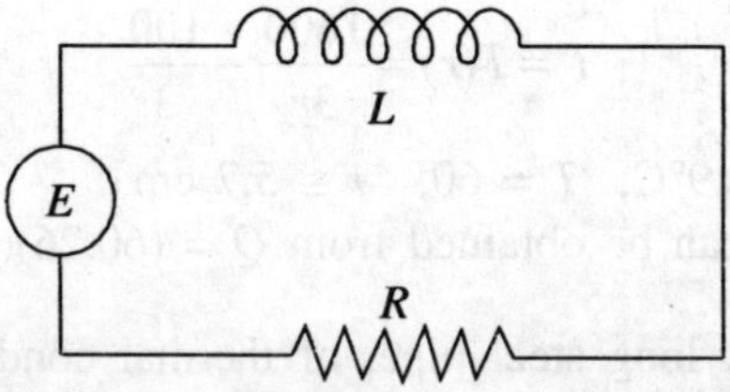

**Fig. 4.13** Simple series circuit containing a resistor and an inductor.

where $L$ and $R$ are the constants known as the inductance and the resistance, respectively. Sometimes the current $i(t)$ is called the response of the system.

The general solution of Eq. (66) is

$$i(t) = \frac{e^{-(R/L)t}}{L}\int e^{(R/L)t}E(t)\,dt + ce^{-(R/L)t} \tag{67}$$

In particular, when $E(t) = E_0$ is a constant, Eq. (67) becomes

$$i(t) = \frac{E_0}{R} + ce^{-(R/L)t} \tag{68}$$

Note that as $t \to \infty$, the second term in Eq. (68) approaches zero. This term is known as the *transient term;* the remaining term is called the *steady state part* of the solution. In this case, $E_0/R$ is called the *steady state current.* For a large time, it is clear that, the circuit is governed by Ohm's law ($E = iR$).

**EXAMPLE 4.35** A 12 V battery is connected to a simple series circuit in which the inductance is ½ H and the resistance is 10 Ω. Determine the current $i$ if $i(0) = 0$.

***Solution*** The differential Eq. (66) for this problem is

$$\frac{1}{2}\frac{di}{dt} + 10i = 12, \quad i(0) = 0$$

and the solution of this equation is

$$i = i(t) = \frac{6}{5} + ce^{-20t}$$

Now $i(0) = 0$ implies $c = -6/5$. Therefore, the response is

$$i = i(t) = \frac{6}{5} - \frac{6}{5}e^{-20t}$$

**EXAMPLE 4.36** A generator having emf 100 V is connected in series with a 10 Ω resistor and an inductor of 2 H. If the switch $k$ is closed at time $t = 0$, obtain a differential equation for the current and determine the current at time $t$.

***Solution*** Let $i$ be the current in amperes flowing as shown in Fig. 4.14. Then

(a) Voltage supplied = 100 V.

(b) Voltage drop across resistance ($Ri$) = $10i$.

(c) Voltage drop across inductor

$$L\frac{di}{dt} = 2\frac{di}{dt}$$

Thus, by Kirchhoff's law, we have

$$100 = 10i + 2\frac{di}{dt} \quad \text{or} \quad \frac{di}{dt} + 5i = 50 \tag{69}$$

which is a first order linear differential equation whose solution is

$$i = 10 + ce^{-5t}$$

Since, the switch $k$ is closed at time $t = 0$, we must have $i = 0$ at $t = 0$ and hence

$$i = 10(1 - e^{-5t})$$

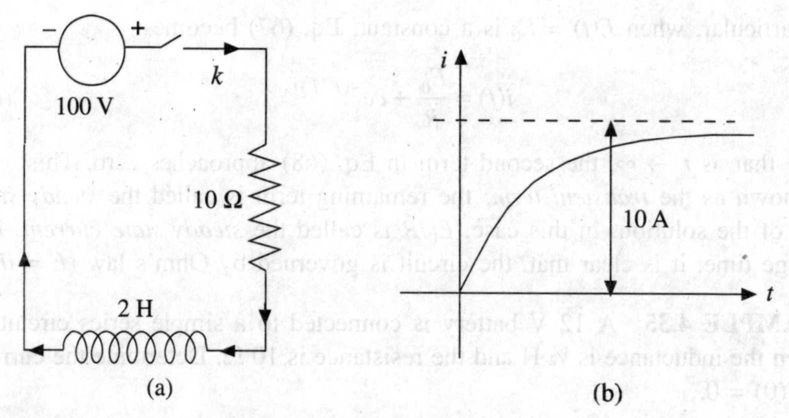


**Fig. 4.14** (a) Simple series circuit; (b) Graph of current $i$ versus time $t$.

**EXAMPLE 4.37** Set up and solve a differential equation for the circuit of the above example if a 100 V generator is replaced by one having an emf of 20 cos 5$t$ V.

***Solution*** Here, Eq. (66) takes the form

$$\frac{di}{dt} + 5i = 10 \cos 5t$$

whose solution is

$$i = \cos 5t + \sin 5t + ce^{-5t}$$

But at $t = 0$, $i = 0$, we have $c = -1$. Thus

$$i = \cos 5t + \sin 5t - e^{-5t}$$

**EXAMPLE 4.38** A decaying emf $E = 200\ e^{-5t}$ is connected in series with a 20 Ω resistor and 0.01 F capacitor. Assuming $Q = 0$ at $t = 0$, find the charge $Q$ and current $i$ at any time. Show that the charge reaches a maximum, calculate it and find out when it is reached.

***Solution*** From Fig. 4.15, we obtain

(a) Voltage supply $E = 200e^{-5t}$.
(b) Voltage drop across resistor $(Ri) = 20i$.
(c) The voltage drop across capacitor is $(Q/c) = Q/0.01 = 100Q$.

Thus, by Kirchhoff's law, we have

$$20i + 100Q = 200e^{-5t}$$

and using $i = dQ/dt$, we get

$$\frac{dQ}{dt} + 5Q = 10e^{-5t}$$

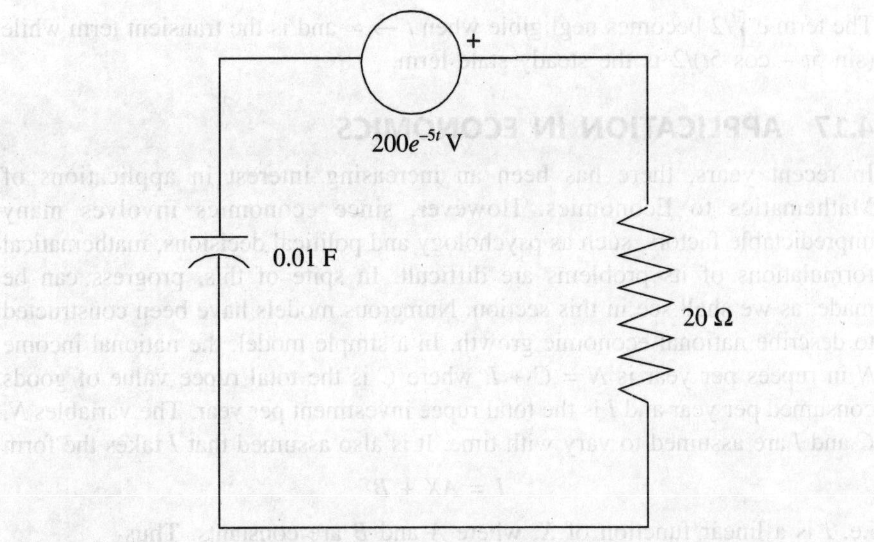


**Fig. 4.15** Simple series circuit.

which has the solution

$$Qe^{5t} = 10t + c$$

But as $Q = 0$ when $t = 0$. Thus, $Q = 10t\, e^{-5t}$. Since, $i = dQ/dt$, we have

$$i = \frac{d}{dt}(10te^{-5t}) = 10e^{-5t} - 50te^{-5t}$$

which gives the current at any time $t$.

To obtain the time $t$ when $Q$ is maximum, put $dQ/dt = 0$ to obtain $t = (1/5)$ s. It can be verified that $Q$ is maximum at this time. The maximum charge is

$$Q = 10 \times \frac{1}{5} e^{-1} \sim 0.74 \text{ C}$$

**EXAMPLE 4.39** In a circuit with no capacitor, the inductance is 4 H, resistance is 20 Ω, and $E = 20 \sin 5t$. Given that $i = 0$ when $t = 0$, find the relationship between $i$ and $t$.

***Solution*** The differential equation for this problem is

$$\frac{di}{dt} + 5i = 5 \sin 5t$$

whose solution is

$$i\, e^{5t} = e^{5t} \frac{\sin 5t - \cos 5t}{2} + c$$

When $i = 0$ at $t = 0$, we find that $c = 1/2$. Hence, the relationship between $i$ and $t$ is

$$i = \frac{\sin 5t - \cos 5t}{2} + \frac{e^{-5t}}{2}$$

The term $e^{-5t}/2$ becomes negligible when $t \to \infty$ and is the transient term while $(\sin 5t - \cos 5t)/2$ is the steady state term.

## 4.17 APPLICATION IN ECONOMICS

In recent years, there has been an increasing interest in applications of Mathematics to Economics. However, since economics involves many unpredictable factors, such as psychology and political decisions, mathematical formulations of its problems are difficult. In spite of this, progress can be made, as we shall see in this section. Numerous models have been constructed to describe national economic growth. In a simple model, the national income $N$ in rupees per year is $N = C + I$, where $C$ is the total rupee value of goods consumed per year and $I$ is the total rupee investment per year. The variables $N$, $C$ and $I$ are assumed to vary with time. It is also assumed that $I$ takes the form

$$I = AX + B$$

i.e. $I$ is a linear function of $X$, where $A$ and $B$ are constants. Thus

$$\frac{dI}{dX} = A, \qquad \frac{dX}{dI} = \frac{1}{A}, \qquad \frac{dX}{dt} = \frac{1}{A}\frac{dI}{dt}$$

The constant $A$ is called the marginal propensity to save. Relating savings to income, it is the fraction of each extra rupee that is saved rather than spent.

Now, suppose $F$ represents the rupee capacity or potential output per year, i.e. $F$ is the value $X$ would have if full payment were realized and maximum use of capital equipment were achieved. It is assumed that the rate of change of $F$ is proportional to the total investment $I$, i.e.

$$\frac{dF}{dt} = kI$$

where $k$ is a positive constant. This is known as the *Domar model*.

The situation in which the productive capacity is fully realized is known as the *economic equilibrium* and it is achieved when the aggregate demand equals potential capacity, or

$$X = F$$

and to satisfy this equation for all $t$, it is necessary that $dX/dt = dF/dt$ for all $t$. From this, we have

$$\frac{1}{A}\frac{dI}{dt} = kI \quad \text{or} \quad \frac{dI}{dt} = (kA)I$$

The solution of this differential equation is $I = I_0 e^{Akt}$, where $I_0$ is the rate of investment at $t = 0$. In this model, we can conclude that investment must grow exponentially to maintain the economic equilibrium. This model is used in Macroeconomics, the branch of Economics dealing with income and output of the whole system.

Now we shall present an application to Microeconomics, which studies individual areas of economic activity.

Let $P$ denote the price in rupees of a single commodity, $Q_j$, the number of items of the commodity in demand per unit time $t$, and $Q_s$ the number of items of the commodity in supply per unit of time. Assume that $Q_j$ and $Q_s$ are linear functions of $P$ so that

$$Q_d = A - PB, \qquad Q_s = -C + PD$$

where $A$, $B$, $C$ and $D$ are positive constants. Here, we have assumed that $Q_d$ is a linear decreasing function of $P$ and $Q_s$ is a linear increasing function of $P$. This model is said to be in equilibrium if $Q_d = Q_s$, i.e. when demand equals supply.

When there is an equilibrium, the variables in the model have no tendency to change. If the variables $Q_d$, $Q_s$ and $P$ change with time, a dynamic analysis is then required. If $P = P(t)$, then we shall find a state of equilibrium when $t \to \infty$.

First, note that if

$$Q_d = A - BP = -C + DP = Q_s$$

then
$$P = \overline{P} = \frac{A+C}{B+D}$$

We call $\overline{P}$ as the *equilibrium price.*

Next, assume that the rate of change of the price $P$ is proportional to the excess demand $Q_d - Q_s$, i.e.

$$\frac{dP}{dt} = K(Q_d - Q_s), \qquad K > 0$$

Thus
$$\frac{dP}{dt} + K(B + D) = K(A + C)$$

which has a solution of the form

$$Pe^{K(B+D)t} = \frac{A+C}{B+D} e^{K(B+D)t} + C_1$$

Setting $P = P(0) = P_0$ at $t = 0$, we get

$$C_1 = P_0 - \frac{A+C}{B+D} = P_0 - \overline{P}$$

and thus
$$P = (P_0 - \overline{P})\, e^{-K(B+D)t} + \overline{P}$$

If $P_0 = \overline{P}$, then $P = \overline{P}$ = constant. If $P_0 \neq \overline{P}$, then

$$\lim_{t\to\infty} P = \overline{P}$$

since $K(B + D) > 0$. As $P(t)$ converges to the equilibrium price $\overline{P}$ as $t \to \infty$, the equilibrium, it is said, is *dynamically stable*; the situation is shown graphically in Fig. 4.16. The model that incorporates the effect of the trends in price on demand and supply has been discussed in Section 6.14.

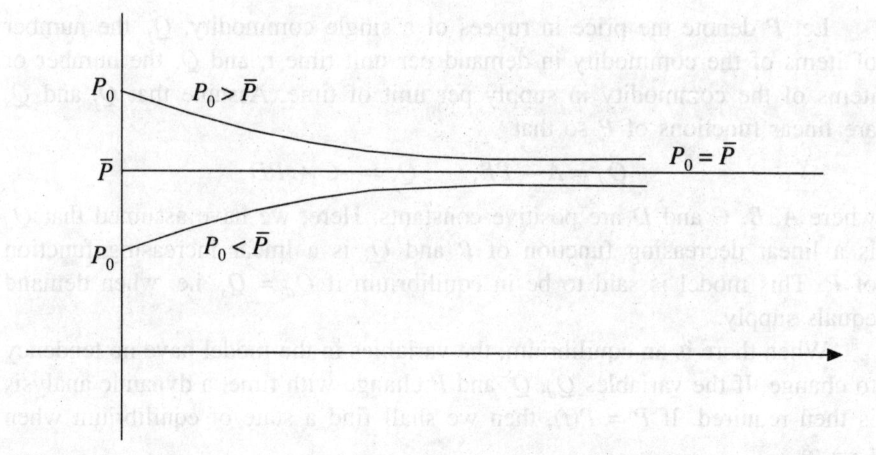


**Fig. 4.16** The price $P$ of a single commodity converges to the equilibrium price $\bar{P}$ as $t \to \infty$—the dynamically stable equilibrium.

## 4.18 THE TRACTRIX (CURVES OF PURSUIT)

Let $C$ be a curve for which the distance to the $y$-axis, measured along the tangent to $C$ from an arbitrary point $P(x, y)$ on $C$, has the constant value $a$. Also assume that $C$ passes through the point $(a, 0)$.

From Fig. 4.17, the slope of $C$ at $P(x, y)$ is given by

$$\frac{dy}{dx} = -\frac{\sqrt{a^2 - x^2}}{x}$$

which on integration yields

$$y = \int -\frac{\sqrt{a^2 - x^2}}{x}\, dx = a \log\left(\frac{a + \sqrt{a^2 - x^2}}{x}\right) - \sqrt{a^2 - x^2} + c_1$$

Setting $x = a$ and $y = 0$, we get $c_1 = 0$. Hence, the equation of the curve $C$ is

$$y = a \log\left(\frac{a + \sqrt{a^2 - x^2}}{x}\right) - \sqrt{a^2 - x^2}$$

The second quadrant branch of the curve $C$ in Fig. 4.17 has the equation

$$y = a \log\left(\frac{a + \sqrt{a^2 - x^2}}{|x|}\right) - \sqrt{a^2 - x^2}$$

and all the four branches are given by

$$y = \pm\left[\, a \log\left(\frac{a + \sqrt{a^2 - x^2}}{|x|}\right) - \sqrt{a^2 - x^2}\,\right]$$

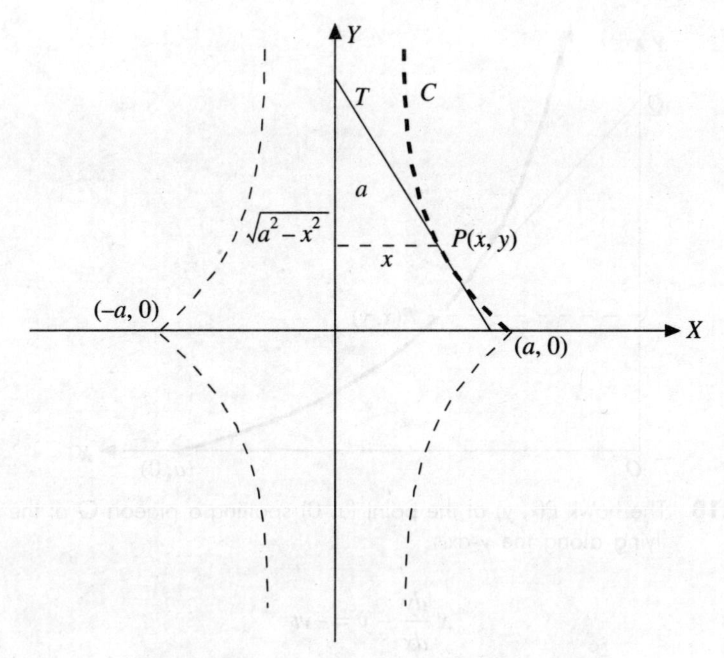


**Fig. 4.17** Four branches of the tractrix.

Equations of the third and fourth quadrant branches are obtained from

$$\frac{dy}{dx} = \frac{\sqrt{a^2 - x^2}}{x}$$

Curve $C$, which was studied by Huygens in 1692, is known as the *tractrix or equitangential curve*. If the $x$- and $y$-axis are introduced on a rough horizontal table, and a string of length $a$ has one end $A$ at the origin and the other end connected to a small weight located at $(a, 0)$, the weight will traverse an approximate tractrix as $A$ moves upward (or downward) along the $y$-axis. This accounts for the name tractrix, derived from the Latin word *tractum* meaning "to drag".

The tractrix has applications in map making, mechanics and in non-Euclidean geometry. Below we shall work out some examples to illustrate the application of tractrix.

**EXAMPLE 4.40** Suppose that a hawk $P$ at point $(a, 0)$ spots a pigeon $Q$ at the origin flying along the $y$-axis at a speed $v$. The hawk immediately flies towards the pigeon at a speed $w$. What will be the flight path of the hawk?

***Solution*** Let $t = 0$ be the time at the instant the hawk starts flying towards the pigeon. After $t$ seconds the pigeon will be at point $Q$ $(0, vt)$ and the hawk at point $P(x, y)$. Since line $PQ$ is tangent to the path (see Fig. 4.18), the slope of the path is

$$\frac{dy}{dx} = \frac{y - vt}{x}$$

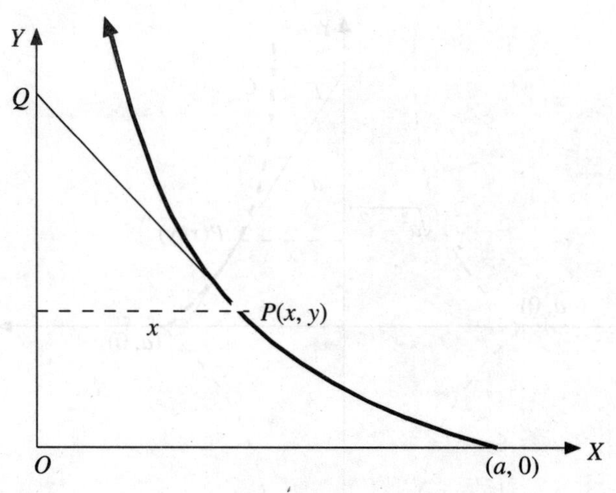


**Fig. 4.18** The hawk $P(x, y)$ at the point $(a, 0)$ spotting a pigeon $Q$ at the origin, flying along the $y$-axis.

or
$$x\frac{dy}{dx} - y = -vt \tag{70}$$

On the other hand, the length of the path travelled by the hawk can be calculated by the formula for arc length as

$$wt = \int ds = \int_x^a \sqrt{1+\left(\frac{dy}{dx}\right)^2}\, dx \tag{71}$$

Solving Eqs. (70) and (71) for $t$ and equating them, we get

$$\frac{y - x\dfrac{dy}{dx}}{v} = \frac{1}{w}\int_x^a \sqrt{1+\left(\frac{dy}{dx}\right)^2}\, dx \tag{72}$$

Differentiation of both sides of Eq. (72) with respect to $x$ yields

$$x\frac{d^2y}{dx^2} = \frac{v}{w}\sqrt{1+\left(\frac{dy}{dx}\right)^2} \tag{73}$$

Setting $p = dy/dx$, Eq. (73) becomes

$$x\frac{dp}{dx} = \frac{v}{w}\sqrt{1+p^2}$$

Separating the variables and integrating, we get

$$\log\left(p + \sqrt{1+p^2}\right) = \frac{v}{w}\log x - c \tag{74}$$

Since, $p = dy/dx = 0$, when $x = a$ (slope of line $PQ = 0$ at $t = 0$), $c = v/w \log a$. Thus, Eq. (74) becomes

$$p + \sqrt{1+p^2} = \left(\frac{x}{a}\right)^{v/w}$$

or

$$p = \frac{dy}{dx} = \frac{1}{2}\left[\left(\frac{x}{a}\right)^{v/w} - \left(\frac{x}{a}\right)^{-v/w}\right] \tag{75}$$

If we assume that the hawk flies faster than the pigeon ($w > v$) then integrate Eq. (75), to obtain

$$y = \frac{a}{2}\left[\frac{(x/a)^{1+(v/w)}}{1+(v/w)} - \frac{(x/a)^{1-(v/w)}}{1-(v/w)}\right] + c \tag{76}$$

Since $y = 0$, when $x = a$, we have

$$c = -\frac{a}{2}\left[\frac{1}{1+[v/w]} - \frac{1}{1-[v/w]}\right] = \frac{avw}{w^2 - v^2}$$

The hawk will catch the pigeon at $x = 0$ and $y = c = avw/w^2 - v^2$. The situation where the hawk flies not faster than the pigeon ($w \le v$) is discussed in Problems 42 and 43.

**EXAMPLE 4.41** A ship A, located at $(a, 0)$, where $a > 0$, sights another ship B at the origin. Ship A pursues ship B at a constant speed $v$ as ship B sails in the positive direction at constant speed $kv$. Where and when does A intercept B if A always sails directly towards B and $0 < k < 1$?

***Solution*** Figure 4.19 shows ship A at $P(x, y)$ and ship B at $Q(0, kvt)$ at time $t$. To find the path of A, we note that

$$\frac{dy}{dx} = \frac{y - kvt}{x - 0}$$

or

$$x\frac{dy}{dx} - y = -kvt$$

Differentiating both sides with respect to $x$, we get

$$x\frac{d^2y}{dx^2} = -kv\frac{dt}{dx}$$

We know that

$$\frac{dt}{dx} = \frac{dt}{ds}\frac{ds}{dx} = -\frac{\sqrt{1+(dy/dx)^2}}{v}$$

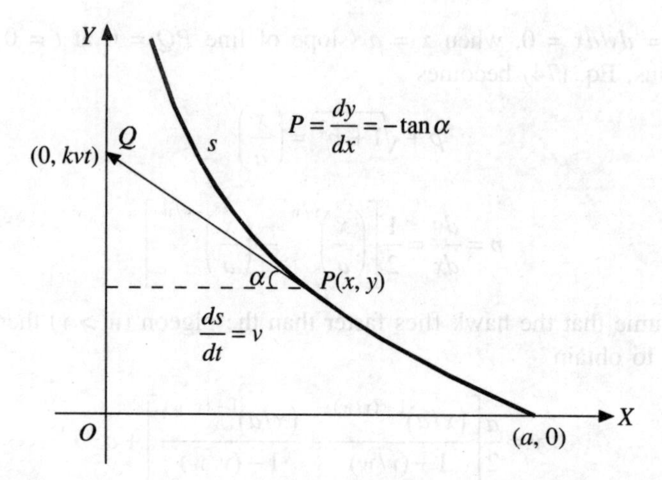


**Fig. 4.19** The ship *A* located at (*a*, 0), *a* > 0 sights another ship *B* at (0, 0).

Thus, the path of *A* is given by the differential equation

$$x\frac{d^2y}{dx^2} = kv\frac{\sqrt{1+(dy/dx)^2}}{v}$$

Letting $p = dy/dx$ and separating the variables, we get

$$x\left(\frac{dp}{dx}\right) = k\sqrt{1+p^2} \quad \text{or} \quad \frac{dp}{\sqrt{1+p^2}} = k\frac{dx}{x}$$

Integrating it, we get

$$\sinh^{-1} p = k \log x + c$$

Now, when $p = 0$ and $x = a$, $c = -k \log a$ and $\sinh^{-1} p = k \log x - \log a = \log (x/a)^k$.

Hence

$$p = \frac{dy}{dx} = \sinh\left[\log\left(\frac{x}{a}\right)^k\right] = \frac{1}{2}\left[\left(\frac{a}{x}\right)^k - \left(\frac{x}{a}\right)^{-k}\right]$$

On integration again, we get

$$y = \frac{a}{2}\left[\frac{(x/a)^{1+k}}{1+k} - \frac{(x/a)^{1-k}}{1-k}\right] + c_1$$

Setting $x = a$ and $y = 0$, we get

$$c_1 = \frac{ak}{1-k^2}$$

When $x = 0$, $y = c_1$ and, therefore, ship *A* intercepts ship *B* at $(0, ak/1 - k^2)$ at time

$$t = \frac{1}{kv}\frac{ak}{1-k^2} = \frac{a}{v(1-k^2)}$$

**EXAMPLE 4.42** A destroyer is in a dense fog, which lifts for an instant, disclosing an enemy submarine on the surface 4 miles away. Suppose that the submarine dives immediately and proceeds at full speed in an unknown direction. What path should the destroyer select to be certain of passing directly over the submarine, if its velocity $v$ is three times that of the submarine?

***Solution*** Suppose that the destroyer has travelled 3 miles towards the place where the submarine was spotted. Then the submarine lies on the circle of radius one mile centred at where it was spotted (Fig. 4.20a), since its velocity is one-third that of the destroyer. As the location of the submarine can easily be described in polar coordinates, we assume that $r = f(\theta)$ be the path the destroyer must follow to be certain of passing over submarine regardless of the direction latter chooses. Then the distance travelled by the submarine to the point where the path will cross is $r - 1$, while that of the destroyer (which is three times longer) is given by the formula of arc length in polar coordinates

$$3(r-1) = \int_0^{\theta} ds = \int_0^{\theta} \sqrt{(dr)^2 + (rd\theta)^2} = \int_0^{\theta} \sqrt{\left(\frac{dr}{d\theta}\right)^2 + r^2}\, d\theta \qquad (77)$$

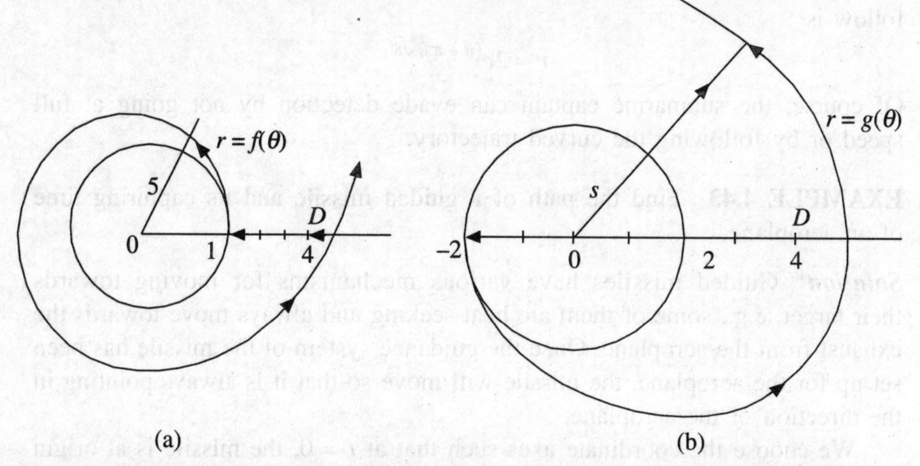


**Fig. 4.20** Path of a destroyer to be certain of passing directly over a submarine: (a) When velocity of the destroyer is three times that of the submarine; (b) When the velocity of the destroyer is six times that of the submarine.

Differentiating both sides of Eq. (77) with respect to $\theta$, we get

$$3\frac{dr}{d\theta}=\sqrt{\left(\frac{dr}{d\theta}\right)^2+r^2} \quad \text{or} \quad 8\left(\frac{dr}{d\theta}\right)^2=r^2$$

Simplifying and separating the variables, we get

$$\frac{dr}{r}=\frac{d\theta}{\sqrt{8}}$$

The solution of this equation is

$$r=ce^{\theta/\sqrt{8}} \tag{78}$$

As $r = 1$ when $\theta = 0$, we have $c = 1$, and the path that the destroyer should follow is the spiral $r=e^{\theta/\sqrt{8}}$ after proceeding 3 miles towards the direction in which the submarine was spotted.

It should be noted that this path is not only the curve, the destroyer could follow. For example, suppose that the destroyer has gone 6 miles towards, where the submarine was spotted (Fig. 4.20b). At this point we can again follow a path $r = g(\theta)$. Since by now the submarine is 2 miles from the origin, the distance travelled by the submarine to where the path will cross is $r - 2$, while the destroyer must go a distance

$$3(r-2)=\int_{-\pi}^{\theta}\left[\sqrt{\left(\frac{dr}{d\theta}\right)^2+r^2}\right]d\theta \tag{79}$$

Equation (79) again leads to the general solution [Eq. (78)], but in this case $r = 2$ when $\theta = -\pi$, so that $c = 2e^{\pi/\sqrt{8}}$. Thus, the spiral, the destroyer must follow is

$$r=2e^{(\theta-\pi)/\sqrt{8}}$$

Of course, the submarine captain can evade detection by not going at full speed or by following the curved trajectory.

**EXAMPLE 4.43** Find the path of a guided missile and its capturing time of an aeroplane.

***Solution*** Guided missiles have various mechanisms for moving towards their target, e.g., some of them are heat seeking and always move towards the exhaust from the aeroplane. Once the guidance system of the missile has been set up for the aeroplane, the missile will move so that it is always pointing in the direction of the aeroplane.

We choose the coordinate axes such that at $t = 0$, the missile is at origin (0, 0) and the aeroplane is at $(a, b)$. Suppose the aeroplane moves parallel to the $x$-axis (Fig. 4.21) with constant speed $V_A$ and the missile has the constant speed $V_M$. Denote the coordinates of the missile by $(x_M, y_M)$.

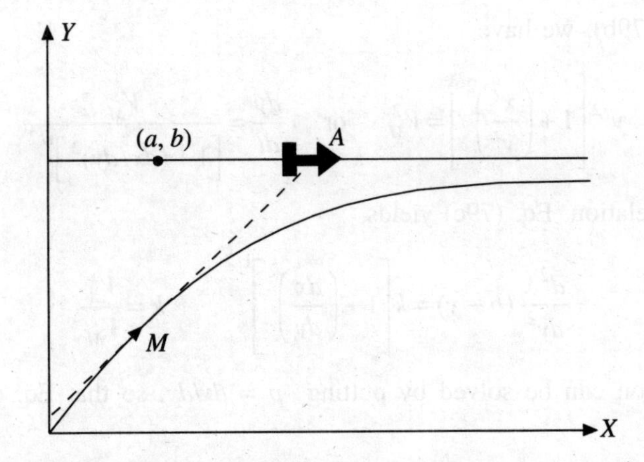


**Fig. 4.21** Path of a guided missile to capture an aeroplane.

The tangent to the missile's path at $M$ will pass through the position of the aeroplane. The equation of the tangent is

$$y - y_M = \frac{dy_M}{dx_M}(x - x_M) = \frac{dy_M / dt}{dx_M / dt}(x - x_M)$$

Now $A(x_A, b)$ lies on this line and $x_A = a + V_A t$. Thus, the above equation yields

$$b - y_M = \frac{y'_M}{x'_M}(a + V_A t - x_M)$$

The path of the missile is thus governed by the differential equation

$$x'(b - y) = y'(a + V_A t - x) \tag{79a}$$

where

$$x'^2 + y'^2 = V_M^2 \tag{79b}$$

with the initial conditions $x(0) = y(0) = 0$. Equation (79a) can be written as

$$\frac{dx}{dy}(b - y) = a + V_A t - x$$

Differentiate it with respect to $t$ to get

$$\frac{d^2x}{dy^2}\frac{dy}{dt}(b - y) = V_A \tag{79c}$$

From Eq. (79b), we have

$$y'^2\left[1+\left(\frac{x'}{y'}\right)^2\right]=V_M^2 \quad \text{or} \quad \frac{dy}{dt}=\frac{V_M}{\left[1+(dx/dy)^2\right]^{1/2}}$$

With this relation, Eq. (79c) yields

$$\frac{d^2x}{dy^2}(b-y)=k\left[1+\left(\frac{dx}{dy}\right)^2\right]^{1/2}, \qquad k=\frac{V_A}{V_M} \tag{79d}$$

This equation can be solved by putting $p = dx/dy$, so that Eq. (79d) takes the form

$$\frac{dp}{dy}(b-y)=k(1+p^2)^{1/2}$$

Now, separate the variables and integrate to get

$$\log\,[p+(1+p^2)^{1/2}] = -k\log\,(b-y)+c$$

Initially, $y = 0$ and $p\ (=dx/dy) = a/b = l$. Thus

$$c = \log\,[l' + (1+l^2)^{1/2}] + k\log b = \log\,(Eb^k)$$

where $E = l + (1 + l^2)^{1/2}$. Hence

$$\log\,[p+(1+p^2)^{1/2}] = \log\,(b-y)^{-k} + \log\,(Eb^k) = \log\left[\frac{Eb^k}{(b-y)^k}\right]$$

or
$$p+(1+p^2)^{1/2} = \frac{Eb^k}{(b-y)^k} \tag{79e}$$

from which we can obtain the value of $p$ as

$$p=\frac{dx}{dy}=\frac{1}{2}\left[\frac{Eb^k}{(b-y)^k}-\frac{(b-y)^k}{Eb^k}\right] \tag{79f}$$

Integrating this equation, we get

$$x=\frac{1}{2}\left[\frac{Eb^k}{(k-1)\,(b-y)^{k-1}}+\frac{(b-y)^{k+1}}{(k+1)\,Eb^k}\right]+c_1$$

(and $k \neq 1$, since we assumed $c < 1$, i.e. $V_A < V_M$).

The constant $c_1$ can be evaluated from the initial condition $x = y = 0$ and

$$c_1=\frac{b[(E^2+1)k+E^2-1]}{2E(1-k^2)} \tag{79g}$$

so that the path of the missile is

$$x = \frac{1}{2}\left[\frac{(b-y)^{k+1}}{(k+1)Eb^k} - \frac{Eb^k(b-y)^{1-k}}{(1-k)}\right] + c_1 \tag{79h}$$

The missile hits the aeroplane when its coordinates are same as that of the aeroplane, i.e. $x = a + V_A t$ and $y = b$. Put these in Eq. (79h) to get

$$a + V_A t = c_1$$

Thus, the capturing time $t$ is

$$t = \frac{(a^2 + b^2)^{1/2} + ak}{V_M(1-k^2)}, \qquad k = \frac{V_A}{V_M} \tag{79i}$$

For more details on tractrix and curves of pursuit, see [26].

## 4.19 PHYSICAL PROBLEMS INVOLVING GEOMETRY

Many types of physical problems are dependent in some way upon geometry. For example, imagine a right circular cylinder, half full of water, rotating with a constant angular velocity about its axis. The shape of the water surface will be determined by the angular velocity of the cylinder. As another example, consider water coming out through a hole at the base of a conical tank. Here the geometrical shape of the container determines the physical behaviour of the water.

In the following illustrative examples we shall consider three physical problems involving geometry; namely, the shape of a reflector, the shape of the water surface in a rotating cylinder, and the flow of water from a tank.

### Curve having a given reflection property

Let $O$ (the origin of any $xy$-coordinate system, Fig. 4.22a), represent the point source of light. Light rays such as $OA$ emerge from $O$, hit the reflector at $A$, and 'bounce off' or are reflected from them on travelling in a straight line. We want to find the shape of the reflector so that all the rays emanating from $O$ 'bounce off' from the reflector parallel to the line $OX$.

Let $CD$ (Fig. 4.22b) be a portion of the reflector and consider any point $P(x, y)$ on it. If $\theta_1$ is the angle of incidence and $\theta_2$ the angle of reflection, then by the principle of optics, $\theta_1 = \theta_2$.* We want to find a relation between the slope $dy/dx$ of the curve (reflector) at $P$ and the coordinates $(x, y)$ of $P$. This has been done as follows: Since, $BP$ (Fig. 4.22b) is parallel to $OX$, $\angle OEP = \theta_2$. Hence, $\angle XOP = \theta_1 + \theta_2 = 2\theta_1$ as $\theta_1 = \theta_2$. The slope of $OP$ is tan $2\theta_1$, but from $\Delta OPF$, it is $y/x$. Hence, $\tan 2\theta_1 = y/x$.

---

*Actually, $90 - \theta_1$, the angle, which ray $OP$ makes with the normal to the arc $CD$ at $P$, is the angle of incidence. Similarly, $90 - \theta_2$, the angle between the reflected ray and the normal, is the angle of reflection. Clearly, if $90 - \theta_1 = 90 - \theta_2$, then $\theta_1 = \theta_2$.

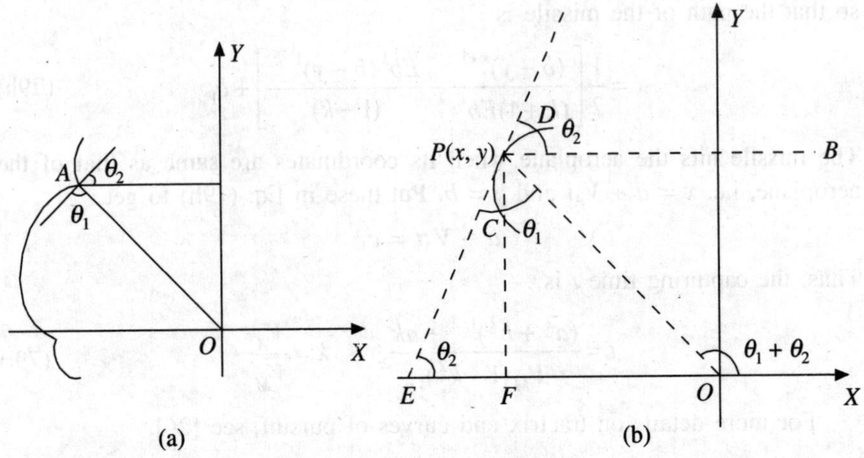


**Fig. 4.22** Curve having a reflection property: (a) A point source of light emerging from O hitting the reflector at A and bouncing off; (b) Schematic diagram of a portion of the reflector.

Also, we have

$$\tan 2\theta_1 = \frac{2\tan\theta_1}{1-\tan^2\theta_1}$$

Since, $\tan\theta_1 = \tan\theta_2 = dy/dx = y'$

$$\frac{2y'}{1-(y')^2} = \frac{y}{x} \tag{80}$$

or

$$\frac{dy}{dx} = \frac{-x \pm \sqrt{x^2+y^2}}{y} \tag{81}$$

which is a homogeneous differential equation. Putting $y = vx$, we get

$$x\frac{dv}{dx} = \frac{-1-v^2 \pm \sqrt{v^2-1}}{v}$$

Separating the variables and integrating, we get

$$\int \frac{dx}{x} = -\int \frac{v\,dv}{v^2+1 \pm \sqrt{v^2+1}}$$

Putting, $v^2 + 1 = u^2$ in the second integral so that $vdv = udu$, we find

$$\log x + c_1 = -\int \frac{du}{u \pm 1} = -\log(u \pm 1) = -\log\left(\sqrt{v^2+1} \pm 1\right)$$

It follows that

$$x\left(\sqrt{v^2+1} \pm 1\right) = c \quad \text{or} \quad \sqrt{x^2+y^2} = c \pm x$$

Therefore,

$$y^2 = \pm 2cx + c^2 \tag{82}$$

For a given value of $c$ ($c \neq 0$), Eq. (82) represents two parabolas symmetric with respect to the $x$-axis (Fig. 4.23). The heavy curve has the equation $y^2 = 2cx + c^2$, $c > 0$; while $y^2 = -2cx + c^2$, $c > 0$ or $y^2 = 2cx + c^2$, $c < 0$ is the equation of the dashed curve. The focus for the family of parabolas is at the origin. In this figure, several light rays have also been shown emanating from the focus and 'bouncing off' the reflector parallel to the $x$-axis.

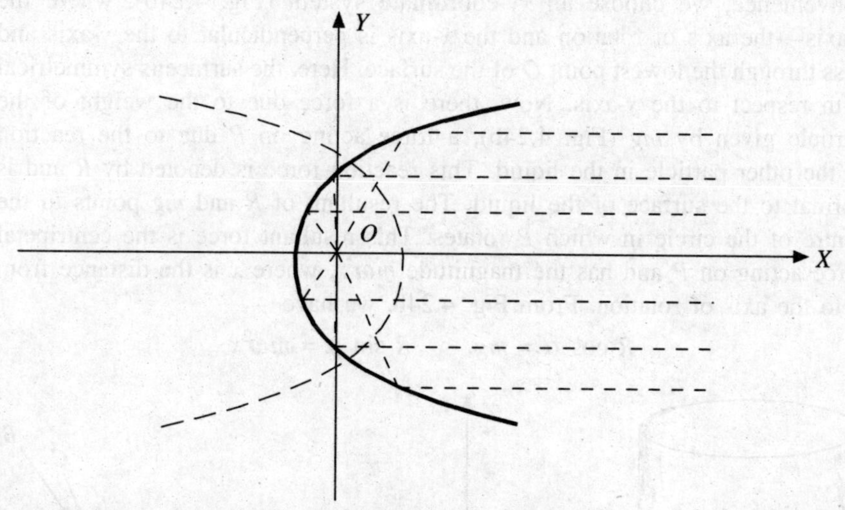


**Fig. 4.23** Two parabolas symmetric with respect to x-axis. Several light rays emanating from the focus and bouncing off the reflector parallel to the x-axis.

If we revolve the parabola about the $x$-axis, we obtain a paraboloid of revolution. An electric bulb placed at the origin sends rays of light 'bouncing off' the reflector to produce a direct beam of light rays which is the most efficient way of lighting. This important property of parabola accounts for the use of parabolic surfaces in telescopes, radar screens, lamps, automobile headlights, television antennas and similar devices.

It may be noted that Eq. (81) can also be solved by writing it as

$$\frac{x\,dx + y\,dy}{\sqrt{x^2 + y^2}} = \pm dx$$

where the left-hand side can be written as $d\left(\sqrt{x^2 + y^2}\right)$. Hence

$$d\left(\sqrt{x^2 + y^2}\right) = \pm dx \quad \text{or} \quad \sqrt{x^2 + y^2} = \pm x + c$$

which leads to the same result as before.

## Rotating fluid

A right circular cylinder having a vertical axis is filled with water and is rotated about its axis with constant angular velocity $\omega$. We then have to find the shape that the water surface takes.

When the cylinder rotates, the surface of the water assumes the shape as in Fig. 4.24a. Consider a particle $P$ of water, of mass $m$, on the surface of the water. When the steady state conditions prevail, this particle will move in a circular path, the circle having its centre on the axis of rotation. For convenience, we choose an $xy$-coordinate system (Fig. 4.24b), where the $y$-axis—the axis of rotation and the $x$-axis is perpendicular to the $y$-axis and pass through the lowest point $O$ of the surface. Here, the surface is symmetrical with respect to the $y$-axis. Now, there is a force due to the weight of the particle given by $mg$ (Fig. 4.24b), a force acting on $P$ due to the reaction of the other particle in the liquid. This reaction force is denoted by $R$ and is normal to the surface of the liquid. The resultant of $R$ and $mg$ points to the centre of the circle in which $P$ rotates. This resultant force is the centripetal force acting on $P$ and has the magnitude $m\omega^2 x$, where $x$ is the distance from $P$ to the axis of rotation. From Fig. 4.24b, we have

$$R \cos \alpha = mg, \qquad R \sin \alpha = m\omega^2 x$$

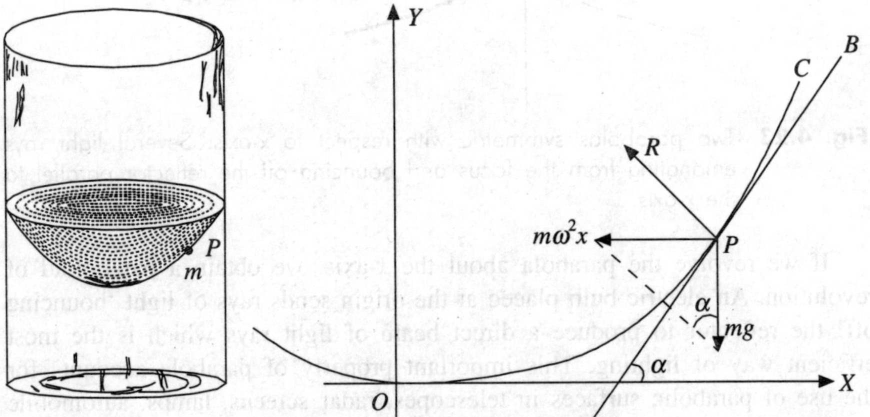

**Fig. 4.24** Rotating fluid: (a) Shape of the surface of water when the cylinder is rotating; (b) Schematic diagram for the curve of a rotating fluid.

Dividing these equations, we get

$$\tan \alpha = \frac{dy}{dx} = \frac{\omega^2 x}{g} \tag{83}$$

Integrating Eq. (83), we obtain

$$y = \frac{\omega^2 x^2}{2g} + c$$

Since $x = 0$, when $y = 0$, we have $c = 0$, and the solution of Eq. (83) becomes

$$y = \frac{\omega^2 x^2}{2g} \tag{84}$$

Hence, in any plane through the $y$-axis, the water level assumes the shape of a parabola. It is a paraboloid of revolution in three dimensions.

## Flow of liquid from a small orifice

The liquid in the vessel shown in Fig. 4.25, flows out through a small sharp-edged orifice. If there were no loss of energy, the speed of the escaping water would be the same as the speed of a freely falling body, namely $\sqrt{2gh}$, where $h$ denotes the height (in ft) of the surface above the orifice at time $t$. Because of the friction and the surface tension, the actual speed has been found (approx.) to be $0.6\sqrt{2gh}$ or $4.8h^{1/2}$ ft/s, when $g = 32$ ft/s$^2$ is the acceleration due to gravity. This is known as *Torricelli's law*. Thus, if the orifice has an area $A$, the fluid leaves the vessel at $4.8Ah^{1/2}$ ft$^3$/s. Hence, if $V$ denotes the volume of liquid in the vessel at time $t$, then

$$\frac{dV}{dt} = -4.8\,Ah^{1/2} \tag{85}$$

The minus sign indicates that $V$ decreases with time $t$.

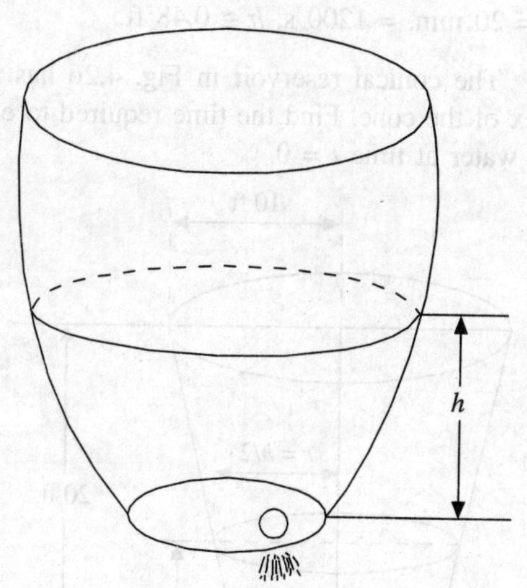


**Fig. 4.25** Flow of liquid through a small sharp edged orifice at the bottom of a vessel.

It may be noted that the shape of the vessel enables us to express $V$ as $V = f(h)$ so that $dV/dt = (df/dh)(dh/dt)$ and the resulting differential equation can be solved to find $h$ in terms of $t$ (or $t$ in terms of $h$).

We shall now apply Eq. (85) to some numerical problems.

**EXAMPLE 4.44** A tank 4 ft deep has a rectangular cross-section 6' × 8'. The tank is initially filled with water, which runs out through an orifice of radius 1" located in the bottom of the tank. Find: (a) the time required for the tank to empty; (b) the time required for half the water to drain out from the tank; and (c) the height of the water in the tank 20 minutes after it starts to drain out.

***Solution*** From Eq. (85), we have

$$\frac{dV}{dt} = -4.8\frac{\pi}{(12)^2}h^{1/2}$$

Since, $V = (6)(8)h$, $dV/dt = 48\,(dh/dt)$, and thus

$$48\frac{dh}{dt} = -4.8\left(\frac{\pi}{144}\right)h^{1/2} \quad \text{or} \quad h^{-1/2}dh = -\frac{\pi}{1440}dt$$

or $$2\sqrt{h} = \frac{-\pi t}{1440} + c$$

Setting $t = 0$ and $h = 4$, we get $c = 4$, and

$$h = \left(2 - \frac{\pi t}{2880}\right)^2 \quad \text{or} \quad t = \frac{2880}{\pi}(2 - \sqrt{h})$$

(a) Setting $h = 0$, we have $t = (5760/\pi)\text{s} \cong 30.6$ min.

(b) Setting $h = 2$, we have $t \cong 9.0$ min.

(c) When $t = 20$ min. $= 1200$ s, $h \cong 0.48$ ft.

**EXAMPLE 4.45** The conical reservoir in Fig. 4.26 has an orifice of 2-in. radius at the vertex of the cone. Find the time required to empty the reservoir if it is filled with water at time $t = 0$.

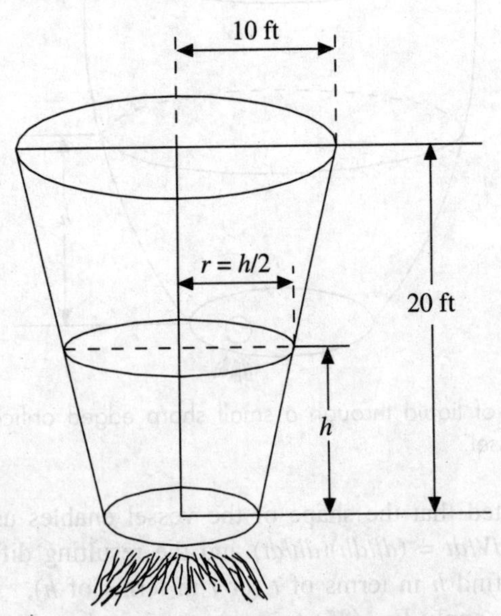


**Fig. 4.26** Flow of liquid through an orifice at the bottom of a conical reservoir.

***Solution*** Since $r = h/2$, we have

$$V = \frac{\pi}{3} r^2 h = \frac{\pi h^3}{12}, \quad \frac{dV}{dt} = \frac{\pi h^2}{4} \frac{dh}{dt}$$

and from Eq. (85), we have

$$\frac{\pi h^2}{4} \frac{dh}{dt} = -4.8\pi \left(\frac{2}{12}\right)^2 h^{1/2}$$

Separating the variables and integrating, we get

$$\int_{20}^{0} h^{3/2} dh = -\int_{0}^{t} \frac{8}{15} dt \quad \text{or} \quad \left(\frac{2}{5} h^{5/2}\right)_{20}^{0} = -\frac{8}{15} t$$

or

$$t = \frac{15}{8} \frac{2}{5} 400\sqrt{20} \text{ s} \cong 22.4 \text{ min}.$$

**Remark.** This model is suitable for the usual type of orifices. For certain applications, the coefficient of $\sqrt{2gh}$ , known as the *coefficient of contraction,* or *discharge coefficient,* may differ from 0.6. A more complicated model is required for other physical assumptions; for example, the air pressure at the liquid level may differ from the air pressure at the level of the orifice.

## 4.20 ORTHOGONAL TRAJECTORIES

If we are given a family of curves (heavy lines in Fig. 4.27), we may think of another family of curves (dashed lines) such that each member of this family cuts each member of the another family at right angles. For example, $AB$ meets several members of the dashed family at right angles at the points

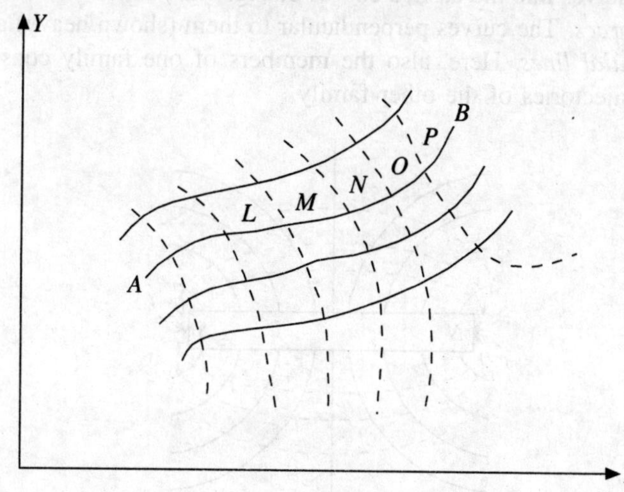


**Fig. 4.27** Two families of curves such that each member of one family intersects each member of other family at right angles.

$L$, $M$, $N$, $O$ and $P$. We say that the families are *mutually orthogonal,* or that either family forms a set of *orthogonal trajectories* of the other family. As an example, consider the family of all circles having their centre at origin (a few such circles appear in Fig. 4.28a). The orthogonal trajectories for this family of circles would be members of the family of straight lines (shown by dashed lines in Fig. 4.28a). Similarly, the orthogonal trajectories of the family of straight lines passing through the origin are circles having centre at the origin.

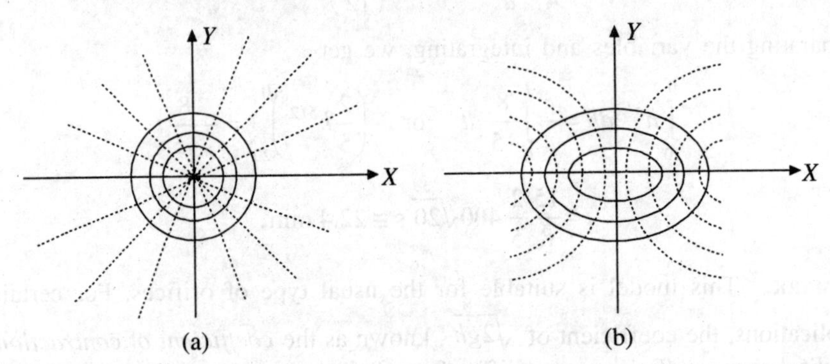


**Fig. 4.28** Orthogonal trajectories: (a) Orthogonal trajectories of the family of circles having centre at origin are the members of the family of straight lines; (b) The family of ellipses and the family of curves orthogonal to them.

As a more complicated illustration, consider the family of ellipses (Fig. 4.28b) and the family of curves orthogonal to them. The curves of one family are the orthogonal trajectories of the other family. Applications of orthogonal trajectories in Physics and Engineering are numerous. As a very simple application, consider Fig. 4.29. Here NS represents a bar magnet, $N$ being its north pole, and $S$ its south pole. If iron fillings are sprinkled around the magnet we find that they arrange themselves like the dashed curves of Fig. 4.29. These curves are called the *lines of forces.* The curves perpendicular to them (shown heavy) are called the *equipotential lines.* Here, also the members of one family constitute the orthogonal trajectories of the other family.

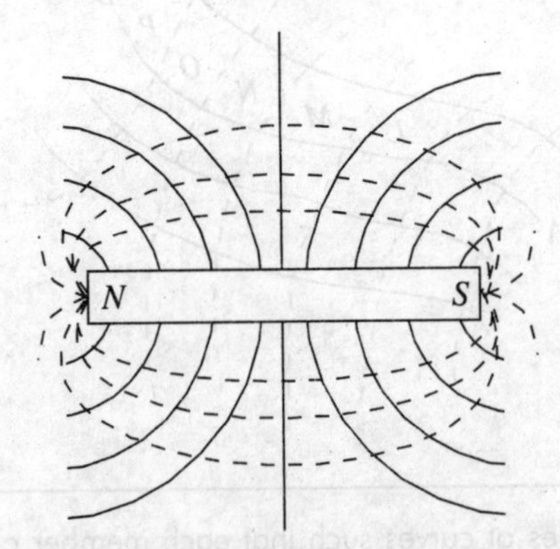


**Fig. 4.29** Lines of forces of a bar magnet.

As another example, consider Fig. 4.30, which represents a weather map so familiar to us. The curves represent *isobars,* which are curves-connecting all cities that report the same barometric pressure at the weather bureau. The orthogonal trajectories of the family of isobars would indicate the general direction of the wind from the higher to low pressure areas. Instead of isobars, Fig. 4.30 could represent *isothermal curves* which are curves connecting points having the same temperature. In such a case, the orthogonal trajectories represent the general direction of heat flow.

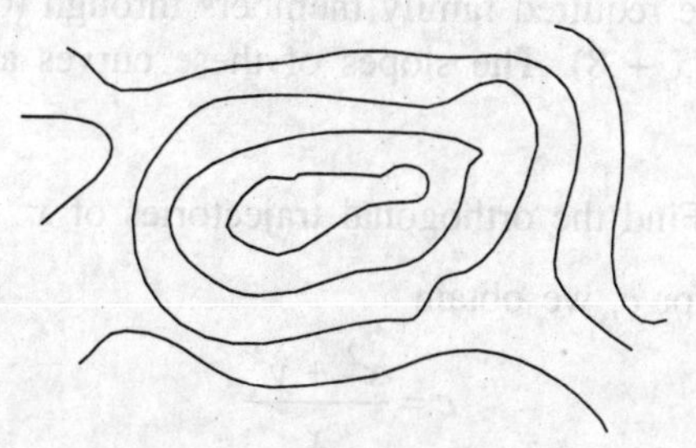

**Fig. 4.30** A weather map showing isobars/isothermal curves.

To find the orthogonal trajectories of a given family of curves, we first find the differential equation

$$\frac{dy}{dx} = f(x, y)$$

which describes the family. The differential equation of the second and orthogonal family is then

$$\frac{dy}{dx} = -\frac{1}{f(x, y)}$$

**EXAMPLE 4.46** Find an equation of the family orthogonal to the family $y = cx$.

***Solution*** Eliminating $c$ from $y = cx$ and $dy/dx = c$, we obtain $dy/dx = y/x$, a differential equation of the family. Hence, a differential equation of the required family is

$$\frac{dy}{dx} = -\frac{1}{y/x} = -\frac{x}{y}$$

Solving this equation, we get $x^2 + y^2 = k^2$, a one-parameter family of circles centred at the origin.

The two families are illustrated in Fig. 4.28a. The domain must not include the origin, since infinitely many members of $y = cx$ pass through that point. The circles are perpendicular to the $x$-axis although none of the circles has a slope at a point at which $y = 0$.

**EXAMPLE 4.47** Find the family orthogonal to the family $y = ce^{-x}$ of exponential curves. Determine the member of each family passing through (0, 4).

***Solution*** From $y' = -ce^{-x}$, we obtain $y' = -y$, a differential equation of the family of exponential curves. Solving $dy/dx = 1/y$ or $ydy = dx$ we obtain $y^2 = 2(x + k)$, a one-parameter family of parabolas. The parabolas are orthogonal to the exponential curves.

Putting $x = 0$ and $y = 4$ in $y = ce^{-x}$ and $y^2 = 2(x + k)$, respectively, we find $c = 4$ and $k = 8$. The required family members through (0, 4) have equations $y = 4e^{-x}$ and $y^2 = 2(x + 8)$. The slopes of these curves at (0, 4) are $-4$ and 1/4, respectively.

**EXAMPLE 4.48** Find the orthogonal trajectories of $x^2 + y^2 = cx$.

***Solution*** Solving for $c$, we obtain

$$c = \frac{x^2 + y^2}{x}$$

Differentiating with respect to $x$, we get

$$y' = \frac{dy}{dx} = \frac{y^2 - x^2}{2xy}$$

The family of orthogonal trajectories thus, have the differential equation

$$\frac{dy}{dx} = \frac{2xy}{x^2 - y^2}$$

which is a homogeneous differential equation and can be solved by putting $y = vx$. The solution is

$$x^2 + y^2 = c_1 y$$

The two orthogonal families are shown in Fig. 4.31. The originally given family is shown by dashed lines.

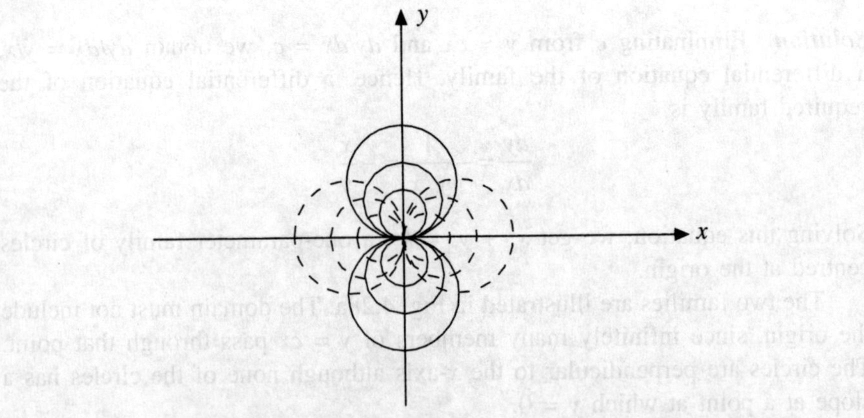


**Fig. 4.31** A family of orthogonal trajectories.

**EXAMPLE 4.49** Find the orthogonal trajectories of the family $y = x + ce^{-x}$ and determine that particular member of each family that passes through (0, 3).

***Solution*** Differentiation of the given equation yields

$$y' = 1 - ce^{-x}$$

Elimination of $c$ gives

$$y' = 1 + x - y$$

Thus, the differential equation for the family of orthogonal trajectories is

$$\frac{dy}{dx} = \frac{-1}{1 + x - y} \quad \text{or} \quad \frac{dx}{dy} + x = y - 1$$

which is a linear differential equation having the solution as

$$xe^y - e^y(y - 2) = c_1$$

Therefore, the required curves passing through (0, 3) are found to be

$$y = x + 3e^{-x}, \qquad x - y + 2 + e^{3-y} = 0$$

**EXAMPLE 4.50** Find the orthogonal trajectories of the family of rectangular hyperbolas $y = c_1/x$.

***Solution*** Differentiation of the given equation and then the elimination of the constant $c_1$ yield the differential equation of the given family

$$\frac{dy}{dx} = -\frac{y}{x}$$

The differential equation of the given orthogonal family is then

$$\frac{dy}{dx} = \frac{x}{y}$$

Solving, we get

$$y^2 - x^2 = c_2$$

where $c_2 = 2c'_2$. The graphs of the two families for various values of $c_1$ and $c_2$ are depicted in Fig. 4.32.

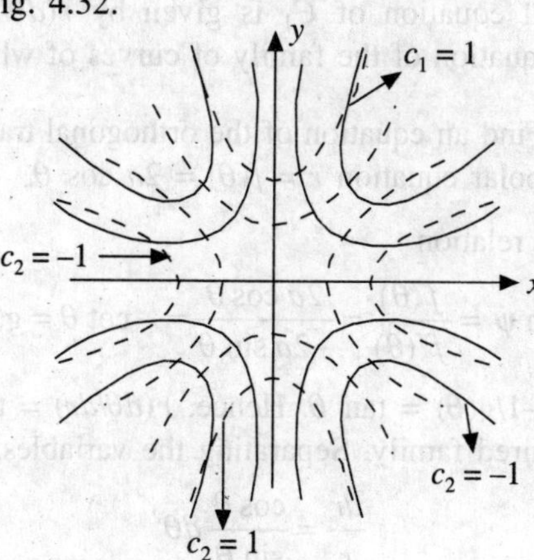


**Fig. 4.32** Orthogonal trajectories of the family of rectangular hyperbolas for different values of the constants.

**Remark.** In finding the orthogonal trajectory of a given family, it is sometimes convenient to use polar coordinates.

In Fig. 4.33, $\psi$ is the angle between the radius vector $OP$ and the tangent to the curve $C$ at $P(r, \theta)$. We know that

$$\tan\psi = \frac{f(\theta)}{f'(\theta)} = \frac{r}{dr/d\theta} = r\frac{d\theta}{dr} = g(\theta)$$

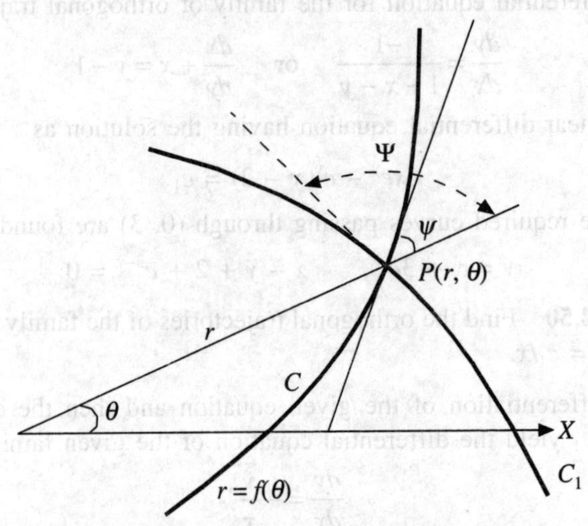


**Fig. 4.33** Schematic diagram of an orthogonal trajectory in polar coordinate.

Denoting the corresponding angle for the curve $C_1$ by $\psi$ and assuming that $C_1$ is perpendicular to $C$ at $P$, we note that

$$\tan\psi = \tan\left(\psi + \frac{\pi}{2}\right) = -\cot\psi = -\frac{1}{\tan\psi}$$

Hence, a differential equation of $C_1$ is given by $r(d\theta/dr) = 1/g(\theta)$. When solved, it gives an equation of the family of curves of which $C_1$ is a member.

**EXAMPLE 4.51** Find an equation of the orthogonal trajectory of the family of circles having a polar equation $r = f(\theta) = 2a\cos\theta$.

***Solution*** From the relation

$$\tan\psi = \frac{f(\theta)}{f'(\theta)} = \frac{2a\cos\theta}{-2a\sin\theta} = -\cot\theta = g(\theta)$$

We obtain $\tan\psi = -1/g(\theta) = \tan\theta$. Hence, $r(d\theta/dr) = \tan\theta$ is a differential equation of the required family. Separating the variables, we get

$$\frac{dr}{r} = \frac{\cos\theta}{\sin\theta}\,d\theta$$

or
$$\log|r| = \log|\sin\theta| + \log 2c$$

or
$$\log\frac{|r|}{2c} = \log|\sin\theta|$$

and
$$r = 2c\sin\theta$$

The required family is a family of circles having centres on the line $\theta = \pi/2$ ($y$-axis).

**EXAMPLE 4.52** Find the orthogonal trajectory of $r = c_1(1 - \sin\theta)$.

***Solution*** For the given curve, we have

$$\frac{dr}{d\theta} = -c_1\cos\theta = -\frac{r\cos\theta}{1-\sin\theta}$$

so that

$$r\frac{d\theta}{dr} = \frac{1-\sin\theta}{\cos\theta} = \tan\psi = g(\theta)$$

Hence

$$r\frac{d\theta}{dr} = \frac{\cos\theta}{1 - r\sin\theta} = \tan\psi$$

Separating the variables, we obtain

$$\frac{dr}{r} = \frac{1-\sin\theta}{\cos\theta}\,d\theta = (\sec\theta - \tan\theta)\,d\theta$$

Integrating, we get

$$\log|r| = \log|\sec\theta + \tan\theta| + \log|\cos\theta| + \log c_2$$

$$= \log|c_2(1 + \sin\theta)|$$

Hence

$$r = c_2(1 + \sin\theta)$$

*Note:* If two families of curves intersect at angle $\alpha \neq \pi/2$, each family is then said to constitute an *isogonal* or *oblique* trajectory of the other family. To find the isogonal trajectory of a given family, we use the formula

$$\tan\alpha = \frac{m_2 - m_1}{1 + m_1 m_2}$$

in place of $m_2 = -1/m_1$. If $dy/dx = f(x, y)$ is the differential equation of the given family, then the differential equation of the isogonal family is

$$\frac{dy}{dx} = \frac{f(x, y) \pm \tan \alpha}{1 \mp f(x, y) \tan \alpha}$$

A family of curves can be *self-orthogonal* in the sense that a member of the orthogonal trajectory is also a member of the original family.

## 4.21 MISCELLANEOUS PROBLEMS IN GEOMETRY

Geometrical problems provide a fertile source of differential equations. We have already seen how differential equations arise in connection with orthogonal trajectories. We shall consider here some other geometrical problems.

**EXAMPLE 4.53** The slope at any point of a curve is $2x + 3y$. If the curve passes through the origin, determine its equation.

***Solution*** The slope at $(x, y)$ is $dy/dx$. Hence

$$\frac{dy}{dx} = 2x + 3y$$

is the required differential equation, which we solve subject to condition $y(0) = 0$. This equation can be written as

$$\frac{dy}{dx} - 3y = 2x$$

and has the integrating factor as $e^{-3x}$. Hence

$$\frac{d}{dx}(ye^{-3x}) = 2xe^{-3x}$$

or

$$ye^{-3x} = \frac{-2xe^{-3x}}{3} - \frac{2e - 3x}{9} + C$$

Since, $y(0) = 0$, $c = 2/9$, we find

$$y = \frac{2}{9}e^{3x} - \frac{2x}{3} - \frac{2}{9}$$

which is the required equation of the curve.

**EXAMPLE 4.54** Find the curve through the point (1, 1) in the $xy$-plane having at each of its points the slope $-y/x$.

***Solution*** The slope at $(x, y)$ is $dy/dx$. Hence

$$\frac{dy}{dx} = -\frac{y}{x}$$

Separating the variables and integrating, the solution is given by

$$y = \frac{c}{x}$$

for every value of the constant $c$. Some of the corresponding curves are shown in Fig. 4.34. As we are looking for the curve which passes through (1, 1), we must have $y = 1$ when $x = 1$. This yields $c = 1$. Hence, the solution of our problem is

$$y = \frac{1}{x}$$

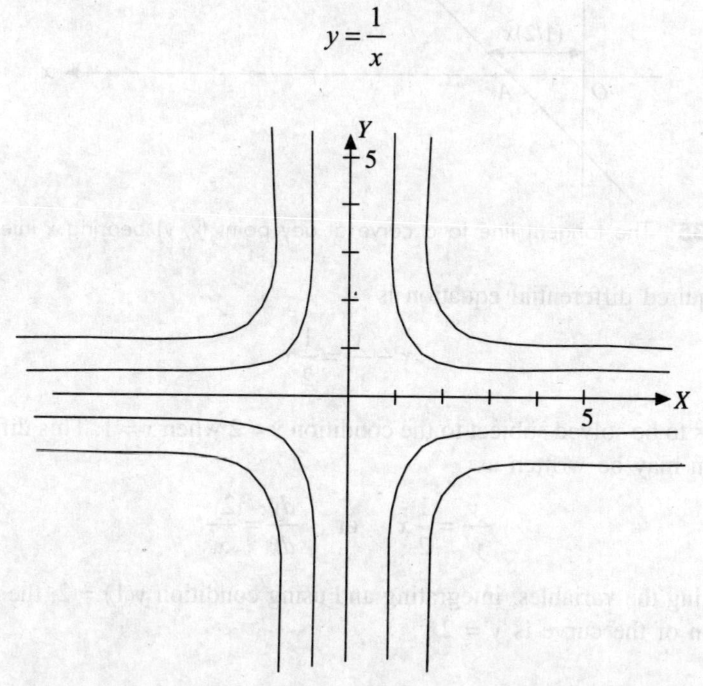


**Fig. 4.34** Graph of the curve $y = c/x$ for some values of the constant c.

**EXAMPLE 4.55** The tangent line to a curve at any point $(x, y)$ on it has its intercept on the $x$-axis always equal to $(1/2)x$. If the curve passes through (1, 2), find its equation.

***Solution*** To solve this example, we must find an expression for the $x$-intercept $OA$ of the tangent line $AP$ to the curve $QPR$ (Fig. 4.35). For this, let $(X, Y)$ be any point on $AP$. Since $y'$ is the slope of the line, its equation is

$$Y - y = y'(X - x)$$

The required intercept is the value of $X$, where $Y = 0$. This is found to be

$$X = x - \frac{y}{y'}$$

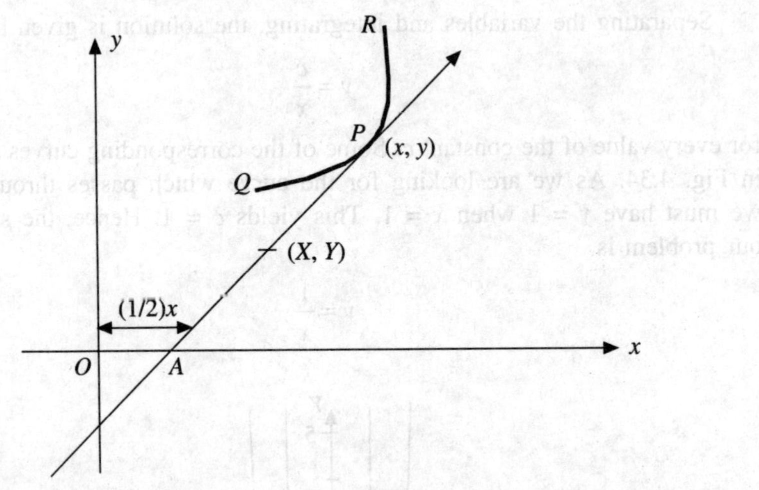


**Fig. 4.35** The tangent line to a curve at any point (x, y) bearing x-intercept.

The required differential equation is

$$x - \frac{y}{y'} = \frac{1}{2}x$$

which is to be solved subject to the condition $y = 2$ when $x = 1$. This differential equation may be written as

$$\frac{y}{y'} = \frac{1}{2}x \quad \text{or} \quad \frac{dy}{dx} = \frac{2y}{x}$$

Separating the variables, integrating and using condition $y(1) = 2$, the required equation of the curve is $y = 2x^2$.

## 4.22 MISCELLANEOUS PROBLEMS IN PHYSICS

**EXAMPLE 4.56** *Escape velocity.* Find the minimum initial velocity of a projectile which is fired in the radial direction from the earth and is supposed to escape from the earth. Neglect the air resistance and the gravitational pull of other celestial bodies.

***Solution*** According to Newton's law of gravitation, the gravitational force as also the acceleration $a$ of the projectile is proportional to $1/r^2$, where $r$ is the distance from the centre of the earth to the projectile. Thus

$$a(r) = \frac{dv}{dt} = \frac{k}{r^2}$$

where $v$ is the velocity and $t$ is the time. Since $v$ is decreasing, $a < 0$ and thus, $k < 0$. Let $R$ be the radius of the earth. When $r = R$, $a = -g$, the acceleration due to gravity at the surface of the earth. Note that the minus sign occurs because $g$ is positive and the action of gravity acts in the negative direction (the direction towards the centre of the earth). Thus

$$-g = a(R) = \frac{k}{R^2}, \quad a(r) = \frac{-gR^2}{r^2}$$

Now, $v = dr/dt$, and the differentiation yields

$$a = \frac{dv}{dt} = \frac{dv}{dr}\frac{dr}{dt} = v\frac{dv}{dr}$$

Separating the variables and integrating, we get

$$vdv = -gR^2 \frac{dr}{r^2}$$

Integrating, we get

$$\frac{v^2}{2} = \frac{gR^2}{r} + c \tag{86}$$

On the surface of the earth, $r = R$ and $v = v_0$, the initial velocity and from Eq. (86), we find that

$$c = \frac{v_0^2}{2} - gR$$

Inserting this value of $c$ in Eq. (86), we get

$$v^2 = \frac{2gR^2}{r} + v_0^2 - 2gR \tag{87}$$

if $v^2 = 0$, then $v = 0$, the projectile stops, and from the physical situation it is clear that the velocity will change from positive to negative so that the projectile will return to the earth. Consequently, we have to choose $v_0$ so large that this cannot happen. If we choose

$$v_0 = \sqrt{2gR} \tag{88}$$

then in Eq. (87) the expression $v_0^2 - 2gR$ is zero and $v^2$ is never zero for any $r$. However, if we choose a smaller value for the initial velocity $v_0$, then $v = 0$ for certain $r$. The velocity $v_0$ in Eq. (88) is called the *escape velocity* from the earth. Since $R = 6372$ km = 3960 miles and $g = 9.8$ m/s$^2$ = 0.0098 km/s$^2$ = 32.17 ft/s$^2$ = 0.00609 miles/s$^2$, we have

$$v_0 = \sqrt{2gR} \cong 11.2 \text{ km/s} \cong 6.95 \text{ miles/s}$$

**Remarks.** 1. The quantity $\sqrt{(2g_1R_1)}$ is the escape velocity of any planet, satellite, or star where $g_1$ and $R_1$ are the acceleration due to gravity and radius. If the radius of such a body is decreased while the mass is unchanged, the escape velocity at the surface of the body increases.

2. Most of the normal stars are maintained in their gaseous, puffed-up state by radiation pressure from within, which is generated by the burning of nuclear fuel. When all the nuclear fuel burns out, the star undergoes a collapse, known as *gravitational collapse,* into a very smaller sphere essentially of the same mass. Depending upon the mass of the star, the crushed, degenerated matter of these collapsed stars can sustain two types of equilibrium. When the mass of the star is less than 1.4 solar masses, then the star is a *white dwarf;* and when the mass of the star lies between the range 1.4 and 3 solar masses, then they are called *neutron stars.* For heavier masses no equilibrium is possible, and the collapse continues until the escape velocity at the surface reaches the velocity of light. Collapsed stars of these nature are completely invisible, since no radiation can ever escape. These stars are the so-called *black holes.*

**EXAMPLE 4.57** *Sky diver.* Suppose that a sky diver falls from rest towards the earth and the parachute opens at an instant, call it $t = 0$, when the sky diver's speed is $v(0) = v_0 = 10.0$ m/s. Find $v(t)$ of the sky diver at any later time $t$. Does $v(t)$ increase indefinitely?

***Solution*** Suppose that the weight of the man plus the equipment is $W = 712$ N, the air-resistance $R$ is proportional to $v^2$, say $R = bv^2$ N, where $b$ is the constant of proportionality and depends mainly upon the parachute. We also assume that $b = 30.0$ Ns$^2$/m$^2$ = 30.0 kg/m.

We now set up the mathematical model (the differential equation) of the problem as follows.

Newton's second law states that

$$\text{mass} \times \text{acceleration} = \text{force}$$

where 'force' means the resultant of the forces acting on the sky diver at any instant. These forces are the weight $W$ and the resistance $R$. The weight $W = mg$, $g = 9.8$ m/s$^2$. Hence, the mass of the man plus equipment is $m = W/g = 72.7$ kg. The air-resistance $R$ acts upward (against the direction of motion), so that the resultant is

$$W - R = mg - bv^2$$

The acceleration is the time derivative of $v$, i.e. $a = dv/dt$. Hence, by Newton's second law

$$m\frac{dv}{dt} = mg - bv^2$$

This is the differential equation of the problem. From the given condition $v(0) = v_0 = 10$, we will now solve this differential equation. This equation can be written as

$$\frac{dv}{dt} = -\frac{b}{m}(v^2 - k^2), \qquad k^2 = \frac{mg}{b}$$

Separating the variables, we get

$$\frac{dv}{v^2 - k^2} = -\frac{b}{m}dt \quad \text{or} \quad \frac{1}{2k}\left(\frac{1}{v-k} - \frac{1}{v+k}\right)dv = -\frac{b}{m}dt$$

Integrating, we get

$$\frac{1}{2k}[\log(v-k) - \log(v+k)] = -\frac{b}{m}t + c_1$$

or

$$\frac{v-k}{v+k} = ce^{-pt}, \qquad p = \frac{2kb}{m}, \qquad c = e^{2kc_1} \tag{89}$$

Solving this for $v$, we have

$$v(t) = k\frac{1 + ce^{-pt}}{1 - ce^{-pt}} \tag{90}$$

Note that as $t \to \infty$, $v(t) \to k$; i.e. $v(t)$ does not increase indefinitely but approaches a limit $k$. This limit is independent of the initial condition $v(0) = v_0$.

We now find $c$ in Eq. (90) such that we obtain the particular solution satisfying the initial condition. From Eq. (89) with $t = 0$, we have

$$c = \frac{v_0 - k}{v_0 + k}$$

With this $c$, Eq. (90) represents the solution we are looking for.

From the given numerical data, we obtain

$$k^2 = \frac{mg}{b} = \frac{W}{b} = \frac{712}{30} = 23.7 \text{ m}^2/\text{s}^2$$

Hence, $k$ = 4.87 m/s. This is the limiting speed. Practically speaking, this is the speed of the sky diver after a sufficiently long time.

To $v(0) = v_0$ = 10 m/s, there corresponds

$$c = \frac{v_0 - k}{v_0 + k} = 0.345$$

Finally

$$p = \frac{2kb}{m} = \frac{2 \times 4.87 \times 30.0}{72.7} = 4.02 \text{ s}^{-1}$$

This altogether yields the result

$$v(t) = 4.87\frac{1 + 0.345\,e^{-4.02t}}{1 - 0.345\,e^{-4.02t}} \quad \text{(see Fig. 4.36)}$$

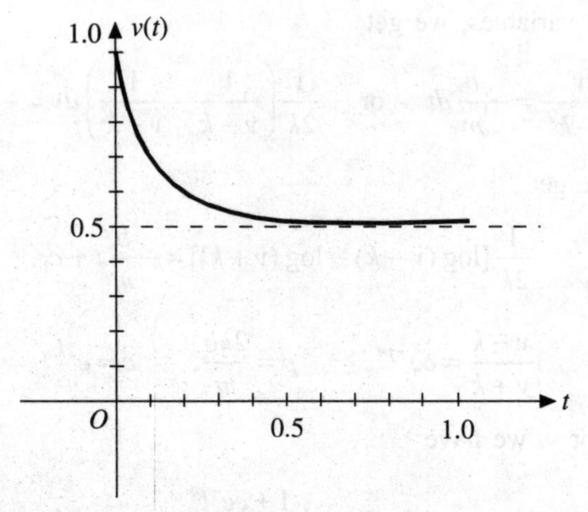


**Fig. 4.36** The speed $v$ of a sky diver at any time $t$.

**EXAMPLE 4.58** A paratrooper (and his parachute) falls from rest. The combined weight of the paratrooper and the parachute is $W$ lb. The parachute has a force acting on it (due to air-resistance) which is proportional to the speed at any instant during the fall. Assuming that the paratrooper falls vertically downward and that the parachute is already open when he takes the jump, describe the ensuing motion.

***Solution*** Draw a physical and force diagram (Fig. 4.37). Assume $A$ to be the origin of the positive $x$-axis. The forces acting are:

(a) Combined weight $W$ acting downward,
(b) Air-resistance $R$ acting upward.

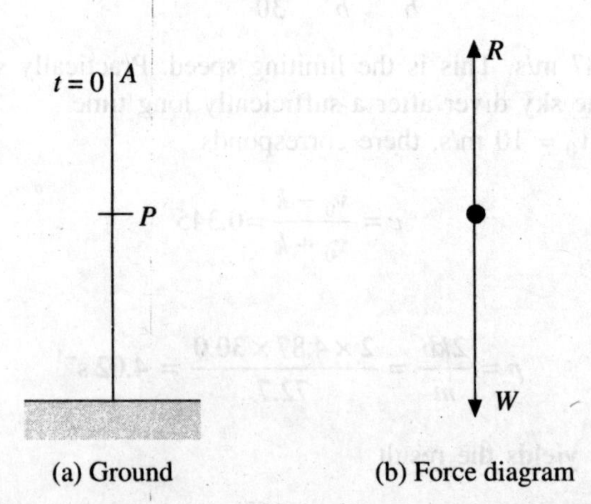


**Fig. 4.37** Schematic diagram for the motion of a paratrooper.

The net force in the positive direction is $W-R$. Since the resistance is proportional to the speed, we have

$$R \propto |v| \qquad \text{or} \qquad R = k\,|v|$$

where $k$ is the constant of proportionality. Since $v$ is always positive, we may write $R = kv$. Hence, the net force is $W - kv$, and from Newton's law, we obtain

$$\frac{W}{g}\frac{dv}{dt} = W - kv$$

Since, the paratrooper starts from rest, $v = 0$ at $t = 0$. Thus, the complete mathematical formulation of the problem is

$$\frac{W}{g}\frac{dv}{dt} = W - kv, \qquad v = 0 \quad \text{at } t = 0$$

Separating the variables and integrating, we get

$$-\frac{W}{k}\log(W - kv) = gt + c_1$$

Since, $v = 0$ at $t = 0$, $c_1 = -w \log W/k$, thus

$$-\frac{W}{k}\log(W - kv) = gt - \frac{W}{k}\log W$$

or

$$\log\left(\frac{W}{W - kv}\right) = \frac{kgt}{W}$$

Hence

$$v = \frac{W}{k}(1 - e^{-kgt/W})$$

As $t \to \infty$, $v \to w/k$, a constant limiting velocity. This accounts for the fact that we notice a parachute travelling at a very nearly uniform speed after a certain length of time has elapsed. We can also determine the distance travelled by the paratrooper as a function of time. From

$$\frac{dx}{dt} = v = \frac{W}{k}(1 - e^{-kgt/W})$$

we have

$$x = \frac{W}{k}\left(t + \frac{W}{gk}e^{-kgt/W}\right) + c_2$$

Note that $x = 0$ at $t = 0$ and thus $c_2 = -W^2/k^2g$. Hence

$$x = \frac{W}{k}\left(t + \frac{W}{gk}e^{-kgt/W} - \frac{W}{gk}\right)$$

## 4.23 MOTION OF A ROCKET

A rocket moves by the backward expulsion of a mass of a gas formed by the burning of the fuel. This ejection of mass has the effect of increasing the forward velocity of the rocket, thus enabling it to continue onward. To consider the motion of rockets, we must treat the motion of objects whose mass is changing. From Newton's second law, we know that the net force acting on an object is equal to the time rate of change in momentum. We will use this in finding the law of motion of a rocket.

Suppose that at time $t$ the total mass of the rocket is $M$, and that at a later time $t + \Delta t$ the mass is $M + \Delta M$, i.e. a mass $-\Delta M$ of gas has been expelled from the back of the rocket (note that the gas expelled in time $\Delta t$ is $-\Delta M$, since $\Delta M$ is a negative quantity). Suppose that the velocity of the rocket relative to earth at time $t$ is $V$ and at time $t + \Delta t$ is $V + \Delta V$, and take the upward direction of the rocket as positive. The expelled gases will have velocity $V + v$ relative to earth, where $v$ is the negative quantity so that $-v$ represents the magnitude of the velocity of the gas relative to the rocket, which for our purpose will be considered constant. The total momentum of the rocket before the loss of gas is $MV$. After the loss of gas, the rocket has a momentum $(M + \Delta M)(V + \Delta V)$, and the gas has the momentum $-\Delta M(V + v)$, so that the total momentum after the loss is $(M + \Delta M)(V + \Delta V) -\Delta M(V + v)$. The change in momentum, i.e. the total momentum after the loss of gas minus the total momentum before the loss is

$$(M + \Delta M)(V + \Delta V) - \Delta M(V + v) - MV = M\Delta V - v\Delta M + \Delta M\Delta V$$

The instantaneous time rate of change in momentum is the limit of the change in momentum divided by $\Delta t$, as $\Delta t \to 0$, i.e.

$$\lim_{\Delta t \to 0}\left(M\frac{\Delta V}{\Delta t} - v\frac{\Delta M}{\Delta t} + \frac{\Delta M}{\Delta t}\Delta V\right) \tag{91}$$

Since, $\Delta M \to 0$, $\Delta V \to 0$,

$$\frac{\Delta M}{\Delta t} \to \frac{dM}{dt} \quad \text{and} \quad \frac{\Delta V}{\Delta t} \to \frac{dV}{dt}$$

as $\Delta t \to 0$, expression (91) becomes

$$M\frac{dV}{dt} - v\frac{dM}{dt}$$

Now, the time rate of change in momentum is the force $F$. Hence

$$F = M\frac{dV}{dt} - v\frac{dM}{dt} \tag{92}$$

is the basic equation for rocket motion.

**EXAMPLE 4.59** A rocket having initial mass $M_0$ g starts radially from the earth's surface. It expels gas at the constant rate of $p$ g/s, at a constant velocity

$q$ cm/s. relative to the rocket, where $p > 0$, $q > 0$. Assuming no external forces are acting on the rocket, find its velocity and the distance travelled at any time.

***Solution*** Referring to the fundamental equation, we have $F = 0$ (as there are no external forces). Since, the rocket loses $p$ g/s, it will lose $pt$ g in $t$ s, and hence its mass after $t$ s is $M = M_0 - pt$. Also, the velocity of the gas relative to the rocket is $v = -q$. Thus, Eq. (92) becomes

$$(M_0 - pt)\frac{dV}{dt} - pq = 0 \quad \text{or} \quad \frac{dV}{dt} = \frac{pq}{M_0 - pt} \tag{93}$$

with the assumed initial condition $V = 0$ at $t = 0$. Integrating Eq. (93), we get

$$V = -q \log (M_0 - pt) + c_1$$

Since $V = 0$ at $t = 0$, $c_1 = q \log M_0$ and

$$V = q \log M_0 - q \log (M_0 - pt) \tag{94}$$

which is the required velocity of the rocket. Let $x$ be the distance which the rocket moves in time $t$ measured from the earth's surface. We have $V = dx/dt$, and Eq. (94) becomes

$$\frac{dx}{dt} = q \log M_0 - q \log (M_0 - pt) = q \log \left(\frac{M_0}{M_0 - pt}\right)$$

from which, after integration and taking $x = 0$ at $t = 0$, we obtain

$$x = qt - \frac{q}{p}(M_0 - pt) \log \frac{M_0}{M_0 - pt} \tag{95}$$

which is the required distance travelled. Note that Eqs. (94) and (95) are valid only for $t < M_0/p$, which is the theoretical limit for the time of flight. The practical limit is much smaller than this.

**EXAMPLE 4.60** A rocket, 80 per cent fuel and 20 per cent structure, discharges fuel at a velocity relative to the rocket of 9000 ft/s. The constant rate of burning of the fuel is 10 per cent of the original fuel/s. Assuming that no external forces are acting, find the velocity and distance travelled at the end of: (a) 2 s, (b) 5 s, and (c) 10 s, if $x = 0$ and $V = 0$ when $t = 0$.

***Solution*** Since, the original amount of the fuel is $0.8\, M_0$, $p = (0.1)(0.8)M_0 = 0.08M_0$. Setting $p = 0.08M_0$, $V = 0$, $q = 9000$. From the above example, we have

$$V(t) = 9000 \log \frac{M_0}{M_0 - 0.08M_0 t} = 9000 \log \frac{1}{1 - 0.08t}$$

and
$$x(t) = 9000t - \frac{9000}{0.08\, M_0}(M_0 - 0.08M_0 t) \log \frac{M_0}{M_0 - 0.08M_0 t}$$

$$= 9000t - 112500(1 - 0.08t) \log \frac{1}{1 - 0.08t}$$

These equations, yields

$$V(2) = 1569 \text{ ft/s}, \qquad V(5) = 4597 \text{ ft/s}, \qquad V(10) = 14{,}485 \text{ ft/s}$$
$$x(2) = 1524 \text{ ft}, \qquad x(5) = 10{,}519 \text{ ft}, \qquad x(10) = 53{,}788 \text{ ft}$$

It is interesting to note that $V(t)$ and $x(t)$ are independent of $M_0$.

## 4.24 FRICTIONAL FORCES

If a body moves on a rough surface, it will experience not only resistance due to air but also another resisting force due to the roughness of the surface. This force is called *friction.* It is known that the frictional force is given by $\mu N$, where $\mu$ is the constant of proportionality known as the *coefficient of friction* and depends on the roughness of the surface, and $N$ is the normal force which the surface acts on the body.

**EXAMPLE 4.61** An object weighing $W$ lb is released from rest at the top of a plane metal slide which is inclined at angle $\theta$ to the horizontal. The resistance due to air is half the velocity and $\mu = 1/4$. If $W = 48$ lb and $\theta = 30°$, then find

(a) the velocity of the object 2 s after it is released,

(b) the velocity of the object when it reaches the bottom if the length of the slide is 24 ft.

***Solution*** Consider the line of motion along the slide. Choose the origin at the top and the positive $x$ direction down the slide (Fig. 4.38). The forces acting on $A$ are (when there is no friction and air resistance):

(i) its weight $W$, acting vertically downward,

(ii) the normal force $N$, exerted by the slide which acts in an upward direction perpendicular to the slide.

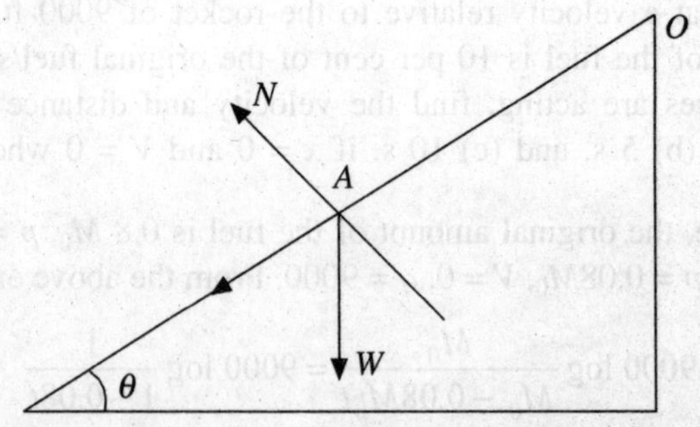


**Fig. 4.38** An object sliding on an inclined plane when the frictional forces are present.

The horizontal and vertical components of $W$ are $W\sin\theta$ and $W\cos\theta$, respectively. The components perpendicular to the slide are in equilibrium and the normal force $N$ is thus, $W\cos\theta$.

Now, when the frictional force and the air-resistance are taken into consideration, the forces acting on $A$, as it moves, are

(I) $f_1$, the component of the weight parallel to the plane, thus $f_1 = W \sin \theta$.

(II) $f_2$, the frictional force which acts in the opposite direction along the slide, thus

$$f_2 = -\mu N = -\mu W \cos \theta$$

(III) $f_3$, the force due to air-resistance, it also acts in the negative direction of the slide and as $v > 0$ from the given condition

$$f_3 = -\frac{1}{2} v$$

Now, from Newton's second law, we have

$$F = ma \quad \text{or} \quad m\frac{dv}{dt} = f_1 + f_2 + f_3$$

or
$$\frac{W}{g}\frac{dv}{dt} = W \sin \theta - \mu W \cos \theta - \frac{1}{2} v \tag{96}$$

which is the required differential of the problem.

From the given data, $W = 48$ lb, $\theta = 30°$, $\mu = 1/4$, $g = 32$, Eq. (96) thus, reduces to

$$\frac{48}{32}\frac{dv}{dt} = 48 \sin 30° - \frac{1}{4} 48 \cos 30° - \frac{1}{2} v$$

or
$$\frac{3}{2}\frac{dv}{dt} = 24 - 6\sqrt{3} - \frac{1}{2} v$$

Separating the variables, integrating and simplifying, we get

$$v = 48 - 12\sqrt{3} - ce^{-t/3}$$

But at $t = 0$, $v = 0$, and $c = 48 - 12\sqrt{3}$. Thus

$$v = (48 - 12\sqrt{3})(1 - e^{-t/3}) \tag{97}$$

Hence

(a) when $t = 2$, $v = 10.2$ ft/s, Integrating Eq. (97), again, we get

$$x = (48 - 12\sqrt{3})(t + 3e^{-t/3}) + c_1$$

As $x = 0$ at $t = 0$, $c_1 = -(48 - 12\sqrt{3})(3)$. Thus, the distance covered in time $t$ is

$$x = (48 - 12\sqrt{3})(t + 3e^{-t/3} - 3) \tag{98}$$

(b) As $x = 24$, Eq. (98) yields

$$3e^{-t/3} = \frac{47 + 2\sqrt{3}}{13} - t$$

and the value of $t$ which satisfies this equation is approximately 2.6. Thus, from Eq. (97), the velocity of the object when it reaches at the bottom is approximately 12.3 ft/s.

## EXERCISES

1. The bacteria in a colony can grow unchecked by the law of exponential growth $y = y_0e^{kt}$. The colony starts with one bacterium and doubles every half hour. How many bacteria will the colony contain at the end of 24 hours? (Under favourable laboratory conditions, the number of cholera bacteria can double every 30 minutes. Of course, in an infected person many of the bacteria are destroyed, but this example helps to explain why a person who feels well in the morning may be dangerously ill by evening.)
2. Bacteria in a certain culture increase at a rate proportional to the number present. If the number doubles in 2 hr., what percentage of the original number will be present at the end of 3 hours?
3. The amount of radioactive isotope $^{14}C$ present in all living organs/matters bears a constant ratio of the amount of the stable isotope $^{12}C$. An analysis of fossil remains of a dinosaur shows that the ratio is only 6.24 per cent of that for living matter. Determine how long ago the dinosaur was living (half-life of $^{14}C$ is approximately 5600 years)?
4. The charcoal from a tree killed in the volcanic eruption that formed a lake contained 44.5% of $^{14}C$ found in living matter. About how old is the lake?
5. Charcoal found in the Lascaux Cave in South France has lost 85.5 per cent of radiocarbon. Find the number of years since the charcoal was living wood. (This result gives an estimate of the age of the famous Lascaux prehistoric painting; the half-life of $^{14}C$ is 5568 years.)
6. If 100 mg of radium is reduced to 90 mg of radium in 200 yr., determine how much radium will remain at the end of 1000 years. Also find the half-life of radium.
7. Show that the half-life $H$ of a radioactive substance can be obtained from two measurements $y_1 = y(t_1)$ and $y_2 = y(t_2)$ of the amount present at times $t_1$ and $t_2$ by the formula $H = [(t_2 - t_1) \log 2]/\log (y_1/y_2)$.
8. Find the time required for money to double when invested at 7 per cent per annum compounded continuously.
9. At what yearly rate of interest compounded continuously does ₹ 1000 increase to ₹ 3000 in 10 years?
10. A cable is wrapped 1.5 turns around a horizontal drum (angle of wrap = $3\pi$). Find the coefficient of friction $\mu$ between the cable and the drum if a pull of 120 lb just supports a load of 840 lb.
11. A copper ball is heated to a temperature of 100°C. At time $t = 0$, it is placed in water which is maintained at a temperature of 30°C and at the end of 3 min., the temperature of the ball is reduced to 70°C. Find the time at which the temperature of the ball reduced to 31°C.

**12.** Water at temperature 10°C takes 5 min. to warm to 20°C at room temperature 40°C. Find the temperature after (i) 20 min., and (ii) 30 min. After how much time will the temperature be 25°C?

**13.** Suppose that a moth ball loses volume by evaporation at a rate proportional to its instantaneous area. If the diameter of the ball decreases from 2 to 1 cm in 3 months, how long will it take until the ball has practically gone (say until its diameter is 1 mm)?

**14.** Use the *Verhulst formula*

$$N = \frac{313{,}400{,}000}{1.5887 + 78.7703\, e^{-0.03134t}}$$

to estimate the population of a country for: (a) 1850 [$t = 160$], (b) 1900 [$t = 110$], and (c) 1950 [$t = 160$]. Also find the limiting population $ab^{-1}$.

**15.** The population of a city grows at a rate equal to the population at any instant. If the population increases 25 per cent in 10 yr., how long will it take the population to double.

**16.** In a model of the changing population $P(t)$ of a community it is assumed that

$$\frac{dP}{dt} = \frac{dB}{dt} - \frac{dD}{dt}$$

where $dB/dt$ and $dD/dt$ are the birth, and death rates, respectively. Solve it for $P(t)$ if $dB/dt = k_1 P$ and $dD/dt = k_2 P$, and analyse the cases $k_1 > k_2$, $k_1 = k_2$, $k_1 < k_2$.

**17.** The population $P(t)$ at any time in a suburb of a city is governed by the initial-value problem $dP/dt = P(10^{-1} - 10^{-7}\, P)$, $P(0) = 5000$, where $t$ is measured in months. What is the limiting value of the population and at what time will the population be equal to one-half of this limiting value?

**18.** Obtain the solution of the differential equation $dP/dt = P(a - b \log P)$ and find the value of the constant $c$ when $P(0) = P_0$.

**19.** In March 1976, the world population reached 4 billion. It was predicted that with an average yearly growth rate of 1.8 per cent, the world population will be 8 billion in 45 years. How does this value compare with that predicted by the model which says that the rate of increase is proportional to the population at any time?

**20.** The average height of a certain plant after 16, 32, and 48 days is 21.6, 43.8, and 54.2 cm, respectively. Assuming that the growth pattern follows the logistic curve, determine: (a) the maximum theoretical height expected, (b) the equation of the logistic curve, and (c) the theoretical heights after 8, 24, and 60 days.

**21.** The following table shows the growth of a colony of bacteria over a number of

| Age ($t$ days) | 0 | 1 | 2 | 3 | 4 | 5 | 6 |
|---|---|---|---|---|---|---|---|
| Area $y$ (cm$^2$) | 1.2 | 3.43 | 9.64 | 19.8 | 27.2 | 33.8 | 37.4 |

days as measured by its size in sq cm: (a) find an equation for the area $y$ in terms of time $t$, (b) using the equation found in (a), compare the computed values of the area with the actual values, (c) what is the theoretical maximum size of the colony?

**22.** The equations of Lotka–Volterra $dy/dt = y(a - bx)$, $dx/dt = x(-c + dy)$, where $a$, $b$, $c$ and $d$ are positive constants, occur in the analysis of the biological balance of two species of animals such as a predator and its prey. Here $x(t)$ and $y(t)$ denote the populations of the two species at any time. Although no explicit solution of the system exist, a solution can be found relating two populations at any time. (Divide the first equation by the second and solve the resulting nonlinear first order equation.)

**23.** The populations of two competing species of animals are described by the nonlinear first order differential equations $dx/dt = k_1x(a - x)$, $dy/dt = k_2xy$. Solve it for $x$ and $y$ in terms of $t$.

**24.** If 5 per cent of the students at a university have a contagious disease and 1 week later a total of 15 per cent have developed the disease, what percentage will have developed it 2, 3, 4 and 5 weeks later, assuming no quarantine?

**25.** A college hostel accommodates 100 students, each of whom is susceptible to a certain virus infection. A simple model of epidemics assumes that during the course of an epidemic the rate of change with respect to time of the number of infected students $I$ is proportional to the number of uninfected students 100-$I$: (a) If at time $t = 0$, a single student becomes infected, show that the number of infected students at any time $t$ is $I = 100\ e^{100kt}/(99 + e^{100kt})$. (b) If $k = 0.01$, when $t$ is measured in days, find the value of the rate of new cases $I_1(t)$ at the end of each day for the first 9 days.

**26.** The supply of food for a certain population is subjected to a seasonal change that affects the growth rate of the population. The differential equation $dx/dt = cx(t) \cos t$, where $c$ is a positive constant, provides a simple model for the seasonal growth of the population. Solve the differential equation in terms of an initial population $x_0$ and the constant $c$. Determine the maximum and minimum populations.

**27.** Of a thousand companies, ten have adopted a new development at time $t = 0$. Given that the number $N(t)$ that have adopted the innovation satisfies the differential equation $dR/dt = 0.0007N(1000 - N)$, find the number $R$ of companies that can be expected to adopt the innovation in 10 years.

**28.** The number of supermarkets $S(t)$ throughout the country that are using a computerized checkout system is described by the initial-value problem $dS/dt = S(1 - 0.0005S)$, $t > 0$, $S(0) = 1$. How many supermarkets are using the computerized method when $t = 10$? How many companies are estimated to adopt the new procedure over a long period of time?

**29.** If glucose is fed intravenously at a constant rate, the change in the overall concentration $G(t)$ of glucose in the blood with respect to time is described by the equation $dG/dt = A/100\ V - aG$. $A$, $V$ and $a$ are positive constants, $A$ being the rate at which glucose is admitted in mg/in, and $V$, the volume of blood in the body in litres (around 5 litres for an adult). $G(t)$ is measured in mg/centilitre. (a) Solve this equation for $G(t)$, using $G_0 = G(0)$. (b) Find the equilibrium value (steady state concentration).

**30.** A tank contains 2000 litres of fluid in which 30 g of salt is dissolved. Brine containing 1 g of salt per litre is then pumped into the tank at a rate of 4 litres/min.; the well-mixed solution is then pumped out at the same rate. Find the number of grams of salt $A(t)$ in the tank at any time.

**31.** A large tank is partially filled with 100 gal of fluid in which 10 lb of salt is dissolved. Brine containing 1/2 lb of salt per gallon is pumped into the tank at a rate of 6 gal/min.; the well-mixed solution is then pumped out at a slower rate of 4 gal/min. Find the number of pounds of salts in the tank after 30 minutes.

**32.** A liquid carries a drug into an organ of volume 500 $cm^3$ at a rate of 10 $cm^3$/s and leaves at the same rate. The concentration of the drug in the entering liquid is 0.08 g/$cm^3$. Assuming that the drug is not present in the organ initially, find (a) the concentration of the drug after 30 s and 120 s, (b) the steady state concentration, (c) how long would it take for the concentration of the drug in the organ to reach 0.04 g/$cm^3$ and 0.06 g/$cm^3$?

**33.** A certain solvent weighs 120 g and dissolves 40 g of a particular solute. Given that 20 g of solute is contained in the solvent at time $t = 0$, find the amount $Q$ of undissolved solute at $t = 4$ hr. if 5 g solute dissolves in one hour. Find also the time required for 10 g solute to dissolve.

**34.** Three grams of substance Y combine with 2 g of substance X to form a 5 g of substance Z. When 90 g of Y are thoroughly mixed with 60 g of X, it is found that in 20 min, 50 g of Z have been formed. How many grams of Z can be formed in 30 min? How long does it take to form 100 g of Z.

**35.** Chemical A is transformed into chemical B. The rate at which B is formed varies directly with the amount of A present at any instant. If 10 lb of A is present initially and 3 lb of A is transformed into B in one hour, find the (a) amount of A transformed after 2, 3, and 4 hours, (b) time in which 75 per cent of chemical A would be transformed to chemical B.

**36.** A sheet of aluminium ($k = 0.49$) is 10 cm thick. One face is kept at 20°C and the other face at 80°C. Assuming that the sheet is sufficiently large for heat flow to be perpendicular to those two faces, find the temperature $T$ in terms of the distance $x$ from the cooler faces. What is the amount of heat transmitted per second across a sq. cm of a section parallel to these faces?

**37.** A hollow spherical glass shell ($k = 0.002$) has an inner radius 6 cm and an outer radius 10 cm. If the inner surface temperature is kept at 50°C and the outer surface at 20°C, what is the temperature $T$ in terms or $r$, the radial distance from the centre of the shell? Find $T$ for $r = 7$, 8 and 9.

**38.** A generator having emf 100 V is connected in series with a 20-Ω resistor and an inductor of 4 H. Given $i = 0$, when $t = 0$, find $i$ in terms of $t$. Find $i$ for $t = 0.2$ s.

**39.** An inductor of 3 H and a 6 Ω resistor are connected in series with a generator having emf 150 $e^{-2t}$ cos 25$t$ V. Find $i$ in terms of $t$ if $i = 20$ when $t = 0$ and find $i$ when $t = 0.5$ s.

**40.** A circuit consists of a 10 Ω resistor and 0.01 F capacitor in series. The charge on the capacitor is 0.05 C. Find the charge and the current flow at time $t$ after the switch is closed.

**41.** Find $I$ in terms of $t$ in the Domar model if $dF/dt = kI$ is replaced by (i) $dF/dt = \sqrt{t}\,I$, (ii) $dF/dt = e^{k_1 I}$.

**42.** Let $v = w$ in the solved Example 4.40. Prove that

$$y = \frac{a}{2}\left\{\frac{1}{2}\left[\left(\frac{x}{a}\right)^2 - 1\right] - \log\frac{x}{a}\right\}$$

so that the hawk will never catch the pigeon. Using Eqs. (71) and (74), show that the distance between the hawk and the pigeon is $(x^2 + a^2)/(2a)$ whenever the hawk is at the point $(x, y)$ on the path. (Thus, the hawk will not come as close as $a/2$ to the pigeon.)

**43.** If $v > w$ in the solved Example 4.40, show that

$$y = \frac{a}{2}\left[\frac{(x/a)^{1+(v/W)} - 1}{1 + (v/W)} + \frac{(a/x)^{(v/W)-1} - 1}{(v/W) - 1}\right]$$

so that the hawk will never catch the pigeon.

**44.** If the destroyer in the solved example 4.42 is only twice as fast as the submarine and the submarine is spotted 3 miles away, find the path now that the destroyer should follow to be certain of passing over the submarine. Assume that both ships execute the same manoeuvres as given in the solved Example 4.42.

**45.** Water 9 ft deep is contained in a tank of rectangular cross-section $5' \times 8'$. The water runs out through a small orifice of radius 1" at the bottom of the tank. Find (a) the time required to empty the tank, (b) the time required to drain half the tank, and (c) the height of the water 20 minutes after it starts to drain.

**46.** A conical tank (see Fig. 4.39) of a circular cross-section whose angle at the apex is 60° has an outlet of cross-sectional area 0.5 cm$^2$. The tank contains water. At time $t = 0$, the outlet is opened and the water flows out. Determine the time when the tank will be empty, assuming that the initial height of the water $h(0) = 1$ m (we can use $dh/dt = -0.600A\sqrt{2g\,h}/B(h)$.

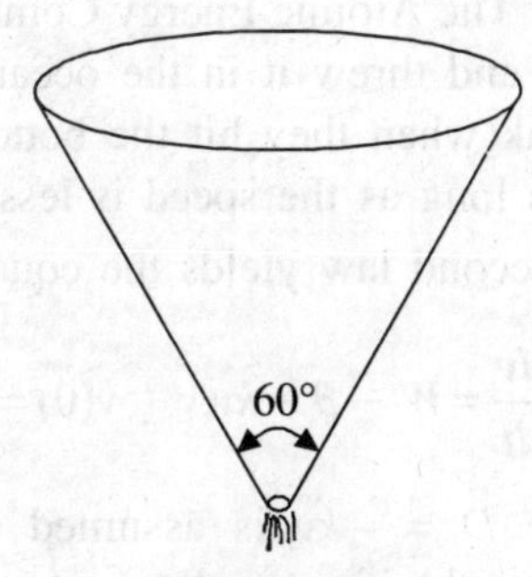


**Fig. 4.39**

**47.** Find the orthogonal trajectories of the family of parabolas opening in the $y$-direction with the vertex at (1, 2).

**48.** Find the members of the orthogonal trajectories for $x + y = ce^y$ which passes through (0, 5).

**49.** Find the orthogonal trajectories of the following family of curves:

(a) $c_1x^2 + y^2 = 1$ (b) $y = c_1e^{-x}$

(c) $y = x/(1 + c_1x)$ (d) $y^3 + 3x^2y = c_1$

(e) $4y + x^2 + 1 + c_1e^{2y} = 0$ (f) $y = 1/(\log c_1x)$

(g) $\sinh y = c_1x$ (h) $x^{1/3} + y^{1/3} = c_1$

**50.** If the sun could be crushed into a smaller sphere with the same mass, estimate what its new radius would have to be in order to increase the escape velocity at its surface to the speed of light.

**51.** *Lambert's law of absorption* states that the absorption of light in a very thin transparent layer is proportional to the thickness of the layer and to the amount incident on that layer. Formulate this in terms of a differential equation and solve it.

**52.** *Boyle-Mariotte's law for ideal gases.* Experiments show that for a gas at low pressure $p$ (and constant temperature $T$), the rate of change of volume $V(p)$ equals $-V/p$. Solve the corresponding differential equation.

**53.** A flywheel of moment of inertia $I$ is rotating with a relatively small constant angular speed $\omega_0$ (radian/s). At some instant $t = 0$, the power is shut off and the motion starts slowing down because of friction. Assuming the friction torque to be proportional to $\sqrt{\omega}$, where $\omega$ is the instantaneous angular speed, and using Newton's second law in *torsional form,* we get, the moment of inertia × angular acceleration = torque. Find $\omega(t)$ and the instant $t_1$ at which the wheel is rotating with $\omega_0/2$ and $t_2$ when it comes to rest.

**54.** Observation shows that the rate of change of atmospheric pressure $p$ with altitude $h$ is proportional to the pressure. Assuming that the pressure at 6000 m (about 18,000 ft) is half of its value $p_0$ at sea level, find the formula for the pressure at any height. [*Hint:* if $y = e^{kx}$ then $y' = ke^{kx} = ky$.]

**55.** *Atomic waste disposal.* The Atomic Energy Commission put atomic wastes into sealed containers and threw it in the ocean. It is important that the containers do not break when they hit the bottom of the ocean. Assume that this is the case as long as the speed is less than 12 m/s (Fig. 4.40). Show that Newton's second law yields the equation of motion

$$m\frac{dv}{dt} = W - B - kv, \qquad v(0) = 0$$

where the drag force $D = -kv$ is assumed to be proportional to $v$. Solve the equation to obtain $v(t)$. Integrate to obtain $y(t)$ such that $y(0) = 0$. Determine the critical time $t_{\text{crit}}$ when the container reaches the critical speed $v_{\text{crit}} = 12$ m/s, assuming that $W = 2254$ N (about 500 lb), $B = 2090$ N (about 460 lb) and $k = 0.637$ kg/s. Show that the container will break if it is dumped at a point where the ocean is deeper than 105 m approximately.

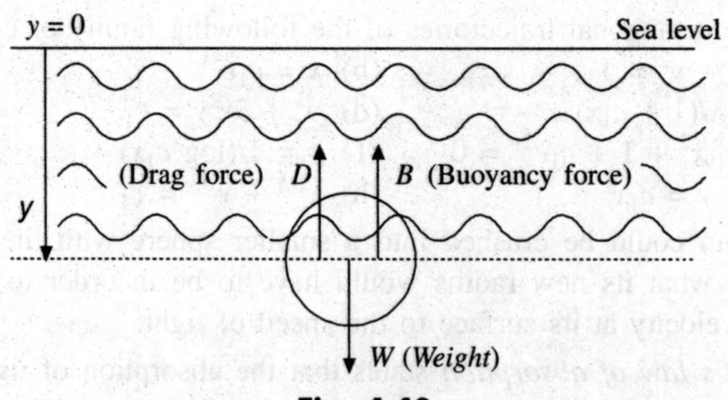


**Fig. 4.40**

CHAPTER 5

# Higher Order Linear Differential Equations

## 5.1 INTRODUCTION

The linear differential equations of the first order were introduced in Section 2.5. In this chapter, we shall discuss the linear differential equations of order greater than one. These equations (as we shall see in Chapter 6) have even greater importance. Motions of pendulums, of elastic strings, of falling bodies, the flow of electric currents and many more such types of problems are related to the solution of linear differential equations of order greater than one.

**Definition 5.1** A linear differential equation of order $n$ is an equation of the form

$$a_n(x)\frac{d^n y}{dx^n} + a_{n-1}(x)\frac{d^{n-1}y}{dx^{n-1}} + \cdots + a_1(x)\frac{dy}{dx} + a_0(x)y = Q(x) \tag{1}$$

where $a_0$, $a_1$, ... , $a_n$ and $Q(x)$ are continuous real functions on a common interval $I$ and $a_n(x) \neq 0$ on $I$. The right-hand side of Eq. (1) is called the *nonhomogeneous term*. If $Q(x)$ is identically zero, then Eq. (1) reduces to

$$a_n(x)\frac{d^n y}{dx^n} + a_{n-1}(x)\frac{d^{n-1}y}{dx^{n-1}} + \cdots + a_1(x)\frac{dy}{dx} + a_0(x)y = 0 \tag{2}$$

and Eq. (2) is called a *homogeneous linear differential equation of order n.* (*Note:* One should not be confused between the term homogeneous as given in Section 2.3 and the one which is used here.)

## 5.2 SOLUTION OF HOMOGENEOUS LINEAR DIFFERENTIAL EQUATIONS OF ORDER *n* WITH CONSTANT COEFFICIENTS

In practice, equations of the form (2), where the coefficients are the functions of $x$ with no restrictions placed on their nature, do not usually have solutions expressible in terms of elementary functions; and even when they do, it is very difficult to find them. However, if each coefficient in Eq. (2) is constant, then the differential Eq. (2) is called a *linear equation with constant coefficients,* and the solution in terms of elementary functions can easily be obtained. Moreover, the linear equations with constant coefficients are of great practical importance as well as of theoretical interest.

In this section, we shall solve the differential equation

$$a_n \frac{d^n y}{dx^n} + a_{n-1} \frac{d^{n-1} y}{dx^{n-1}} + \cdots + a_1 \frac{dy}{dx} + a_0 y = 0 \tag{3}$$

where $a_0, a_1, \ldots, a_n$ are all constants and $a_n \neq 0$.

Suppose that a possible solution of Eq. (3) is

$$y = e^{mx} \tag{4}$$

Since

$$\frac{dy}{dx} = me^{mx}, \qquad \frac{d^2 y}{dx^2} = m^2 e^{mx}, \qquad \frac{d^n y}{dx^n} = m^n e^{mx}$$

Equation (3) takes the form

$$a_n m^n e^{mx} + a_{n-1} m^{n-1} e^{mx} + \cdots + a_1 m e^{mx} + a_0 e^{mx} = 0 \tag{5}$$

As $e^{mx} \neq 0$ for all $m$ and $x$, we can divide Eq. (5) by it to get

$$a_n m^n + a_{n-1} m^{n-1} + \cdots + a_1 m + a_0 = 0 \tag{6}$$

Now, each value of $m$ for which Eq. (6) holds will make $y = e^{mx}$ a solution of Eq. (3). But Eq. (6) is an algebraic equation in $m$ of degree $n$ and, therefore, by the fundamental theorem of algebra, it has at least one and not more than $n$ distinct roots. We denote these roots by $m_1, m_2, \ldots, m_n$, where $m$'s need not all be distinct; then each function

$$y_1 = e^{m_1 x}, \qquad y_2 = e^{m_2 x}, \ldots, \qquad y_n = e^{m_n x}$$

is a solution of Eq. (3).

Equation (6) is called the *auxiliary equation* (A.E.) or the *characteristic equation* (C.E.) of Eq. (3) and can easily be obtained from this equation by simply replacing $y'$ with $m$, $y''$ with $m^2$ and so on, and $y^{(n)}$ with $m^n$.

While solving the auxiliary equation, the following three cases may occur:

1. All the roots are distinct and real.
2. All the roots are real but some are repeating.
3. All the roots are imaginary.

We shall discuss all these three possibilities separately.

*Case I.* If the $n$ roots $m_1, m_2, \ldots, m_n$ of A.E. (6) are distinct, then $n$ solutions of Eq. (3) are

$$y_1 = e^{m_1 x}, \qquad y_2 = e^{m_2 x}, \ldots, \qquad y_n = e^{m_n x}$$

But these $n$ solutions are different and linearly independent and thus, the general solution of Eq. (3) is

$$y_c = y = c_1 e^{m_1 x} + c_2 e^{m_2 x} + \cdots + c_n e^{m_n x} \quad (7)$$

Here, $y_c$ is known as the *complementary function.*

**EXAMPLE 5.1** Solve $\dfrac{d^3y}{dx^3} + 6\dfrac{d^2y}{dx^2} + 11\dfrac{dy}{dx} + 6y = 0$.

***Solution*** Here, the A.E. is

$$m^3 + 6m^2 + 11m + 6 = 0 \quad \text{or} \quad (m + 1)(m + 2)(m + 3) = 0$$

which gives

$$m = -1, -2, -3$$

Therefore, the general solution is

$$y = c_1 e^{-x} + c_2 e^{-2x} + c_3 e^{-3x}$$

**EXAMPLE 5.2** Solve $\dfrac{d^2y}{dx^2} - 3\dfrac{dy}{dx} + 2y = 0$ with $y = 0$, $x = 0$ and $\dfrac{dy}{dx} = 0$.

***Solution*** The A.E. is $m^2 - 3m + 2 = 0$ which gives $m = 1, 2$, and solution is

$$y = c_1 e^x + c_2 e^{2x} \quad (8)$$

Given that $y = 0$ when $x = 0$, Eq. (8) yields

$$0 = c_1 + c_2 \quad (9)$$

Also, from Eq. (8)

$$\frac{dy}{dx} = c_1 e^x + 2c_2 e^{2x}$$

Using $dy/dx = 0$ when $x = 0$, we have

$$0 = c_1 + 2c_2 \quad (10)$$

Solving Eqs. (9) and (10), we get $c_1 = c_2 = 0$, and the solution (8) thus becomes $y = 0$.

**EXAMPLE 5.3** Solve $\dfrac{d^3y}{dx^3} - \dfrac{d^2y}{dx^2} - 6\dfrac{dy}{dx} = 0$.

***Solution*** The A.E. is $m^3 - m^2 - 6m = 0$, which gives

$$m(m^2 - m - 6) = 0 \quad \text{or} \quad m = 0, -2, 3$$

Thus, the solution is

$$y = c_1 e^{0x} + c_2 e^{-2x} + c_3 e^{3x} = c_1 + c_2 e^{-2x} + c_3 e^{3x}$$

*Case II.* If the characteristic Eq. (6) has a root $m = a$, which repeats $n$ times, then the general solution of Eq. (3) is

$$y = (c_1 + c_2x + c_3x^2 + \cdots + c_nx^{n-1})e^{ax} \tag{11}$$

If A.E. (6) has $k$ roots each equal to $m_1$ and the remaining $(n - k)$ roots are all different, then the solution of Eq. (3) is

$$y = (c_1x^{k-1} + c_2x^{k-2} + \ldots + c_{k-1}x + c_k)e^{m_1x} + c_{k+1}e^{m_{k+1}x} + \ldots + c_ne^{m_nx} \tag{12}$$

**EXAMPLE 5.4** Solve $\dfrac{d^3y}{dx^3} - 3\dfrac{dy}{dx} + 2y = 0$.

***Solution*** The A.E. is

$$m^3 - 3m + 2 = 0$$

which gives $m = 1, 1, -2$, and the solution is

$$y = (c_1 + c_2x)e^x + c_3e^{-2x}$$

**EXAMPLE 5.5** Solve $16\dfrac{d^2y}{dx^2} + 24\dfrac{dy}{dx} + 9y = 0$.

***Solution*** The A.E. is $16m^2 + 24m + 9 = 0$. Solving, we get

$$(4m + 3)^2 = 0 \quad \text{or} \quad m = -\frac{3}{4}, -\frac{3}{4}$$

The solution is

$$y = (c_1x + c_2)e^{-(3/4)x}$$

**EXAMPLE 5.6** Solve $\dfrac{d^2y}{dx^2} - 2a\dfrac{dy}{dx} + a^2y = 0$.

***Solution*** Here, the A.E. is $m^2 - 2am + a^2 = 0$ and has a double root $m = a$. The solution is

$$y = (c_1x + c_2)e^{ax}$$

*Case III.* If the constant coefficients in Eq. (6) are real, then any imaginary root it may have must occur in conjugate pairs. Thus, if $\alpha + i\beta$ is one root, then $\alpha - i\beta$ must be another root. If $\alpha \pm i\beta$ are the two imaginary roots of a characteristic equation of a second order linear differential equation, then the solution is

$$\begin{aligned} y &= Ae^{(\alpha+i\beta)x} + Be^{(\alpha-i\beta)x} \\ &= Ae^{\alpha x}e^{i\beta x} + Be^{\alpha x}e^{-i\beta x} \\ &= e^{\alpha x}(Ae^{i\beta x} + Be^{-i\beta x}) \end{aligned} \tag{13}$$

But $e^{i\beta x} = \cos \beta x + i \sin \beta x$, $e^{-i\beta x} = \cos \beta x - i \sin \beta x$. Substitute these in the above equation and simplify to get

$$y = e^{\alpha x}[(A + B) \cos \beta x + i(A - B) \sin \beta x]$$
$$= e^{\alpha x}(c_1 \cos \beta x + c_2 \sin \beta x) \qquad (14)$$

where $c_1 = A + B$, $c_2 = (A - B)i$ are the new constants.

If $\alpha + i\beta$ and $\alpha - i\beta$ each occurs twice as a root, then the general solution is

$$y = e^{\alpha x}[(c_1 + c_2 x) \cos \beta x + (c_3 + c_4 x) \sin \beta x] \qquad (15)$$

**EXAMPLE 5.7** Solve $\dfrac{d^2 y}{dx^2} + 4y = 0$.

***Solution*** The A.E. $m^2 + 4 = 0$ has a pair of roots $m = \pm 2i$. The solution is

$$y = e^{0x}(c_1 \cos 2x + c_2 \sin 2x) = c_1 \cos 2x + c_2 \sin 2x$$

**EXAMPLE 5.8** Solve $\dfrac{d^4 y}{dx^4} + 8\dfrac{d^2 y}{dx^2} + 16y = 0$.

***Solution*** The A.E. $m^4 + 8m^2 + 16 = 0$ has a double root $m = \pm 2i$. The solution is

$$y = (c_1 + c_2 x) \cos 2x + (c_3 + c_4 x) \sin 2x$$

**EXAMPLE 5.9** Solve $\dfrac{d^3 y}{dx^3} + y = 0$.

***Solution*** The A.E. is $m^3 + 1 = 0$ or $(m + 1)(m^2 - m + 1) = 0$, which gives

$$m = -1, \quad \frac{1 \pm \sqrt{3}\, i}{2}$$

and the solution is

$$y = c_1 e^{-x} + e^{(1/2)x}\left(c_2 \cos \frac{\sqrt{3}}{2} x + c_3 \sin \frac{\sqrt{3}}{2} x\right)$$

**Example 5.10** Solve $\left(\dfrac{dy}{dx} - y\right)^2 \left(\dfrac{d^2 y}{dx^2} + y\right)^2 = 0$.

***Solution*** The A.E. is $(m - 1)^2 (m^2 + 1)^2 = 0$, which gives $m = 1, 1, \pm i, \pm i$. The solution is

$$y = (c_1 + c_2 x)\, e^x + (c_3 + c_4 x) \cos x + (c_5 + c_6 x) \sin x$$

## 5.3 SOLUTION OF NONHOMOGENEOUS LINEAR DIFFERENTIAL EQUATIONS WITH CONSTANT COEFFICIENTS BY MEANS OF POLYNOMIAL OPERATORS

The general solution of a nonhomogeneous linear differential equation

$$a_n \frac{d^n y}{dx^n} + a_{n-1} \frac{d^{n-1} y}{dx^{n-1}} + \cdots + a_1 \frac{dy}{dx} + a_0 y = Q(x)$$

where $a_n \neq 0$, $Q(x) \neq 0$ and $a_0, a_1, \ldots, a_n$ are constants, is

$$y = y_c + y_p$$

Here, $y_c$ and $y_p$ are, respectively, known as the complementary function (C.F.) and *particular integral* (P.I.) or *particular solution*. It was seen in the earlier section that when $Q(x) = 0$, the general solution is $y = y_c$, and we had also discussed the method of finding the complementary function. It only remains to find the particular integral. We shall be concerned here mainly with the cases where $Q(x)$ consists of such terms as $b$, $x^k$, $e^{ax}$, $\sin ax$, $\cos ax$ and a finite number of combinations of such terms, where $a$ and $b$ are constants and $k$ is a positive integer.

Before going into the details of the procedures of finding the P.I. of a given differential equation, we shall give some definitions about the polynomial and differential operators and their related properties (without proof).

**Definition 5.2** A mathematical device by means of which we can convert one function into another is known as an *operator*. For example, the operation of differentiation is an operator as it converts a differentiable function $f(x)$ into a new function $f'(x)$. The letter $D$, which we shall use to denote the differentiation, is called the *differential operator*. Hence, if $y$ is an $n$th order differentiable function, then

$$D^0 y = y, \qquad Dy = y', \qquad D^2 y = y'', \ldots, \qquad D^n y = y^n \tag{16}$$

**Definition 5.3** If $P(D)$ is a polynomial operator of order $n$ defined by

$$P(D) = a_0 + a_1 D + a_2 D^2 + \cdots + a_n D^n, \qquad a_n \neq 0 \tag{17}$$

and $y$ is an $n$th order differentiable function, then

$$P(D)y = (a_0 + a_1 D + \cdots + a_n D^n)y = a_0 y + a_1 Dy + \cdots + a_n D^n y \tag{18}$$

From Eq. (16), we have

$$P(D)y = a_n y^n + a_{n-1} y^{n-1} + \cdots + a_1 y' + a_0 y \tag{19}$$

Using this equation, the linear nonhomogeneous differential equation of order $n$ with constant coefficients can be written as

$$a_n y^n + a_{n-1} y^{n-1} + \cdots + a_1 y' + a_0 y = Q(x), \qquad a_n \neq 0 \tag{20}$$

or

$$P(D)y = Q(x)$$

**Property 5.1** If $P(D)$ is a polynomial operator (17) and $y_1$, $y_2$ are two $n$th order differentiable functions, then

$$P(D)\,(b_1 y_1 + b_2 y_2) = b_1 P(D) y_1 + b_2 P(D) y_2 \tag{21}$$

where $b_1$ and $b_2$ are constants. In general, we have

$$P(D)\,(b_1 y_1 + b_2 y_2 + \cdots + b_n y_n) = b_1 P(D) y_1 + b_2 P(D) y_2 + \cdots + b_n P(D) y_n \tag{22}$$

**REMARKS.** 1. An operator which satisfies Eq. (21) is called a *linear operator*. Hence, the polynomial operator [Eq. (17)] is linear.

2. From the linearity of the polynomial operator, one can prove the following:

(a) If $y_1, y_2, \ldots, y_n$ are $n$ solutions of the homogeneous linear equation $P(D)y = 0$, then $y_c = y = c_1y_1 + c_2y_2 + \cdots + c_ny_n$ is also a solution.

(b) If $y_c$ is a solution of the homogeneous linear equation $P(D)y = 0$ and $y_p$ is a particular solution of the nonhomogeneous equation $P(D)y = Q(x)$, then $y = y_c + y_p$ is a solution of $P(D)y = Q(x)$.

3. If $y_1, y_2, \ldots, y_n$ constitute the solution of the respective equations

$$P(D)y_1 = Q_1(x), \qquad P(D)y_2 = Q_2(x), \ldots, \qquad P(D)y_n = Q_n(x)$$

then $y = y_1 + y_2 + \cdots + y_n$ is a solution of

$$P(D)y = Q_1(x) + Q_2(x) + \cdots + Q_n(x).$$

This is known as the *principle of superposition.*

**Property 5.2** The sum of the two polynomial operators $P_1(D)$ and $P_2(D)$ is defined as

$$(P_1 + P_2)y = P_1y + P_2y$$

**Property 5.3** The product of a function $g(x)$ by a polynomial operator $P(D)$ is defined by

$$[g(x)P(D)]y = g(x)\ [P(D)y]$$

**Property 5.4** The product of two polynomial operators $P_1(D)$ and $P_2(D)$ is defined as

$$[P_1(D)P_2(D)]y = P_1(D)\ [P_2(D)]y$$

**Property 5.5** The polynomial operator also satisfies the following:

(1) $P_1(D)[P_2(D)P_3(D)] = [P_1(D)P_2(D)]P_3(D)$ (associative law for multiplication).

(2) $P_1(D)[P_2(D) + P_3(D)] = P_1(D)P_2(D) + P_1(D)P_3(D)$ (distributive law for multiplication)

**Property 5.6** If $P(D)$ is defined by Eq. (17), then

$$P(D) = a_n(D - x_1)(D - x_2) \ldots (D - x_n) \tag{23}$$

where $x_1, x_2, \ldots, x_n$ are the real or imaginary roots of A.E. (6) of $P(D)y = 0$. That is to say, a polynomial operator with constant coefficients can be factored just as if it were an ordinary polynomial.

**REMARK.** If $P(D)$ is a polynomial operator given by Eq. (17), then

$$P(D) = P_1(D)P_2(D)$$

where $P_1(D)$ and $P_2(D)$ may be composite factors of $P(D)$, i.e. $P_1$ and $P_2$ may be products of factors of Eq. (23).

**Property 5.7** The polynomial operator satisfies the commutative law for multiplication, viz.,

$$(D - x_1)(D - x_2) = (D - x_2)(D - x_1)$$

**Definition 5.4** If $P(D)$ is a polynomial defined by Eq. (17), then

$$P(D + a) = a_n(D + a)^n + \cdots + a_1(D + a) + a_0$$

where $a$ is a constant.

**Property 5.8** (*Exponential shift property*) If $P(D)$ is a polynomial operator defined by Eq. (17) and $f(x)$ is an $n$th order differentiable function of $x$, then

$$P(D)(fe^{ax}) = e^{ax}P(D + a)f \tag{24}$$

where $a$ is a constant.

**REMARKS.** 1. $(D - a)^n(fe^{ax}) = e^{ax}D^nf$ (25)

2. If $c$ is a constant, then

$$P(D)(ce^{ax}) = ce^{ax}P(a) \tag{26}$$

We are now in a position to give a method of solving Eq. (1), when the coefficients are constant, by making use of the polynomial operator. The procedure is illustrated by means of the following example:

**EXAMPLE 5.11** Solve $y''' + 2y'' - y' - 2y = e^{2x}$. (27)

***Solution*** In the operator notation, Eq. (27) can be written as

$$(D^3 + 2D^2 - D - 2)y = e^{2x} \tag{28}$$

or

$$(D - 1)(D + 1)(D + 2)y = e^{2x} \tag{29}$$

Let

$$u = (D + 1)(D + 2)y \tag{30}$$

Then Eq. (29) becomes $(D - 1)u = e^{2x}$, which is a linear differential equation. The solution is

$$u = e^{2x} + c_1e^x$$

Putting this relation in Eq. (30), we get

$$(D + 1)\ (D + 2)y = e^{2x} + c_1e^x \tag{31}$$

Let

$$(D + 2)y = v \tag{32}$$

Then, Eq. (31) becomes $(D + 1)v = e^{2x} + c_1e^x$, which is a linear equation, and its solution is

$$v = \frac{1}{3}e^{2x} + \frac{c_1}{2}e^x + c_2e^{-x}$$

Putting this equation in (32), we obtain

$$Dy + 2y = \frac{1}{3}e^{2x} + \frac{1}{2}c_1e^x + c_2e^{-x}$$

which, again, is a linear differential equation with the solution as

$$y = \frac{1}{12}e^{2x} + \frac{c_1}{6}e^x + c_2e^{-x} + c_3e^{-2x}$$

Here $(c_1/6)e^x + c_2e^{-x} + c_3e^{-2x}$ is the complementary function and $(1/12)e^{2x}$ is the particular integral (free from constants).

In general, if the nonhomogeneous linear differential equation

$$a_ny^n + a_{n-1}y^{n-1} + \cdots + a_1y' + a_0 = Q(x), \qquad a_n \neq 0 \tag{33}$$

of order $n$ is written as

$$(D - x_1)(D - x_2) \ldots (D - x_n)y = Q(x) \tag{34}$$

where $x_1, x_2, \ldots, x_n$ are the roots of A.E., then a general solution can be obtained as follows:

Let
$$u = (D - x_2) \ldots (D - x_n)y \tag{35}$$

Then, Eq. (34) takes the form

$$(D - x_1)u = Q(x) \tag{36}$$

which is a linear equation in $u$. Find its solution and put it in Eq. (35) to get

$$(D - x_2)(D - x_3) \ldots (D - x_n)y = u(x) \tag{37}$$

Let
$$v = (D - x_3) \ldots (D - x_n)y \tag{38}$$

Then, Eq. (37) becomes

$$(D - x_2)v = u(x) \tag{39}$$

an equation linear in $v$. Find its solution and put it in Eq. (38) to get

$$(D - x_3)(D - x_4) \ldots (D - x_n)y = v(x)$$

Repeat this process an additional $(n - 2)$ times to get the solution for $y$.

**Inverse operation**

**Definition 5.5** Let $P(D)y = Q(x)$, where $P(D)$ is the polynomial operator defined in Eq. (17) and $Q(x)$ is the function consisting only of such terms as $b$, $x^k$, $e^{ax}$, $\sin ax$, $\cos ax$ and a finite number of combination of these terms, where $a$, $b$ are constants and $k$ is a positive integer. The inverse operator of $P(D)$, written as $P^{-1}(D)$ or $1/P(D)$, is then defined as an operator which, when operating on $Q(x)$, will give the particular integral $y_p$ of $P(D)y = Q(x)$ that contains no constant multiples of a term in C.F., i.e.

$$P^{-1}(D)Q(x) = y_p \qquad \text{or} \qquad y_p = \frac{1}{P(D)}Q(x)$$

**Remarks.** 1. From Definition 5.5, we conclude that $D^{-n}Q(x)$ will mean the integration of $Q(x)$ $n$ times by ignoring constants of integration.

2. Also, if $P(D)y = 0$, then

$$y_p = P^{-1}(D)(0) = 0, \quad \text{or} \quad y_p = \frac{1}{P(D)}(0) = 0$$

**Property 5.9** $P(D)\,[P^{-1}(D)Q(x)] = Q(x)$

The method of Example 5.11 and the discussion that follows this example for finding the P.I. is a longer method. In practice, we have special methods for finding the P.I. As already mentioned, the function $Q(X)$ of $P(D)y = Q(X)$ may contain only such terms as $b$, $x^k$, $e^{ax}$, $\sin ax$, $\cos ax$ or a finite number of combinations of these terms. We shall now consider each of them.

### 5.3.1 When $Q(x) = bx^k$ and $P(D) = D - a_0$, $a_0 \neq 0$

Here, $P(D)y = Q(x)$ becomes

$$(D - a_0)y = bx^k$$

Then

$$y = \frac{bx^k}{D - a_0} = \frac{1}{-a_0\left(1 - \dfrac{D}{a_0}\right)}(bx^k) = -\frac{1}{a_0}\left(1 + \frac{D}{a_0} + \frac{D^2}{a_1^2} + \cdots + \frac{D^k}{a_0^k}\right) bx^k$$

One should not go beyond the $D^k/a_0^k$ terms as $D^{k+1}\, x^k = 0$. After differentiating the above equation, we get

$$y_p = y = -\frac{b}{a_0}\left(x^k + \frac{kx^{k-1}}{a_0} + \cdots + \frac{k!}{a_0^k}\right), \qquad a_0 \neq 0$$

In general, if

$$P(D)y = (a_nD^n + \cdots + a_1D + a_0)y = bx^k \tag{40}$$

then

$$\text{P.I.} = y_p = y = \frac{1}{P(D)}\,bx^k$$

$$= \frac{1}{a_0\left(1 + \dfrac{a_1}{a_0}D + \dfrac{a_2}{a_0}D^2 + \cdots + \dfrac{a_n}{a_0}D^n\right)}\,bx^k$$

$$= \frac{b}{a_0}(1 + b_1D + \cdots + b_kD^k)x^k \tag{41}$$

where $(1 + b_1D + b_2D^2 + \cdots + b_kD^k)/a_0$ is the series expansion of $1/P(D)$ obtained by ordinary division.

If $k = 0$, then Eq. (40) becomes $P(D)y = b$ and from Eq. (41), we have

$$\text{P.I.} = y_p = y = \frac{1}{P(D)} b = \frac{b}{a_0}, \qquad a_0 \neq 0 \tag{42}$$

**EXAMPLE 5.12** Solve $\dfrac{d^2 y}{dx^2} - 2\dfrac{dy}{dx} - 3y = 5$.

***Solution*** The A.E. is $m^2 - 2m - 3 = 0$ and the roots are $-1$ and 3. Thus

$$\text{C.F.} = y_c = c_1 e^{-x} + c_2 e^{3x}$$

Also

$$P(D) = D^2 - 2D - 3$$

with $a_0 = -3$ and $b = 5$. Hence

$$\text{P.I.} = y_p = \frac{b}{P(D)} = \frac{b}{a_0} = -\frac{5}{3}$$

Therefore, the complete solution is

$$y = y_c + y_p = c_1 e^{-x} + c_2 e^{3x} - \frac{5}{3}$$

**EXAMPLE 5.13** Solve $(D^2 - 4)y = x^2$.

***Solution*** The A.E. is $m^2 - 4 = 0$ which gives $m = \pm 2$. Thus

$$\text{C.F.} = y_c = c_1 e^{2x} + c_2 e^{-2x}$$

Also

$$\text{P.I.} = y_p = \frac{1}{P(D)} x^2$$

$$= \frac{1}{D^2 - 4} x^2$$

$$= -\frac{1}{4\left(1 - \dfrac{D^2}{4}\right)} x^2$$

$$= -\frac{1}{4}\left(1 - \frac{D^2}{4}\right)^{-1} x^2$$

$$= -\frac{1}{4}\left(1 + \frac{D^2}{4} + \cdots\right) x^2$$

$$= -\frac{1}{4}\left(x^2 + \frac{1}{2}\right)$$

Therefore, the required solution is

$$y = y_c + y_p = c_1 e^{2x} + c_2 e^{-2x} - \frac{1}{4}\left(x^2 + \frac{1}{2}\right)$$

**EXAMPLE 5.14** Solve $(D^2 + 2D + 1)y = 2x + x^2$.

***Solution*** The A.E. is $m^2 + 2m + 1 = 0$ having roots as $-1, -1$. The complementary function is thus

$$y_c = (c_1 x + c_2)\, e^{-x}$$

Also

$$\text{P.I.} = y_p = \frac{1}{P(D)}(2x + x^2)$$

$$= \frac{1}{(D+1)^2}(2x + x^2)$$

$$= (D + 1)^{-2}\,(2x + x^2)$$

$$= (1 - 2D + 3D^2 + \cdots)(2x + x^2)$$

$$= x^2 - 2x + 2$$

Hence, the complete solution is

$$y = y_c + y_p = (c_1 x + c_2)\, e^{-x} + x^2 - 2x + 2$$

**EXAMPLE 5.15** Solve $(D^3 - 2D + 4)y = x^4 + 3x^2 - 5x + 2$.

***Solution*** The A.E. is

$$m^3 - 2m + 4 = 0$$

or

$$(m + 2)(m^2 - 2m + 2) = 0$$

or

$$m = -2,\ 1 \pm i.$$

Thus

$$\text{C.F.} = y_c = c_1 e^{-2x} + e^x(c_2 \cos x + c_3 \sin x)$$

Also

$$\text{P.I.} = y_p = \frac{1}{D^3 - 2D + 4}(x^4 + 3x^2 - 5x + 2)$$

$$= \frac{1}{4\left(1 - \frac{1}{2}D + \frac{1}{4}D^3\right)}(x^4 + 3x^2 - 5x + 2)$$

$$= \frac{1}{4}\left[1 - \frac{1}{2}\left(D + \frac{1}{2}D^3\right)\right]^{-1}(x^4 + 3x^2 - 5x + 2)$$

$$= \frac{1}{4}\left[1 + \frac{1}{2}\left(D - \frac{1}{2}D^3\right) + \frac{1}{4}\left(D - \frac{1}{2}D^3\right)^2\right.$$

$$\left. + \frac{1}{8}\left(D - \frac{1}{2}D^3\right)^2 + \ldots\right](x^4 + 3x^2 - 5x + 2)$$

Performing the indicated differentiation, we get, after simplification, the relation

$$y_p = \frac{1}{4}\left(x^4 + 2x^3 + 6x^2 - 5x - \frac{7}{2}\right)$$

Hence, the required solution is

$$y = y_c + y_p = c_1e^{-2x} + e^x(c_2 \cos x + c_3 \sin x) + \frac{1}{4}\left(x^4 + 2x^3 + 6x^2 - 5x - \frac{7}{2}\right)$$

### 5.3.2 When $Q(x) = bx^k$ and $P(D) = a_nD^n + a_{n-1}D^{n-1} + \cdots + a_1D$

In this case, $a_0 = 0$, and $D$ is a factor of $P(D)$. Therefore, by Property 5.6, we can write

$$P(D) = D(a_nD^{n-1} + \cdots + a_2D + a_1)$$

where $a_1 \neq 0$. If both $a_0 = 0$ and $a_1 = 0$, then $D^2$ is a factor of $P(D)$ so that $P(D)$ can be written as

$$P(D) = D^2(a_nD^{n-2} + \cdots + a_3D + a_2)$$

In general, if $D^r$ is a factor of $P(D)$, then $P(D)y = bx^k$ has the form

$$P(D)y = D^r(a_nD^{n-r} + \cdots + a_{r+1}D + a_r)y = bx^k,\ a_r \neq 0$$

Thus, by definition,

$$\text{P.I.} = y_p = \frac{1}{D^r(a_nD^{n-r} + \cdots + a_{r+1}D + a_r)}bx^k \qquad (43)$$

**EXAMPLE 5.16** Solve $(D^3 - D^2 - 6D)y = x^2 + 1$.

***Solution*** The A.E. $m^3 - m^2 - 6m = 0$ has the roots 0, 3 and –2. Thus

$$\text{C.F.} = y_c = c_1 + c_2e^{3x} + c_3e^{-2x}$$

Also

$$\text{P.I.} = y_p = \frac{1}{D^3 - D^2 - 6D}(x^2 + 1)$$

$$= -\frac{1}{6D}\left(1 + \frac{D}{6} - \frac{D^2}{6}\right)^{-1}(x^2 + 1)$$

$$= -\frac{1}{6D}\left[1-\left(\frac{D}{6}-\frac{D^2}{6}\right)+\left(\frac{D}{6}-\frac{D^2}{6}\right)^2+\cdots\right](x^2+1)$$

$$= -\frac{1}{6D}\left(1+x^2-\frac{1}{3}x+\frac{7}{18}\right)$$

$$= -\frac{1}{6}D^{-1}\left(1+x^2-\frac{1}{3}x+\frac{7}{18}\right)$$

$$= -\frac{1}{6}\left(\frac{25}{18}x+\frac{x^3}{3}-\frac{x^2}{6}\right)$$

Therefore, the complete solution is

$$y = y_c + y_p = c_1 + c_2e^{3x} + c_3e^{-2x} - \frac{1}{6}\left(\frac{25}{18}x+\frac{x^3}{3}-\frac{x^2}{6}\right)$$

**EXAMPLE 5.17** Solve $\frac{d^3y}{dx^3}+3\frac{d^2y}{dx^2}+2\frac{dy}{dx}=x^2$.

***Solution*** The A.E. $m^3 + 3m^2 + 2m = 0$ can be written as $m(m + 1)(m + 2) = 0$ so that the roots are 0, –1 and –2, and the complementary function is

$$y_c = c_1 + c_2e^{-x} + c_3e^{-2x}$$

Now

$$\text{P.I.} = y_p = \frac{1}{D^3+3D^2+2D}x^2$$

$$= \frac{1}{D(D+1)(D+2)}x^2$$

$$= \frac{1}{2D}\left[(1+D)^{-1}\left(1+\frac{D}{2}\right)^{-1}\right]x^2$$

$$= \frac{1}{2D}(1-D+D^2+\cdots)\left(1-\frac{D}{2}+\frac{D^2}{4}+\cdots\right)x^2$$

$$= \frac{1}{2D}\left(x^2-3x+\frac{7}{2}\right)$$

$$= \frac{1}{2}\left(\frac{x^3}{3} - \frac{3x^2}{2} + \frac{7}{2}x\right)$$

$$= \frac{1}{12}(2x^3 - 9x^2 + 21x)$$

Therefore, the complete solution is

$$y = y_c + y_p = c_1 + c_2e^{-x} + c_3e^{-2x} + \frac{1}{12}(2x^3 - 9x^2 + 21x)$$

### 5.3.3 When $Q(x) = be^{ax}$

In this case, $P(D)y = Q(x)$ becomes $P(D)y = be^{ax}$. The particular integral here is

$$y_p = \frac{1}{P(D)} be^{ax} = \frac{be^{ax}}{P(a)}, \quad P(a) \neq 0 \tag{44}$$

Note that $a$ in $P(a)$ is same as $a$ in $e^{ax}$. The case when $P(a) = 0$ will be discussed later.

**EXAMPLE 5.18** Solve $(D^2 - 2D + 5)y = e^{-x}$.

***Solution*** The A.E. is $m^2 - 2m + 5 = 0$, and the roots are $-1 \pm 2i$. Thus

$$\text{C.F.} = y_c = e^{-x}(c_1 \cos 2x + c_2 \sin 2x)$$

Now

$$\text{P.I.} = \frac{1}{D^2 - 2D + 5} e^{-x} = \frac{1}{(-1)^2 - 2(-1) + 5} e^{-x} = \frac{1}{8} e^{-x}$$

The required solution is

$$y = e^{-x}(c_1 \cos 2x + c_2 \sin 2x) + \frac{1}{8}e^{-x}$$

**EXAMPLE 5.19** Solve $(D^3 - D^2 - 4D + 4)y = e^{3x}$.

***Solution*** The characteristic equation is

$$m^3 - m^2 - 4m + 4 = 0 \quad \text{or} \quad (m - 1)(m^2 - 4) = 0 \quad \text{or} \quad m = 1, 2, -2.$$

Thus,

$$\text{C.F.} = c_1e^{x} + c_2e^{2x} + c_3e^{-2x}$$

and

$$\text{P.I.} = \frac{e^{3x}}{D^3 - D^2 - 4D + 4} = \frac{e^{3x}}{(3)^3 - (3)^2 - 4(3) + 4} = \frac{e^{3x}}{10}$$

Therefore, the general solution is

$$y = \text{C.F.} + \text{P.I.} = c_1 e^{x} + c_2 e^{2x} + c_3 e^{-2x} + \frac{1}{10} e^{3x}$$

## 5.3.4 When $Q(x) = b \sin ax$ or $b \cos ax$

We know that

$$e^{iax} = \cos ax + i \sin ax \tag{45}$$

Consequently, P.I. can be obtained from Eq. (44) by using Eq. (45).

Alternatively, if $P(-a^2) \neq 0$, then

$$y_p = \frac{1}{P(D^2)} \sin ax = \frac{1}{P(-a^2)} \sin ax \tag{46}$$

$$y_p = \frac{1}{P(D^2)} \cos ax = \frac{1}{P(-a^2)} \cos ax \tag{47}$$

**Remark.** If $y_p = \frac{1}{P(D)} \sin ax$, then put $-a^2$ for $D^2$, $a^4$ for $D^4$, $-a^6$ for $D^6$ and so on, in $P(D)$ to calculate $y_p$.

The above method fails when $P(-a^2) = 0$ and in such a case, we proceed as follows:

From Eq. (45), we have

$$\frac{1}{D^2 + a^2} \sin ax = \text{Im} \frac{1}{D^2 + a^2} e^{iax}$$

$$\frac{1}{D^2 + a^2} \cos ax = \text{Re} \frac{1}{D^2 + a^2} e^{iax}$$

Now

$$\begin{aligned}
\frac{1}{D^2 + a^2} e^{iax} &= \frac{1}{(D - ia)(D + ia)} e^{iax} \\
&= \frac{1}{(D - ia)} \frac{e^{iax}}{2ia} \\
&= \frac{x}{2ia} e^{iax} \\
&= \frac{x}{2ia} (\cos ax + i \sin ax) \\
&= \frac{x}{2a} (\sin ax - i \cos ax)
\end{aligned}$$

Equating the real and imaginary parts, we have

$$\frac{1}{D^2 + a^2} \cos ax = \frac{x}{2a} \sin ax \tag{48}$$

$$\frac{1}{D^2 + a^2} \sin ax = -\frac{x}{2a} \cos ax \tag{49}$$

**EXAMPLE 5.20** Solve $(D^2 - 3D + 2)y = 3 \sin 2x$. (50)

***Solution*** The A.E. $m^2 - 3m + 2 = 0$ has the roots 1, 2. Thus

$$\text{C.F.} = y_c = c_1 e^x + c_2 e^{2x}$$

Now, using Eq. (45), the imaginary part of a particular integral of

$$(D^2 - 3D + 2)y = 3e^{2ix} \tag{51}$$

will be a solution of Eq. (50). Hence, $P(D) = D^2 - 3D + 2$ and by Eq. (44), with $b = 3$, and $a = 2i$, we have

$$y_p = \frac{3e^{2ix}}{P(D)} = \frac{3e^{2ix}}{P(2i)}$$

$$= \frac{3e^{2ix}}{(2i)^2 - 3(2i) + 2}$$

$$= \frac{3(1-3i)}{-20}(e^{2ix})$$

$$= -\frac{3}{20}(1-3i)(\cos 2x + i \sin 2x)$$

$$= -\frac{3}{20}[(\cos 2x + 3 \sin 2x) + i(\sin 2x - 3 \cos 2x)]$$

The imaginary part of this equation is

$$y_p = \frac{3}{20}(3 \cos 2x - \sin 2x)$$

which is the required particular integral. Therefore, the complete solution of the given differential equation is

$$y = y_c + y_p = c_1 e^x + c_2 e^{2x} + \frac{3}{20}(3 \cos 2x - \sin 2x)$$

**EXAMPLE 5.21** Solve $(D^3 + D^2 - D - 1)y = \cos 2x$.

***Solution*** The A.E. is $m^3 + m^2 - m - 1 = 0$. Solving, we get $m = 1, -1, -1$. Thus

$$y_c = c_1 e^x + (c_2 + c_3 x)e^{-x}$$

$$y_p = \frac{\cos 2x}{D^3 + D^2 - D - 1} = \frac{1}{D(-2^2) + (-2^2) - D - 1} \cos 2x$$

$$= \frac{\cos 2x}{-5(D+1)} = \frac{-1}{5} \frac{(D-1)}{(D^2 - 1)} \cos 2x$$

$$= -\frac{1}{5} \frac{(D-1)}{(-2^2) - 1} \cos 2x = \frac{1}{25}(D-1) \cos 2x$$

$$= \frac{1}{25}(-2 \sin 2x - \cos 2x)$$

The general solution, therefore, is

$$y = y_c + y_p = c_1 e^x + (c_2 + c_3 x)e^{-x} - \frac{1}{25}(2 \sin 2x + \cos 2x)$$

**EXAMPLE 5.22** Solve $(D^3 + 1)y = \cos 2x$.

***Solution*** The A.E. $m^3 + 1 = 0$ has the roots

$$-1, \frac{1}{2} \pm i\frac{\sqrt{3}}{2}.$$

Thus,

$$y_c = c_1 e^{-x} + e^{x/2}\left(c_2 \cos \frac{\sqrt{3}}{2}x + c_3 \sin \frac{\sqrt{3}}{2}x\right)$$

Also

$$y_p = \frac{1}{D^3 + 1} \cos 2x$$

$$= \frac{1}{D(-2^2) + 1} \cos 2x$$

$$= \frac{1}{1 - 4D} \cos 2x$$

$$= \frac{1 + 4D}{(1 - 16D^2)} \cos 2x$$

$$= \frac{1 + 4D}{1 - 16(-2^2)} \cos 2x$$

$$= \frac{1}{65}(1 + 4D) \cos 2x$$

$$= \frac{1}{65}(\cos 2x - 8 \sin 2x)$$

Therefore, the required solution is

$$y = y_c + y_p = c_1e^{-x} + e^{x/2}\left(c_2 \cos\frac{\sqrt{3}}{2}x + c_3 \sin\frac{\sqrt{3}}{2}x\right) + \frac{1}{65}(\cos 2x - 8\sin 2x)$$

**EXAMPLE 5.23** Solve $(D^2 + 4)y = \cos 2x$.

***Solution*** The A.E. $m^2 + 4 = 0$ gives $m = \pm 2i$. Thus

$$\text{C.F.} = c_1 \cos 2x + c_2 \sin 2x$$

and

$$\text{P.I.} = \frac{1}{D^2 + a^2}\cos 2x = \frac{1}{D^2 + 4}\cos 2x = \frac{x}{2(2)}\sin 2x \qquad \text{[from Eq. (48)]}$$

Hence, the required solution is

$$y = \text{C.F.} + \text{P.I.} = c_1 \cos 2x + c_2 \sin 2x + \frac{x}{4}\sin 2x$$

**EXAMPLE 5.24** Solve $(D^4 - 1)y = \sin x$.

***Solution*** The A.E. $m^4 - 1 = 0$ has the roots $\pm 1$, $\pm i$. Thus

$$y_c = c_1e^x + c_2e^{-x} + c_3 \cos x + c_4 \sin x$$

$$y_p = \frac{1}{D^4 - 1}\sin x$$

$$= \frac{1}{(D^2 - 1)(D^2 + 1)}\sin x$$

$$= \frac{1}{(-1-1)(D^2 + 1)}\sin x$$

$$= -\frac{1}{2}\left[-\frac{x}{2(1)}\cos x\right]$$

Therefore, the general solution is

$$y = y_c + y_p = c_1e^x + c_2e^{-x} + c_3 \cos x + c_4 \sin x + \frac{1}{4}x \cos x$$

### 5.3.5 When $Q(x) = e^{ax}V$, where $V$ is a function of $x$

In this case

$$y_p = \frac{1}{P(D)}e^{ax}V = e^{ax}\frac{1}{P(D + a)}V \tag{52}$$

**EXAMPLE 5.25** Solve $(D^2 - 2D + 1)y = e^x x^2$.

***Solution*** Here

$$\text{C.F.} = (c_1 + c_2 x)e^x$$

and

$$\text{P.I.} = \frac{1}{D^2 - 2D + 1} e^x x^2$$

$$= e^x \frac{1}{(D+1)^2 - 2(D+1) + 1} x^2$$

$$= e^x \frac{1}{D^2} x^2$$

$$= e^x D^{-2} x^2$$

$$= \frac{e^x x^4}{12}$$

Therefore, the required solution is

$$y = (c_1 + c_2 x)\, e^x + \frac{1}{12} e^x x^4$$

**EXAMPLE 5.26** Solve $(D^2 + 4D - 12)y = (x - 1)e^{2x}$.

***Solution*** The A.E. $m^2 + 4m - 12 = 0$ has the roots 2 and –6. Thus,

$$y_c = c_1 e^{2x} + c_2 e^{-6x}$$

$$y_p = \frac{(x-1)\, e^{2x}}{P(D)} = e^{2x} \frac{1}{(D+2)^2 - 4(D+2) - 12}(x - 1)$$

$$= e^{2x} \frac{1}{D^2 + 8D}(x - 1)$$

$$= e^{2x} \frac{1}{8D}\left(1 + \frac{1}{8}D\right)^{-1} (x - 1)$$

$$= \frac{e^{2x}}{8D}\left(1 - \frac{1}{8}D + \frac{1}{64}D^2 + \cdots\right)(x - 1)$$

$$= \frac{e^{2x}}{8} D^{-1}\left(x - \frac{9}{8}\right)$$

$$= \frac{1}{64} e^{2x} (4x^2 - 9x)$$

Therefore, the required solution is

$$y = c_1 e^{2x} + c_2 e^{-6x} + \frac{1}{64} e^{2x}(4x^2 - 9x)$$

**Example 5.27** Solve $(D^2 - 2D + 5)y = e^{2x} \sin x$.

***Solution*** The A.E. $m^2 - 2m + 5 = 0$ has the roots $1 \pm 2i$. Thus

$$y_c = e^x(c_1 \cos 2x + c_2 \sin 2x)$$

and

$$y_p = \frac{e^{2x} \sin x}{P(D)} = e^{2x} \frac{1}{(D+2)^2 - 2(D+2) + 5} \sin x$$

$$= e^{2x} \frac{1}{D^2 + 2D + 5} \sin x$$

$$= e^{2x} \frac{1}{-1^2 + 2D + 5} \sin x$$

$$= e^{2x} \frac{1}{2D + 4} \sin x$$

$$= e^{2x} \frac{1}{4D^2 - 16} (2D - 4) \sin x$$

$$= -\frac{e^{2x}}{20} (2D - 4) \sin x$$

$$= -\frac{e^{2x}}{10} (\cos x - 2 \sin x)$$

Therefore, the general solution is

$$y = y_c + y_p = e^x(c_1 \cos 2x + c_2 \sin 2x) - \frac{1}{10} e^{2x}(\cos x - \sin 2x)$$

### 5.3.6 When $Q(x) = be^{ax}$ and $P(a) = 0$

In this case, $(D - a)$ is a factor of $P(D)$. Suppose that $(D - a)^n$ is a factor of $P(D)$ and hence, we can write $P(D) = (D - a)^n f(D)$, $f(D) \neq 0$, and then

$$y_p = \frac{1}{P(D)} be^{ax} = \frac{1}{(D-a)^n f(D)} be^{ax} = \frac{bx^n e^{ax}}{n!\, f(a)}, \qquad f(a) \neq 0 \qquad (53)$$

**EXAMPLE 5.28** Solve $(D^2 + 6D + 9)y = 2e^{-3x}$.

***Solution*** Here

$$y_c = (c_1 + c_2x)e^{-3x}$$

and

$$y_p = \frac{2e^{-3x}}{D^2 + 6D + 9} = \frac{2e^{-3x}}{(D-3)^2 + 6(D-3) + 9}(1) = 2e^{-3x}\frac{1}{D^2}(1) = 2e^{-3x}\left(\frac{1}{2}x^2\right)$$

The required solution is, therefore

$$y = y_c + y_p = (c_1 + c_2x)\, e^{-3x} + x^2e^{-3x}$$

**EXAMPLE 5.29** Solve $(D^2 - 4D + 4)y = 8(x^2 + e^{2x} + \sin 2x)$.

***Solution*** Here

$$\text{C.F.} = (c_1 + c_2x)e^{2x}$$

$$\text{P.I.} = \frac{Q(x)}{P(D)} = \frac{8(x^2 + e^{2x} + \sin 2x)}{(D-2)^2}$$

$$= \frac{8x^2}{(D-2)^2} + \frac{8e^{2x}}{(D-2)^2} + \frac{8 \sin 2x}{(D-2)^2}$$

$$= 2\frac{1}{(1-1/2D)^2}x^2 + 8e^{2x}\frac{1}{(D+2-2)^2}(1) + 8\frac{1}{(-4-4D+4)}\sin 2x$$

$$= 2\left(1 - \frac{1}{2}D\right)^{-2} x^2 + 8e^{2x}D^{-2}(1) - 2D^{-1}\sin 2x$$

$$= 2\left(1 + D + \frac{3}{4}D^2 + \cdots\right)x^2 + \frac{8}{2}e^{2x}x^2 + \cos 2x$$

$$= 2x^2 + 4x + 3 + 4x^2e^{2x} + \cos 2x$$

Here

$$y = \text{C.F.} + \text{P.I.} = (c_1 + c_2x)e^{2x} + 2x^2 + 4x + 4x^2e^{2x} + \cos 2x + 3$$

**EXAMPLE 5.30** Solve $(D^2 + 4D + 4)y = e^{2x} - e^{-2x}$.

***Solution*** Here

$$\text{C.F.} = (c_1 + c_2x)e^{2x}$$

and $$\text{P.I.} = \frac{1}{D^2+4D+4}(e^{2x} - e^{-2x})$$

$$= \frac{1}{D^2+4D+4}e^{2x} - \frac{1}{D^2+4D+4}e^{-2x}$$

$$= \frac{1}{2^2+4(2)+1}e^{2x} - e^{-2x}\frac{1}{(D-2)^2+4(D-2)+4}(1)$$

$$= \frac{1}{16}e^{2x} - e^{-2x}D^{-2}(1)$$

$$= \frac{1}{16}e^{2x} - \frac{1}{2}x^2e^{-2x}$$

Therefore, the required solution is

$$y = (c_1 + c_2x)e^{2x} + \frac{1}{16}e^{2x} - \frac{1}{2}x^2e^{-2x}$$

### 5.3.7 When *Q(x) = xV*, where *V* is any function of *x*

Here

$$y_p = \frac{1}{P(D)}(xV)$$

$$= x\frac{1}{P(D)}V - \frac{P'(D)}{[P(D)]^2}V \qquad (54)$$

**EXAMPLE 5.31** Solve $\dfrac{d^2y}{dx^2} - 2\dfrac{dy}{dx} + y = x\sin x$.

***Solution*** Here

$$y_c = (c_1 + c_2x)\,e^x$$

$$y_p = \frac{1}{D^2-2D+1}x\sin x$$

$$= x\frac{1}{D^2-2D+1}\sin x - \frac{2D-2}{(D^2-2D+1)^2}\sin x$$

$$= x\frac{1}{-1-2D+1}\sin x - \frac{2D-2}{(-1-2D+1)^2}\sin x$$

$$= -\frac{1}{2}xD^{-1}\sin x - \frac{1}{4}D^{-2}(2D-2)\sin x$$

$$= \frac{1}{2}x\cos x - \frac{1}{2}D^{-2}(\cos x - \sin x)$$

$$= \frac{1}{2}(x\cos x + \cos x - \sin x)$$

Thus, the general solution is

$$y = y_c + y_p = (c_1 + c_2x)e^x + \frac{1}{2}(x\cos x + \cos x - \sin x)$$

**Example 5.32** Solve $\dfrac{d^2y}{dx^2} + 2\dfrac{dy}{dx} + y = x\cos x$.

***Solution*** Here

$$y_c = (c_1 + c_2x)e^{-x}$$

$$y_p = \frac{1}{D^2 + 2D + 1}x\cos x$$

$$= x\frac{1}{D^2 + 2D + 1}\cos x - \frac{2D+2}{(D^2 + 2D + 1)^2}\cos x$$

$$= \frac{x}{2}D^{-1}\cos x - \frac{1}{2D^2}(D+1)\cos x$$

$$= \frac{x}{2}\sin x - \frac{1}{2}D^{-2}(-\sin x + \cos x)$$

$$= \frac{x}{2}\sin x - \frac{1}{2}(\sin x - \cos x)$$

Hence, the general solution is

$$y = (c_1 + c_2x)e^{-x} + \frac{1}{2}(x\sin x - \sin x + \cos x)$$

**EXAMPLE 5.33** Solve $(D^2 - 2D + 1)y = xe^x \sin x$.

***Solution*** Here

$$y_c = (c_1 + c_2x)e^x$$

$$y_p = \frac{1}{D^2 - 2D + 1}xe^x\sin x = e^x\frac{1}{(D+1)^2 - 2(D+1) + 1}x\sin x$$

$$= e^xD^{-2}(x\sin x)$$

Integrating by parts, we get

$$y_p = e^x(-x \sin x - 2 \cos x)$$

Thus, the general solution is

$$y = (c_1 + c_2x)\, e^x - e^x(x \sin x + 2 \cos x)$$

**EXAMPLE 5.34** Solve $(D^2 + 1)\, y = x^2 \sin 2x$.

***Solution*** Here

$$y_c = c_1 \cos x + c_2 \sin x$$

$$y_p = \frac{1}{D^2+1} x^2 \sin 2x$$

$$= \text{Im} \frac{1}{D^2+1} x^2 e^{2ix}$$

$$= \text{Im of } e^{2ix} \frac{1}{(D+2i)^2+1} x^2$$

$$= \text{Im of } e^{2ix} \frac{1}{-3\left(1-\frac{4}{3}iD-\frac{1}{3}D^2\right)} x^2$$

$$= \text{Im of } \frac{e^{2ix}}{-3}\left[1-\left(\frac{4iD+D^2}{3}\right)\right]^{-1} x^2$$

$$= \text{Im of } \frac{e^{2ix}}{-3}\left(1+\frac{4iD}{3}-\frac{13}{9}D^2+\dots\right)x^2$$

$$= \text{Im of } \frac{e^{2ix}}{-3}\left[x^2+\frac{4i2x}{3}-\frac{13}{9}(2)\right]$$

$$= \text{Im of } \frac{1}{-3}(\cos 2x + i \sin 2x)\left[\left(x^2-\frac{26}{9}\right)+i\frac{8}{3}x\right]$$

$$= -\frac{1}{3}\frac{8}{3}x \cos 2x - \frac{1}{3}\left(x^2-\frac{26}{9}\right)\sin 2x$$

Therefore, the complete solution is

$$y = c_1 \cos x + c_2 \sin x - \frac{1}{27}[24x \cos 2x + (9x^2 - 26) \sin 2x]$$

## 5.4 METHOD OF UNDETERMINED COEFFICIENTS

We know that the general solution of the differential equation

$$a_n \frac{d^n y}{dx^n} + a_{n-1} \frac{d^{n-1} y}{dx^{n-1}} + \cdots + a_1 \frac{dy}{dx} + a_0 y = Q(x) \tag{55}$$

where $a_n \neq 0$ and $Q(x) \neq 0$, is

$$y = y_c + y_p \tag{56}$$

where $y_c$, the complementary function, is the general solution of the related homogeneous equation of (55), and $y_p$ is the particular integral of (55). In Sections 5.2 and 5.3, we have discussed the methods of finding $y_c$ and $y_p$, respectively. We shall now give yet another method of finding $y_p$ of Eq. (55). This method is known as the *method of undetermined coefficients.* Here, $Q(x)$ can only contain terms such as $b$, $x^k$, $e^{ax}$, $\sin ax$, $\cos ax$ and a finite number of combinations of such terms. To find $y_p$ by this method, it is necessary to compare the terms of $Q(x)$ in Eq. (55) with those of $y_c$. In the process of comparison, different possibilities may occur. We shall consider each of these and explain them with the help of solved examples.

*Case I: No term of $Q(x)$ in Eq. (55) is the same as a term of $y_c$.* In this case, $y_p$ will be a linear combination of the terms in $Q(x)$ and all its linearly independent derivatives. (For a discussion of linear dependence, see [13]).

**EXAMPLE 5.35** Solve $(D^2 + 4D + 4)y = 4x^2 + 6e^x$. (57)

***Solution*** Here

$$y_c = (c_1 + c_2 x)\, e^{-2x} \tag{58}$$

Since, $Q(x) = 4x^2 + 6e^x$ has no term common with $y_c$, $y_p$ will be a linear combination of $Q(x)$ and all its linearly independent derivatives (which are, neglecting the constant coefficients, $x^2$, $x$, $1$, $e^x$). Hence

$$y_p = Ax^2 + Bx + C + De^x \tag{59}$$

where $A$, $B$, $C$, $D$ are to be determined. Differentiate Eq. (59) twice to get

$$y'_p = 2Ax + B + De^x \tag{60}$$

$$y''_p = 2A + De^x \tag{61}$$

For Eq. (59) to be a solution of Eq. (57), we make use of Eqs. (59)–(61) in Eq. (57) to get

$$2A + De^x + 4(2Ax + B + De^x) + 4(Ax^2 + Bx + C + De^x) = 4x^2 + 6e^x$$

Simplifying and equating the coefficients of like terms in the two members of the above equation, we get

$$4A = 4, \qquad 8A + 4B = 0$$

$$2A + 4B + 4C = 0, \qquad 9D = 6$$

Solutions of these equations are

$$A = 1, \qquad B = -2, \qquad C = \frac{3}{2}, \qquad D = \frac{2}{3}$$

Substituting these values in Eq. (59), we obtain

$$y_p = x^2 - 2x + \frac{2}{3}e^x + \frac{3}{2}$$

Hence, the general solution of Eq. (57) is

$$y = y_c + y_p = (c_1 + c_2 x)e^{-2x} + x^2 - 2x + \frac{2}{3}e^x + \frac{3}{2}$$

**EXAMPLE 5.36** Solve $(D^2 + 2D + 5)y = 12e^x - 34 \sin 2x$. (62)

***Solution*** Here

$$y_c = e^{-x}(c_1 \cos 2x + c_2 \sin 2x)$$

$$y_p = Ae^x + B \sin 2x + C \cos 2x \tag{63}$$

Also,

$$y'_p = Ae^x + 2B \cos 2x - 2C \sin 2x \tag{64}$$

$$y''_p = Ae^x - 4B \sin 2x - 4C \cos 2x \tag{65}$$

From Eqs. (63)–(65), Eq. (62) takes the form

$$8Ae^x + (B - 4C) \sin 2x + (4B + C) \cos 2x = 12e^x - 34 \sin 2x$$

Equating the coefficients of like terms on the two sides of this equation, we get

$$8A = 12, \qquad B - 4C = -34, \qquad 4B + C = 0$$

Thus, $A = 3/2$, $B = -2$, $C = 8$ and Eq. (63) now reduces to

$$y_p = \frac{3}{2}e^x - 2 \sin 2x + 8 \cos 2x$$

Therefore, the general solution of Eq. (62) is

$$y = y_c + y_p = e^{-x}(c_1 \cos 2x + c_2 \sin 2x) + \frac{3}{2}e^x - 2 \sin 2x + 8 \cos 2x$$

*Case II: When $Q(x)$ in (55) contains a term which is $x^k$ times a term $f(x)$ of $y_c$, where $k$ is zero or a positive integer.* Here, the particular integral, $y_p$, of Eq. (55) will be a linear combination of $x^{k+1}f(x)$ and all its linearly independent derivatives (ignoring the constant coefficients). If in addition, $Q(x)$ contains terms which correspond to Case I, then the proper terms required by this case must be included in $y_p$.

**EXAMPLE 5.37** Solve $(D^2 - 3D + 2)y = 2x^2 + 3e^{2x}$.

***Solution*** Here, $y_c = c_1e^x + c_2e^{2x}$. Comparing the right-hand side of the given equation with $y_c$, we observe that $Q(x)$ contain $e^{2x}$ which (neglecting the constant coefficients) is $x^0$ times the same terms in $y_c$. Hence, for this term, $y_p$ must contain a linear combination of $x^{0+1}e^{2x}$ and all its linearly independent derivatives. Also, $Q(x)$ has the term $x^2$ which belongs to Case I. For this term, $y_p$ must include a linear combination of it and all its linearly independent derivatives, we can neglect the function $e^{2x}$ as it already appeared in $y_c$. Therefore

$$y_p = Ax^2 + Bx + C + Dxe^{2x}$$

Differentiate it twice to get

$$y'_p = 2Ax + B + 2Dxe^{2x} + De^{2x}$$
$$y''_p = 2A + 4Dxe^{2x} + 4De^{2x}$$

Substituting the values of $y_p$, $y'_p$ and $y''_p$ in the given differential equation, we get

$$2Ax^2 + (2B - 6A)x + (2A - 3B + 2C) + De^{2x} = 2x^2 + 3e^{2x}$$

Equating the coefficients of like terms on the two sides of this equation and solving the resulting equations, we obtain

$$A = 1, \quad B = 3, \quad C = \frac{7}{2}, \quad D = 3$$

Putting these values in $y_p$, we have

$$y_p = x^2 + 3x + \frac{7}{2} + 3xe^{2x}$$

Therefore, the complete solution is

$$y = y_c + y_p = c_1e^x + c_2e^{2x} + x^2 + 3x + 3xe^{2x} + \frac{7}{2}$$

**EXAMPLE 5.38** Solve $(D^2 - 3D + 2)y = xe^{2x} + \sin x$.

***Solution*** Here

$$y_c = c_1 e^x + c_2 e^{2x}$$

$$y_p = Ax^2 e^{2x} + Bxe^{2x} + C \sin x + D \cos x$$

Find $y'_p$ and $y''_p$. Substituting the values of $y_p$, $y'_p$ and $y''_p$ in the given differential equation, and equating the coefficients of like terms on both sides of the resulting equation, we get, after simplification,

$$A = \frac{1}{2}, \quad B = -1, \quad C = \frac{1}{10}, \quad D = \frac{3}{10}$$

Thus, $$y_p = \frac{1}{2}x^2 e^{2x} - xe^{2x} + \frac{1}{10}\sin x + \frac{3}{10}\cos x$$

Therefore, the general solution of the given differential equation is

$$y = y_c + y_p = c_1 e^x + c_2 e^{2x} + \frac{1}{2}x^2 e^{2x} - xe^{2x} + \frac{1}{10}\sin x + \frac{3}{10}\cos x$$

*Case: III* If

(i) the A.E. of Eq. (55) has an $r$ multiple root, and
(ii) $Q(x)$ contains a term which (neglecting the constant coefficients) is $x^k f(x)$, $f(x)$ is a term in $y_c$ and is obtained from the $r$ multiple root; then, $y_p$ will be a linear combination of $x^{k+r} f(x)$ and all its linearly independent derivatives. If, in addition, $Q(x)$ contains terms that belong to Cases I and II, then the proper terms, which these cases demand, must also be included in $y_p$.

**EXAMPLE 5.39** Solve $(D^2 + 4D + 4)y = 3xe^{-2x}$.

***Solution*** Here

$$y_c = c_1 e^{-2x} + c_2 xe^{-2x}$$

Note that the A.E. has a multiple root $m = -2$, and $Q(x)$ contains the term $xe^{-2x}$ which is $x$ times the term $e^{-2x}$ in $y_c$, and that this term in $y_c$ came from a multiple root. Hence, $r = 2$ and $k = 1$. Therefore, $y_p$ must be a linear combination of $x^3 e^{-2x}$ and all its linearly independent derivatives. In obtaining this linear combination, we can neglect $e^{-2x}$ and $xe^{-2x}$ as they are already in $y_c$. Thus

$$y_p = Ax^3 e^{-2x} + Bx^2 e^{-2x}$$

Find $y'_p$, $y''_p$. Substituting these values of $y_p$, $y'_p$ and $y''_p$ in the given differential equation, we get

$$6Axe^{-2x} + 2Be^{-2x} = 3xe^{-2x}$$

which gives $A = 1/2$, $B = 0$. Hence

$$y_p = \frac{1}{2}x^3 e^{-2x}$$

Therefore, the required solution is

$$y = y_c + y_p = c_1 e^{-2x} + c_2 x e^{-2x} + \frac{1}{2}x^3 e^{-2x}$$

## 5.5 METHOD OF VARIATION OF PARAMETERS

In Section 5.4, we have solved the linear differential equation

$$a_n \frac{d^n y}{dx^n} + a_{n-1}\frac{d^{n-1} y}{dx^{n-1}} + \ldots + a_1 \frac{dy}{dx} + a_0 y = Q(x) \tag{66}$$

where

(i) the coefficients are constant, and
(ii) $Q(x)$ is a function which has a finite number of linearly independent derivatives.

Now, is it possible to remove either or both of these conditions? In this section, we shall answer this and then solve Eq. (66).

In respect of (i), there are only few linear equations having nonconstant coefficients whose solutions can be written in terms of the elementary function and the standard methods for solving them are available. Later (Section 5.6), we shall give a method by which we can solve a second order linear differential equation with nonconstant coefficients provided that we know one solution.

Regarding (ii), it is possible to solve Eq. (66), even when $Q(x)$ has an infinite number of linearly independent derivatives. The method for solving Eq. (66) in such a situation is discussed now, and is known as the method of *variation of parameters*.

For the sake of convenience, we consider a second order linear differential equation with constant coefficients

$$a_2 \frac{d^2 y}{dx^2} + a_1 \frac{dy}{dx} + a_0 y = Q(x), \qquad a_2 \neq 0 \tag{67}$$

where $Q(x)$ is a continuous function of $x$ and is nonzero. The related homogeneous equation is

$$a_2 \frac{d^2 y}{dx^2} + a_1 \frac{dy}{dx} + a_0 y = 0 \tag{68}$$

Now, find the two linearly independent solutions $y_1$ and $y_2$ of Eq. (68), and form

$$y_p(x) = u(x)y_1(x) + v(x)y_2(x) \tag{69}$$

where $u(x)$ and $v(x)$ are functions of $x$ which are to be determined.

Differentiating Eq. (69) twice, we get

$$y'_p = uy'_1 + u'y_1 + vy'_2 + v'y_2 = (uy'_1 + vy'_2) + (u'y_1 + v'y_2) \quad (70)$$

$$y''_p = (uy''_1 + uy''_2) + (u'y'_1 + v'y'_2) + (u'y_1 + v'y_2)' \quad (71)$$

From Eqs. (69)–(71), we see that $y_p$ is a solution of Eq. (67) if

$$u(a_2y''_1 + a_1y'_1 + a_0y_1) + v(a_2y''_2 + a_1y'_2 + a_0y_2) + a_2(u'y'_1 + v'y'_2)$$
$$+ a_2(u'y_1 + v'y_2)' + a_1(u'y_1 + v'y_2) = Q(x) \quad (72)$$

As $y_1$ and $y_2$ are the solutions of Eq. (68), the quantities in the first two parentheses of Eq. (72) are equal to zero. The remaining terms will be equal to $Q(x)$ if we choose $u$ and $v$ such that

$$u'y_1 + v'y_2 = 0, \qquad u'y'_1 + v'y'_2 = \frac{Q(x)}{a_2} \quad (73)$$

Solving Eq. (73), we get

$$u' = \frac{\begin{vmatrix} 0 & y_2 \\ \dfrac{Q(x)}{a_2} & y_2' \end{vmatrix}}{\begin{vmatrix} y_1 & y_2 \\ y_1' & y_2' \end{vmatrix}}, \qquad v' = \frac{\begin{vmatrix} y_1 & 0 \\ y_1' & \dfrac{Q(x)}{a_2} \end{vmatrix}}{\begin{vmatrix} y_1 & y_2 \\ y_1' & y_2' \end{vmatrix}} \quad (74)$$

Integrating Eqs. (74), we can find $u$ and $v$. Substituting these values in Eq. (69), we get a particular integral of Eq. (67).

**REMARKS.** 1. If the nonhomogeneous linear differential equation is of the order greater than 2, then it can be shown that

$$y_p = u_1y_1 + u_2y_2 + \cdots + u_ny_n$$

where $y_1, y_2, \ldots, y_n$ are the $n$ independent solution of the related homogeneous equation.

2. This method can also be used when $Q(x)$ has a finite number of linearly independent derivatives. This method can also be applicable for higher order differential equations (cf., pp 229).

**EXAMPLE 5.40** Solve

$$(D^2 - 3D + 2)y = \sin e^{-x}. \quad (75)$$

***Solution*** Here

$$y_c = c_1e^x + c_2e^{2x}$$

Therefore, the two linearly independent solutions of the related homogeneous equation of Eq. (75) are

$$y_1 = e^x, \quad y_2 = e^{2x} \tag{76}$$

From Eq. (73), we have

$$u'e^x + v'e^{2x} = 0, \qquad u'e^x + v'(2e^{2x}) = \sin e^{-x}$$

Solving these, we get

$$u' = -e^{-x} \sin e^{-x}, \qquad v' = e^{-2x} \sin e^{-x}$$

Therefore,

$$u = \int (\sin e^{-x})(-e^{-x})\, dx, \qquad v = -\int e^{-x} \sin e^{-x}(-e^{-x})\, dx$$

or $$u = -\cos e^{-x}, \qquad v = -\sin e^{-x} + e^{-x} \cos e^{-x} \tag{77}$$

From Eqs. (76) and (77), Eq. (69) becomes

$$y_p = -e^{2x} \sin e^{-x}$$

Therefore, the required solution is

$$y = y_c + y_p = c_1 e^x + c_2 e^{2x} - e^{2x} \sin e^{-x}$$

**EXAMPLE 5.41** Solve $(D^2 + 4D + 4)y = 3xe^{-2x}$.

***Solution*** We have already solved this example (see Example 5.39) by the method of undetermined coefficients. Here

$$y_c = c_1 e^{-2x} + c_2 x e^{-2x}$$

Thus, the two linearly independent solutions of the related homogeneous equations are

$$y_1 = e^{-2x}, \qquad y_2 = xe^{-2x}$$

From Eq. (73), we obtain

$$u'e^{-2x} + v'xe^{-2x} = 0$$

or $$u'(-2e^{-2x}) + v'(-2xe^{-2x} + e^{-2x}) = 3xe^{-2x}$$

Solving for $u'$ and $v'$ and then integrating, we get

$$u = -x^3, \qquad v = \frac{3}{2}x^2$$

Hence $$y_p = uy_1 + vy_2 = -x^3 e^{-2x} + \frac{3}{2}x^2(xe^{-2x})$$

Thus, the general solution is

$$y = y_c + y_p = c_1 e^{-2x} + c_2 x e^{-2x} + \frac{1}{2}x^3 e^{-2x}$$

which is the same as in Example 5.39.

## 5.6 LINEAR DIFFERENTIAL EQUATIONS WITH NONCONSTANT COEFFICIENTS

Consider the general linear differential equation

$$f_n(x)\frac{d^n y}{dx^n} + f_{n-1}(x)\frac{d^{n-1} y}{dx^{n-1}} + \dots + f_1(x)\frac{dy}{dx} + f_0(x)y = Q(x) \tag{78}$$

and its related homogeneous equation

$$f_n(x)\frac{d^n y}{dx^n} + f_{n-1}(x)\frac{d^{n-1}y}{x^{n-1}} + \cdots + f_1(x)\frac{dy}{dx} + f_0(x)y = 0 \tag{79}$$

where $f_n(x), f_{n-1}(x), \ldots, f_1(x), f_0(x)$ and $Q(x)$ are each continuous functions of $x$ and $f_n(x) \neq 0$.

In most of the cases, the solution of Eq. (78) will not be expressible in terms of elementary functions. When the coefficients are constant, the methods for finding the solutions of Eqs. (78) and (79) are known. Here we shall give a method for solving Eqs. (78) and (79) when the coefficients $f_n(x)$ are not constant. We consider a second order differential equation

$$f_2(x)\frac{d^2 y}{dx^2} + f_1(x)\frac{dy}{dx} + f_0(x)y = Q(x) \tag{80}$$

and its related homogeneous equation

$$f_2(x)\frac{d^2 y}{dx^2} + f_1(x)\frac{dy}{dx} + f_0(x)y = 0 \tag{81}$$

Suppose that a nontrivial solution ($y_1 \neq 0$) of Eq. (81) is known. The method by means of which we can obtain a second independent solution of Eq. (81), as well as a particular integral of Eq. (80) is called the *reduction of order method.*

Let $y_2(x)$ be a second solution of Eq. (81) and have the form

$$y_2(x) = y_1(x)\int u(x)dx \tag{82}$$

where $u(x)$ is a function of $x$ to be determined. Differentiating Eq. (82) twice, we get

$$y'_2(x) = y_1 u + y'_1 \int u(x)dx \tag{83}$$

$$y''_2(x) = y_1 u' + y'_1 u + y'_1 u + y''_1 \int u(x)dx \tag{84}$$

Putting the values of $y_2$, $y'_2$, $y''_2$ in Eq. (81) and simplifying, we obtain

$$[f_2(x)y''_1 + f_1(x)y'_1 + f_0(x)y_1]\int u(x)\,dx + f_2(x)y_1 u' + [2f_2(x)y'_1 + f_1(x)y_1]u = 0 \tag{85}$$

As $y_1$ is a solution of Eq. (81), the quantity in the first bracket of Eq. (85) is zero, and Eq. (85) thus reduces to

$$f_2(x)y_1 u' + [2f_2(x)y'_1 + f_1(x)y_1]u = 0 \tag{86}$$

Multiplying Eq. (86) by $dx/[uf_2(x)y_1]$, we have

$$\frac{du}{u} + 2\frac{dy_1}{y_1} = -\frac{f_1(x)}{f_2(x)}dx$$

Integration yields

$$\log u + 2\log y_1 = -\int \frac{f_1(x)}{f_2(x)}\,dx$$

or

$$\log (uy_1^2) = -\int \frac{f_1(x)}{f_2(x)}\,dx$$

or

$$u = \frac{\exp\left[-\int \frac{f_1(x)}{f_2(x)}\,dx\right]}{y_1^2} \tag{87}$$

Substituting this value of $u$ in Eq. (82), we get the required solution.

**EXAMPLE 5.42** Solve $x^2y'' - xy' + y = 0$. Given $y_1 = x$ as a solution.

***Solution*** Here, $f_2(x) = x^2$, $f_1(x) = -x$ and from Eq. (87), we have

$$u = \frac{\exp\left[-\int \frac{-x}{x^2}\,dx\right]}{x^2} = \frac{\exp(\log x)}{x^2} = x^{-1}$$

Thus

$$y_2 = y_1\int u(x)dx = x\int x^{-1}dx = x\log x$$

Therefore, the required solution is

$$y = c_1x + c_2x\log x$$

**EXAMPLE 5.43** Given that $y = x$ is a solution of

$$x^2\frac{d^2y}{dx^2} + x\frac{dy}{dx} - y = 0, \qquad x \neq 0$$

Find the general solution of

$$x^2\frac{d^2y}{dx^2} + x\frac{dy}{dx} - y = x.$$

***Solution*** Given $y_1(x) = x$. Thus

$$y_2(x) = y_1(x)\int u(x)dx = x\int u(x)dx \tag{88}$$

Differentiating it twice, we get

$$y'_2 = xu + \int u(x)dx, \qquad y''_2 = xu' + u + u$$

Substituting the values of $y_2$, $y'_2$, $y''_2$ in the given differential equation and simplifying, we obtain

$$u' + \frac{3}{x}u = x^{-2}$$

which is a linear equation and has the solution

$$u = \frac{x^{-1}}{2} + cx^{-3}$$

Put this value of $u$ in Eq. (88) to obtain

$$y = \frac{x}{2}\log x - c\frac{x^{-1}}{2} + c_2 x$$

which is equivalent to

$$y = c_1 x^{-1} + c_2 x + \frac{x}{2}\log x$$

as the required solution.

## 5.7 THE CAUCHY–EULER EQUATION

An equation of the form

$$a_n x^n \frac{d^n y}{dx^n} + a_{n-1} x^{n-1} \frac{d^{n-1} y}{dx^{n-1}} + \dots + a_1 x \frac{dy}{dx} + a_0 y = Q(x) \qquad (89)$$

where $a_0, a_1, \dots, a_{n-1}, a_n$ are constant, is called a *Cauchy–Euler equation* of order $n$. To know the solution of such an equation, we make some suitable substitution so that Eq. (89) may reduce to an equation for which the methods of solution are known. Such a substitution is $x = e^t$. This substitution reduces Eq. (89) to a linear equation with constant coefficients. To see how this can be done, we consider the second order Cauchy-Euler equation as

$$a_2 x^2 \frac{d^2 y}{dx^2} + a_1 x \frac{dy}{dx} + a_0 y = Q(x) \qquad (90)$$

Take $x = e^t$. Then $\log x = t$,

$$\frac{dy}{dx} = \frac{dy}{dt}\frac{dt}{dx} = \frac{1}{x}\frac{dy}{dt}$$

$$\frac{d^2 y}{dx^2} = \frac{1}{x}\frac{d}{dx}\left(\frac{dy}{dt}\right) + \frac{dy}{dt}\frac{d}{dx}\left(\frac{1}{x}\right) = \frac{1}{x^2}\left(\frac{d^2 y}{dt^2} - \frac{dy}{dt}\right)$$

Thus $$x\frac{dy}{dx} = \frac{dy}{dt}, \qquad x^2 \frac{d^2 y}{dx^2} = \frac{d^2 y}{dt^2} - \frac{dy}{dt}$$

and hence from Eq. (90), we obtain

$$a_2\left(\frac{d^2 y}{dt^2} - \frac{dy}{dt}\right) + a_1 \frac{dy}{dt} + a_0 y = Q(e^t)$$

or $$A_2 \frac{d^2 y}{dt^2} + A_1 \frac{dy}{dt} + A_0 y = R(t) \qquad (91)$$

where $A_2 = a_2$, $A_1 = a_1 - a_2$, $A_0 = a_0$ and $R(t) = Q(e^t)$. Equation (91) is a linear differential equation with constant coefficients and can be solved by already known methods.

In a similar way, we can solve Eq. (89).

**EXAMPLE 5.44** Solve $x^2 \dfrac{d^2y}{dx^2} + x\dfrac{dy}{dx} - 4y = x^2$.

***Solution*** Putting $x = e^t$, or $\log x = t$ and $D_1 = d/dt$, the given differential equation can be written as

$$[D_1(D_1 - 1) + D_1 - 4]y = e^{2t}$$

or

$$(D_1^2 - 4)y = e^{2t}$$

Here, A.E. has the roots $m = \pm 2$. Thus

$$\text{C.F.} = y_c = c_1e^{2t} + c_2e^{-2t} = c_1x^2 + c_2x^{-2}$$

and

$$\text{P.I.} = y_p = \frac{1}{D_1^2 - 4}e^{2t}$$

$$= e^{2t}\frac{1}{(D_1 + 2)^2 - 4}(1)$$

$$= e^{2t}\frac{1}{D_1^2 + 4D_1}(1)$$

$$= e^{2t}\frac{1}{4D_1}\left(1 + \frac{D_1}{4}\right)^{-1}(1)$$

$$= e^{2t}\frac{1}{4D_1}(1)$$

$$= \frac{1}{4}e^{2t}\,t$$

$$= \frac{1}{4}x^2 \log x$$

Hence, the required solution is

$$y = y_c + y_p = c_1x^2 + c_2x^{-2} + \frac{1}{4}x^2 \log x$$

**EXAMPLE 5.45** Solve $x^2D^2y - xDy - 3y = x^2 \log x$.

***Solution*** Putting $x = e^t$, or, $\log x = t$ and $D_1 = d/dt$, the given equation becomes

$$(D_1^2 - 2D_1 - 3)y = te^{2t}$$

Here

$$y_c = c_1e^{-t} + c_2e^{3t} = c_1x^{-1} + c_2x^3$$

$$y_p = \frac{1}{D_1^2 - 2D_1 - 3}te^{2t}$$

$$= e^{2t}\frac{1}{(D_1+2)^2 - 2(D_1+2) - 3}(t)$$

$$= e^{2t}\frac{1}{-3}\left(1 - \frac{2}{3}D_1 - \frac{1}{3}D_1^2\right)^{-1}(t)$$

$$= -\frac{1}{3}e^{2t}\left(1 + \frac{2}{3}D_1 + \ldots\right)(t)$$

$$= -\frac{1}{3}e^{2t}\left(t + \frac{2}{3}\right)$$

$$= -\frac{1}{3}x^2\left(\log x + \frac{2}{3}\right)$$

The required solution, therefore, is

$$y = y_c + y_p = c_1x^{-1} + c_2x^3 - \frac{1}{3}x^2\left(\log x + \frac{2}{3}\right)$$

**EXAMPLE 5.46** Solve $x^3D^3y + 3x^2D^2y + xDy + y = x + \log x$.

***Solution*** Let $x = e^t$, and $D_1 = d/dt$ so that the given equation takes the form

$$[D_1(D_1 - 1)(D_1 - 2) + 3D_1(D_1 - 1) + D_1 + 1]y = e^t + t$$

or

$$(D_1^3 + 1)y = e^t + t$$

The A.E. $m_1^3 + 1 = 0$ has the roots

$$-1 \text{ and } \frac{1}{2} \pm i\frac{\sqrt{3}}{2}$$

Thus

$$y_c = c_1e^{-t} + e^{t/2}\left[c_2\cos\left(\frac{\sqrt{3}}{2}t\right) + c_2\sin\left(\frac{\sqrt{3}}{2}t\right)\right]$$

$$= c_1x^{-1} + \sqrt{x}\left[c_2\cos\left(\frac{\sqrt{3}}{2}\log x\right) + c_3\sin\left(\frac{\sqrt{3}}{2}\log x\right)\right]$$

and $y_p = \dfrac{1}{D_1^3+1}e^t + \dfrac{1}{D_1^3+1}t = \dfrac{1}{1+1}e^t + (1+D_1^3)^{-1}(t) = \dfrac{1}{2}e^t + t = \dfrac{1}{2}x + \log x$

Therefore, the required solution is

$$y = c_1 x^{-1} + \sqrt{x}\left[c_2 \cos\left(\frac{\sqrt{3}}{2}\log x\right) + c_3 \sin\left(\frac{\sqrt{3}}{2}\log x\right)\right] + \frac{1}{2}x + \log x$$

## 5.8 LEGENDRE'S LINEAR EQUATION

An equation of the form

$$k_n(ax+b)^n \frac{d^n y}{dx^n} + k_{n-1}(ax+b)^{n-1}\frac{d^{n-1}y}{dx^{n-1}} + \cdots + k_0 y = Q(x) \qquad (92)$$

where $k_0, k_1, \ldots, k_n$ are constants and $Q(x)$ is a function of $x$, is called *Legendre's linear equation.*

Such equations can be reduced to linear equations with constant coefficients by the substitution

$$ax + b = e^t, \qquad t = \log(ax+b)$$

Then, if $D = \dfrac{d}{dt}$,

$$\frac{dy}{dx} = \frac{dy}{dt}\frac{dt}{dx} = \frac{a}{ax+b}\frac{dy}{dt}$$

or

$$(ax+b)\frac{dy}{dx} = a\frac{dy}{dt} = aDy$$

$$\frac{d^2y}{dx^2} = \frac{d}{dx}\frac{a}{ax+b}\frac{dy}{dt} = \frac{a^2}{(ax+b)^2}\left(\frac{d^2y}{dt^2} - \frac{dy}{dt}\right)$$

i.e.

$$(ax+b)^2\frac{d^2y}{dx^2} = a^2 D(D-1)y$$

Similarly

$$(ax+b)^3\frac{d^3y}{dx^3} = a^3 D(D-1)(D-2)$$

and so on.

By making these replacements in Eq. (92), we get a linear equation with constant coefficients.

**Example 5.47** Solve $(1+x)^2\dfrac{d^2y}{dx^2} + (1+x)\dfrac{dy}{dx} + y = 2\sin[\log(1+x)]$.

***Solution*** Put $1 + x = e^t$, $t = \log(1+x)$ and $D = d/dt$. Then the given equation becomes

$$(D^2+1)y = 2\sin t$$

which is a linear equation with constant coefficients. Here

$$\text{C.F.} = y_c = c_1 \cos t + c_2 \sin t$$
$$\text{P.I.} = y_p = -t \cos t$$

Hence, the complete solution is

$$y = y_c + y_p = c_1 \cos t + c_2 \sin t - t \cos t$$

Therefore, the solution of the given equation is

$$y = c_1 \cos \log (1 + x) + c_2 \sin \log (1 + x) - \log (1 + x) \cos \log (1 + x)$$

## 5.9 MISCELLANEOUS DIFFERENTIAL EQUATIONS

In this section, we study some other important types of differential equations which require special methods for their solutions and have a number of applications.

**Equations of the form $d^2y/dx^2 = f(x)$**

Integrating with respect to $x$, we get

$$\frac{dy}{dx} = \int f(x)\, dx + c = F(x) \quad \text{(say)}$$

Again integrating, we have

$$y = \int F(x)dx + c_1$$

as the required solution.

In general, the solution of a differential equation of the form $d^ny/dx^n = f(x)$ is obtained by integrating it $n$ times successively.

**EXAMPLE 5.48** Solve $\dfrac{d^2y}{dx^2} = xe^x$.

***Solution*** Integrating the given equation, we get

$$\frac{dy}{dx} = xe^x - \int e^x dx + c_1 = (x - 1)e^x + c_1$$

Integrate again to obtain

$$y = (x - 2)e^x + c_1 x + c_2$$

which is the required solution.

**Equations of the form $d^2y/dx^2 = f(y)$**

Multiplying both sides by $2(dy/dx)$, we have

$$2\frac{dy}{dx}\frac{d^2y}{dx^2} = 2\frac{dy}{dx} f(y)$$

Integrating with respect to $x$, we get

$$\left(\frac{dy}{dx}\right)^2 = 2\int f(y)dy + c = F(y) \quad \text{(say)}$$

or
$$\frac{dy}{dx} = \sqrt{F(y)}$$

Now, separating the variables and integrating, we obtain

$$\int \frac{dy}{\sqrt{F(y)}} = x + c_1$$

which gives the required solution. Such equations frequently occur in dynamics.

**EXAMPLE 5.49** Solve $d^2y/dx^2 = 2(y^3 + y)$ under the condition $y = 0$, $dy/dx = 1$, when $x = 0$.

***Solution*** Multiplying the given equation by $2(dy/dx)$, we get

$$2\frac{dy}{dx}\frac{d^2y}{dx^2} = 4(y^3 + y)\frac{dy}{dx}$$

Integrating with respect to $x$, we get

$$\left(\frac{dy}{dx}\right)^2 = y^4 + 2y^2 + c \tag{93}$$

As $dy/dx = 1$ for $y = 0$, we find $c = 1$ and Eq. (93) becomes

$$\left(\frac{dy}{dx}\right)^2 = y^4 + 2y^2 + 1 = (y^2 + 1)^2$$

or
$$\frac{dy}{dx} = y^2 + 1$$

Now, separate the variables and integrate to get

$$\tan^{-1}y = x + c_1 \tag{94}$$

As $y = 0$ for $x = 0$, we obtain $c_1 = 0$, and Eq. (94) becomes

$$\tan^{-1}y = x \qquad \text{or} \qquad y = \tan x$$

which is the required solution.

## 5.10 DIFFERENTIAL EQUATIONS FOR SPECIAL FUNCTIONS

In this section, we list below some special types of differential equations which find their vast applications in many branches of Physics and Engineering. For more details, see [2].

(i) The differential equation

$$x^2y'' + xy' + (x^2 - k^2)y = 0 \tag{95}$$

where $k$ is a positive constant or zero, is called *Bessel's equation,* the simplest nontrivial linear ordinary differential equation with an irregular singularity. The solution of Eq. (95) (known as *Bessel's function*) appears in a wide variety of problems in physics. For example, the separation of the Helmholtz or wave equation in circular cylindrical or in spherical polar coordinates leads to some form of Eq. (95). Moreover, Eq. (95) leads to the theory of asymptotic expansion and the Bessel function is used to describe the motion of planets (Kepler's equation).

(ii) The linear differential equation

$$(1 - x^2)y'' - 2xy' + k(k + 1)y = 0 \tag{96}$$

where $k$ is a real constant and occurs in many physical problems, is known as *Legendre equation.* When the electrostatic potential is expressed in spherical polar coordinates, the Legendre polynomials [solution of Eq. (96)] are conveniently used. Legendre polynomials satisfy the parity or reflection property that plays an important role in quantum mechanics. Legendre polynomials also have series representation which enables us to give a better description of earth's gravitational potential, and thus, confirms the pear-shaped deformation of earth.

(iii) The differential equation

$$y'' - 2xy' + 2ky = 0 \tag{97}$$

where $k$ is real, is known as the *Hermite equation.* The most important single application of the Hermite polynomial [solution of Eq. (97)] is the description of a quantum mechanical simple harmonic oscillator.

(iv) The differential equation

$$xy'' + (1 - x)y' + ky = 0 \tag{98}$$

where $k$ is real, is called the *Laguerre equation.* One of the most important applications of Laguerre polynomials [solution of Eq. (98)] is the Schrödinger wave equation for the hydrogen atom.

(v) The differential equation

$$(x - 1)(x + 1)y'' + xy' - k^2y = 0 \tag{99}$$

is known as the *Chebyshev* (or *Tschebyscheff*) *equation.* The solution of this equation has many useful applications in numerical computations.

(vi) The differential equation

$$x(1 - x)y'' + [c - (a + b + 1)x]y' - aby = 0 \tag{100}$$

where $a$, $b$, $c$ are constants, is known as *Gauss's* or *hypergeometric equation,* and the solution of this equation is called the *hypergeometric function.*

It is interesting to note that the solutions of Eqs. (95)–(100) play an important role in the development of special functions—an exciting branch of mathematics.

## 5.11 SERIES SOLUTION OF A DIFFERENTIAL EQUATION—FROBENIUS METHOD

In Sections 5.2–5.5, we have discussed the methods for solving a linear differential equation with constant coefficients, while a method for solving a linear differential equation with non-constant coefficient was given in Section 5.6. Here, we shall describe yet another method for solving a linear differential equation when the coefficients are functions of the independent variable.

Consider a second-order homogeneous linear differential equation

$$f_2(x)\frac{d^2y}{dx^2} + f_1(x)\frac{dy}{dx} + f_0(x)y = 0 \qquad (101)$$

(cf. Eq. 81), where $f_2(x)$, $f_1(x)$ and $f_0(x)$ are polynomials.

A number of problems of physics and engineering involve differential equations of the form (101) in which $f_2(x)$, $f_1(x)$ and $f_0(x)$ are polynomials. For example, the Bessel's equation (95), Legendre's equation (96), Hermite equation (97), Laguerre equation (98) and hypergeometric equation (100), etc., involve polynomials; and it is for this reason as well as for the sake of convenience in calculations, we assume that the functions $f_2(x)$, $f_1(x)$ and $f_0(x)$ in Eq. (101) are polynomials. But the method of solution that we shall develope here shall also be applicable to other class of functions.

We know that the general solutions of the differential equations

$$D^2y - 3Dy + 2y = 0 \qquad \text{and} \qquad D^2y + 4y = 0$$

by the method of Section 5.2, respectively, are

$$y = c_1e^x + c_2e^{2x} \qquad \text{and} \qquad y = c_1 \cos 2x + c_2 \sin 2x \qquad (102)$$

Using the series expressions of $e^x$, $\sin x$ and $\cos x$, the solutions in Eq. (102), respectively, take the forms

$$y = c_1\left(1 + x + \frac{x^2}{2!} + \cdots\right) + c_2\left[1 + 2x + \frac{(2x)^2}{2!} + \cdots\right] \qquad (103)$$

and
$$y = c_1\left[1 - \frac{(2x)^2}{2!} + \cdots\right] + c_2\left[2x - \frac{(2x)^3}{3!} + \cdots\right] \qquad (104)$$

Also, consider the differential equation

$$x^2y'' - xy' + y = 0$$

Put $x = e^t$, then using the method of Section 5.7, the solution of this differential equation is

$$y = (c_1 + c_2t)\, e^t = c_1\, x + c_2x \log x$$

Using the Taylor series expansion of log $x$, this solution may be expressed as

$$y = c_1x + c_2x\left[(x-1) - \frac{(x-1)^2}{2} + \frac{(x-1)^3}{3} - \frac{(x-1)^4}{4} + \cdots\right] \quad (105)$$

Equations (103)–(105) suggest that the general solution of linear differential equations can also be written in terms of a series. In this section, we shall present an effective method for determining the solution of Eq. (101) in terms of a power series. This method is due to the German mathematician Georg Frobenius (1849–1917), who published it in 1873 and is known as *Frobenius method*. The general theory of this method is rather complicated, however, the working rule for such a method is given by the following steps:

1. Let the solution of Eq. (101) be of the form

$$y = x^k(c_0 + c_1x + c_2x^2 + \cdots + c_mx^m + \cdots) = \sum_{m=0}^{\infty} c_mx^{k+m}, \qquad c_0 \neq 0 \quad (106)$$

2. Differentiate Eq. (106) with respect to $x$ to get

$$y' = \sum_{m=0}^{\infty} c_m(k+m)x^{k+m-1},\; y'' = \sum_{m=0}^{\infty} c_m(k+m)(k+m-1)x^{k+m-2}$$

so that Eq. (101) reduces to

$$f_2(x)\sum_{m=0}^{\infty} c_m(k+m)(k+m-1)x^{k+m-2} + f_1(x)\sum_{m=0}^{\infty} c_m(k+m)x^{k+m-1}$$

$$+ f_0(x)\sum_{m=0}^{\infty} c_mx^{k+m} = 0 \quad (107)$$

which is an equation in powers of $x$.

3. Equate the coefficients of lowest powers of $x$ in Eq. (107), to zero, we get a quadratic equation in $k$. This quadratic equation is known as *indicial equation*. Find the roots of this indicial equation, we then have the following possibilities:
   (i) The roots are not equal and not differ by an integer.
   (ii) The roots are not equal, differ by an integer and make a coefficient of $y$ indeterminate.
   (iii) The roots are not equal, differ by an integer and make a coefficient of $y$ infinite.
   (iv) The roots are equal.

4. Equate the coefficients of general power of $x$ (e.g. $x^{k+m}$ or $x^{k+m-1}$, whichever may be lower), in the identity obtained from Eq. (107), to zero, we get a relation between the coefficients $c_m$ and $c_{m-1}$ or, $c_m$ and $c_{m-2}$, etc. Such a relation is known as a *recurrence relation*.
5. If there is a recurrence relation between $c_m$ and $c_{m-2}$, then $c_1$ is determined by equating the coefficients of next higher power of $x$ (than already used in obtaining the indicial equation) to zero. While, if there is a recurrence relation between $c_m$ and $c_{m-1}$, then we can omit this step.
6. The solution of the given differential equation can now be obtained by substituting the values of the coefficients $c_m$ etc., as calculated from the steps 4 and 5, in Eq. (106).

It may be noted that for a given differential equation there is a solution corresponding to each root of the indicial equation. Thus, if $k_1$ and $k_2$ are the roots of the indicial equation, then with each $k_1$ and $k_2$ there is associated a solution $y_1$ and $y_2$, respectively. Hence, the general solution of Eq. (101) is

$$y = Ay_1 + By_2$$

where $A$ and $B$ are arbitrary constants.

From the above discussion it is clear that the solution of Eq. (101) depends upon the nature of the roots of the indicial equation and thus we shall now discuss, with the help of examples, the different possibilities mentioned in step 3. As we proceed to discuss all these possibilities, some other necessary informations about the working rule shall also be mentioned.

***Case (i): Roots are not equal and not differ by an integer***

**EXAMPLE 5.50** Solve $2x^2y'' - xy' + (1 - x^2)y = 0$.

***Solution*** The given differential equation is

$$2x^2y'' - xy' + (1 - x^2)y = 0 \tag{108}$$

Let the solution of Eq. (108) be of the form

$$y = x^k(c_0 + c_1x + c_2x^2 + \cdots + c_mx^m + \cdots) = \sum_{m=0}^{\infty} c_mx^{k+m}, c_0 \neq 0 \tag{109}$$

then

$$y' = \sum_{m=0}^{\infty} c_m(k+m)x^{k+m-1}, \qquad y'' = \sum_{m=0}^{\infty} c_m(k+m)(k+m-1)x^{k+m-2}$$

Substitute these values of $y$, $y'$ and $y''$ in Eq. (108), we get

$$\sum_{m=0}^{\infty} c_m[2(k+m)(k+m-1) - (k+m) + 1]x^{k+m} - \sum_{m=0}^{\infty} c_mx^{k+m+2} = 0 \tag{110}$$

Taking $m = 0$ in Eq. (110), the lowest power of $x$ is $x^k$; equate its coefficients to zero, we get

$$(2k - 1)(k - 1) = 0 \tag{111}$$

(as $c_0 \neq 0$). Equation (111) is the indicial equation having the roots as $k_1 = 1$ and $k_2 = 1/2$ (here, the roots are not equal and do not differ by an integer).

Equation (110) can also be written as

$$\sum_{m=0}^{\infty} \{c_m[2(k+m)(k+m-1)-(k+m)+1]-c_{m-2}\}x^{k+m} = 0$$

Now in this equation, equate the coefficients of $x^{k+m}$ to zero, we get the recurrence relation

$$c_m = \frac{c_{m-2}}{(2k+2m-1)(k+m-1)} \tag{112}$$

To get the value of $c_1$, put $m = 1$ in Eq. (110) and equate the coefficients of lower degree term of $x$ to zero, we get

$$c_1[k(2k + 1)] = 0$$

Since $k \neq 0$, we have

$$c_1 = 0 \tag{113a}$$

Thus from Eq. (112), we have

$$c_3 = c_5 = c_7 = \cdots = 0 \tag{113b}$$

Also, from Eq. (112), for $m = 2, 4, \ldots$, we have

$$c_2 = \frac{c_0}{(2k+3)(k+1)} \tag{114}$$

$$c_4 = \frac{c_2}{(2k+7)(k+3)} \tag{115}$$

and so on. From Eqs. (113)–(115), Eq. (109) thus becomes

$$y = c_0 x^k \left[1 + \frac{x^2}{(2k+3)(k+1)} + \cdots\right] \tag{116}$$

so that when $k = 1$ and $k = 1/2$, Eq. (116), respectively, leads to

$$y_1 = c_0 x\left(1 + \frac{x^2}{2\cdot 5} + \cdots\right) \quad \text{and} \quad y_2 = c_0 x^{1/2}\left(1 + \frac{x^2}{2\cdot 3} + \cdots\right)$$

Hence, the complete solution of Eq. (108) is

$$y = Ay_1 + By_2 = c_0 x\left(1 + \frac{x^2}{2\cdot 5} + \cdots\right) + c_0 x^{1/2}\left(1 + \frac{x^2}{2\cdot 3} + \cdots\right)$$

where $A$ and $B$ are arbitrary constant and they are equal to $c_0$.

**EXAMPLE 5.51** Solve $4xy'' + 2y' + y = 0$.

***Solution*** The given differential equation is

$$4xy'' + 2y' + y = 0 \tag{117}$$

Assume that Eq. (117) has a solution of the form

$$y = x^k(c_0 + c_1x + c_2x^2 + \cdots + c_mx^m + \cdots) = \sum_{m=0}^{\infty} c_mx^{k+m}, \qquad c_0 \neq 0 \tag{118}$$

then

$$y' = \sum_{m=0}^{\infty} c_m(k+m)x^{k+m-1}, \qquad y'' = \sum_{m=0}^{\infty} c_m(k+m)(k+m-1)x^{k+m-2}$$

Substitute the values of $y$, $y'$ and $y''$ in Eq. (117), we get

$$\sum_{m=0}^{\infty} c_m[4(k+m)(k+m-1)+2(k+m)]x^{k+m-1} + \sum_{m=0}^{\infty} c_mx^{k+m} = 0 \tag{119}$$

Put $m = 0$ in Eq. (119), the lowest power of $x$ is $x^{k-1}$; equate its coefficients to zero, we get

$$k(4k - 2) = 0 \tag{120}$$

(as $c_0 \neq 0$). Equation (120) is the indicial equation having roots as $k_1 = 0$ and $k_2 = 1/2$, which are not equal and do not differ by an integer.

Equation (119) can also be written as

$$\sum_{m=0}^{\infty} \{c_m[4(k+m)(k+m-1)+2(k+m)] + c_{m-1}\}x^{k+m-1} = 0$$

To get a recurrence relation, equate the coefficients of $x^{k+m-1}$, in this equation to zero, we have

$$c_m = -\frac{c_{m-1}}{4(k+m)(k+m-1)+2(k+m)} \tag{121}$$

Thus, when $m = 1, 2, 3, \ldots$, Eq. (121) leads to

$$c_1 = -\frac{c_0}{(4k+2)(k+1)} \tag{122}$$

$$c_2 = \frac{c_0}{(4k+2)(4k+6)(k+1)(k+2)} \tag{123}$$

and so on. Now from Eqs. (122) and (123), Eq. (118) reduces to

$$y = c_0x^k\left[1 - \frac{x}{(4k+2)(k+1)} + \frac{x^2}{(4k+2)(4k+6)(k+1)(k+2)} + \cdots\right] \tag{124}$$

For $k = 0$, Eq. (124) leads to

$$y_1 = c_0\left(1 - \frac{x}{2!} + \frac{x^2}{4!} + \cdots\right) \tag{125}$$

while for $k = 1/2$, Eq. (124) yields

$$y_2 = c_0 x^{1/2}\left(1 - \frac{x}{3!} + \frac{x^2}{5!} + \cdots\right) \tag{126}$$

Therefore, from Eq. (125) and (126) the general solution of Eq. (117) is

$$y = Ay_1 + By_2 = c_0\left(1 - \frac{x}{2!} + \frac{x^2}{4!} + \cdots\right) + c_0 x^{1/2}\left(1 - \frac{x}{3!} + \frac{x^2}{5!} + \cdots\right)$$

***Case (ii): Roots are not equal, differ by an integer and make a coefficient of y indeterminate***

In this case let $k_1$ and $k_2$ be the two roots of the indicial equation and are such that $k_1 < k_2$, and if one of the coefficients of $y$ becomes indeterminate by putting $k = k_2$ (say) then the complete solution is obtained by taking $k = k_2$ in $y$. The solution corresponding to $k = k_1$ in $y$ has to be rejected as it contains a numerical multiple of one of the series appearing in the solution corresponding to $k = k_2$.

**EXAMPLE 5.52** Solve $(1 - x^2)y'' - xy' + 4y = 0$.

***Solution*** The given differential equation is

$$(1 - x^2)y'' - xy' + 4y = 0 \tag{127}$$

Let the series solution of Eq. (127) be

$$y = x^k(c_0 + c_1 x + c_2 x^2 + \cdots + c_m x^m + \cdots) = \sum_{m=0}^{\infty} c_m x^{k+m}, \qquad c_0 \neq 0 \tag{128}$$

then

$$y' = \sum_{m=0}^{\infty} c_m (k+m) x^{k+m-1}, \qquad y'' = \sum_{m=0}^{\infty} c_m (k+m)(k+m-1) x^{k+m-2}$$

and Eq. (127) becomes

$$\sum_{m=0}^{\infty} c_m (k+m)(k+m-1) x^{k+m-2} - \sum_{m=0}^{\infty} c_m (k+m+2)(k+m-2) x^{k+m} = 0 \tag{129}$$

Put $m = 0$ in Eq. (129) and equate the coefficients of lowest power of $x$ (i.e. $x^{k-2}$) to zero, we get

$$k(k - 1) = 0 \tag{130}$$

(as $c_0 \neq 0$). The roots of the indicial equation (130) are $k = 0, 1$, which are not equal and differ by an integer.

For the recurrence relation, consider Eq. (129) and equate the coefficients of $x^{k+m-2}$ to zero so that we have

$$c_m = \frac{k + m - 4}{k + m - 1} c_{m-2} \tag{131}$$

To obtain the value of $c_1$, put $m = 1$ in Eq. (129) and equate the coefficients of the lowest power of $x$ to zero, we get

$$c_1 k(k + 1) = 0 \tag{132}$$

But from Eq. (130), $k = 0$ and Eq. (132) thus make $c_1$ indeterminate. Moreover, from Eq. (131), $c_2$, $c_3$, $c_4$, ... can be expressed in terms of $c_0$ and $c_3$, $c_5$, ... if $c_1$ is assumed to be finite. Thus for $k = 0$, Eq. (131) reduces to

$$c_m = \frac{m-4}{m-1} c_{m-2}$$

which leads to

$$c_2 = -2c_0, \qquad c_3 = -\frac{1}{2}c_1, \qquad c_4 = 0, \qquad c_5 = -\frac{1}{8}c_1, \qquad c_6 = 0, \ldots \tag{133}$$

Now using $k = 0$ (one of the roots of indicial equation) and Eq. (133) in Eq. (128), we get

$$y = c_0(1 - 2x^2) + c_1\left(x - \frac{1}{2}x^3 - \frac{1}{8}x^5 + \cdots\right) \tag{134}$$

Now corresponding to the other root $k = 1$, from Eqs. (131) and (128), the solution is

$$y = c_0\left(x - \frac{1}{2}x^3 - \frac{1}{8}x^5 + \cdots\right) \tag{135}$$

The solution (135) is simply a constant multiple of one of the series in the solution (134), so we reject it.

Hence, the required solution is given by Eq. (134). It contains two arbitrary constants $c_0$ and $c_1$.

**EXAMPLE 5.53** Solve $y'' - xy' + y = 0$.

***Solution*** Given that

$$y'' - xy' + y = 0 \tag{136}$$

Let the series solution of Eq. (127) be of the form

$$y = x^k(c_0 + c_1 x + c_2 x^2 + \cdots + c_m x^m + \cdots) = \sum_{m=0}^{\infty} c_m x^{k+m}, \qquad c_0 \neq 0 \tag{137}$$

then

$$y' = \sum_{m=0}^{\infty} c_m (k + m) x^{k+m-1}, \qquad y'' = \sum_{m=0}^{\infty} c_m (k + m)(k + m - 1) x^{k+m-2}$$

Equation (136) now takes the form

$$\sum_{m=0}^{\infty} c_m (k + m)(k + m - 1) x^{k+m-2} - \sum_{m=0}^{\infty} c_m (k + m - 1) x^{k+m} = 0 \tag{138}$$

Put $m = 0$ in Eq. (138) and equate the coefficients of the lowest power of $x$ (i.e. $x^{k-2}$) to zero, we get

$$k(k-1) = 0, \qquad c_0 \neq 0 \tag{139}$$

as the indicial equation whose roots are $k = 0, 1$.

For the recurrence relation, consider Eq. (138) and equate the coefficients of $x^{k+m-2}$ to zero, so that we have

$$c_m = \frac{k+m-3}{(k+m)(k+m-1)} c_{m-2} \tag{140}$$

Now put $m = 1$ in Eq. (138) and equate the coefficients of the lowest power of $x$ to zero, we get

$$c_1 k(k+1) = 0$$

so that when $k = 0$ this equation makes $c_1$ indeterminate. Also, from Eq. (140), assuming $c_1$ to be finite, we have (for $k = 0$)

$$c_m = \frac{m-3}{m(m-1)} c_{m-2} \tag{141}$$

Equation (141) now leads to

$$c_2 = -\frac{1}{2}c_0,\ c_3 = 0,\ c_4 = -\frac{1}{24}c_0,\ c_5 = 0,\ c_6 = \frac{3}{30}\left(-\frac{1}{24}\right)c_0, \ldots \tag{142}$$

Therefore, from Eqs. (137) and (142), the required solution (neglecting the solution corresponding to the root $k = 1$) is

$$y = c_1 x + c_0\left(1 - \frac{x^2}{2!} - \frac{x^4}{4!} - 3\frac{x^6}{6!} - \cdots\right)$$

***Case (iii): Roots are not equal, differ by an integer and make a coefficient of y infinite***

Here, let $k_1$ and $k_2$ be two roots such that $k_1 < k_2$ differing by an integer and if for $k = k_2$ some of the coefficients of $y$ becomes infinite, we then replace $c_0$ by $a_0(k - k_2)$. This leads to a solution corresponding to the root $k = k_2$. For the root $k = k_1$ the solution is not linearly independent (if the ratio of the two series in the solution is constant, the two solutions are not linearly independent). The linearly independent solution is given by $(\partial y/\partial k)_{k=k_2}$.

Thus, for this case, the general solution of the given differential equation is

$$y = A(y)_{k=k_2} + B\left(\frac{\partial y}{\partial k}\right)_{k=k_2}$$

where $(y)_{k=k_2}$ and $(\partial y/\partial k)_{k=k_2}$ are the two linearly independent solutions. $A$ and $B$ are arbitrary constants.

**EXAMPLE 5.54** Solve $x(1 - x)y'' - 3xy' - y = 0$.

***Solution*** The given equation can be written as

$$(x - x^2)y'' - 3xy' - y = 0 \tag{143}$$

Let the series solution of Eq. (143) be of the form

$$y = x^k(c_0 + c_1x + c_2x^2 + \cdots + c_mx^m + \cdots) = \sum_{m=0}^{\infty} c_m x^{k+m}, \qquad c_0 \neq 0 \tag{144}$$

then

$$y' = \sum_{m=0}^{\infty} c_m(k + m)x^{k+m-1}, \qquad y'' = \sum_{m=0}^{\infty} c_m(k + m)(k + m - 1)x^{k+m-2}$$

and Eq. (143) reduces to

$$\sum_{m=0}^{\infty} c_m[(k + m)(k + m - 1)]x^{k+m-1} - \sum_{m=0}^{\infty} c_m(k + m + 1)^2 x^{k+m} = 0 \tag{145}$$

Put $m = 0$ in Eq. (145) and equate the coefficients of the lowest power of $x$ to zero, we get the indicial equation

$$k(k - 1) = 0, \qquad c_0 \neq 0 \tag{146}$$

which has the roots as $k = 0, 1$.

For the recurrence relation, consider Eq. (145) and equate the coefficients of $x^{k+m-1}$ to zero, we have

$$c_m = \frac{k+m}{k+m-1} c_{m-1} \tag{147}$$

Taking $m = 1, 2, 3, \ldots$ in Eq. (147), we get

$$c_1 = \frac{k+1}{k} c_0, \qquad c_2 = \frac{k+2}{k} c_0, \qquad c_3 = \frac{k+3}{k} c_0 \tag{148}$$

and so on. From Eq. (148), Eq. (144) leads to

$$y = c_0 x^k \left(1 + \frac{k+1}{k} x + \frac{k+2}{k} x^2 + \frac{k+3}{k} x^3 + \cdots\right) \tag{149}$$

so that when $k = 0$ (one of the roots of the indicial equation), the coefficient on the right-hand side of Eq. (149) becomes infinite. Thus, we make the substitution $c_0 = a_0(k - k_2) = a_0 k$ in Eq. (149) to get

$$y = a_0 x^k[k + (k + 1)x + (k + 2)x^2 + (k + 3)x^3 + \cdots] \tag{150}$$

Thus, corresponding to $k = 0$, the solution of Eq. (143), from Eq. (150) is

$$(y)_{k=0} = a_0(x + 2x^2 + 3x^3 + \cdots) \tag{151}$$

The solution, corresponding to $k = 1$, from Eq. (149) is

$$y = c_0(x + 2x^2 + 3x^3 + \cdots) \tag{152}$$

The solution (152) is not linearly independent. To have a linearly independent solution of Eq. (143), differentiate Eq. (150) partially with respect to $k$ (using logarithmic differentiation), so that we have

$$\frac{\partial y}{\partial k} = a_0 x^k \log x[k + (k + 1)x + (k + 2)x^2 + (k + 3)x^3 + \cdots]$$
$$+ a_0 x^k (1 + x + x^2 + \cdots) \tag{153}$$

Thus, for $k = 0$, Eq. (153) becomes

$$\left(\frac{\partial y}{\partial k}\right)_{k=0} = a_0 \log x(x + 2x^2 + 3x^3 + \cdots) + a_0(1 + x + x^2 + \cdots)$$

which from (151) reduces to

$$\left(\frac{\partial y}{\partial k}\right)_{k=0} = (y)_{k=0} \log x + a_0(1 + x + x^2 + \cdots) \tag{154}$$

Therefore, the required solution of the given differential equation is

$$y = A(y)_{k=0} + B\left(\frac{\partial y}{\partial k}\right)_{k=0}$$

where $(y)_{k=0}$ and $(\partial k/\partial k)_{k=0}$ are given by Eqs. (151) and (154), respectively and $A$ and $B$ are arbitrary constants.

**EXAMPLE 5.55** Solve $(x^2 - x)y'' + xy' - y = 0$.

***Solution*** The given equation can be written as

$$x^2y'' - xy'' + xy' - y = 0 \tag{155}$$

Let the series solution of Eq. (155) be of the form

$$y = x^k(c_0 + c_1x + c_2x^2 + \cdots + c_mx^m + \cdots) = \sum_{m=0}^{\infty} c_m x^{k+m}, \qquad c_0 \neq 0 \tag{156}$$

then

$$y' = \sum_{m=0}^{\infty} c_m x^{k+m-1}, \quad y'' = \sum_{m=0}^{\infty} c_m(k + m)(k + m - 1)x^{k+m-2}$$

and Eq. (155) reduces to

$$\sum_{m=0}^{\infty} c_m[(k + m)^2 - 1]\, x^{k+m} - \sum_{m=0}^{\infty} c_m(k + m)(k + m - 1)x^{k+m-1} = 0 \tag{157}$$

Put $m = 0$ in Eq. (157) and equate the coefficients of the lowest power of $x$ to zero, we get the indicial equation

$$k(k - 1) = 0, \qquad c_0 \neq 0$$

which has the roots as $k = 0, 1$.

From Eq. (157), the recurrence relation is

$$c_m = \frac{k+m-2}{k+m-1} c_{m-1} \tag{158}$$

When $m = 1, 2, 3, \ldots$, Eq. (158) leads to

$$c_1 = \frac{k-1}{k} c_0, \quad c_2 = \frac{k-1}{k+1} c_0, \quad c_3 = \frac{k-1}{k+2} c_0 \tag{159}$$

and so on. From Eq. (159), Eq. (156) reduces to

$$y = c_0 x^k \left(1 + \frac{k-1}{k} x + \frac{k-1}{k+1} x^2 + \frac{k-1}{k+2} x^3 + \cdots \right) \tag{160}$$

so that when $k = 0$ (one of the roots of the indicial equation), the coefficient on the right-hand side of Eq. (160) becomes infinite. Thus, we put $c_0 = a_0(k - k_2) = a_0 k$ in Eq. (160) to get

$$y = a_0 x^k \left[ k + (k-1)x + \frac{k(k-1)}{k+1} x^2 + \frac{k(k-1)}{k+2} x^3 + \cdots \right] \tag{161}$$

Thus, corresponding to $k = 0$, the solution of the given equation, from Eq. (161), is

$$(y)_{k=0} = -a_0 x \tag{162}$$

To get another linearly independent solution, differentiate Eq. (161) partially with respect to $k$ (using logarithmic differentiation), and we have

$$\frac{\partial y}{\partial k} = y \log x + a_0 x^k \left[ 1 + x + \frac{(2k-1)(k+1) - (k^2 - k)}{(k+1)^2} x^2 + \cdots \right]$$

so that

$$\left(\frac{\partial y}{\partial k}\right)_{k=0} = -a_0 x \log x + a_0 \left(1 + x - x^2 + \cdots\right) \tag{163}$$

using Eq. (162) for $y$, which is another solution of the given equation. Therefore, the complete solution of Eq. (155) is

$$y = A(y)_{k=0} + B\left(\frac{\partial y}{\partial k}\right)_{k=0}$$

where $(y)_{k=0}$ and $(\partial y/\partial k)_{k=0}$ are given by Eqs. (162) and (163), respectively and $A$ and $B$ are arbitrary constants.

***Case (iv): Roots are equal***

If the roots of the indicial equation are $k = k_1 = k_2$ then the two independent solutions of the given differential equation are $(y)_{k=k1}$ and $(\partial y/\partial k)_{k=k2}$.

**EXAMPLE 5.56** Solve $xy'' + y' + xy = 0$.

***Solution*** The given equation is

$$xy'' + y' + xy = 0 \tag{164}$$

If a series solution of Eq. (164) is of the form

$$y = x^k(c_0 + c_1x + c_2x^2 + \cdots + c_mx^m + \cdots) = \sum_{m=0}^{\infty} c_mx^{k+m}, \qquad c_0 \neq 0 \tag{165}$$

then

$$y' = \sum_{m=0}^{\infty} c_m(k+m)x^{k+m-1}, \qquad y'' = \sum_{m=0}^{\infty} c_m(k+m)(k+m-1)x^{k+m-2}$$

so that Eq. (164) reduces to

$$\sum_{m=0}^{\infty} c_m[(k+m)^2]x^{k+m-1} + \sum_{m=0}^{\infty} c_mx^{k+m+1} = 0 \tag{166}$$

Put $m = 0$ in Eq. (166) and equate the coefficients of the lowest power of $x$ to zero, we get $k^2 = 0$, which leads to $k = 0, 0$ as $c_0 \neq 0$. Thus the roots are equal.

From Eq. (166), the recurrence relation is

$$c_m = -\frac{c_{m-2}}{(k+m)^2} \tag{167}$$

Put $m = 1$ in Eq. (166) and equate the coefficient of $x^k$ to zero, we get $c_1(k+1)^2 = 0$ which leads to $c_1 = 0$ and thus from Eq. (167), we have

$$c_3 = c_5 = c_7 = \cdots = 0 \tag{168}$$

Put $m = 2, 4, \ldots$ in Eq. (167), we get

$$c_2 = -\frac{c_0}{(k+2)^2}, \qquad c_4 = \frac{c_0}{(k+2)^2(k+4)^2} \tag{169}$$

and so on. From Eqs. (168) and (169), Eq. (165) takes the form

$$y = c_0x^k\left[1 - \frac{x^2}{(k+2)^2} + \frac{x^4}{(k+2)^2(k+4)^2} + \cdots\right] \tag{170}$$

Thus, corresponding to $k = 0$, the solution of the given Eq. (164), from Eq. (170) is

$$(y)_{k=0} = c_0\left(1 - \frac{x^2}{2^2} + \frac{x^4}{2^2 \cdot 4^2} + \cdots\right) \tag{171}$$

To get another solution, differentiate Eq. (170) partially with respect to $k$ (using logarithmic differentiation), and we have

$$\frac{\partial y}{\partial k} = y \log x + c_0x^k\left[-\frac{-2}{(k+2)^3}x^2 + \frac{-2 \cdot 2(k+3)}{(k+2)^3(k+4)^3}x^4 + \cdots\right] \tag{172}$$

where $y$ is given by Eq. (170).

When $k = 0$, Eq. (172) reduces to

$$\left(\frac{\partial y}{\partial k}\right)_{k=0} = c_0\left(1-\frac{x^2}{2^2}+\frac{x^4}{2^2\cdot 4^2}+\cdots\right)\log x + c_0\left(\frac{2}{2^3}x^2-\frac{2\cdot 2\cdot 3}{2^3\cdot 4^3}x^4+\cdots\right)$$

$$= c_0\left(1-\frac{x^2}{2^2}+\frac{x^4}{2^2\cdot 4^2}+\cdots\right)\log x$$

$$+c_0\left[\frac{x^2}{2^2}-\frac{x^4}{2^2\cdot 4^2}\left(1+\frac{1}{2}\right)+\frac{x^6}{2^2\cdot 4^2\cdot 6^2}\left(1+\frac{1}{2}+\frac{1}{3}\right)+\cdots\right]$$

or
$$\left(\frac{\partial y}{\partial k}\right)_{k=0} = (y)_{k=0}\log x - c_0\sum_{m=1}^{\infty}(-1)^m\frac{H_m x^{2m}}{2^{2m}(m!)^2} \tag{173}$$

where

$$H_n = 1+\frac{1}{2}+\frac{1}{3}+\cdots+\frac{1}{n}=\sum_{m=1}^{n}\frac{1}{m}$$

Equation (173) represents another solution of the given equation. Therefore, the complete solution of the given equation is

$$y = A(y)_{k=0} + B\left(\frac{\partial y}{\partial k}\right)_{k=0}$$

where $(y)_{k=0}$ and $(\partial y/\partial k)_{k=1}$ are given by Eqs. (171) and (173), respectively and $A$ and $B$ are arbitrary constants.

**REMARK.** Equation (164) is known as *Bessel's equation of order zero*. If $c_0 = 1$, then from Eqs. (171) and (173), we have

$$J_0(x) = \sum_{m=0}^{\infty}\frac{(-1)^m x^{2m}}{2^{2m}(m!)^2} \tag{174}$$

and
$$y_0 = J_0(x)\log x - \sum_{m=0}^{\infty}(-1)^m\frac{H_m x^{2m}}{2^{2m}(m!)^2} \tag{175}$$

Equations (174) and (175) are, respectively, known as *Bessel's functions of first and second kind of order zero*.

**EXAMPLE 5.57** Solve $(x - x^2)y'' + (1 - x)y' - y = 0$.

***Solution*** Let the solution of the given equation be of the form

$$y = x^k(c_0 + c_1x + c_2x^2 + \cdots + c_mx^m + \cdots) = \sum_{m=0}^{\infty} c_m x^{k+m}, \quad c_0 \neq 0 \tag{176}$$

then $y' = \sum_{m=0}^{\infty} c_m(k + m)x^{k+m-1}, \quad y'' = \sum_{m=0}^{\infty} c_m(k + m)(k + m - 1)x^{k+m-2}$

and the given equation can be expressed as

$$\sum_{m=0}^{\infty} c_m(k+m)^2 x^{k+m-1} - \sum_{m=0}^{\infty} c_m[(k+m)^2+1]x^{k+m} = 0 \tag{177}$$

which leads to

$$c_0 k^2 = 0, \qquad c_0 \neq 0$$

Thus, the indicial equation $k^2 = 0$ has equal roots as $k = 0, 0$. Also from Eq. (177), the recurrence relation is

$$c_m = \frac{(k+m-1)^2+1}{(k+m)^2} c_{m-1} \tag{178}$$

When $m = 1, 2, 3, \ldots$, Eq. (178) leads to

$$c_1 = \frac{k^2+1}{(k+1)^2} c_0, \qquad c_2 = \frac{[(k+1)^2+1](k^2+1)}{(k+1)^2(k+2)^2} c_0$$

and so on. Thus, from Eq. (176), we have

$$y = c_0 x^k \left[1 + \frac{k^2+1}{(k+1)^2} x + \frac{(k^2+2k+2)(k^2+1)}{(k+1)^2(k+2)^2} x^2 + \cdots\right] \tag{179}$$

so that for $k = 0$, one solution of the given equation is

$$(y)_{k=0} = c_0 \left(1 + x + \frac{2}{2^2} x^2 + \cdots\right) \tag{180}$$

Now, differentiating Eq. (179) partially with respect to $k$ (using logarithmic differentiation), we get

$$\frac{\partial y}{\partial k} = y \log x + c_0 x^k \left[\frac{2k(k+1)^2 - 2(k+1)(k^2+1)}{(k+1)^4} x + \cdots\right] \tag{181}$$

Thus, using Eqs. (180) and (181), the second solution of the given equation, corresponding to $k = 0$, is

$$\left(\frac{\partial y}{\partial k}\right)_{k=0} = c_0 \left(1 + x + \frac{2}{2^2} x^2 + \cdots\right) \log x + c_0(-2x - x^2 + \cdots) \tag{182}$$

Therefore, the complete solution of the given equation is

$$y = A(y)_{k=0} + B\left(\frac{\partial y}{\partial k}\right)_{k=0}$$

where $(y)_{k=0}$ and $(\partial y/\partial k)_{k=0}$ are given by Eqs. (180) and (182), respectively and $A$ and $B$ are arbitrary constants.

## 5.12 BESSEL, LEGENDRE AND HYPERGEOMETRIC EQUATIONS AND THEIR SOLUTIONS

In Section 5.10, we have seen that Eqs. (95), (96) and (100) have a wide range of applications in several branches of physics and engineering. The solution of these equations leads to an interesting branch of mathematics, known as *special functions*. This section is devoted to find the solution of these equations.

***Bessel equation***

An equation of the form

$$x^2y'' + xy' + (x^2 - n^2)y = 0 \tag{183}$$

is known as *Bessel equation* of order $n$, where $n$ is a real number. The solution of Eq. (183) is known as *Bessel function*.

A systematic analysis of the solution of Eq. (183) was made by the German mathematician F.W. Bessel (1784–1846). He was the first mathematician who measured the distance to a star. He also made significant contributions in the fields of astronomy, geodesy and classical mechanics. Bessel's equations have a number of applications in physical and engineering sciences and, in particular, are very effective in solving the problems involving circular geometry, the temperature distribution in circular plates, flow of electricity in cylinders, the vibration of membranes, etc.

We shall now find a solution of Eq. (183) when $2n$ is non-integral. Let a solution of Eq. (183) be of the form

$$y = x^k(c_0 + c_1x + c_2x^2 + \cdots + c_mx^m + \cdots) = \sum_{m=0}^{\infty} c_m x^{k+m}, \qquad c_0 \neq 0 \tag{184}$$

then

$$y' = \sum_{m=0}^{\infty} c_m(k+m)x^{k+m-1}, \qquad y'' = \sum_{m=0}^{\infty} c_m(k+m)(k+m-1)x^{k+m-2}$$

and Eq. (183) takes the form

$$\sum_{m=0}^{\infty} c_m[(k+m)(k+m-1) + (k+m) - n^2]x^{k+m} + \sum_{m=0}^{\infty} c_m x^{k+m+2} = 0 \tag{185}$$

From Eq. (185), the indicial equation is

$$k^2 - n^2 = 0, \qquad c_0 \neq 0 \tag{186}$$

having the roots as $k_1 = n$ and $k_2 = -n$ (here, the roots are not equal and do not differ by an integer).

From Eq. (185), the recurrence relation is

$$c_m = -\frac{c_{m-2}}{(k+m+n)(k+m-n)} \tag{187}$$

Since Eq. (187) is a recurrence relation between $c_m$ and $c_{m-2}$, we shall use Step 5 of Section 5.11 to calculate $c_1$. We begin with Eq. (185) and write it as

(take $m = 1$)

$$c_1[(k+1)k + (k+1) - n^2]\, x^{k+1} + c_1 x^{k+2} = 0$$

Now equate the coefficient of the lowest degree term of $x$ to zero so that we have

$$c_1[(k+1+n)(k+1-n)] = 0$$

which shows that $c_1 = 0$ as $k = \pm n$. Thus from Eq. (187), we get

$$c_1 = c_3 = c_5 = \cdots = 0 \tag{188}$$

Now taking $m = 2, 4, \ldots$ in Eq. (187), we get

$$c_2 = \frac{c_0}{(k+2+n)(k+2-n)} \tag{189}$$

$$c_4 = \frac{c_0}{(k+2+n)(k+2-n)(k+4+n)(k+4-n)} \tag{190}$$

and so on. From Eqs. (188)–(190), Eq. (184) reduces to

$$y = c_0 x^k \left[1 - \frac{x^2}{(k+2+n)(k+2-n)} + \frac{x^4}{(k+2+n)(k+2-n)(k+4+n)(k+4-n)} + \cdots\right] \tag{191}$$

Thus for $k = n$, Eq. (191) becomes

$$y_1 = c_0 x^n \left[1 + (-1)\frac{x^2}{2^2 \cdot 1!(n+1)} + (-1)^2 \frac{x^4}{2^4 \cdot 2!(n+1)(n+2)} + \cdots\right] \tag{192}$$

If we take

$$c_0 = \frac{1}{2^n \Gamma(n+1)}$$

then Eq. (192) takes the form

$$y_1 = \frac{x^n}{2^n \Gamma(n+1)}\left[1 + (-1)\frac{x^2}{2^2 \cdot 1!(n+1)} + (-1)^2 \frac{x^4}{2^4 \cdot 2!(n+1)(n+2)} + \cdots\right]$$

$$= \frac{x^n}{2^n \Gamma(n+1)} \sum_{m=0}^{\infty} (-1)^m \frac{x^{2m}}{2^{2m}\, m!(n+1)(n+2)\cdots(n+m)}$$

or

$$y_1 = \sum_{m=0}^{\infty} (-1)^m \left(\frac{x}{2}\right)^{n+2m} \frac{1}{m!\,\Gamma(n+m+1)} \equiv J_n(x) \tag{193}$$

Similarly, for $k = -n$, Eq. (192) leads to

$$y_2 = \sum_{m=0}^{\infty} (-1)^m \left(\frac{x}{2}\right)^{-n+2m} \frac{1}{m!\,\Gamma(-n+m+1)} \equiv J_{-n}(x) \tag{194}$$

Hence the general solution of Bessel's equation is

$$y = AJ_n(x) + BJ_{-n}(x) \tag{195}$$

where $A$ and $B$ are arbitrary constants.

The solution $J_n(x)$, as given by Eq. (193), is known as *Bessel's function of first kind of order n.*

**REMARK.** 1. If $n$ is an integer, then $\Gamma(n+m+1)=(n+m)!$ and Eq. (193) becomes

$$J_n(x)=\sum_{m=0}^{\infty}(-1)^m \frac{1}{m!(n+m)!}\left(\frac{x}{2}\right)^{n+2m} \tag{193a}$$

so that for $n = 0$ and 1 respectively, Eq. (193 a), leads to

$$J_0(x)=\sum_{m=0}^{\infty}(-1)^m \frac{1}{(m!)^2}\left(\frac{x}{2}\right)^{2m}$$

$$=1-\frac{x^2}{2^2}+\frac{x^4}{2^2\cdot 4^2}-\frac{x^6}{2^2\cdot 4^2\cdot 6^2}+\cdots \tag{193b}$$

and

$$J_1(x)=\sum_{m=0}^{\infty}(-1)^m \frac{1}{m!(m+1)!}\left(\frac{x}{2}\right)^{2m+1}$$

$$=\frac{x}{2}-\frac{x^3}{2^2\cdot 4}+\frac{x^5}{2^2\cdot 4^2\cdot 6}-\cdots \tag{193c}$$

$J_0(x)$ and $J_1(x)$ are known as Bessel functions of order 0 and 1, respectively. Moreover, $J_0(0) = 1$ and $J_1(0) = 0$.

2. Due to the presence of negative and positive terms in $J_0(x)$ and $J_1(x)$, the graphs of $J_0(x)$ and $J_1(x)$ oscillate and decay fast as $x \to \infty$.

***Legendre equation***

An equation of the form

$$(1 - x^2)y'' - 2xy' + n(n + 1)y = 0 \tag{196}$$

where $n$ is real number, is known as *Legendre's equation* of order $n$. The solution of Eq. (196) is known as *Legendre function* or *polynomial.*

Equation (196) was given by A.M. Legendre (1752–1833), a French mathematician, as the outcome of his studies of the attraction of spheroids. Legendre is best remebered for his work in the field of elliptic integrals and theory of numbers. Legendre's equation has a number of applications in the physical problems involving spherical geometry and gravitation.

We shall now find a series solution of Eq. (196). Let a series solution of this equation be of the form

$$y = x^k(c_0 + c_1x + c_2x^2 + \cdots + c_mx^m + \cdots) = \sum_{m=0}^{\infty} c_m x^{k+m}, \qquad c_0 \neq 0 \tag{197}$$

then

$$y' = \sum_{m=0}^{\infty} c_m (k+m) x^{k+m-1}, \quad y'' = \sum_{m=0}^{\infty} c_m (k+m)(k+m-1) x^{k+m-2}$$

so that Eq. (196) reduces to

$$\sum_{m=0}^{\infty} c_m[(k+m)(k+m-1]x^{k+m-2} - \sum_{m=0}^{\infty} c_m[(k+m-n)(k+m+n+1)]x^{k+m} = 0 \tag{198}$$

From Eq. (198), the indicial equation (as $c_0 \neq 0$) is

$$k(k-1) = 0 \tag{199}$$

and the roots of this equation are $k = 0, 1$, which are not equal and differ by an integer.

The recurrence relation, from Eq. (198), is

$$c_m = \frac{(k+m-2-n)(k+m+n-1)}{(k+m)(k+m-1)} c_{m-2} \tag{200}$$

To get the value of $c_1$, put $m = 1$ in Eq. (198) and equate the coefficients of the lowest power of $x$ to zero, we get

$$c_1 k(k+1) = 0 \tag{201}$$

When $k = 0$, $c_1$ is indeterminate from Eq. (201). But for $k = 0$, Eq. (200) leads to

$$c_m = \frac{(m-2-n)(m+n-1)}{m(m-1)} c_{m-2} \tag{202}$$

so that

$$c_2 = -\frac{n(n+1)}{2} c_0$$

$$c_3 = \frac{(1-n)(2+n)}{3 \cdot 2} c_1$$

$$c_4 = -\frac{n(n+1)(n+3)(2-n)}{4 \cdot 3 \cdot 2} c_0$$

$$c_5 = \frac{(1-n)(3-n)(2+n)(4+n)}{5 \cdot 4 \cdot 3 \cdot 2} c_1$$

and so on. Thus, from Eq. (197), we have

$$y = c_0 \left[ 1 - \frac{n(n+1)}{2!} x^2 - \frac{n(n+1)(n+3)(2-n)}{4!} x^4 + \cdots \right]$$
$$+ c_1 \left[ x + \frac{(1-n)(2+n)}{3!} x^3 + \frac{(1-n)(3-n)(2+n)(4+n)}{5!} x^5 + \cdots \right] \tag{203}$$

which is the required solution of the Legendre's equation (196), and is expressed in the *ascending powers* of $x$. The solution corresponding to the other root $k = 1$, is simply a constant multiple of one of the series in the solution (203), so we reject it.

It is also possible to obtain the solution of Eq. (196) in terms of the *descending powers* of $x$. This form of solution is more important (due to its applications to physical problems) than the solution given by Eq. (203). For such a solution, we assume that Eq. (196) has a series solution of the form

$$y = x^k(c_0 + c_1x^{-1} + c_2x^{-2} + \cdots + c_mx^{-m} + \cdots) = \sum_{m=0}^{\infty} c_m x^{k-m}, \quad c_0 \neq 0 \tag{204}$$

then

$$y' = \sum_{m=0}^{\infty} c_m(k - m)x^{k-m-1}, \quad y'' = \sum_{m=0}^{\infty} c_m(k - m)(k - m - 1)x^{k-m-2}$$

so that Eq. (196) reduces to

$$\sum_{m=0}^{\infty} c_m(k - m)(k - m - 1)x^{k-m-2} - \sum_{m=0}^{\infty} c_m(k - m - n)(k - m + n + 1)x^{k-m} = 0 \tag{205}$$

Put $m = 0$ in Eq. (205) and equate the coefficients of the highest power of $x$ to zero (since the series solution is assumed to be in the descending powers of $x$), we get the indicial equation as

$$c_0(k - n)(k + n + 1) = 0 \tag{206}$$

and since $c_0 \neq 0$, the roots of the indicial equation are

$$k = n, -(n + 1) \tag{206a}$$

Now, put $m = 1$ in Eq. (205) and equate the coefficients of $x^{k-1}$ to zero, we get

$$c_1(k - 1 - n)(k + n) = 0$$

which yields $c_1 = 0$ as neither $(k + n)$ nor $(k - n - 1)$ is zero due to Eq. (206a).

To get a recurrence relation, consider Eq. (205) so that

$$c_m = \frac{(k - m + 2)(k - m + 1)}{(k - m - n)(k - m + n + 1)} c_{m-2} \tag{207}$$

As $c_1 = 0$, Eq. (207) leads to

$$c_3 = c_5 = c_7 = \cdots = 0$$

We thus have

*Case (i):* When $k = n$

From Eq. (207), we have

$$c_m = \frac{(n - m + 2)(n - m + 1)}{(-m)(-m + 2n + 1)} c_{m-2}$$

so that

$$c_2 = \frac{n(n-1)}{-2(2n-1)}c_0 \qquad c_4 = \frac{n(n-1)(n-2)(n-3)}{4 \cdot 2(2n-1)(2n-3)}c_0$$

and so on. Thus from Eq. (204), a solution of Legendre's equation (196) (for $k = n$) is

$$y = c_0\left[x^n - \frac{n(n-1)}{2(2n-1)}x^{n-2} + \frac{n(n-1)(n-2)(n-3)}{2 \cdot 4(2n-1)(2n-3)}x^{n-4} + \cdots\right] \quad (208)$$

*Case (ii)*: When $k = -(n + 1)$

From Eq. (207), we have

$$c_m = \frac{(-n-m+1)(-n-m)}{(-m)(-m-2n-1)}c_{m-2}$$

so that

$$c_2 = \frac{(n+1)(n+2)}{2(2n+3)}c_0, \quad c_4 = \frac{(n+1)(n+2)(n+3)\,(n+4)}{2 \cdot 4(2n+3)(2n+5)}c_0$$

and so on. Thus from Eq. (204), another solution of Legendre's equation (196) (for $k = -n - 1$) is

$$y = c_0\left[\frac{x^{-n-1}(n+1)(n+2)}{2(2n+3)}x^{-n-3} + \frac{(n+1)(n+2)(n+3)\,(n+4)}{2 \cdot 4(2n+3)(2n+5)}x^{-n-5} + \cdots\right] \quad (209)$$

If $n$ is a positive integer and

$$c_0 = \frac{1 \cdot 3 \cdot 5 \cdots (2n-1)}{n!}$$

then solution (208) can be expressed as

$$y = \frac{1 \cdot 3 \cdot 5 \cdots (2n-1)}{n!}\left[x^n - \frac{n(n-1)}{2(2n-1)}x^{n-2} + \frac{n(n-1)(n-2)(n-3)}{2 \cdot 4(2n-1)(2n-3)}x^{n-4} + \cdots\right] \quad (210)$$

which is a terminating series. If we denote solution (210) by $P_n(x)$, then

$$P_n(x) = \sum_{m=0}^{n/2}(-1)^m \frac{(2n-2m)!}{2^n\, m!(n-2m)!(n-m)!}x^{n-2m} \quad (211)$$

where

$$\left(\frac{n}{2}\right) = \begin{cases} \frac{n}{2}, & \text{if } n \text{ is even} \\ \frac{n-1}{2}, & \text{if } n \text{ is odd} \end{cases}$$

$P_n(x)$ given by Eq. (211) is one of the solution (for $k = n$) of Legendre's equation and is known as *Legendre's function of first kind.*

Further, if $n$ is a positive integer and

$$c_0 = \frac{n!}{1\cdot 3\cdot 5\cdots(2n+1)}$$

and let $Q_n(x)$ denotes the solution (209), then

$$Q_n(x) = \frac{n!}{1\cdot 3\cdot 5\cdots(2n+1)}\left[x^{-n-1} + \frac{(n+1)(n+2)}{2(2n+3)}x^{-n-3} + \frac{(n+1)(n+2)(n+3)(n+4)}{2\cdot 4(2n+3)(2n+5)}x^{-n-5} + \cdots\right] \quad (212)$$

Here, $Q_n(x)$ is a non-terminating infinite series as $n$ is positive; and is known as *Legendre's function of second kind.*

Hence, the most general solution of Legendre's equation (196) is

$$y = AP_n(x) + BQ_n(x)$$

where $P_n(x)$ and $Q_n(x)$ are, respectively, given by Eqs. (211) and (212) and $A$ and $B$ are arbitrary constants.

**REMARK.** From Eq. (210), by taking $n = 0, 1, 2, 3, \ldots$, we get the first few Legendre polynomials as follows:

$$P_0(x) = 1, P_1(x) = x, P_2(x) = \frac{1}{2}(3x^2 - 1), P_3(x) = \frac{1}{2}(5x^3 - 3x)$$

$$P_4(x) = \frac{1}{8}(35x^4 - 30x^2 + 3), P_5(x) = \frac{1}{8}(63x^5 - 70x^3 + 15x),$$

etc. Such types of Legendre polynomials can be used to express a given polynomial in terms of Legendre polynomials (cf., Example 5.72).

***Hypergeometric equation***

A differential equation of the form

$$x(1 - x)y'' + [c - (a + b + 1)x]\, y' - aby = 0 \quad (213)$$

where $a$, $b$ and $c$ are constants, is known as a *hypergeometric equation.* A particular solution of this equation is known as *hypergeometric function.*

Equation (213) has a wide range of applications in problems of mathematical physics, for example, this equation can be used to obtain a solution of Schrodinger equation for hydrogen atom.

Now, we shall obtain a series solution of Eq. (213). Let a series solution of this equation be of the form

$$y = x^k(c_0 + c_1x + c_2x^2 + \cdots + c_mx^m + \cdots) = \sum_{m=0}^{\infty} c_m x^{k+m}, \qquad c_0 \neq 0 \quad (214)$$

then

$$y' = \sum_{m=0}^{\infty} c_m (k+m) x^{k+m-1}, \qquad y'' = \sum_{m=0}^{\infty} c_m (k+m)(k+m-1) x^{k+m-2}$$

so that Eq. (213) reduces to

$$\sum_{m=0}^{\infty} c_m (k+m)(k+m-1+c) x^{k+m-1} - \sum_{m=0}^{\infty} c_m (k+m+a)(k+m+b) x^{k+m} = 0 \tag{215}$$

From Eq. (215), the indicial equation (as $c_0 \neq 0$) is

$$k(k - 1 + c) = 0$$

and the roots of this equation are $k = 0, 1 - c$.

The recurrence relation, from Eq. (215), is

$$c_m = \frac{(k+m-1+a)(k+m-1+b)}{(k+m)(k+m-1+c)} c_{m-1} \tag{216}$$

For $m = 1, 2, 3, \ldots$, Eq. (216) leads to

$$c_1 = \frac{(k+a)(k+b)}{(k+1)(k+c)} c_0$$

$$c_2 = \frac{(k+a)(k+b)(k+1+a)(k+1+b)}{(k+1)(k+2)(k+c)(k+1+c)} c_0 \tag{217}$$

and so on. Thus, from Eqs. (217) and (214), a solution of the given equation, when $k = 0$, is

$$y = c_0 \left[ 1 + \frac{ab}{c} x + \frac{ab(1+a)(1+b)}{1 \cdot 2 \cdot c(1+c)} x^2 + \cdots \right] \tag{218}$$

while, for $k = 1 - c$, another solution of the given equation, from Eqs. (217) and (214), is

$$y = c_0 x^{1-c} \left[ 1 + \frac{(1-c+a)(1-c+b)}{1 \cdot (1-c)} x + \frac{(1-c+a)(1-c+b)(2-c+a)(2-c+b)}{1 \cdot 2 \cdot (2-c)(3-c)} x^2 + \cdots \right] \tag{219}$$

Put

$$a' = 1 - c + a, \qquad b' = 1 - c + b, \qquad c' = 2 - c \tag{220}$$

then Eq. (219) takes the form

$$y = c_0 x^{1-c} \left[ 1 + \frac{a'b'}{1 \cdot c'} x + \frac{a'b'(1+a')(1+b')}{1 \cdot 2 \cdot c'(1+c')} x^2 + \cdots \right] \tag{221}$$

If we take $c_0 = A$ in Eq. (218) and $c_0 = B$ in Eq. (221), then the general solution of the given equation is

$$y = A\left(1 + \frac{ab}{c}x + \cdots\right) + Bx^{1-c}\left(1 + \frac{a'b'}{c'}x + \cdots\right) \tag{222}$$

where $a'$, $b'$ and $c'$ are given by Eq. (220).

**REMARKS.** 1. If we take $c_0 = 1$ in Eq. (218), then the series on the right-hand side of Eq. (218) is known as *hypergeometric series*. Equation (218) can also be expressed as

$$y = \sum_{n=0}^{\infty} \frac{(a)_n (b)_n}{n!(c)_n} x^n = {}_2F_1(a, b; c; x) \tag{223}$$

where $(a)_n = a(a + 1) \cdots (a + n - 1)$.

Equation (223) is a particular solution of Eq. (213) and is known as *hypergeometric function*.

2. When $a = 1$ and $b = c$, then Eq. (223) leads to

$$y = {}_2F_1(1, b; b; x) = 1 + x + x^2 + x^3 + \cdots$$

which is a geometric series and thus the series in Eq. (223) is known as *hypergeometric series*.

## 5.13 MISCELLANEOUS SOLVED EXAMPLES

In this section, we shall solve some more problems based on the different sections of this chapter, and we have

**EXAMPLE 5.58** Solve the following differential equations:

(i) $(D^3 - D)y = 2x + 1$

(ii) $(D^3 + 2D^2 + D)y = x^2 + x$

***Solution*** (i) Here

$$y_c = c_1 + c_2 e^{-x} + c_3 e^x$$

and

$$\begin{aligned} y_p &= \frac{1}{D^3 - D}(2x + 1) \\ &= -\frac{1}{D}(1 - D^2)^{-1}(2x + 1) \\ &= -\frac{1}{D}(1 + D^2 + \ldots)(2x + 1) \\ &= -(x^2 + x) \end{aligned}$$

Therefore, the required solution is

$$y = c_1 + c_2 e^{-x} + c_3 e^{x} - x^2 - x$$

(ii) Here

$$y_c = c_1 + (c_2 + c_3 x)e^{-x}$$

and

$$y_p = \frac{1}{D^3 + 2D^2 + D}(x^2 + x)$$

$$= \frac{1}{D}\{1 + (2D + D^2)\}^{-1}(x^2 + x)$$

$$= \frac{1}{D}(x^2 - 3x + 4)$$

$$= \frac{x^3}{3} - \frac{3x^2}{2} + 4x$$

Therefore, the required solution is

$$y = c_1 + (c_2 + c_3 x)e^{-x} + \frac{x^3}{3} - \frac{3x^2}{2} + 4x$$

**EXAMPLE 5.59** Solve the following differential equations

(i) $(D^2 + D - 6)y = 5e^{-3x}$

(ii) $(4D^2 - 4D + 1)y = e^{x/2}$

***Solution*** (i) Here

$$y_c = c_1 e^{-3x} + c_2 e^{2x}$$

Also, $f(-3) = 0$ and $f'(-3) \neq 0$. Thus

$$y_p = \frac{1}{(D+3)(D-2)} 5e^{-3x}$$

$$= 5(D+3)^{-1}[(D-2)^{-1} e^{-3x}]$$

$$= 5(D+3)^{-1}[(-3-2)^{-1} e^{-3x}]$$

$$= -(D+3)^{-1} e^{-3x} \cdot 1$$

$$= -e^{-3x}(D - 3 + 3)^{-1} \cdot 1$$

$$= -e^{-3x} D^{-1}(1) = -xe^{-3x}$$

Therefore, the required solution is

$$y = c_1 e^{-3x} + c_2 e^{2x} - xe^{-3x}$$

(ii) Here

$$y_c = (c_1 + c_2 x)e^{x/2}$$

Also, $f\left(\frac{1}{2}\right) = 0$. Thus

$$y_p = (2D-1)^{-2} e^{x/2}$$

$$= e^{x/2}[2(D+\frac{1}{2})-1]^{-2}(1)$$

$$= \frac{1}{4} e^{x/2} D^{-2}(1) = \frac{1}{8} x^2 e^{x/2}$$

Therefore, the solution is

$$y = (c_1 + c_2 x)e^{x/2} + \frac{1}{8} x^2 e^{x/2}$$

**EXAMPLE 5.60** Solve:

(i) $(D^3 - 3D^2 + 4D - 2)y = \cos x$

(ii) $(D^2 - 4D - 5)y = 3\cos(4x+3)$

(iii) $(D^3 - D^2 + 4D - 4)y = \sin 3x$

***Solution*** (i) Here

$$y_c = c_1 e^x + e^x (c_2 \cos x + c_3 \sin x)$$

and

$$y_p = \frac{1}{D^3 - 3D^2 + 4D - 2} \cos x$$

$$= \frac{1}{(-1)D - 3(-1) + 4D - 2} \cos x$$

$$= \frac{1}{3D+1} \cos x = \frac{(3D-1)}{(3D+1)(3D-1)} \cos x$$

$$= \frac{3D-1}{9D^2 - 1} \cos x = \frac{3D-1}{9(-1)-1} \cos x$$

$$= -\frac{1}{10}(3D-1)\cos x = \frac{1}{10}(3\sin x + \cos x)$$

Therefore, the required solution is

$$y = c_1 e^x + e^x (c_2 \cos x + c_3 \sin x) + \frac{1}{10}(3\sin x + \cos x)$$

(ii) Here

$$y_c = c_1 e^{-x} + c_2 e^{5x}$$

and $$y_p = \frac{1}{D^2-4D-5} 3\cos(4x+3)$$

$$= -\frac{3}{4D+21}\cos(4x+3)$$

$$= -\frac{3(4D-21)}{(4D+21)(4D-21)}\cos(4x+3)$$

$$= -\frac{3(4D-21)}{16D^2-441}\cos(4x+3) = \frac{3(4D-21)}{697}\cos(4x+3)$$

$$= -\frac{3}{697}[16\sin(4x+3) - 21\cos(4x+3)]$$

Therefore, the solution is

$$y = c_1e^{-x} + c_2e^{5x} - \frac{3}{697}[16\sin(4x+3) - 21\cos(4x+3)]$$

(iii) Here

$$y_c = c_1e^x + c_2\cos 2x + c_3\sin 2x$$

and $$y_p = \frac{1}{D^3 - D^2 + 4D - 4}\sin 3x = -\frac{1}{5}\frac{1}{D-1}\sin 3x$$

$$= -\frac{1}{5}\frac{(D+1)}{(D-1)(D+1)}\sin x = -\frac{1}{5}\frac{D+1}{D^2-1}\sin 3x$$

$$= \frac{1}{50}(3\cos 3x + \sin 3x)$$

Therefore, the required solution is

$$y = c_1e^x + c_2\cos 2x + c_3\sin 2x + \frac{1}{50}(3\cos 3x + \sin 3x)$$

**EXAMPLE 5.61** Solve:

(i) $(D^2 - 3D + 2)y = xe^{3x} + \sin 2x$

(ii) $(D^2 + 2)y = x^2e^{3x} + e^x\cos 2x$

(iii) $(D^4 - 1)y = e^x\cos x$

(iv) $(D^2 - 5D + 6)y = (x + e^x)x$

***Solution*** (i) Here

$$y_c = c_1e^x + c_2e^{2x}$$

and $$y_p = \frac{1}{D^2-3D+2}xe^{3x} + \frac{1}{D^2-3D+2}\sin 2x$$

$$= e^{3x}\frac{1}{(D+3)^2 - 3(D+3) + 2}x + \frac{1}{-2^2 - 3D + 2}\sin 2x$$

$$= e^{3x} \frac{1}{D^2 + 3D + 2} x - \frac{1}{3D + 2} \sin 2x$$

$$= e^{3x} \frac{1}{2\left(1 + \frac{3}{2}D + \frac{1}{2}D^2\right)} x - \frac{3D - 2}{(3D + 2)(3D - 2)} \sin 2x$$

$$= \frac{1}{2} e^{3x} [1 + \left(\frac{3}{2}D + \frac{1}{2}D^2\right)]^{-1} x - \frac{(3D - 2)}{9D^2 - 4} \sin 2x$$

$$= \frac{1}{2} e^{3x} \left(1 - \frac{3}{2}D\right) x + \frac{1}{40}(3D - 2) \sin 2x$$

$$= \frac{1}{2} e^{3x} \left(x - \frac{3}{2}\right) + \frac{1}{20}(3\cos 2x - 2\sin 2x)$$

Therefore, the solution is

$$y = c_1 e^x + c_2 e^{2x} + \frac{1}{2} e^{3x} (x - \frac{3}{2}) + \frac{1}{20}(3\cos 2x - 2\sin 2x)$$

(ii) Here,

$$y_c = c_1 \cos\sqrt{2}x + c_2 \sin\sqrt{2}x$$

and
$$y_p = \frac{1}{D^2 + 2} x^2 e^{3x} + \frac{1}{D^2 + 2} e^x \cos 2x$$

$$= e^{3x} \frac{1}{(D + 3)^2 + 2} x^2 + e^x \frac{1}{(D + 1)^2 + 2} \cos 2x$$

$$= e^{3x} \frac{1}{D^2 + 6D + 11} x^2 + e^x \frac{1}{D^2 + 2D + 3} \cos 2x$$

$$= \frac{1}{11} e^{3x} [1 + (\frac{6}{11}D + \frac{1}{11}D^2)]^{-1} x^2 + e^x \frac{1}{2D - 1} \cos 2x$$

$$= \frac{1}{11} e^{3x} [1 - \frac{6}{11}D + \frac{25}{121}D^2)] x^2 + e^x \frac{2D + 1}{4D^2 - 1} \cos 2x$$

$$= \frac{1}{11} e^{3x} (x^2 - \frac{12}{11}x + \frac{25}{121}) - e^x \frac{1}{17}(\cos 2x - 4\sin 2x)$$

Therefore, the required solution is

$$y = c_1 \cos\sqrt{2}x + c_2 \sin\sqrt{2}x + \frac{1}{11} e^{3x} (x^2 - \frac{12}{11}x + \frac{25}{121}) - e^x \frac{1}{17}(\cos 2x - 4\sin 2x)$$

(iii) Here

$$y_c = c_1e^{-x} + c_2e^x + c_3\cos x + c_4\sin x$$

and
$$y_p = \frac{1}{D^4 - 1}e^x\cos x = e^x\frac{1}{(D+1)^4 - 1}\cos x$$

$$= e^x\frac{1}{D^4 + 4D^3 + 6D^2 + 4D}\cos x = -\frac{1}{5}e^x\cos x$$

Therefore, the required solution is

$$y = c_1e^{-x} + c_2e^x + c_3\cos x + c_4\sin x - \frac{1}{5}e^x\cos x$$

(iv) Here

$$y_c = c_1e^{2x} + c_2e^{3x}$$

and
$$y_p = \frac{1}{D^2 - 5D + 6}x^2 + e^x\frac{1}{D^2 - 5D + 6}x$$

$$= \frac{1}{6}\left[1 - \left(\frac{5}{6}D - \frac{1}{6}D^2\right)\right]^{-1}x^2 + e^x\frac{1}{(D+1)^2 - 5(D+1) + 6}x$$

$$= \frac{1}{6}\left[1 + \left(\frac{5}{6}D - \frac{1}{6}D^2\right) + \left(\frac{5}{6}D - \frac{1}{6}D^2\right)^2 + \cdots\right]x^2 + e^x\frac{1}{(D^2 - 3D + 2)}x$$

$$= \frac{1}{6}\left(x^2 + \frac{5}{3}x - \frac{1}{3} + \frac{25}{18}\right) + \frac{1}{2}e^x\left[1 - \left(\frac{3}{2}D - \frac{1}{2}D^2\right)\right]^{-1}x$$

$$= \frac{1}{108}(18x^2 + 30x + 19) + \frac{1}{4}e^x(2x - 3)$$

Therefore, the solution is

$$y = c_1e^{2x} + c_2e^{3x} + \frac{1}{108}(18x^2 + 30x + 19) + \frac{1}{4}e^x(2x - 3)$$

**EXAMPLE 5.62** Solve:

(i) $(D^2 - 1)y = xe^x\sin x$
(ii) $(D^2 - 4D + 4)y = 8x^2e^{3x}\sin 2x$

***Solution*** (i) Here

$$y_c = c_1e^{-x} + c_2e^x$$

and $$y_p = \frac{1}{D^2 - 1} xe^x \sin x$$

$$= x\frac{1}{D^2-1}e^x \sin x - \frac{2D}{(D^2-1)^2}e^x \sin x \qquad \text{[from Eq. (54)]}$$

$$= xe^x \frac{1}{(D+1)^2 - 1}\sin x - 2D\left[e^x \frac{1}{\{(D+1)^2-1\}^2}\sin x\right]$$

$$= xe^x \frac{1}{D^2+2D}\sin x - 2D\left[e^x \frac{1}{(D^2+2D)^2}\sin x\right]$$

$$= xe^x \frac{1}{2D-1}\sin x - 2D\left[e^x \frac{1}{(2D-1)^2}\sin x\right]$$

$$= xe^x \frac{(2D+1)}{(4D^2-1)}\sin x - 2D\left[e^x \frac{(2D+1)^2}{(4D^2-1)^2}\sin x\right]$$

$$= -\frac{1}{5}xe^x(2\cos x + \sin x) - 2D\left[\frac{1}{25}e^x(4D^2+4D+1)\sin x\right]$$

$$= -\frac{1}{25}e^x[2(1+5x)\cos x + 5x - 14\sin x]$$

Therefore, the required solution is

$$y = c_1e^{-x} + c_2e^x - \frac{1}{25}e^x[2(1+5x)\cos x + 5x - 14\sin x]$$

(ii) Here

$$y_c = (c_1 + c_2x)e^{2x}$$

and $$y_p = \frac{1}{D^2 - 4D + 4}8x^2e^{2x}\sin 2x$$

$$= 8e^{2x}\frac{1}{(D+2)^2 - 4(D+2) + 4}x^2 \sin 2x$$

$$= 8e^{2x}\frac{1}{D^2}(x^2 \sin 2x)$$

$$= 8e^{2x}(-2x^2 \sin 2x + 3\sin 2x - 4x\cos 2x)$$

(after integrating by parts and simplification). Therefore, the required solution is

$$y = (c_1 + c_2x)e^{2x} + 8e^{2x}(-2x^2 \sin 2x + 3\sin 2x - 4x\cos 2x)$$

**EXAMPLE 5.63** Solve $y'' + 16y = 32\sec 2x$ by variation of parameter method.

***Solution*** Here

$$y_c = c_1 \cos 4x + c_2 \sin 4x \tag{224}$$

Let

$$y_1 = \cos 4x,\ y_2 = \sin 4x \tag{225}$$

then from Eq. (74), after simplification, we get

$$u' = -16\sin 2x,\ v' = 8(2\cos 2x - \sec 2x)$$

which leads to

$$u = 8\cos 2x,\ v = 8\sin 2x - 4\log(\sec 2x + \tan 2x) \tag{226}$$

Using Eqs. (225) and (226) in Eq. (69), we get

$$y_p = 8\cos 2x \cos 4x + [8\sin 2x - 4\log(\sec 2x + \tan 2x)]\sin 4x$$

Therefore, the required solution is

$$y = c_1 \cos 4x + c_2 \sin 4x + 8\cos 2x \cos 4x + [8\sin 2x - 4\log(\sec 2x + \tan 2x)]\sin 4x$$

**REMARK.** It may be noted the above method can also be applied to the differential equations with variable coefficients. This method can also be extended to differential equations of any order. Consider, for example, a third order differential equation as

$$a_3 \frac{d^3y}{dx^3} + a_2 \frac{d^2y}{dx^2} + a_1 \frac{dy}{dx} + a_0 y = Q(x),\ a_3 \neq 0 \tag{227}$$

where $Q(x)$ is a continuous function of $x$ and is non-zero. Here

$$y_c = c_1 y_1(x) + c_2 y_2(x) + c_3 y_3(x) \tag{228}$$

and

$$y_p = u(x) y_1(x) + v(x) y_2(x) + w(x) y_3(x) \tag{229}$$

Similar to Eq. (73), we have

$$\begin{aligned} u'y_1 + v'y_2 + w'y_3 &= 0 \\ u'y_1' + v'y_2' + w'y_3' &= 0 \\ u'y_1'' + v'y_2'' + w'y_3'' &= q(x),\ q(x) = \frac{Q(x)}{a_3} \end{aligned} \tag{230}$$

Solving Eq. (230) for $u$, $v$ and $w$ and using Eqs. (228) and (229), the solution of Eq. (227) is

$$y = y_c + y_p$$

**EXAMPLE 5.64** Solve $y''' - 6y'' + 11y' - 6y = e^{-x}$.

***Solution*** Here

$$y_c = c_1 e^x + c_2 e^{2x} + c_3 e^{3x}$$

and

$$y_1 = e^x,\ y_2 = e^{2x},\ y_3 = e^{3x}$$

$$y_1' = e^x, y_2' = 2e^{2x}, y_3' = 3e^{3x}$$

$$y_1'' = e^x, y_2'' = 4e^{2x}, y_3'' = 9e^{3x}$$

Thus, Eq. (230) leads to

$$u'e^x + v'e^{2x} + w'e^{3x} = 0$$

$$u'e^x + 2v'e^{2x} + 3w'e^{3x} = 0$$

$$u'e^x + 4v'e^{2x} + 9w'e^{3x} = e^{-x}$$

Solving these equations, we have

$$u' = \frac{1}{2}e^{-2x}, v' = -e^{-3x}, w' = \frac{1}{2}e^{-4x}$$

so that

$$u = -\frac{1}{4}e^{-2x}, v = \frac{1}{3}e^{-3x}, w = -\frac{1}{8}e^{-4x}$$

Thus, from Eq. (229), we have

$$y_p = -\frac{1}{24}e^{-x}$$

Therefore, the solution is

$$y = c_1e^x + c_2e^{2x} + c_3e^{3x} - \frac{1}{24}e^{-x}$$

**EXAMPLE 5.65** Solve $y'' - 4y' + 13y = 12e^{2x}\sin 3x$ by the method of undetermined coefficients.

***Solution*** Here

$$y_c = c_1e^{2x}\cos 3x + c_2e^{2x}\sin 3x$$

and $Q(x) = 12e^{2x}\sin 3x$. Since $e^{2x}\sin 3x$ appears in $y_c$ and $Q(x)$, therefore we choose

$$y_p = xe^{2x}(A\cos 3x + B\sin 3x)$$

Now substituting the values of $y_p, y_p'$ and $y_p''$ in the given differential equation and comparing the coefficients on both sides, we get after simplification $A = -2$ and $B = 0$. Thus

$$y_p = -2xe^{2x}\cos x$$

Therefore, the required solution is

$$y = c_1e^{2x}\cos 3x + c_2e^{2x}\sin 3x - 2xe^{2x}\cos x$$

**EXAMPLE 5.66** Solve $y''' - 6y'' + 12y' - 8y = 12e^{2x} + 27e^{-x}$ by the method of undetermined coefficients.

***Solution*** Here

$$y_c = c_1e^{2x} + c_2xe^{2x} + c_3x^2e^{2x}$$

and $Q(x) = 12e^{2x} + 27e^{-x}$. As $Q(x)$ contains a term $e^{2x} = x^0e^{2x}$ which is in $y_c$ due to the multiple root, $y_p$ will contain terms $x^3e^{2x}$ and all its linearly independent derivatives. Thus

$$y_p = Ax^3e^{2x} + Be^{-x}$$

Now find the values of $y_p', y_p''$ and $y_p'''$, substitute them in the given differential equation and after comparing the coefficients on both sides of the resulting equation, we obtain $A = 2$ and $B = -1$. Thus

$$y_p = 2x^3e^{2x} - e^{-x}$$

Therefore, the required solution is

$$y = (c_1 + c_2x + c_3x^2)e^{2x} + 2x^3e^{2x} - e^{-x}$$

**EXAMPLE 5.67** Using Frobenius method, solve the following differential equations:

(i) $2x^2y'' + xy' - (1 + x)y = 0$
(ii) $2x^2y'' + xy' - (1 + x^2)y = 0$

***Solution*** (i) Assume that the given differential equation has a solution of the form

$$y = \sum_{m=0}^{\infty} c_m x^{k+m}, \quad c_0 \neq 0 \tag{231}$$

then

$$y' = \sum_{m=0}^{\infty} (k+m)c_m x^{k+m-1}, y'' = \sum_{m=0}^{\infty} (k+m)(k+m-1)c_m x^{k+m-2} \tag{232}$$

From Eqs. (231) and (232), the given equation reduces to

$$\sum_{m=0}^{\infty} [2(k+m)+1](k+m-1)c_m x^{k+m} - \sum_{m=0}^{\infty} c_m x^{k+m+1} = 0 \tag{233}$$

Put $m = 0$ in Eq. (233) and equate the coefficient of lowest power of $x$ to zero, we get the indicial equation

$$(2k+1)(k-1) = 0$$

whose roots are $k = 1, -1/2$ which are different and do not differ by an integer. From Eq. (233), the recurrence relation is

$$c_m = \frac{1}{(2k+2m+1)(k+m-1)} c_{m-1} \tag{234}$$

For $m=1,2,3,\ldots$, Eq. (234) leads to

$$c_1 = \frac{1}{k(2k+3)} c_0, c_2 = \frac{1}{k(2k+3)(2k+5)(k+1)} c_1$$

etc. Substituting these values in Eq. (231), we get

$$y = c_0 x^k \left[ 1 + \frac{1}{k(2k+3)} x + \frac{1}{k(k+1)(2k+3)(2k+5)} x^2 + \cdots \right] \quad (235)$$

Put $k = 1$ and take $c_0 = A$ in Eq. (235), we get

$$y = Ax\left(1 + \frac{1}{5}x + \frac{1}{70}x^2 + \cdots\right) = Ay_1 \quad (236)$$

Put $k = -1/2$ and take $c_0 = B$ in Eq. (235), we get

$$y = Bx^{-1/2}\left(1 - x - \frac{1}{2}x^2 + \cdots\right) = By_2 \quad (237)$$

From Eqs. (236) and (237), the required solution is

$$y = Ay_1 + By_2$$

(ii) Let a series solution of the given equation be of the form given by Eq. (231). Substituting the values of $y, y'$ and $y''$ from Eqs. (231) and (232) in the given equation, we get after simplification

$$\sum_{m=0}^{\infty} [2(k+m)(k+m-1) + (k+m) - 1] c_m x^{k+m} - \sum_{m=0}^{\infty} c_m x^{k+m+2} = 0 \quad (238)$$

Put $m = 0$ in Eq. (238) and equate the coefficient of lowest power of $x$ to zero, we get the indicial equation

$$2k^2 - k - 1 = 0$$

whose roots are $k = 1, -1/2$, which are different and do not differ by an integer.

Put $m = 1$ in Eq. (238) and equate the coefficient of lowest power of $x$ to zero, we get

$$c_1(2k^2 + 3k) = 0$$

For $k = 1, -1/2$, this equation leads to $c_1 = 0$. Thus, Eq. (238) can be expressed as

$$\sum_{m=2}^{\infty} [2(k+m)(k+m-1) + (k+m) - 1] c_m x^{k+m} - \sum_{m=0}^{\infty} c_m x^{k+m+2} = 0$$

Put $i = m+2$ in the second term of this equation, we get

$$\sum_{m=2}^{\infty} [2(k+m)(k+m-1) + (k+m) - 1] c_m x^{k+m} - \sum_{i=2}^{\infty} c_{i-2} x^{k+i} = 0$$

As $i$ is a dummy variable, this equation can be written as

$$\sum_{m=2}^{\infty}\{[2(k+m)(k+m-1)+(k+m)-1]c_m - c_{m-2}\}x^{k+m} = 0$$

which leads to

$$c_m = \frac{1}{(k+m)(2k+2m-1)-1}c_{m-2} \tag{239}$$

Since $c_{-1} = 0$, Eq. (239) leads to $c_{-3} = c_{-5} = c_{-7} = ... = 0$.
For $k = 1$, Eq. (239) leads to

$$c_m = \frac{1}{(m+1)(2m+1)-1}c_{m-2}$$

so that

$$c_2 = \frac{1}{14}c_0,\ c_4 = \frac{1}{616}c_0, \ldots.$$

while for $k = -1/2$, Eq. (239) leads to

$$c_m = \frac{1}{\left(m-\dfrac{1}{2}\right)(2m-2)-1}c_{m-2}$$

so that

$$c_2 = \frac{1}{2}c_0,\ c_4 = \frac{1}{40}c_0, \ldots$$

Thus, the two solutions, from Eq. (231) are

$$y = Ax\left(1+\frac{1}{14}x^2+\frac{1}{616}x^4+\cdots\right) = Ay_1,\ (A = c_0)$$

and

$$y = Bx^{-1/2}\left(1+\frac{1}{2}x^2+\frac{1}{40}x^4+\cdots\right) = By_2,\ (B = c_0)$$

Therefore, the required solution is

$$y = Ay_1 + By_2$$

**EXAMPLE 5.68** Using Frobenius method, solve the following differential equations:

(i) $x^2y'' + 6xy' + (x^2+6)y = 0$

(ii) $x^2y'' - xy' - \left(x^2+\dfrac{5}{4}\right)y = 0$

***Solution*** (i) Let a series solution of the given equation be of the form given by Eq. (231). Now, substitute the values of $y, y'$ and $y''$ from Eqs. (231) and (232) in the given differential equation, we get after simplification

$$\sum_{m=0}^{\infty}[(k+m)(k+m+5)+6]c_m x^{k+m} - \sum_{m=0}^{\infty}c_m x^{k+m+2} = 0 \tag{240}$$

Put $m = 0$ in Eq. (240) and equate the coefficient of lowest power of $x$ to zero, we get

$$k^2 + 5k + 6 = 0$$

which is the indicial equation having the roots as $k = -2, -3$. Here the roots are different and differ by an integer. Take $m = m + 2$ in the first term of Eq. (240) and equate the coefficients of $x^{k+m+2}$ to zero, we get the following recurrence relation

$$c_{m+2} = -\frac{1}{(k+m+2)(k+m+7)+6}c_m \tag{241}$$

Also, by taking $m = 1$ in Eq. (240) and equating the coefficient of lowest power of $x$ to zero, we get

$$c_1[(k+1)(k+6)+6] = 0$$

For $k = -2$, this equation leads to $c_1 = 0$, while for $k = -3$, $c_1$ is indeterminate. Assuming $c_1$ to be finite, for $k = -3$, Eq. (241) leads to

$$c_{m+2} = -\frac{1}{(m-1)(m+4)+6}c_m \tag{242}$$

so that when $m = 0, 1, 2, \ldots$, this equation yields

$$c_2 = -\frac{1}{2}c_0,\ c_3 = -\frac{1}{6}c_1,\ c_4 = \frac{1}{24}c_0,\ c_5 = \frac{1}{120}c_1, \ldots$$

Thus, the solution, for $k = -3$ is

$$y = x^k(c_0 + c_1x + c_2x^2 + c_3x^3 + \cdots)$$

$$= x^{-3}\left[c_0\left(1 - \frac{x^2}{2!} + \frac{x^4}{4!} + \cdots\right) + c_1\left(x - \frac{x^3}{3!} + \frac{x^5}{5!} + \cdots\right)\right]$$

$$= c_0\left(\frac{\cos x}{x^3}\right) + c_1\left(\frac{\sin x}{x^3}\right) \tag{243}$$

While for $k = -2$ and $m = 0, 1, 2, \ldots$, Eq. (242) leads to

$$c_1 = 0, c_2 = -\frac{1}{6}c_0, c_3 = 0, c_4 = \frac{1}{120}c_0, c_5 = 0, \ldots$$

so that from Eq. (231), the solution is

$$y = c_0x^{-2}\left(1 - \frac{x^2}{3!} + \frac{x^4}{5!} + \cdots\right) = c_0x^{-3}\left(x - \frac{x^3}{3!} + \frac{x^5}{5!} + \cdots\right)$$

This solution has already appeared as one of the terms of the solution given by Eq. (243), so we reject this solution. Therefore, the required solution is given by Eq. (243).

(ii) Let a series solution of the given equation be of the form given by Eq. (231). Now, substitute the values of $y, y'$ and $y''$ from Eqs. (231) and (232) in the given differential equation, we get after simplification

$$\sum_{m=0}^{\infty}\left[(k+m)(k+m-2)-\frac{5}{4}\right]c_m x^{k+m}-\sum_{m=0}^{\infty}c_m x^{k+m+2}=0 \tag{244}$$

Put $m = 0$ in this equation and equate the coefficient of lowest power of $x$ to zero, we get the indicial equation

$$k^2-2k-\frac{5}{4}=0$$

whose roots are $k = -1/2, 5/2$. These roots are different and differ by an integer. Now put $m = 1$ in Eq. (244) and equate the coefficient of lowest power of $x$ to zero, we get

$$\left[(k+1)(k-1)-\frac{5}{4}\right]c_1=0$$

which leads to $c_1 = 0$ as $[(k + 1)(k - 1) - (5/4)] \neq 0$ for both $k = -1/2$ and $k = 5/2$. Also, from Eq. (244), the recurrence relation is

$$c_m=\frac{1}{(k+m)(k+m-2)-\dfrac{5}{4}}c_{m-2} \tag{245}$$

Put $m = 3$ and $k = -1/2$ in Eq. (245), we get $c_3 = 0/0$ which shows that for the root $k = -1/2$ one of the coefficients of the solution becomes indeterminate, thus choosing $c_3$ to be finite we have to find the solution. Eq. (245) for $k = -1/2$ leads to

$$c_m=\frac{1}{m(m-3)}c_{m-2},\ m\neq 3$$

and we thus have

$$c_2=-\frac{1}{2}c_0,\ c_4=-\frac{1}{2\cdot 4}c_0,\ c_5=\frac{1}{10}c_3,\ c_6=-\frac{1}{2\cdot 3\cdot 4\cdot 6}c_0,\ c_7=\frac{1}{280}c_3,\ldots$$

Therefore, from Eq. (231), the required solution is

$$y=c_0x^{-1/2}\left[1-\frac{x^2}{2!}-\frac{1}{2!}\frac{x^4}{4}+\cdots\right]+c_3x^{5/2}\left[1+\frac{1}{5}\frac{x^2}{2}+\frac{1}{70}\frac{x^4}{4}+\cdots\right] \tag{246}$$

The solution corresponding to the roots $k = 5/2$ has to be rejected as it is already contained in Eq. (246).

**EXAMPLE 5.69** Using Frobenius method, solve the following differential equations:

(i) $x^2y''+xy'+(x^2-1)y=0$

(ii) $(x^2+x)y''+3xy'+y=0$

***Solution*** (i) Let a series solution of the given equation be of the form given by Eq. (231). Substitute the values of $y$, $y'$ and $y''$, from Eqs. (231) and (232), the given differential equation reduces to

$$\sum_{m=0}^{\infty}(k+m-1)(k+m+1)c_m x^{k+m} + \sum_{m=0}^{\infty} c_m x^{k+m+2} = 0 \tag{247}$$

From Eq. (247), the indicial equation is

$$(k-1)(k+1) = 0$$

and the roots are $k = -1$, 1 which are different and differ by an integer. Also from Eq. (247) for $m = 1$, we have $k(k + 2)c_1 = 0$ which shows that $c_1 = 0$ for both the values of $k$. Further, from Eq. (247), the recurrence relation is

$$c_m = -\frac{1}{(k+m-1)(k+m+1)}c_{m-2} \tag{248}$$

Since $c_1 = 0$, Eq. (248) leads to $c_3 = c_5 = c_7 = \ldots = 0$; while for $m = 2, 4, 6, \ldots$, Eq. (248) leads to

$$c_2 = -\frac{1}{(k+1)(k+3)}c_0,\ c_4 = \frac{1}{(k+1)(k+3)^2(k+5)}c_0, \ldots$$

Using these values in Eq. (231), we get

$$y = c_0 x^k\left[1 - \frac{1}{(k+1)(k+3)}x^2 + \frac{1}{(k+1)(k+3)^2(k+5)}x^4 + \cdots\right] \tag{249}$$

For $k = -1$, the coefficients in the above equation become infinite, so we use the substitution $c_0 = a_0(k - k_2) = a_0(k + 1)$ in Eq. (249) to get

$$y = a_0 x^k\left[(k+1) - \frac{1}{k+3}x^2 + \frac{1}{(k+3)^2(k+5)}x^4 + \cdots\right] \tag{250}$$

Thus, corresponding to the root $k = -1$, the solution [from Eq. (250)] is

$$(y)_{k=-1} = a_0 x^{-1}\left[-\frac{1}{2}x^2 + \frac{1}{16}x^4 + \cdots\right] \tag{251}$$

To get another solution, differentiate Eq. (250) partially with respect to $k$ so that we have

$$\frac{\partial y}{\partial k} = a_0 x^k \log x\left[(k+1) - \frac{1}{k+3}x^2 + \frac{1}{(k+3)^2(k+5)}x^4 + \cdots\right]$$

$$+ a_0 x^k\left[1 + \frac{1}{(k+3)^2}x^2 - \left\{\frac{2}{(k+3)^3(k+5)} + \frac{1}{(k+3)(k+5)^2}\right\}x^4 + \cdots\right] \tag{252}$$

Put $k = 1$ in Eq. (250), we get

$$y = -2^2 a_0 x^{-1}\left(-\frac{1}{2}x^2 + \frac{1}{16}x^4 + \cdots\right)$$

which is not linearly independent. To get the linearly independent solution, put $k = -1$ in Eq. (252), we have

$$\left(\frac{\partial y}{\partial k}\right)_{k=-1} = (y)_{k=-1} \log x + a_0 x^{-1}\left[1 + \frac{1}{2}x^2 - \frac{1}{2^2 \cdot 4}\left(\frac{2}{2} + \frac{1}{4}\right)x^4 + \cdots\right] \quad (253)$$

Therefore, the required solution is

$$y = A(y)_{k=-1} + B\left(\frac{\partial y}{\partial k}\right)_{k=-1}$$

(ii) Let a series solution of the given differential equation be of the form given by Eq. (231), then using the values of $y$, $y'$ and $y''$, the given differential equation leads to

$$\sum_{m=0}^{\infty} c_m[(k+m)(k+m-1) + 3(k+m) + 1]1x^{k+m}$$

$$+\sum_{m=0}^{\infty} c_m[(k+m)(k+m-1)]x^{k+m-1} = 0 \quad (254)$$

Put $m = 0$ in Eq. (254) and equate the coefficient of lowest degree term to zero, we get

$$c_0 k(k-1) = 0$$

As $c \neq 0$, the roots are $k = 0, 1$. From Eq. (254), the recurrence relation is

$$c_{m+1} = -\frac{k+m+1}{k+m}c_m,\ m \geq 0 \quad (255)$$

For $m = 0, 1, 2, \ldots$, we thus have

$$c_1 = -\frac{k+1}{k}c_0,\ c_2 = \frac{k+2}{k}c_0,\ c_3 = -\frac{k+3}{k}c_0, \ldots$$

The solution is

$$y = c_0 x^k\left[1 - \frac{k+1}{k}x + \frac{k+2}{k}x^2 - \frac{k+3}{k}x^3 + \cdots\right] \quad (256)$$

For $k = 0$ (one of the root of the indicial equation), the coefficients in Eq. (256) become infinite. Thus, we make the substitution

$$c_0 = a_0(k - k_2) = a_0 k, (a_0 \neq 0)$$

in Eq. (256) so that

$$y = a_0 x^k[k - (k+1)x + (k+2)x^2 - (k+3)x^3 + \cdots] \quad (257)$$

Thus, corresponding to the root $k = 0$, from Eq. (257) the solution is

$$(y)_{k=0} = -a_0 x[1 - 2x + 3x^2 + \cdots] \quad (258)$$

while, the solution corresponding to the root $k = 1$, from Eq. (257) is

$$(y)_{k=1} = a_0 x[1 - 2x + 3x^2 - 4x^3 + \cdots]$$

which is not linearly independent. To get the linearly independent solution, differentiate Eq. (257) partially with respect to $k$ so that we have

$$\frac{\partial y}{\partial k} = a_0 x^k \log x[k-(k+1)x+(k+2)x^2+\cdots] + a_0 x^k [1-x+x^2-x^3+\cdots]$$

which, for $k = 0$, leads to

$$\left(\frac{\partial y}{\partial k}\right)_{k=0} = (y)_{k=0} \log x + a_0[1-x+x^2-x^3+\cdots] \tag{259}$$

Therefore, the required solution is

$$y = A(y)_{k=0} + B\left(\frac{\partial y}{\partial k}\right)_{k=0}$$

where $(y)_{k=0}$ and $\left(\frac{\partial y}{\partial k}\right)_{k=0}$ are, respectively, given by Eqs. (258) and (259).

**EXAMPLE 5.70** Show that

(i) $J_{1/2}(x) = \sqrt{\frac{2}{\pi x}} \sin x$

(ii) $J_{-(1/2)}(x) = \sqrt{\frac{2}{\pi x}} \cos x$

(iii) $[J_{1/2}(x)]^2 + [J_{-(1/2)}(x)]^2 = \frac{2}{\pi x}$

***Solution*** (i) From Eq. (193), we have

$$J_n(x) = \frac{x^n}{2^n \Gamma(n+1)}\left[1 - \frac{1}{2\cdot 2(n+1)}x^2 + \frac{1}{2\cdot 4\cdot 2^2(n+1)(n+2)}x^4 + \cdots\right] \tag{260}$$

Take $n = 1/2$ in Eq. (260), we get after simplification

$$J_{\frac{1}{2}}(x) = \frac{2\sqrt{x}}{\sqrt{2\pi}}\frac{1}{x}\left[x - \frac{1}{3!}x^3 + \frac{1}{5!}x^5 + \cdots\right] = \sqrt{\frac{2}{\pi x}} \sin x$$

(ii) Take $n = -(1/2)$ in Eq. (260), we get after simplification

$$J_{-\frac{1}{2}}(x) = \frac{x^{-1/2}}{2^{-1/2}\Gamma\left(\frac{1}{2}\right)}\left[1 - \frac{1}{2!}x^2 + \frac{1}{4!}x^4 + \cdots\right] = \sqrt{\frac{2}{\pi x}} \cos x$$

(iii) From (i) and (ii) , we get the required result.

**EXAMPLE 5.71** Express $P(x) = 4P_3(x) + 3P_2(x) + 2P_1(x) + P_0(x)$ as a polynomial in $x$, where $P_n(x)$ is the Legendre polynomial of order $n$.

***Solution*** From Eq. (210), we have

$$P_0(x)=1,\ P_1(x)=x,\ P_2(x)=\frac{1}{2}(3x^2-1),\ P_3(x)=\frac{1}{2}(5x^3-3x)$$

Substituting these values in the given expression, we get

$$P(x)=\frac{1}{2}(20x^3+9x^2-8x-1)$$

**EXAMPLE 5.72** Express $f(x)=x^4+2x^3+2x^2-x-3$ in terms of Legendre polynomials.

***Solution*** From Eq. (210), we have

$$P_0(x)=1, P_1(x)=x, P_2(x)=\frac{1}{2}(3x^2-1), P_3(x)=\frac{1}{2}(5x^3-3x),$$

$$P_4(x)=\frac{1}{8}(35x^4-30x^2+3)$$

These equations lead to

$$1=P_0(x), x=P_1(x), x^2=\frac{1}{3}[2P_2(x)+P_0(x)], x^3=\frac{1}{5}[2P_3(x)+3P_1(x)],$$

$$x^4=\frac{1}{35}[8P_4(x)+20P_2(x)+7P_0(x)]$$

Substituting these values in the given expression, we get

$$f(x)=\frac{8}{35}P_4(x)+\frac{4}{5}P_3(x)+\frac{40}{21}P_2(x)+\frac{1}{5}P_1(x)-\frac{224}{105}P_0(x)$$

which is the required result.

## EXERCISES

Solve the following differential equations:

**1.** $\dfrac{d^2y}{dx^2}+(a+b)\dfrac{dy}{dx}+aby=0.$ **2.** $\dfrac{d^3y}{dx^3}-2\dfrac{d^2y}{dx^2}+4\dfrac{dy}{dx}-8y=0.$

**3.** $\dfrac{d^3y}{dx^3}-\dfrac{d^2y}{dx^2}-\dfrac{dy}{dx}-2y=0.$ **4.** $(D^3-D^2-D+1)y=0.$

**5.** $(D^3-4D^2+5D-2)y=0.$ **6.** $(D^4-2D^3+2D^2-2D+1)y=0.$

**7.** $(D^2-2D+5)y=0$ given that $y=0$ and $dy/dx=4$ when $x=0$.

**8.** $(D^4-4D^3+8D^2-8D+4)y=0.$

Solve the following differential equations using Example 5.11:

**9.** $y''-y'-2y=e^x.$ **10.** $y''+3y'+2y=12e^x.$

Find the particular integral of the following differential equations using methods 1–7 of Section 5.3:

**11.** $y'' + 3y' + 2y = 4.$

**12.** $y'' + y' + y = x^2.$

**13.** $D^2y - 3Dy + 2y = x.$

**14.** $\dfrac{d^2y}{dx^2} + \dfrac{dy}{dx} = x^2 + 2x.$

**15.** $\dfrac{d^3y}{dx^3} - \dfrac{d^2y}{dx^2} = 2x^3.$

**16.** $y^{iv} - y''' + y'' = 6.$

**17.** $y'' + 3y' + 2y = 12e^x.$

**18.** $y'' + y = 3e^{-2x}.$

**19.** $\dfrac{d^2y}{dx^2} - y = \sin x.$

**20.** $\dfrac{d^2y}{dx^2} - y = \cos x.$

**21.** $\dfrac{d^2y}{dx^2} - 3\dfrac{dy}{dx} + 2y = 3\sin x.$

**22.** $4\dfrac{d^2y}{dx^2} - 5\dfrac{dy}{dx} = x^2e^x.$

**23.** $\dfrac{d^2y}{dx^2} + \dfrac{dy}{dx} + y = 3x^2e^x.$

**24.** $\dfrac{d^3y}{dx^3} + 3\dfrac{d^2y}{dx^2} + 3\dfrac{dy}{dx} + y = e^{-x}(2 - x^2).$

**25.** $y'' - y = 2e^x.$

**26.** $y'' + y' - 2y = 3e^{-2x}.$

**27.** $y^{iv} - 3y''' - 6y'' + 28y' - 24y = e^{2x}.$

**28.** $y'' - 2y' + y = 7e^x.$

**29.** $y'' + 3y' + 2y = 8 + 6e^x + 2\sin x.$

**30.** $y'' + 3y' + 2y = 2(e^{-2x} + x^2).$

**31.** $y^{v} + 2y''' + y' = 2x + \sin x + \cos x.$

**32.** $\dfrac{dy}{dx} - 3y = x^3 + 3x - 5.$

Solve the following differential equations using methods 1–7 of Section 5.3:

**33.** $(D^2 + a^2)y = \tan ax.$

**34.** $(D^4 - m^4)y = \cos mx + \cosh mx.$

**35.** $(D^3 - D^2 - D + 1)y = x(e^{-x} + 1).$

**36.** $(D^2 + 1)y = xe^{2x}.$

**37.** $(D^2 + 1)y = e^{-x} + \cos x + x^3 + e^x\cos x.$

**38.** $(D^3 + 3D^2 + 3D + 1)y = e^{-x}.$

**39.** $(D^4 + D^2 + 16)y = 16x^2 + 256.$

**40.** $(D^2 + 1)(D^2 + 4)y = \cos\dfrac{x}{2}\cos\dfrac{3x}{2}.$

**41.** $(D^5 - D)y = 12e^x + 8\sin x - 2x.$

**42.** $(D^2 - 4D + 4)y = x^2 + e^x + \sin 2x.$

**43.** $(D^2 + 1)y = \cos x + xe^{2x} + e^x\sin x.$

Solve the following differential equations by the method of undetermined coefficients:

**44.** $\dfrac{d^2y}{dx^2} + 3\dfrac{dy}{dx} + 2y = 12e^x.$

**45.** $\dfrac{d^2y}{dx^2} + 3\dfrac{dy}{dx} + 2y = \sin x.$

**46.** $\dfrac{d^2y}{dx^2}+\dfrac{dy}{dx}+y=x^2.$

**47.** $(D^2 - 2D - 8)y = 9xe^x + 10e^{-x}.$

**48.** $(D^2 - 3D)y = 2e^{2x} \sin x.$

**49.** $(D^4 - 2D^2 + 1)y = x - \sin x.$

**50.** $(D^2 + D)y = x^2 + 2x.$

Using the method of variation of parameter, solve the following equations:

**51.** $y'' + 3y' + 2y = 12e^x.$

**52.** $y'' + 2y' + y = x^2e^{-x}.$

**53.** $y'' + y = 4x \sin x.$

**54.** $y'' + 2y' + y = e^{-x} \log x.$

**55.** $y'' - 2y' + y = e^x \log x.$

**56.** $y''' + 4y' = \sec 2x.$

**57.** $y''' - 6y'' + 12y' - 8y = \dfrac{e^{2x}}{x}.$

Use the reduction of order method to find the solution of the following equation; one solution of the homogeneous equation is given:

**58.** $y'' - \dfrac{2}{x}y' + \dfrac{2}{x^2}y = 0, \qquad y_1 = x.$

**59.** $(2x^2 + 1)y'' - 4xy' + 4y = 0, \qquad y_1 = x.$

**60.** $y'' - \dfrac{2}{x}y' + \dfrac{2}{x^2}y = x \log x, \qquad y_1 = x.$

**61.** $x^2y'' + xy' - y = x^2e^{-x}, \qquad y_1 = x.$

Solve the following differential equations:

**62.** $x^2\dfrac{d^2y}{dx^2} - x\dfrac{dy}{dx} + y = 2 \log x.$

**63.** $x^4\dfrac{d^3y}{dx^3} + 2x^3\dfrac{d^2y}{dx^2} - x^2\dfrac{dy}{dx} + xy = 1.$

**64.** $x^2\dfrac{d^2y}{dx^2} - x\dfrac{dy}{dx} + 2y = x \log x.$

**65.** $x^2\dfrac{d^2y}{dx^2} + 2x\dfrac{dy}{dx} - 20y = (x+1)^2.$

**66.** $x^2\dfrac{d^2y}{dx^2} + 4x\dfrac{dy}{dx} + 2y = e^x.$

**67.** $(2x+3)^2\dfrac{d^2y}{dx^2} - 2(2x+3)\dfrac{dy}{dx} - 12y = 6x.$

**68.** $(x+3)^2\dfrac{d^2y}{dx^2} - 4(x+3)\dfrac{dy}{dx} + 6y = \log(x+3).$

**69.** $\dfrac{d^2y}{dx^2} = x^2 \sin x.$

**70.** $\dfrac{d^3y}{dx^3} = x + \log x.$

**71.** $\dfrac{d^2y}{dx^2} = 3\sqrt{y}, \qquad y = 1, \qquad \dfrac{dy}{dx} = 2,$ when $x = 0.$

**72.** $\dfrac{d^2y}{dx^2}=\dfrac{36}{y^2}, \quad y=8, \quad \dfrac{dy}{dx}=0, \quad \text{when } x=0.$

Find the series solution of the following differential equations:

**73.** $9x(1-x)y'' - 12y' + 4y = 0.$

**74.** $(2x + x^3)y'' - y' - 6xy = 0.$

**75.** $2x(1-x)y'' + (1-x)y' + 3y = 0.$

**76.** $2x^2y'' + 3xy' - (x^2+1)y = 0.$

**77.** $(1-x^2)y'' + 2xy' + y = 0.$

**78.** $(2+x^2)y'' + xy' + (1+x)y = 0.$

**79.** $(1-x^2)y'' + 2y = 0.$

**80.** $(1-x^2)y'' - 2xy' + 2y = 0$ (Legendre's equation of order one)

**81.** $x^2y'' + xy' + (x^2-4)y = 0$ (Bessel's equation of order 2)

**82.** $x(1-x)y'' - (1+3x)y' - y = 0.$

**83.** $(x + x^2 + x^3)y'' + 3x^2y' - 2y = 0.$

**84.** Express Eq. (161) as

$$y = a_0x^k\left[k + \frac{k(k-1)}{k}x + \frac{k(k-1)}{k+1}x^2 + \cdots\right]$$

and obtain one solution $y_1$ for $k = 1$ in the above equation and the other solution $(\partial y/\partial k)_{k=0}$.

Solve the following differential equations using Frobenius method:

**85.** $(x - x^2)y'' + (1-5x)y' - 4y = 0.$

**86.** $xy'' + (1+x)y'' + 2y = 0.$

**87.** $(x^3 + x^2)y'' + xy' - 2xy = 0.$

**88.** $4(x^4 - x^2)y'' + 8x^3y' - y = 0.$

**89.** $2xy'' + y' - 2y = 0.$

**90.** $(x^3 - x)y'' + (8x^2 - 2)y' + 12xy = 0.$

**91.** $xy'' + 3y' - x^2y = 0.$

**92.** $2x^2y'' + (x^2 - x)y' + y = 0.$

**93.** $2xy'' + (1+x)y' - 2y = 0.$

**94.** $2x(x+1)y'' + 3(x+1)y' - y = 0$

**95.** $2x^2(1-x)y'' - x(1+7x)y' + y = 0.$

**96.** $2xy'' + 5(1-2x)y' - 5y = 0.$

**97.** $2x(x+3)y'' - 3(x+1)y' + 2y = 0.$

CHAPTER 6

# Applications of Higher-order Differential Equations

Higher order differential equations, which were introduced in Chapter 5, have numerous important applications. In particular, second order differential equations with constant coefficients have a number of applications in Physics, Electrical and Mechanical Engineering, Medical Sciences and Economics. This chapter is devoted to the applications of higher order differential equations to such disciplines.

## 6.1 RECTILINEAR MOTION (SIMPLE HARMONIC MOTION)

The motion of a particle moving in a straight line is described by the following differential equation:

$$F\left(t, x, \frac{dx}{dt}, \frac{d^2x}{dt^2}\right) = 0 \tag{1}$$

The general solution of Eq. (1) contains two arbitrary constants, which can be obtained from the initial conditions

$$t = 0, \qquad x = x_0, \qquad t = b, \qquad \frac{dx}{dt} = v_0 \tag{2}$$

The resulting particular solution is given by the displacement function $x = h(t)$. The domain of $h$ is the time interval. The differential Eq. (1) and the initial conditions (2) furnish a mathematical model for the physical situation.

**EXAMPLE 6.1** A particle moves on the $x$-axis with an acceleration $a = 6t - 4$ ft/s$^2$. Find the position and velocity of the particle at $t = 3$, if the particle is at origin and has a velocity 10 ft/s when $t = 0$.

***Solution*** Here, at $t = 0$, $x = 0$, $v = 10$.

Now
$$a = \frac{dv}{dt} = \frac{d^2x}{dt^2} = 6t - 4$$

which gives
$$v = 3t^2 - 4t + c_1$$

and thus $c_1 = 10$. Also
$$v = \frac{dx}{dt} = 3t^2 - 4t + 10 \tag{3}$$

which on integration yields
$$x = t^3 - 2t^2 + 10t + c_2$$

From the initial conditions, $c_2 = 0$. Thus
$$x = t^3 - 2t^2 + 10t \tag{4}$$

Putting $t = 3$ in Eqs. (3) and (4), we get
$$(v)_{t=3} = 25 \text{ ft/s}, \qquad (x)_{t=3} = 39 \text{ ft}$$

**EXAMPLE 6.2** A ball is released from a balloon which is 192 ft above the ground and rising at 64 ft/s. Find the distance over which the ball continues to rise and the time elapsed before it strikes the ground.

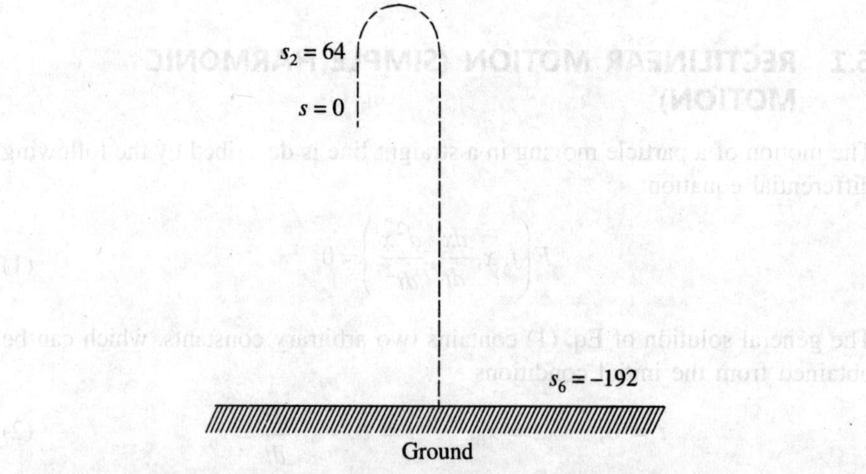


**Fig. 6.1** Motion of a ball released from a balloon.

***Solution*** Let the displacement $s$ of the ball be measured from the point at which the ball is released and $s$ be positive upward (Fig. 6.1). Also, let the time $t$ be measured from the instant the ball is released. Assume that the ball has the constant acceleration $a = dv/dt = -32$ ft/s$^2$. As the velocity decreases with time, the acceleration, known as the *acceleration due to gravity*, is negative. The initial velocity of the ball is the same as that of the balloon and thus we have to solve the initial value problem

$$a = \frac{dv}{dt} = -32; \quad \begin{Bmatrix} t = 0 \\ v = 64 \end{Bmatrix}, \quad \begin{Bmatrix} t = 0 \\ s = 0 \end{Bmatrix}$$

Integrating this, we get $v = -32t + c$, which yields $c = 64$, and hence

$$v = \frac{ds}{dt} = -32t + 64$$

Integrating again, we get

$$s = -16t^2 + 64t + k$$

and thus $k = 0$. Hence

$$s = -16t^2 + 64t$$

The ball continues to rise until $v = -32t + 64 = 0$, or $t = 2$ s. The distance travelled in the first two seconds is

$$s_2 = [-16t^2 + 64t]_{t=2} = 64 \text{ ft}$$

To find the time before the ball strikes the ground, we get $s = -192$ and solve for $t$

$$-192 = -16t^2 + 64t \quad \text{or} \quad (t - 6)(t + 2) = 0$$

Since $t \geq 0$, the time is 6 s.

In the above examples, Eq. (1) has a particularly simple form $d^2x/dt^2 = f(t)$. We shall now consider problems whose solutions involve various other forms of Eq. (1).

In Fig. 6.2, a small block of mass $m$ is attached to one end of a spring, and the other end of the unstretched spring is attached to a fixed wall. The block is pulled to the right, stretching the spring $x_0$ ft, and then released from rest (with initial velocity zero). The block moves on a smooth plane (frictionless).

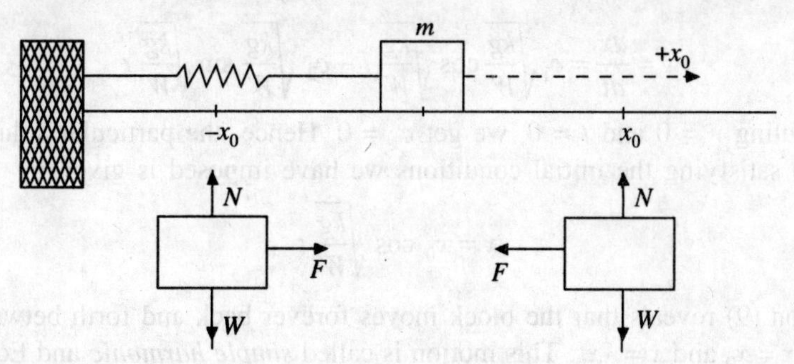


**Fig. 6.2** Movement of a block of mass $m$ attached to one end of a spring; the other end of the unstretched spring is attached to a fixed wall.

According to Hooke's law, the force (in lb) that stretches or compresses a spring is proportional to the change in length of the spring. This law is applicable to all elastic materials, provided the extensions (or compression) do not exceed the so-called elastic limit of the material. Denote the magnitude

$|\mathbf{F}|$ of this force by $k\,|x|$; $k > 0$ is known as the *spring constant*. The force $\mathbf{F} = (kx)\hat{i}$ is called the *restoring force*, since it always acts towards the origin and tends to restore the block to its original (or equilibrium) position at $x = 0$. Note that, $kx > 0$ when $x > 0$; $kx = 0$ when $x = 0$; and $kx < 0$ when $x < 0$.

The force system acting on the block when the block is to the right (and also to the left) of the equilibrium position is shown in Fig. 6.2. The vector **W** represents the force the earth exerts on the block, and vector **N** represents the force the plane exerts on the block. Each of the vectors **W** and **N**, magnitude W, and the weight of the block is measured in pounds.

By Newton's second law, the unbalanced force acting on the block in the $x$-direction equals the mass of the block times its acceleration in the $x$-direction. Since the unbalanced force in the $x$-direction has the value $-kx$,

$$m\frac{d^2x}{dt^2} = -kx \tag{5}$$

where $x$ is positive, negative or zero.

The mass $m$ is measured in slugs and $m = W/g$ ($g = 32$ ft/s$^2$). The differential equation governing the motion of the block can be written as

$$\frac{d^2x}{dt^2} + \frac{kg}{W}x = 0 \tag{6}$$

which is a linear differential equation with constant coefficients. The A.E. $m^2 + (kg/W) = 0$ has roots $\pm i\sqrt{gk/W}$ and the complete solution of Eq. (6) is

$$x = c_1 \sin\sqrt{\frac{kg}{W}}\,t + c_2 \cos\sqrt{\frac{kg}{W}}\,t \tag{7}$$

Substituting $t = 0$ and $x = x_0$, we get $c_2 = x_0$.

Differentiation of Eq. (7) with respect to $t$ yields

$$v = \frac{dx}{dt} = c_1\sqrt{\frac{kg}{W}} \cos\sqrt{\frac{kg}{W}}\,t - c_2\sqrt{\frac{kg}{W}} \sin\sqrt{\frac{kg}{W}}\,t \tag{8}$$

Substituting $v = 0$ and $t = 0$, we get $c_1 = 0$. Hence, the particular solution of Eq. (6) satisfying the initial conditions we have imposed is given by

$$x = x_0 \cos\sqrt{\frac{kg}{W}}\,t \tag{9}$$

Equation (9) reveals that the block moves forever back and forth between the points $x = x_0$ and $x = -x_0$. This motion is called *simple harmonic* and Eq. (6) is the *differential equation of the simple harmonic motion.* The time required for the block to go from $x = x_0$ to $x = -x_0$ and back again is called the *period* of the motion. It is equal to the fundamental period of the periodic function given by

$$\psi(t) = \cos\sqrt{\frac{kg}{W}}\,t$$

and has the value

$$T = \frac{\pi}{\sqrt{kg/W}} = 2\pi\sqrt{\frac{W}{kg}}\ \text{s} \tag{10}$$

The reciprocal of $T$ denotes the number of complete oscillations or cycles of the block per second and is called the *frequency* of the motion. The graph of Eq. (9) is shown in Fig. 6.3.

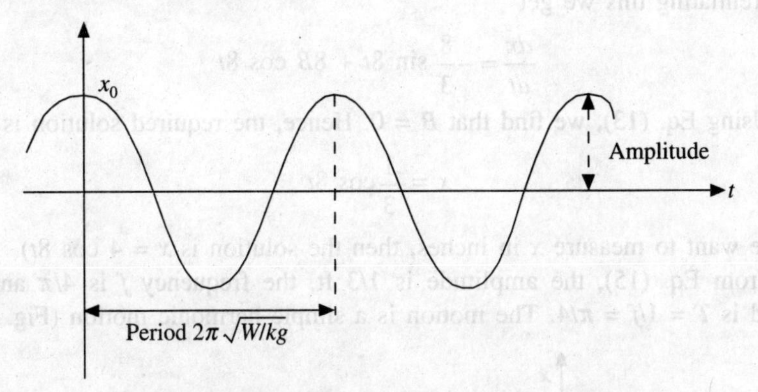


**Fig. 6.3** Graph of a simple harmonic motion.

**EXAMPLE 6.3** Experimentally it is found that a 6 lb weight stretches a certain spring 6 inches. If the weight is pulled 4 in. below the equilibrium position and released: (i) set up the differential equation and associated conditions describing the motion; (ii) find the position of the weight as a function of time; (iii) find the amplitude, period and frequency of motion; (iv) determine the position, velocity and acceleration of the weight ½ s after it has been released.

***Solution*** By Hooke's law, $F = kx$, or $6 = (½)k$ (as 6 in. = ½ ft), i.e. $k = 12$. The differential equation of the present example is

$$\frac{d^2x}{dt^2} + \frac{(12)(32)}{6}x = 0$$

or

$$\frac{d^2x}{dt^2} + 64x = 0 \tag{11}$$

At $t = 0$, the weight is 4 in. below the equilibrium position. We thus have

$$x = \frac{1}{3}\ \text{ft} \qquad \text{at } t = 0 \tag{12}$$

Also, since the weight is released (i.e. it has zero velocity) at $t = 0$

$$\frac{dx}{dt} = 0 \qquad \text{at} \quad t = 0 \tag{13}$$

The auxiliary equation for (11) is $m^2 + 64 = 0$ and has roots $\pm 8i$. Hence, the solution of Eq. (11) is

$$x = A \cos 8t + B \sin 8t \tag{14}$$

From Eq. (12), we have $A = 1/3$, so that

$$x = \frac{1}{3} \cos 8t + B \sin 8t$$

Differentiating this we get

$$\frac{dx}{dt} = -\frac{8}{3} \sin 8t + 8B \cos 8t$$

Using Eq. (13), we find that $B = 0$. Hence, the required solution is

$$x = \frac{1}{3} \cos 8t \tag{15}$$

(If we want to measure $x$ in inches, then the solution is $x = 4 \cos 8t$).

From Eq. (15), the amplitude is 1/3 ft, the frequency $f$ is $4/\pi$ and the period is $T = 1/f = \pi/4$. The motion is a simple harmonic motion (Fig. 6.4).

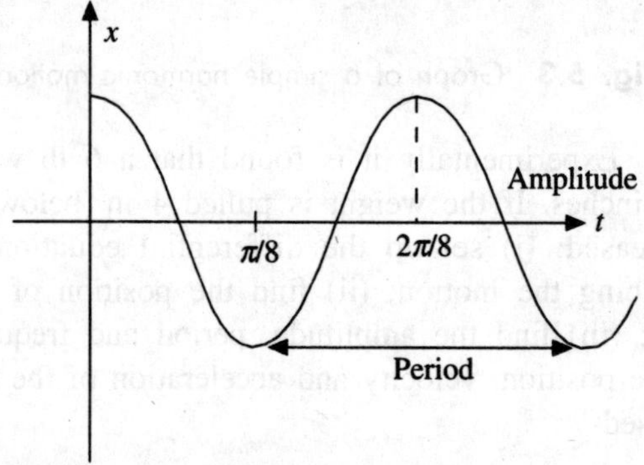


**Fig. 6.4** Graph of a simple harmonic motion illustrating Example 6.3.

Differentiate, now Eq. (15) with respect to $x$, to get

$$v = \frac{dx}{dt} = -\frac{8}{3} \sin 8t$$

$$a = \frac{d^2x}{dt^2} = -\frac{64}{3} \cos 8t$$

Put $t = 1/2$ and use 4 radians $= 4(180/\pi)$ degrees $\cong 229°$ to find

$$x = \frac{1}{3}(-0.656) = -0.219$$

$$v = -\frac{8}{3}(0.755) = 2.01$$

$$a = -\frac{64}{3}(-0.656) = 14.0$$

**EXAMPLE 6.4** In the above example, the weight is pulled 4 in. below the equilibrium position and is then given a downward velocity of 2 ft/s, instead of being released. Find the amplitude, period and frequency of the motion.

***Solution*** Here, the differential equation is

$$\frac{d^2x}{dt^2} + 64x = 0 \tag{16}$$

and the initial conditions are

$$x = \frac{1}{3}, \qquad \frac{dx}{dt} = 2 \quad \text{at} \quad t = 0 \tag{17}$$

The general solution of Eq. (16) is

$$x = A \cos 8t + B \sin 8t$$

From the first condition of Eq. (17), $A = 1/3$. Hence, $x = (1/3)\cos 8t + B \sin 8t$. Differentiating it with respect to $t$, we get

$$\frac{dx}{dt} = -\frac{8}{3} \sin 8t + 8B \cos 8t$$

and using the second condition of Eq. (17), we find $B = 1/4$. The required solution is

$$x = \frac{1}{3} \cos 8t + \frac{1}{4} \sin 8t \tag{18}$$

If $x$ is measured in inches, then Eq. (18) becomes

$$x = 4 \cos 8t + 3 \sin 8t \tag{19}$$

It is often useful to write Eq. (18) in an equivalent form by making use of the identity*

$$\left.\begin{aligned} a \cos \omega t + b \sin \omega t &= \sqrt{a^2 + b^2} \sin (\omega t + \phi) \\ \text{where} \quad \sin \phi = \frac{a}{\sqrt{a^2 + b^2}}, &\qquad \cos \phi = \frac{b}{\sqrt{a^2 + b^2}} \end{aligned}\right\} \tag{20}$$

---

* This is easy to prove, since

$$\sqrt{a^2 + b^2} \sin (\omega t + \phi) = \sqrt{a^2 + b^2} (\sin \omega t \cos \phi + \cos \omega t \sin \phi)$$

$$= \sqrt{a^2 + b^2} \left( \sin \frac{b\omega t}{\sqrt{a^2 + b^2}} + \cos \frac{a\omega t}{\sqrt{a^2 + b^2}} \right)$$

$$= a \cos \omega t + b \sin \omega t$$

($\phi$ is called the phase angle). From Eq. (20), Eq. (18) becomes

$$x = \sqrt{\left(\frac{1}{3}\right)^2 + \left(\frac{1}{4}\right)^2} \sin(8t + \phi) = \frac{5}{12} \sin(8t + \phi) \tag{21}$$

where $\sin \phi = 4/5$, $\cos \phi = 3/5$, and $\phi = 53°8'$, or 0.9274 radian. Hence, Eq. (21) becomes

$$x = \frac{5}{12} \sin(8t + 0.9274) \tag{22}$$

if $x$ is in feet, and, if $x$ is in inches, then

$$x = 5 \sin(8t + 0.9274) \tag{23}$$

The graph of Eq. (22) is shown in Fig. 6.5.

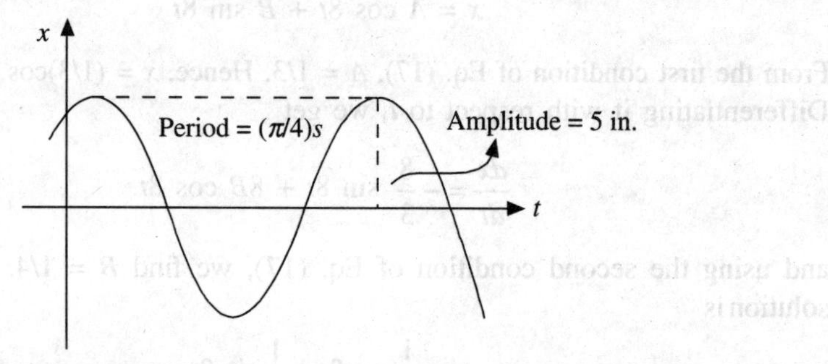


**Fig. 6.5** Graph of a simple harmonic motion illustrating Example 6.4.

The amplitude is 5 in. (or 5/12 ft), the period is $2\pi/8 = (\pi/4)$ s, and the frequency is $(4/\pi)$ cycles/s.

In general, if a motion can be described by an equation of the form $x = A \sin(\omega t + \phi)$, then

$$\left.\begin{aligned} &\text{Amplitude} = A, \qquad \text{Period } T = \frac{2\pi}{\omega} \\ &\text{Frequency} = f = \frac{1}{T} = \frac{\omega}{2\pi} \qquad (\text{or } \omega = 2\pi f) \end{aligned}\right\} \tag{24}$$

The solution of Eq. (6) can also be obtained as follows:

Let $\omega^2 = kg/W$. The acceleration can be written as

$$\frac{d^2x}{dt^2} = \frac{dv}{dt} = \frac{dv}{dx}\frac{dx}{dt} = v\frac{dv}{dx}$$

Equation (6) now takes the form

$$v\frac{dv}{dx} + \omega^2 x = 0$$

Separating the variables and integrating, we get

$$\frac{v^2}{2} = \frac{A^2\omega^2}{2} - \frac{\omega^2 x^2}{2} \quad \text{or} \quad v = \frac{dx}{dt} = \pm\omega\sqrt{A^2 - x^2}$$

Separating the variables and integrating again, we get

$$\sin^{-1}\frac{x}{A} = \pm\omega t + \phi$$

and $\quad x = A \sin(\pm\omega t + \phi) = A \cos\phi \sin(\pm\omega t) + A \sin\phi \cos(\pm\omega t)$

which is of the form (7)

$$x = c_1 \sin \omega t + c_2 \cos \omega t$$

**EXAMPLE 6.5** A particle starts from rest, a distance 10 cm from a fixed point $O$. It moves along a horizontal straight line towards $O$ under the influence of an attractive force at $O$. This force at any time varies as the distance of the particle from $O$. If the acceleration of the particle is 9 cm/s$^2$ directed towards $O$ when the particle is 1 cm from $O$, describe the motion.

***Solution*** Assume that the particle starts at $A$ (Fig. 6.6) at $t = O$. Take the fixed point $O$ as the origin of the coordinate system and choose $OA$ as the positive

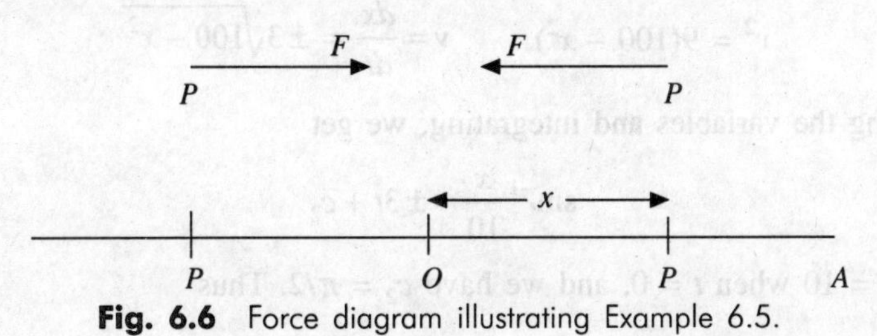


**Fig. 6.6** Force diagram illustrating Example 6.5.

direction. Let $P$ be the position of the particle at any time. Since the magnitude of the force of attraction towards point $O$ is proportional to the distance from point $O$. We have, from Newton's law, the relation

$$m\frac{d^2x}{dt^2} = -kx$$

or
$$\frac{d^2x}{dt^2} = -\frac{k}{m}x \tag{25}$$

Since, the acceleration is 9 cm/s$^2$ directed towards $O$ when the particle is 1 cm from $O$, we have $x = 1$, $a = d^2x/dt^2 = -9$. Hence, from (25) $k/m = 9$. Thus

$$\frac{d^2x}{dt^2} = -9x \tag{26}$$

As we have a second order differential equation, two conditions are required.

Since, the particle starts from rest 10 cm from $O$, we have $x = 10$, $v = 0$ at $t = 0$. Thus

$$\left.\begin{aligned} &\frac{d^2x}{dt^2} = -9x, \qquad x = 10 \\ &\frac{dx}{dt} = 0 \qquad \text{at} \quad t = 0 \end{aligned}\right\} \tag{27}$$

Let $dx/dt = v$, so that

$$\frac{d^2x}{dt^2} = \frac{dv}{dt} = \frac{dv}{dx}\frac{dx}{dt} = v\frac{dv}{dx}$$

and Eq. (26) takes the form

$$v\frac{dv}{dx} = -9x$$

Separating the variables and integrating, we get

$$v^2 = -9x^2 + c_1$$

Since $v = 0$, when $x = 10$, $c_1 = 900$. Thus

$$v^2 = 9(100 - x^2), \qquad v = \frac{dx}{dt} = \pm 3\sqrt{100 - x^2}$$

Separating the variables and integrating, we get

$$\sin^{-1}\frac{x}{10} = \pm 3t + c_2$$

Since, $x = 10$ when $t = 0$, and we have $c_2 = \pi/2$. Thus

$$\sin^{-1}\frac{x}{10} = \frac{\pi}{2} \pm 3t$$

or
$$x = 10 \cos 3t \tag{28}$$

The graph of $x$ versus $t$ is shown in Fig. 6.7. It shows that the particle starts at $x = 10$ when $t = 0$, then proceeds through $O$ to the place $x = -10$, from where it returns to $O$ again, passes through, and goes to $x = 10$. The cycle then repeats over and over again. This behaviour is similar to that of the bob of a pendulum swinging back and forth and is an example of a simple harmonic motion.

Here, amplitude is 10 cm, and from the graph, the period is $2\pi/3$. Another way to see that the period $2\pi/3$ without the graph is to determine when the particle is at an extremity of its path, for example, when $x = 10$. From Eq. (28), it is seen that this will occur when

$$\cos 3t = 1, \quad \text{or} \quad 3t = 0, 2\pi, 4\pi, \ldots$$

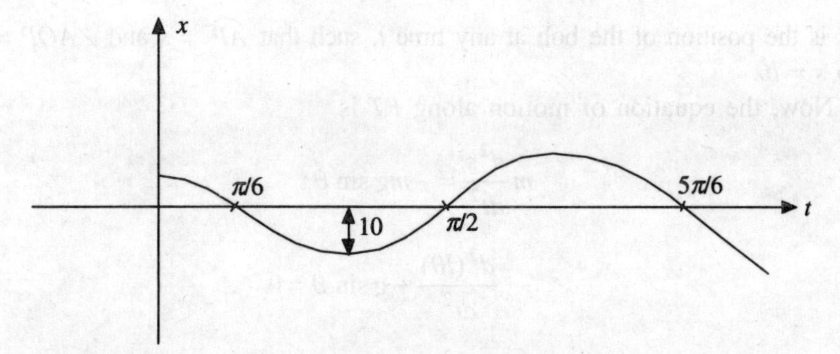


**Fig. 6.7** The simple harmonic motion illustrating Example 6.5.

Hence, the first time that $x = 10$ is $t = 0$, the next time $t = 2\pi/3$, the next time $t = 4\pi/3$, and so on. The difference between successive time is $2\pi/3$, which is the period. The frequency is $3/2\pi$ cycles/s.

REMARK. A dynamical system which vibrates with simple harmonic motion is called a *harmonic oscillator.*

## 6.2 THE SIMPLE PENDULUM

A simple pendulum consists of a particle of weight $W$ (bob) supported by a straight rod or piece of string of length $l$. The particle is free to oscillate in a vertical plane; the mass of the particle is assumed to be concentrated at a point; and the weight of the rod is assumed to be negligible.

Let $O$ be the fixed point (Fig. 6.8) and $A$ be the position of the bob initially.

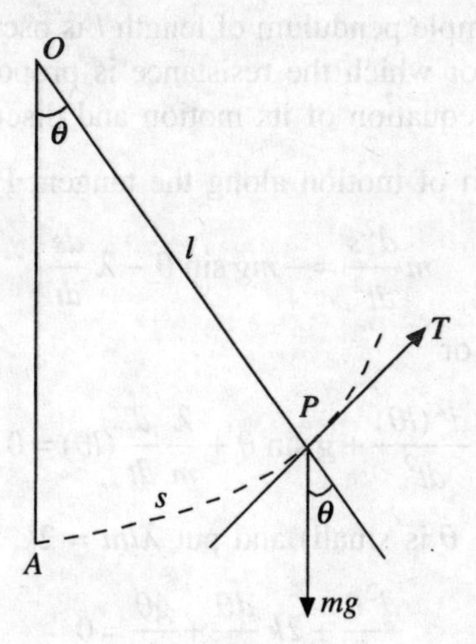


**Fig. 6.8** Schematic diagram for a simple pendulum.

If $P$ is the position of the bob at any time $t$, such that $\widehat{AP} = s$ and $\angle AOP = \theta$, then $s = l\theta$.

Now, the equation of motion along $PT$ is

$$m\frac{d^2s}{dt^2} = -mg\sin\theta$$

or
$$\frac{d^2(l\theta)}{dt^2} + g\sin\theta = 0$$

or
$$\frac{d^2\theta}{dt^2} + \frac{g}{l}\sin\theta = 0$$

or
$$\frac{d^2\theta}{dt^2} + \frac{g}{l}\left(\theta - \frac{\theta^3}{3!} + \cdots\right) = 0$$

or
$$\frac{d^2\theta}{dt^2} + \frac{g}{l}\theta = 0 \text{ (for a first approx.)}$$

The auxiliary equation has the roots $\pm i\sqrt{g/l}$, and the solution is

$$\theta = c_1\cos\sqrt{\frac{g}{l}}\,t + c_2\sin\sqrt{\frac{g}{l}}\,t$$

Therefore, the motion of a simple pendulum is simple harmonic and the time of an oscillation is $2\pi\sqrt{l/g}$.

The movement of the bob from one end to the other constitutes half an oscillation and is known as a *beat* or a *swing*. The time of one beat is $\pi\sqrt{l/g}$.

**EXAMPLE 6.6** A simple pendulum of length $l$ is oscillating through a small angle $\theta$ in a medium for which the resistance is proportional to the velocity. Obtain the differential equation of its motion and discuss the motion.

***Solution*** The equation of motion along the tangent PT (see Fig. 6.8) is

$$m\frac{d^2s}{dt^2} = -mg\sin\theta - \lambda\frac{ds}{dt}$$

where $\lambda$ is a constant, or

$$\frac{d^2(l\theta)}{dt^2} + g\sin\theta + \frac{\lambda}{m}\frac{d}{dt}(l\theta) = 0$$

Replace $\sin\theta$ by $\theta$ (as $\theta$ is small) and put $\lambda/m = 2k$, to obtain

$$\frac{d^2\theta}{dt^2} + 2k\frac{d\theta}{dt} + \frac{g\theta}{l} = 0$$

which is the required differential equation.

The auxiliary equation has the roots $-k \pm i\sqrt{W^2 - k^2}$, where $W = g/l$. The oscillatory motion of the bob is possible only when $k < W$. The solution of the present differential equation is

$$\theta = e^{-kt}\left(c_1 \cos \sqrt{W^2 - k^2}\; t + c_2 \sin \sqrt{W^2 - k^2}\; t\right)$$

which gives the vibratory motion of period $2\pi / \sqrt{W^2 - k^2}$.

**EXAMPLE 6.7** A mass weighing 2 lb stretches a spring 6 inches. At $t = 0$, the mass is released from a point 8 in. below the equilibrium position with an upward velocity of 4/3 ft/s. Determine the function $x(t)$ which describes the subsequent motion.

***Solution*** From $W = mg$, we have $m = 2/32 = 1/16$ slug. Also from Hooke's law, $2 = k(1/2)$ and $k = 4$ lb/ft. Hence, the differential equation

$$\frac{d^2x}{dt^2} + \frac{kg}{W}x = 0$$

becomes

$$\frac{d^2x}{dt^2} + \frac{(4)(32)}{2}x = 0$$

or

$$\frac{d^2x}{dt^2} + 64x = 0 \tag{29}$$

The initial displacement and velocity are

$$x(0) = \frac{2}{3}, \quad \left(\frac{dx}{dt}\right)_{t=0} = -\frac{4}{3}$$

where the negative sign in the second equation is a consequence of the fact that the mass is given an initial velocity in the negative or upward direction.

The general solution of Eq. (29) is

$$x(t) = c_1 \cos 8t + c_2 \sin 8t \tag{30}$$

Applying the initial conditions to Eq. (30), we get $c_1 = 2/3$, and

$$x(t) = \frac{2}{3}\cos 8t + c_2 \sin 8t$$

Now

$$\frac{dx}{dt} = -\frac{16}{3}\sin 8t + 8c_2 \cos 8t$$

which, from the initial conditions, yields $c_2 = -1/6$. Thus, the equation of motion is

$$x(t) = \frac{2}{3}\cos 8t - \frac{1}{6}\sin 8t$$

From Eq. (20), we have

$$x(t) = A \sin (8t + \phi)$$

The amplitude is $A = \sqrt{(2/3)^2 + (-1/6)^2} = 0.69$ ft. Also, the phase angle $\phi$ defined by

$$\tan \phi = \frac{a}{b} = \frac{2/3}{-1/6} = -4$$

or, $\tan^{-1}(-4) = \phi = -1.326$ radians (the range of the inverse tangent is $-\pi/2 < \tan^{-1} x < \pi/2$). Unfortunately, this angle is located in the fourth quadrant and, therefore, contradicts the fact that $\sin \phi > 0$ and $\cos \phi < 0$. Hence, we must take $\phi$ to be the second quadrant angle

$$\phi = \pi + (-1.326)$$

Thus, we have

$$x(t) = \frac{\sqrt{17}}{6} \sin (8t + 1.816)$$

The motion in Fig. 6.2 is not a pure simple harmonic motion since some friction must act on the block and eventually the block will come at rest. Before considering the effect of friction, we present an example showing one way in which a pure simple harmonic motion can be induced. This kind of motion appears frequently in applications.

**EXAMPLE 6.8** In Fig. 6.9, the point $P(x, y)$ moves around the circle of radius $r$ at constant speed $|\mathbf{V}|$. Assuming that $P$ starts at $P_0$ and moves counter-clockwise around the circle, show that the projection $P(x, 0)$ of $P(x, y)$ undergoes a simple harmonic motion on the $x$-axis.

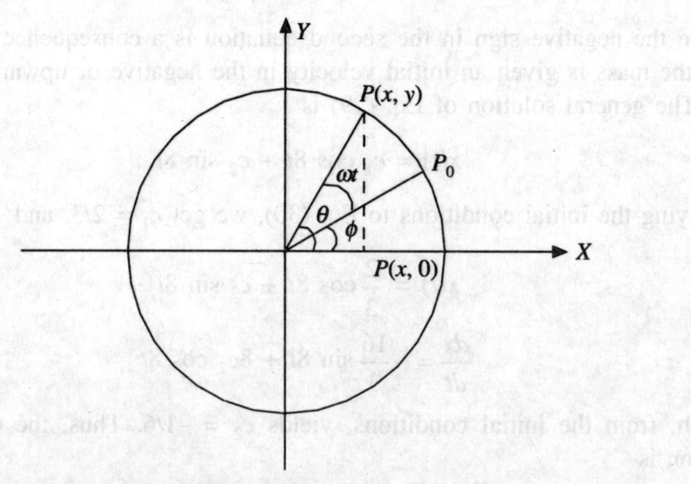


**Fig. 6.9** Movement of a point around a circle of radius *r*.

***Solution*** Let $S$ denote the arc length from $P_0$ to $P$. Then

$$S = r(\theta - \phi) = r\theta - r\phi$$

$$\frac{dS}{dt} = |\mathbf{V}| = r\frac{d\theta}{dt} = \text{constant}$$

Thus, $d\theta/dt = \omega$ is constant and $\theta = \omega t + \phi$ since $\phi = \theta$ when $t = 0$. We now have

$$x = r\cos(\omega t + \phi)$$

$$v = \frac{dx}{dt} = -r\omega \sin(\omega t + \phi)$$

$$a = \frac{dv}{dt} = \frac{d^2x}{dt^2} = -r\omega^2 \cos(\omega t + \theta)$$

Since, $d^2x/dt^2 = -\omega^2 x$, we have

$$x = c_1 \sin \omega t + c_2 \cos \omega t$$

$$v = \frac{dx}{dt} = c_1\omega \cos \omega t - c_2\omega \sin \omega t$$

and $P(x, 0)$ undergoes a simple harmonic motion. If $x = r$ and $v = 0$ when $t = 0$ then $c_2 = r$, $c_1 = 0$, and $x = r\cos \omega t$. This situation would prevail when $\phi = 0$.

## 6.3 DAMPED MOTION

The discussion of free harmonic motion is somewhat unrealistic since the motion described by Eq. (6) assumes that no retarding forces (frictional forces, air resistance) are acting on the moving mass. Unless the mass is suspended in a perfect vacuum, there will be at least a resisting force due to the surrounding medium. For example, as Fig. 6.10 shows, the mass $m$ could be suspended in a viscous medium or connected to a dashpot damping device.

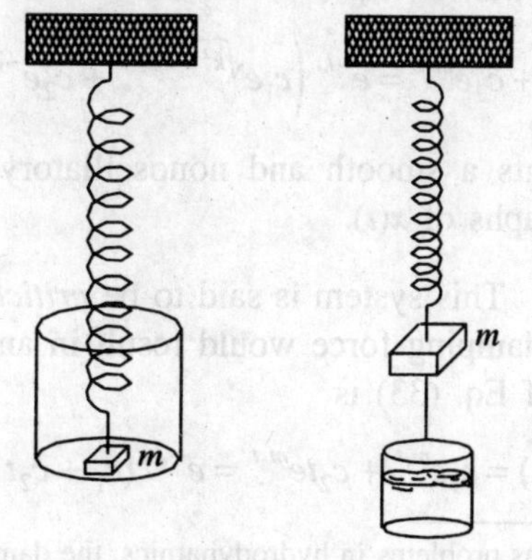


**Fig. 6.10** Dashpot damping device.

In the study of mechanics, damping forces acting on a body are considered to be proportional to a power of the instantaneous velocity; in particular, we shall assume throughout the subsequent discussion that this force is given by a constant multiple of $dx/dt$.* When no other external forces are impressed on the system, it follows from Newton's second law that

$$m\frac{d^2x}{dt^2} = -kx - \beta\frac{dx}{dt} \tag{31}$$

where $\beta$ is a positive constant and the negative sign is due to the fact that the damping force acts in a direction opposite to the motion.

Dividing Eq. (31) by $m$, the differential equation for *free, damped motion* is then

$$\frac{d^2x}{dt^2} + \frac{\beta}{m}\frac{dx}{dt} + \frac{k}{m}x = 0 \tag{32}$$

or

$$\frac{d^2x}{dt^2} + 2\lambda\frac{dx}{dt} + \omega^2 x = 0 \tag{33}$$

where

$$2\lambda = \frac{\beta}{m} = \frac{\beta g}{W}, \qquad \omega^2 = \frac{k}{m} = \frac{kg}{W} \tag{34}$$

The symbol $2\lambda$ is used only for algebraic convenience since the auxiliary equation is $m^2 + 2\lambda m + \omega^2 = 0$, and the corresponding roots are then

$$m_1 = -\lambda + \sqrt{\lambda^2 - \omega^2}, \qquad m_2 = -\lambda - \sqrt{\lambda^2 - \omega^2}$$

We can now distinguish three possible cases depending upon the algebraic sign of $\lambda^2 - \omega^2$. Since, each factor will contain the damping factor $e^{-\lambda t}$, $\lambda > 0$, the displacement of the mass will become negligible for large time.

*Case I* $\lambda^2 - \omega^2 > 0$: In this situation the system is said to be *overdamped* since the damping coefficient $\beta$ is large as compared to the spring constant $k$. The corresponding solution of Eq. (33) is

$$x(t) = c_1 e^{m_1 t} + c_2 e^{m_2 t} = e^{-kt}\left(c_1 e^{\sqrt{k^2 - \omega^2}\, t} + c_2 e^{-\sqrt{k^2 - \omega^2}\, t}\right) \tag{35}$$

This equation represents a smooth and nonoscillatory motion. Figure 6.11 shows two possible graphs of $x(t)$.

*Case II* $\lambda^2 - \omega^2 = 0$: This system is said to be *critically damped* since any slight decrease in the damping force would result in an oscillatory motion. The general solution of Eq. (33) is

$$x(t) = c_1 e^{m_1 t} + c_2 t e^{m_2 t} = e^{-\lambda t}(c_1 + c_2 t) \tag{36}$$

---

* In many instances, such as problems in hydrodynamics, the damping force is proportional to $(dx/dt)^2$.

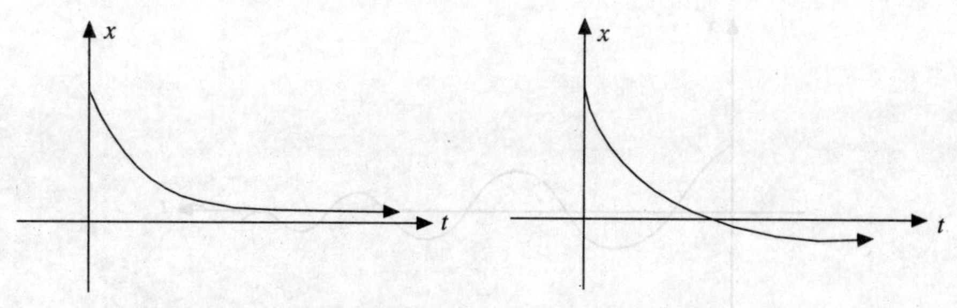


**Fig. 6.11** Graph of a smooth and non-oscillatory motion—overdamped motion.

Some graphs of the typical motion are depicted in Fig. 6.12.

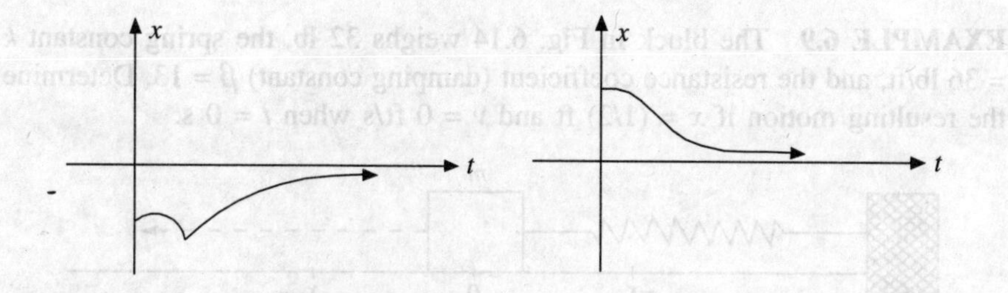


**Fig. 6.12** Critically damped motion.

Notice that the motion is quite similar to that of an overdamped system. It is also apparent from Eq. (36) that the mass can pass through the equilibrium position at the most once.*

*Case III* $\lambda^2 - \omega^2 < 0$: In this case, the system is said to be *underdamped* since the damping coefficient is small as compared to the spring constant. The roots $m_1$ and $m_2$ are now complex:

$$m_1 = -\lambda + \left(\sqrt{\omega^2 - \lambda^2}\right)i, \qquad m_2 = -\lambda - \left(\sqrt{\omega^2 - \lambda^2}\right)i$$

and the general solution of Eq. (33) is

$$x(t) = e^{-\lambda t}\left[c_1 \cos \sqrt{\omega^2 - \lambda^2}\, t + c_2 \sin \sqrt{\omega^2 - \lambda^2}\, t\right] \tag{37}$$

As indicated in Fig. 6.13, the motion described by Eq. (37) is oscillatory but, because of the coefficient $e^{-\lambda t}$, the amplitude of vibration approaches zero as $t \to \infty$.

* An examination of the derivatives of Eqs. (35) and (36) would show that these functions can have atmost one relative maximum or minimum for $t > 0$.

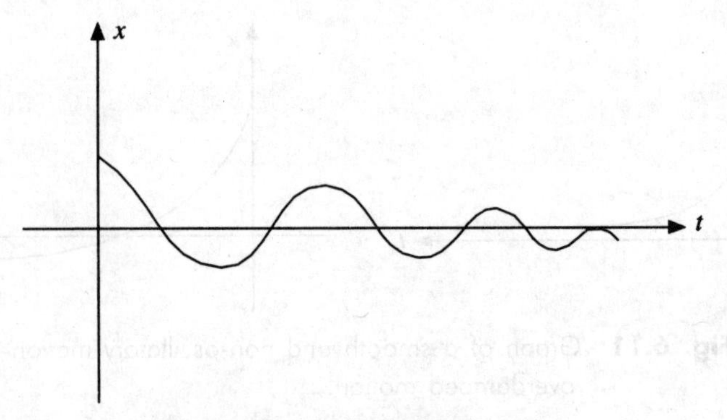


**Fig. 6.13** Underdamped motion.

**EXAMPLE 6.9** The block in Fig. 6.14 weighs 32 lb, the spring constant $k$ = 36 lb/ft, and the resistance coefficient (damping constant) $\beta = 13$. Determine the resulting motion if $x = (1/2)$ ft and $v = 0$ ft/s when $t = 0$ s.

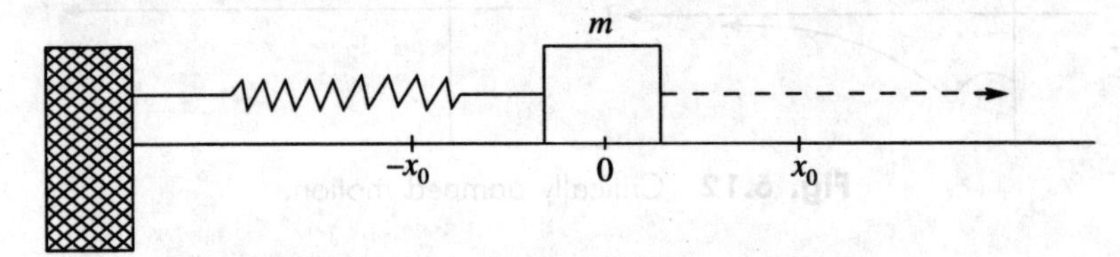


**Fig. 6.14** Movement of a block attached to a spring.

***Solution*** Here, Eq. (32) becomes

$$\frac{d^2x}{dt^2} + 13\frac{dx}{dt} + 36x = 0$$

The auxiliary equation has the roots $-4$ and $-9$ and the differential equation has the solution

$$x = c_1e^{-4t} + c_2e^{-9t} \tag{38}$$

The velocity $v = dx/dt$ is given by

$$\frac{dx}{dt} = -4c_1e^{-4t} - 9c_2e^{-9t} \tag{39}$$

Substituting $x = 1/2$ and $t = 0$ in Eq. (38) and $v = 0$, $t = 0$ in Eq. (39), we find that $c_1 = 9/10$ and $c_2 = -4/10$. Thus, the displacement $x$ is given by

$$x = \frac{9e^{-4t} - 4e^{-9t}}{10} = \frac{e^{-4t}}{10}(9 - 4e^{-5t})$$

It is easily seen that $x$ is never zero and that $x \to 0$ as $t \to \infty$. The motion is overdamped and the displacement time curve is shown in Fig. 6.15.

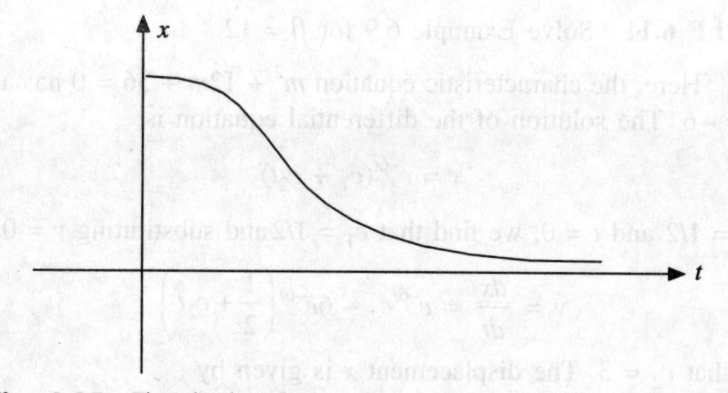


**Fig. 6.15** The displacement time curve for overdamped motion.

**EXAMPLE 6.10** In Example 6.3 assume that a damping force in pounds numerically equal to 3.75 times the instantaneous velocity is taken into account. Find $x$ as a function of $t$.

***Solution*** Taking into account the damping force $-3.75\ (dx/dt)$ in the differential equation of Example 6.3, we find

$$\frac{6}{32}\frac{d^2x}{dt^2} = -12x - 3.75\frac{dx}{dt}$$

or

$$\frac{d^2x}{dt^2} + 20\frac{dx}{dt} + 64x = 0 \tag{40}$$

The initial conditions are, as before,

$$x = \frac{1}{3} \quad \text{at } t = 0, \qquad \frac{dx}{dt} = 0 \quad \text{at } t = 0 \tag{41}$$

The auxiliary equation for (40) has roots $-4$, $-16$. Hence

$$x = c_1 e^{-4t} + c_2 e^{-16t}$$

Using conditions (41), we find

$$x = \frac{4}{9}e^{-4t} - \frac{1}{9}e^{-16t}$$

The graph appears as in Fig. 6.16. It is seen that no oscillations occur; the weight has so much damping that it just gradually returns to the equilibrium position without passing through it. The motion is overdamped.

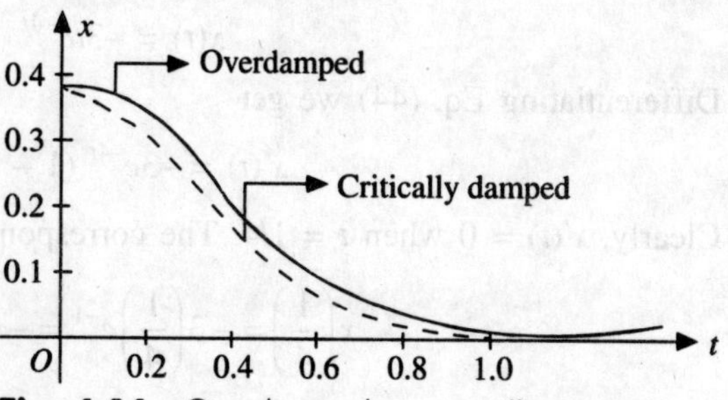


**Fig. 6.16** Overdamped motion illustrating Example 6.10.

**EXAMPLE 6.11** Solve Example 6.9 for $\beta = 12$.

***Solution*** Here, the characteristic equation $m^2 + 12m + 36 = 0$ has a repeated root $m = -6$. The solution of the differential equation is

$$x = e^{-6t}(c_1 + c_2 t)$$

Using $x = 1/2$ and $t = 0$, we find that $c_1 = 1/2$ and substituting $v = 0$, $t = 0$ in

$$v = \frac{dx}{dt} = e^{-6t}c_2 - 6e^{-6t}\left(\frac{1}{2} + c_2 t\right)$$

we find that $c_2 = 3$. The displacement $x$ is given by

$$x = e^{-6t}\left(\frac{1}{2} + 3t\right)$$

and the displacement-time curve resembles that of Fig. 6.15 except that $x$ approaches zero more rapidly. The motion is critically damped.

**EXAMPLE 6.12** An 8-lb weight stretches a spring 2 ft. Assuming that a damping force numerically equals to two times the instantaneous velocity acts on the system, determine the equation of motion if the weight is released from the equilibrium position with an upward velocity of 3 ft/s.

***Solution*** From Hooke's law, we have $k = 4$ lb/ft, and from $m = W/g$, we have $m = 1/4$ slug. Thus, the differential equation of motion is

$$\frac{d^2x}{dt^2} + 8\frac{dx}{dt} + 16x = 0 \tag{42}$$

The auxiliary equation $m^2 + 8m + 16 = 0$ has the roots $m_1 = m_2 = -4$ and the solution is

$$x(t) = c_1 e^{-4t} + c_2 t e^{-4t} \tag{43}$$

The initial conditions

$$x(0) = 0, \qquad \left(\frac{dx}{dt}\right)_{t=0} = -3$$

demand that $c_1 = 0$ and $c_2 = -3$. Thus, the equation of motion is

$$x(t) = -3te^{-4t} \tag{44}$$

Differentiating Eq. (44) we get

$$x'(t) = -3e^{-4t}\ (1 - 4t)$$

Clearly, $x'(t) = 0$ when $t = 1/4$. The corresponding extreme displacement is

$$x\left(\frac{1}{4}\right) = -3\left(\frac{1}{4}\right)e^{-1} = -0.276 \text{ ft}$$

As shown in Fig. 6.17, we interpret this value to mean that the weight reaches a maximum height of 0.276 ft above the equilibrium position.

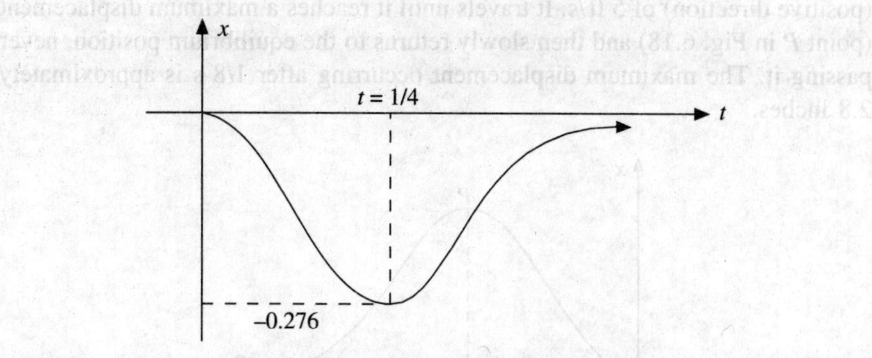


**Fig. 6.17** Displacement time curve.

**EXAMPLE 6.13** Instead of 3.75 in Example 6.10, use 3 and find $x$ as a function of $t$.

***Solution*** Equation (40) becomes

$$\frac{d^2x}{dt^2} + 16\frac{dx}{dt} + 64x = 0 \tag{45}$$

The roots of the auxiliary equation are –8 and –8, and the solution of Eq. (45) is

$$x(t) = c_1 e^{-8t} + c_2 t e^{-8t}$$

and using the initial conditions (41), this becomes

$$x(t) = \frac{1}{3}e^{-8t} + \frac{8}{3}te^{-8t}$$

The graph appears as in Fig. 6.16 (dashed lines) and is to be compared with the curve (solid line) of the figure in Example 6.10.

It is interesting to see what would happen if the initial conditions in the above example were changed. It will be clear that such a modification could not have possibly changed overdamped or critically damped motion into an oscillatory motion. However, some characteristics of the motion may be changed.

**EXAMPLE 6.14** Assume the differential equation of Example 6.13, but change the initial conditions to $x = 0$, $dx/dt = 5$ at $t = 0$.

***Solution*** The solution of the differential Eq. (45) is

$$x = c_1 e^{-8t} + c_2 t e^{-8t}$$

Using the given initial conditions, we find

$$x = 5te^{-8t}$$

The graph is depicted in Fig. 6.18. To interpret this motion, observe that initially the weight is at equilibrium position and given a velocity downward (positive direction) of 5 ft/s. It travels until it reaches a maximum displacement (point $P$ in Fig. 6.18) and then slowly returns to the equilibrium position, never passing it. The maximum displacement occurring after 1/8 s is approximately 2.8 inches.

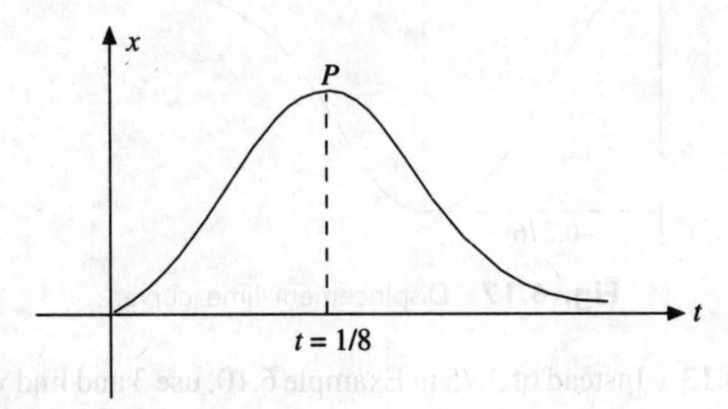


**Fig. 6.18** Displacement time curve for Example 6.14.

**EXAMPLE 6.15** A 16 lb weight is attached to a 5 ft long spring. At equilibrium the spring measures 8.2 ft. If the weight is pushed up and released from rest at a point 2 ft above the equilibrium position, find the displacement $x(t)$ if it is further known that the surrounding medium offers a resistance numerically equal to the instantaneous velocity.

***Solution*** The elongation of the spring after the weight is attached is 8.2 – 5= 3.2 ft, and thus from Hooke's law, $k$ = 5 lb/ft. Also, we have $m$ = 16/32 = 1/2 slug, so that the differential equation is

$$\frac{d^2x}{dt^2} + 2x + 10x = 0 \tag{46}$$

The roots of the auxiliary equation are $-1 \pm 3i$, and the solution of Eq. (46) is

$$x(t) = e^{-t}(c_1 \cos 3t + c_2 \sin 3t) \tag{47}$$

From the given conditions

$$x(0) = -2, \qquad \left(\frac{dx}{dt}\right)_{t=0} = 0$$

we find that $c_1 = -2$ and $c_2 = -2/3$. Thus

$$x(t) = e^{-t}\left(-2\cos 3t - \frac{2}{3}\sin 3t\right)$$

The solution of Eq. (47) with initial conditions can also be written in the form [making use of identity (20)]

$$x = x(t) = \frac{2}{3}\sqrt{10}\, e^{-t} \sin(3t + 4.391)$$

The graph of this function is shown in Fig 6.19 and the motion is underdamped.

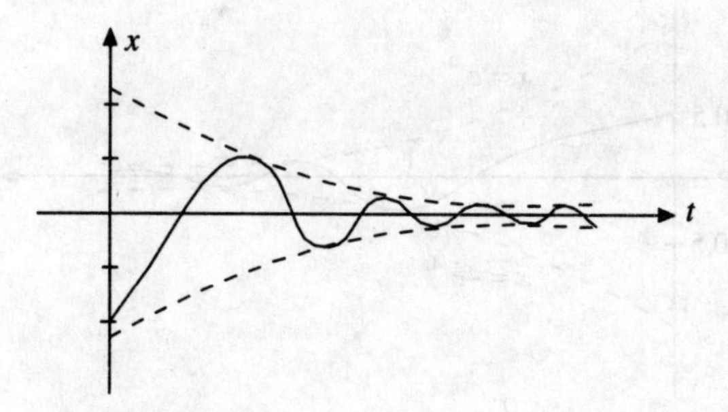


**Fig. 6.19** Underdamped motion.

**EXAMPLE 6.16** Solve Example 6.9 for $\beta = 8$.

***Solution*** The characteristic equation $m^2 + 8m + 36 = 0$ has roots $-4 \pm 2\sqrt{5}\,i$. The solution of the differential equation is

$$x = e^{-4t}(c_1 \sin 2\sqrt{5}t + c_2 \cos 2\sqrt{5}t).$$

Setting $t = 0$ and $x = 1/2$, we find $c_2 = 1/2$. Substituting $v = 0$ and $t = 0$ into

$$v = \frac{dx}{dt} = e^{-4t}(2\sqrt{5}\,c_1 \cos 2\sqrt{5}\,t - \sqrt{5} \sin 2\sqrt{5}\,t)$$

$$-4e^{-4t}\left(c_1 \sin 2\sqrt{5}\,t + \frac{1}{2} \cos 2\sqrt{5}\,t\right)$$

we see that $c_1 = 1/\sqrt{5}$. Thus, the displacement $x$ is given by

$$x = e^{-4t}\left(\frac{1}{\sqrt{5}} \sin 2\sqrt{5}\,t + \frac{1}{2} \cos 2\sqrt{5}\,t\right) \tag{48}$$

The motion is underdamped, the factor $e^{-4t}$ is the *damping factor* and the factor $[(1/\sqrt{5}) \sin 2\sqrt{5}t + (1/2) \cos 2\sqrt{5}t]$ is the *harmonic factor*. Figure 6.20 shows that the displacement time curve is drawn by first sketching the curve $f(t) = \pm e^{-4t}$, and that the height of the displacement curve is numerically equal to the value of the damping factor when the harmonic factor is numerically equal to 1, and zero when the harmonic factor has the value zero. It is also helpful to write Eq. (48) in the form

$$x = 0.3\sqrt{5}\, e^{-4t} \sin (2\sqrt{5}\,t + \phi)$$

where $\cos \phi = 2/3$ and $\sin \phi = \sqrt{5}/3$. Although the motion is not periodic, the quantity $2\pi/(2\sqrt{5}) = \pi/\sqrt{5}$ is termed as *quasi-period.*

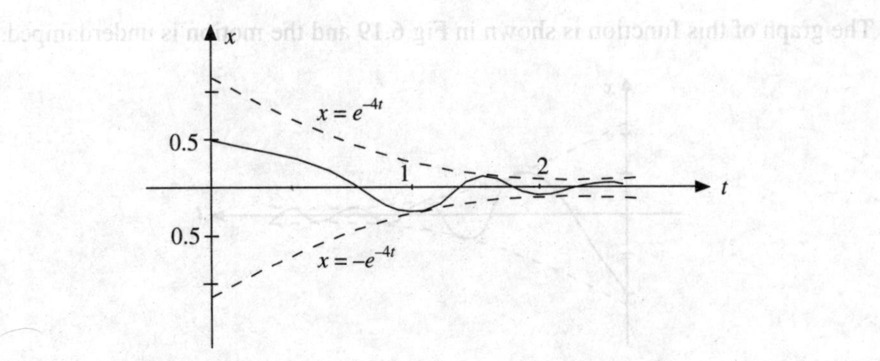


**Fig. 6.20** Displacement time curve for Example 6.16.

**EXAMPLE 6.17** Assume that a damping force (given in pounds) numerically equal to 1.5 times the instantaneous velocity in ft/s acts on the weight shown in Example 6.3: (a) set up the differential equation, and (b) find the position $x$ of the weight as a function of time $t$.

***Solution*** Taking into account the damping force $-1.5\,(dx/dt)$, the differential equation is

$$\frac{d^2x}{dt^2} + 8\frac{dx}{dt} + 64x = 0 \tag{49}$$

The initial conditions are

$$x(0) = \frac{1}{3}, \quad \left(\frac{dx}{dt}\right)_{t=0} = 0$$

The auxiliary equation for (49) has the roots $-4 \pm 4\sqrt{3}\,i$, and so the solution is

$$x = e^{-4t}(c_1 \cos 4\sqrt{3}\,t + c_2 \sin 4\sqrt{3}\,t)$$

which, using the initial conditions, becomes

$$x = \frac{1}{9}e^{-4t}(3\cos 4\sqrt{3}\,t + \sqrt{3}\sin 4\sqrt{3}\,t) \tag{50}$$

From Eq. (20) this may be written as

$$x = \frac{2\sqrt{3}}{9}e^{-4t}\sin\left(4\sqrt{3}\,t + \frac{\pi}{3}\right) \tag{51}$$

The graph of Eq. (51), shown in Fig. 6.21, lies between the graphs of $x = (2\sqrt{3}/9)\,e^{-4t}$ and $x = -(2\sqrt{3}/9)\,e^{-4t}$ (shown as dashed lines in the figure), since sine varies between $-1$ and $+1$.

The motion is underdamped (damped oscillatory or damped vibratory). It may be noted that Eq. (51) has the form

$$x = \tilde{A}(t)\sin(\omega t + \phi) \tag{52}$$

where $\tilde{A}(t) = (2\sqrt{3}/9)\,e^{-4t}$, $\omega = 4\sqrt{3}$ and $\phi = \pi/3$. The quasi-period is $2\pi/\omega = 2\pi/4\sqrt{3}$.

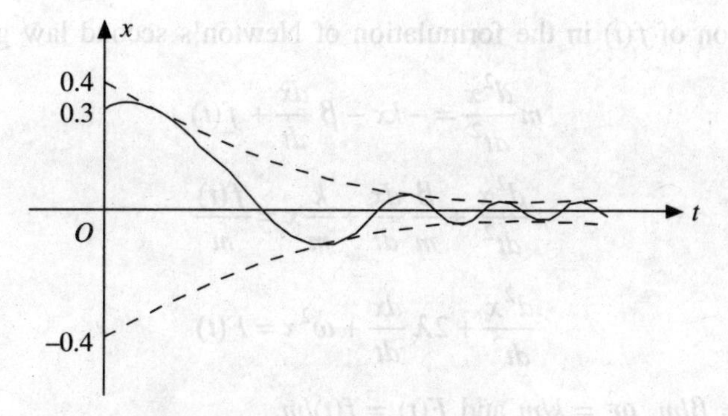


**Fig. 6.21** Underdamped motion.

By analogy with the undamped case, $\tilde{A}(t)$ is called the *damped amplitude* or *time-varying amplitude*. It is seen that the amplitude decreases with time, thus agreeing with our experience. One fact that should be noted is that the frequency with damping is less than that without damping. This is plausible since one would expect opposition to the motion to increase the time for a complete cycle. The undamped frequency, i.e. with $\beta = 0$ is often called *natural frequency*. It is of great importance in the phenomenon of resonance, which will be discussed later.

## 6.4 FORCED MOTION

Here, we take into consideration an external force acting on a vibrating spring. For example, $f(t)$ could represent a driving (impressed) force causing an oscillatory vertical motion of the support of the spring (Fig. 6.22).

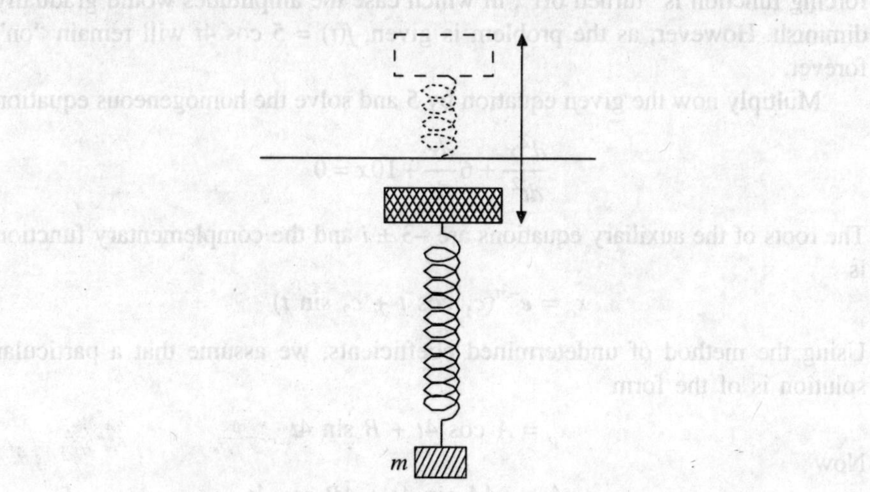


**Fig. 6.22** Oscillatory motion of the spring.

The inclusion of $f(t)$ in the formulation of Newton's second law gives

$$m\frac{d^2x}{dt^2} = -kx - \beta\frac{dx}{dt} + f(t) \tag{53}$$

or

$$\frac{d^2x}{dt^2} + \frac{\beta}{m}\frac{dx}{dt} + \frac{k}{m}x = \frac{f(t)}{m} \tag{54}$$

or

$$\frac{d^2x}{dt^2} + 2\lambda\frac{dx}{dt} + \omega^2 x = F(t) \tag{55}$$

where $2\lambda = \beta/m$, $\omega^2 = k/m$ and $F(t) = f(t)/m$.

To solve the latter nonhomogeneous equation, we can employ either the method of undetermined coefficients or variation of parameters.

**EXAMPLE 6.18** Interpret and solve the initial value problem

$$\frac{1}{5}\frac{d^2x}{dt^2} + 1.2\frac{dx}{dt} + 2x = 5\cos 4t$$

$$x(0) = \frac{1}{2}, \quad \left(\frac{dx}{dt}\right)_{t=0} = 0$$

***Solution*** We can interpret the problem to represent a vibrational system consisting of a mass ($m$ = 1/5 slug) attached to a spring ($k$ = 2 lb/ft). The mass is released from rest 1/2 unit (ft) below the equilibrium position. Although the motion is damped ($\beta$ = 1.2), the system is also being driven by an external periodic force ($T = \pi/2$ s) beginning at $t = 0$. Intuitively we would expect that even with damping the system will remain in motion until such time as the forcing function is "turned off", in which case the amplitudes would gradually diminish. However, as the problem is given, $f(t) = 5\cos 4t$ will remain "on" forever.

Multiply now the given equation by 5 and solve the homogeneous equation

$$\frac{d^2x}{dt^2} + 6\frac{dx}{dt} + 10x = 0$$

The roots of the auxiliary equations are $-3 \pm i$ and the complementary function is

$$x_c = e^{-3t}(c_1 \cos t + c_2 \sin t)$$

Using the method of undetermined coefficients, we assume that a particular solution is of the form

$$x_p = A\cos 4t + B\sin 4t$$

Now

$$x_p' = -4A\sin 4t + 4B\cos 4t$$

$$x_p'' = -16A\cos 4t - 16B\sin 4t$$

so that

$$x''_p + 6x'_p + 10x_p = 25 \cos 4t$$

gives

$$(-6A + 24B) \cos 4t + (-24A - 6B) \sin 4t = 25 \cos 4t$$

The resulting system of equations

$$-6A + 24B = 25, \qquad -24A - 6B = 0$$

yield $A = -25/102$ and $B = 50/51$. Thus

$$x_p = -\frac{25}{102} \cos 4t + \frac{50}{51} \sin 4t$$

and the general solution of the given equation is

$$x = x_c + x_p = e^{-3t}(c_1 \cos t + c_2 \sin t) - \frac{25}{10} \cos 4t + \frac{50}{51} \sin 4t$$

Using the initial conditions, we find $c_1 = 38/51$ and $c_2 = -86/51$. Therefore, the equation of motion is

$$x(t) = e^{-3t}\left(\frac{38}{51} \cos t - \frac{86}{51} \sin t\right) - \frac{25}{102} \cos 4t + \frac{50}{51} \sin 4t$$

**EXAMPLE 6.19** In Example 6.17, assume that a periodic force (external) given by $f(t) = 24 \cos 8t$ is acting. Find $x$ in terms of $t$, using the initial conditions given there.

***Solution*** The differential equation is

$$\frac{6}{32} \frac{d^2x}{dt^2} = -12x - 1.5 \frac{dx}{dt} + 24 \cos 8t$$

or

$$\frac{d^2x}{dt^2} + 8 \frac{dx}{dt} + 64x = 128 \cos 8t \tag{56}$$

and the initial conditions are $x = 1/3$, $dx/dt$ at $t = 0$. The complementary solution of Eq. (56) is

$$x_c = e^{-4t}(c_1 \cos 4\sqrt{3}\,t + c_2 \sin 4\sqrt{3}\,t)$$

If we assume that a particular solution is $A \sin 8t + B \cos 8t$, we find $A = 2$, $B = 0$. Hence, the general solution of Eq. (56) is

$$x = e^{-4t}(c_1 \cos 4\sqrt{3}\,t + c_2 \sin 4\sqrt{3}\,t) + 2 \sin 8t$$

From the initial conditions, we have $c_1 = 1/3$, $c_2 = -11\sqrt{3}/9$, and thus

$$x = \frac{e^{-4t}}{9}(3 \cos 4\sqrt{3}\,t - 11\sqrt{3} \sin 4\sqrt{3}\,t) + 2 \sin 8t \tag{57}$$

The graph of Eq. (57) appears as in Fig. 6.23. It will be observed that the terms in Eq. (57) involving $e^{-4t}$ become negligible when $t$ is large. These terms are called *transient terms* and are significant only when $t$ is near zero. These transient terms in the solution, when they are significant, are sometimes called the *transient solution*. When the transient terms are negligible, the term 2 sin 8$t$ remains. This is called the *steady state term* or *steady state solution*, since it indicates the behaviour of the system when the conditions have become steady. It is seen that the steady state curve (shown as dashed lines in Fig. 6.23) is periodic and has the same period as that of the applied external force.

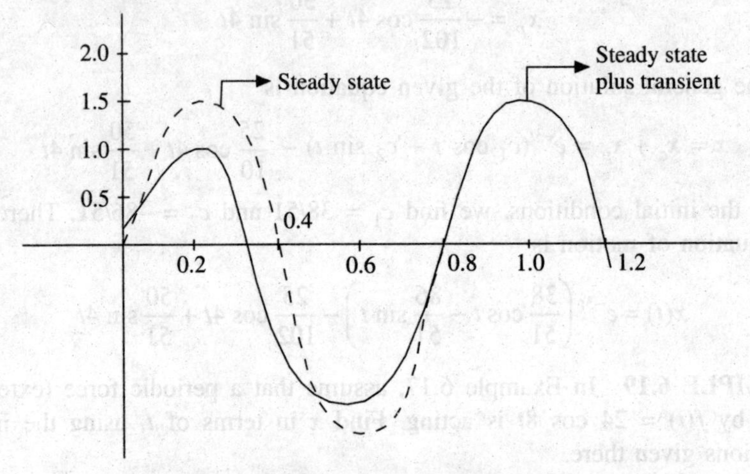


**Fig. 6.23** Displacement time curve.

When $f$ is periodic, such as $f(t) = f_0 \sin \gamma t$ or $f(t) = f_0 \cos \gamma t$, the general solution of Eq. (55) consists of

$$x(t) = \text{transient} + \text{steady state}$$

**EXAMPLE 6.20** Solve Example 6.16 for an impressed force $F(t) = 72 \cos 6t$ acting on the block.

***Solution*** The motion is governed by the equation

$$\frac{d^2x}{dt^2} - 8\frac{dx}{dt} + 36x = 72 \cos 6t$$

Using the method of undetermined coefficients, the displacement $x$ is given by

$$x = e^{-4t}(c_1 \sin 2\sqrt{5}\, t + c_2 \cos 2\sqrt{5}\, t) + \frac{3}{2} \sin 6t$$

From the initial conditions, we find $c_1 = -7/2\sqrt{5}$ and $c_2 = 1/2$.

The terms in the solution that involve $e^{-4t}$ have an appreciable effect on the motion only for small values of $t$, since they contribute very little to the value of $x$ when $t$ is large. These are transient terms and (3/2) sin 6$t$ is the steady state solution. The steady state solution is periodic and has the same period as that of the impressed force. The motion is approximately a simple harmonic motion of period $\pi/3$ and amplitude 3/2.

There is another way of viewing Eq. (55). In this approach, $F(t)$ is regarded as an input to the physical system governed by the differential equation and $x(t)$, the solution of the differential equation, is regarded as the output, or the response of the system to the input $f(t)$. In the problems we consider, for most input $f(t)$, the output $x(t)$ will be determined uniquely except for the transient solution which depends on the initial conditions.

The input-output point of view is very fruitful in many applications. It is often used, for example, in analysing systems. (For more details on input-output analysis, see [10].)

**EXAMPLE 6.21** The block shown in Fig. 6.24 weighs $W$ lb when it is held at rest by the spring and its position is the equilibrium position. Find its displacement from the equilibrium position if it is pulled $x_0$ ft below the equilibrium position and released from rest.

***Solution*** Let $x$ denote the displacement of the block, measured positive downward from the equilibrium position. If $d$ is the stretch in the spring when the block is in equilibrium position ($x$ = 0) and $k$, the spring constant, then from Hooke's law, we have $W = kd$.

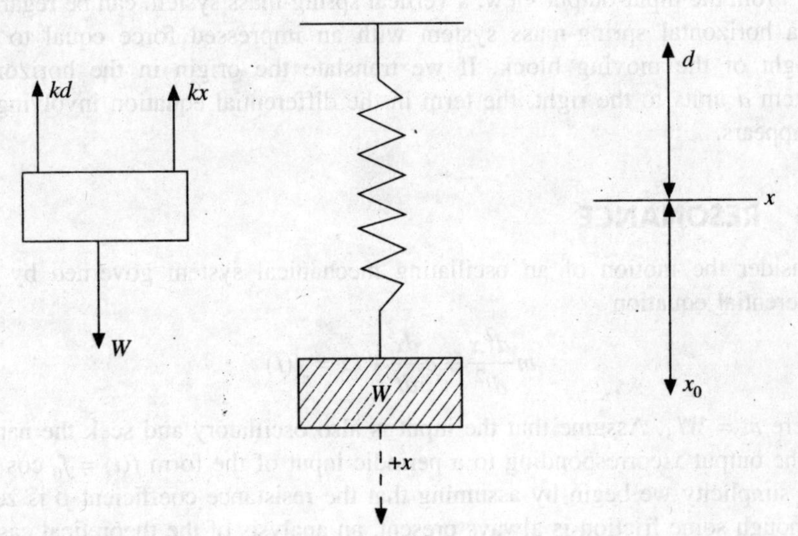


**Fig. 6.24** Oscillatory motion of the spring-force diagram.

When the block is $x$ ft from the equilibrium position, the spring exerts a force of $-k(d + x) = -kd - kx = -W - kx$ in pounds in the $x$-direction. Since, the earth pulls on the block with a force of $W$ lb, from Newton's second law, we have

$$\frac{W}{g}\frac{d^2x}{dt^2} = W + (-W - kx)$$

or

$$\frac{d^2x}{dt^2} + \frac{kg}{W}x = 0 \tag{58}$$

Using the initial conditions, $x = x_0$ and $v = dx/dt = 0$ at $t = 0$, we find that the displacement is given by

$$x = x_0 \cos\sqrt{\frac{kg}{W}}\,t$$

The block undergoes a simple harmonic motion about the point $x = 0$. If the impressed force $f(t)$ is applied to the block, then Eq. (58) becomes

$$\frac{d^2x}{dt^2} + \frac{kg}{W}x = \frac{g}{W}f(t)$$

and to solve it we proceed as in Example 6.20.

If the resistance force is also acting, then the differential equation is

$$\frac{d^2x}{dt^2} + \frac{cg}{W}\frac{dx}{dt} + \frac{kg}{W}x = \frac{g}{W}f(t)$$

From the input-output view, a vertical spring-mass system can be regarded as a horizontal spring-mass system with an impressed force equal to the weight of the moving block. If we translate the origin in the horizontal system $d$ units to the right, the term in the differential equation involving $W$ disappears.

## 6.5 RESONANCE

Consider the motion of an oscillating mechanical system governed by the differential equation

$$m\frac{d^2x}{dt^2} + c\frac{dx}{dt} + kx = f(t)$$

where $m = W/g$. Assume that the input is also oscillatory and seek the nature of the output $x$ corresponding to a periodic input of the form $f(t) = f_0 \cos \omega t$. For simplicity we begin by assuming that the resistance coefficient $\beta$ is zero. Although some friction is always present, an analysis of the theoretical case $c = 0$ is instructive and suggests what we might expect for small values of $c$. Let

$$\omega_0 = \sqrt{\frac{k}{m}} = \sqrt{\frac{kg}{W}} \tag{59}$$

We first note that the differential equation

$$\frac{d^2x}{dt^2} + \omega_0^2 x = \frac{f_0}{m}\cos\omega t \tag{60}$$

has complementary function given by

$$x_c = c_1 \cos \omega_0 t + c_2 \sin \omega_0 t$$

If the frequency $\omega/2\pi$ of the input is the same as the natural frequency $\omega_0/2\pi$ of the system, we try to find a particular solution of the form $x_p = t(c_1 \cos \omega_0 t + c_2 \sin \omega_0 t)$. We find (see also Problem 23) that $x_p = (f_0/2m\omega_0)t \sin \omega_0 t$. Then Eq. (60) has the complete solution given by

$$x = x_c + x_p = c_1 \cos \omega_0 t + c_2 \sin \omega_0 t + \frac{f_0}{2m\,\omega_0} t \sin \omega_0 t$$

Whereas the input has the constant amplitude $f_0$, the values of the output $x$ are unbounded. This is seen by observing that as $t \to \infty$ along a sequence for which

$$\sin \omega_0 t = \pm 1, \qquad \left| \frac{f_0}{2m\omega_0} t \sin \omega_0 t \right| \to \infty$$

The absolute value of $c_1 \cos \omega_0 t + c_2 \sin \omega_0 t$ does not exceed $\sqrt{c_1^2 + c_2^2}$. The graph of

$$x_p = \frac{f_0}{2m\omega_0} t \sin g\,\omega_0 t$$

is depicted in Fig. 6.25.

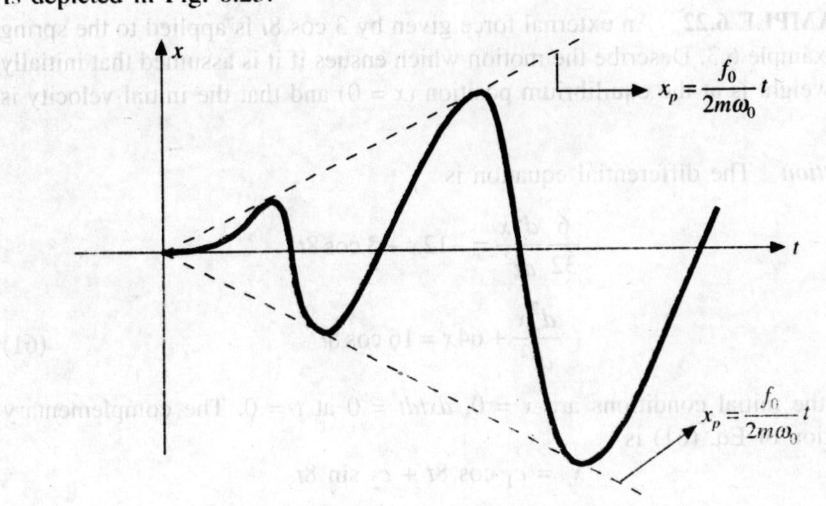


**Fig. 6.25** Pure or undamped resonance.

The phenomenon we have described is known as *pure* or *undamped resonance* and is also termed as *sympathetic vibration.* It is approximated when $\beta$ is small and the input and the natural frequencies coincide. The forcing function is then said to be in *resonance* with the system. Resonance can be both harmful and useful and must be considered in designing the systems. Bridges, cars, ships, aeroplanes, etc. are vibrating systems and can be affected adversely by resonance. A body oscillating on a spring will move farther and farther from the equilibrium position when resonance occurs. The spring may eventually break or Hooke's law will no longer be true. Soldiers crossing a bridge often break step so that they will not generate a frequency equal to the natural frequency of the bridge, thus avoiding the collapse of the bridge. The famous example of the resonance phenomenon is the disaster of the Facome Bridge (for a detailed discussion of this, see [3]). When a tree is uprooted by having a periodic force applied to produce swaying, the resonant effect is being used to advantage. A rocking motion applied to a car often succeeds in getting the car out of a hole or a ditch. Desirable amplification effects are often produced in electric circuits by resonant effects. Acoustic vibrations can be equally destructive as large mechanical vibrations. Operatic and jazz singers sometimes take pride in their ability to inflict destruction on the lowly water glass. The sounds from organs and piccolos have been known to crack windows.

> When the trumpet sounds a single blast: when the earth with all its mountains is lifted up and with one mighty crash is shattered into dust—on that day the dread event will come to pass. The sky will rent asunder; so that day it will be frail.
>
> (Al-Quran, Chapter 69—The Truth: Verse 13–16).

**EXAMPLE 6.22** An external force given by 3 cos 8$t$ is applied to the spring of Example 6.3. Describe the motion which ensues if it is assumed that initially the weight is at the equilibrium position ($x = 0$) and that the initial velocity is zero.

***Solution*** The differential equation is

$$\frac{6}{32}\frac{d^2x}{dt^2} = -12x + 3\cos 8t$$

or

$$\frac{d^2x}{dt^2} + 64x = 16\cos 8t \tag{61}$$

and the initial conditions are $x = 0$, $dx/dt = 0$ at $t = 0$. The complementary solution of Eq. (61) is

$$x_c = c_1 \cos 8t + c_2 \sin 8t$$

For a particular solution, we assume that

$$x_p = t(A \cos 8t + B \sin 8t)$$

By the method of undetermined coefficients, we find that $A = 0$, $B = 1$, and the general solution is

$$x = x_c + x_p = c_1 \cos 8t + c_2 \sin 8t + t \sin 8t$$

Using the initial conditions, we get $c_1 = c_2 = 0$. Hence

$$x = t \sin 8t \tag{62}$$

The graph of Eq. (62) lies between the graphs of $x = t$ and $x = -t$ as shown in Fig. 6.26. It is seen that the oscillations build up without limit. Naturally, the spring is bound to break within a short time.

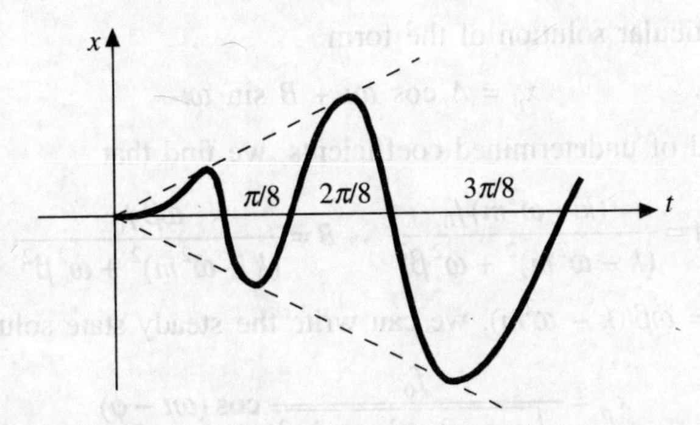


**Fig. 6.26** Displacement time curve.

In this example, it should be noted that the damping was neglected and resonance occurred because *the frequency of the applied force (external) was equal to the natural frequency of the undamped system.* This is a general principle. In the case where damping occurs, the oscillations do not build up without limit but nevertheless become very large. Resonance in this case occurs when the frequency of the applied external force is slightly smaller than the natural frequency of the system (For further discussion of this, see Problem 23).

An interesting oscillation takes place when $\omega$ and $\omega_0$ are unequal but differ only slightly, i.e. when the input and natural frequencies are nearly equal.

If $x = 0$ and $dx/dt = 0$ when $t = 0$, the differential Eq. (60) has the solution

$$x = \frac{f_0}{m(\omega_0^2 - \omega^2)}(\cos \omega t - \cos \omega_0 t) \tag{63}$$

Applying the identity

$$\cos a - \cos b = 2 \sin \frac{a+b}{2} \sin \frac{a-b}{2}$$

Equation (63) can be written as

$$x = \frac{2 f_0}{m(\omega_0^2 - \omega^2)} \sin \left[\frac{1}{2}(\omega_0 + \omega)t\right] \sin \left[\frac{1}{2}(\omega_0 - \omega)t\right] \tag{64}$$

Equation (63) gives the displacement $x$ as the superposition of two sinusoidal waves whose periods differ slightly. Since the period of $\sin [\frac{1}{2}(\omega_0 - \omega)t]$ is much larger than the period of sin $[\frac{1}{2}(\omega_0 + \omega)t]$, it is evident from Eq. (64) that the superposition in Eq. (63) results in a displacement-time curve of the type given in Fig. 6.40 (see Problem 22).

Now consider the case in which the resistance coefficient $\beta$ in equation

$$\frac{W}{g}\frac{d^2x}{dt^2} + \beta\frac{dx}{dt} + kx = f(t)$$

will have complex roots $a \pm ib$ where $a < 0$, $b > 0$, and the transient solution will have the form

$$x_c = e^{at}(c_1 \cos bt + c_2 \sin bt)$$

Assume a particular solution of the form

$$x_p = A \cos \omega t + B \sin \omega t \tag{65}$$

By the method of undetermined coefficients, we find that

$$A = \frac{(k - \omega^2 m) f_0}{(k - \omega^2 m)^2 + \omega^2 \beta^2}, \quad B = \frac{\omega \beta f_0}{(k - \omega^2 m)^2 + \omega^2 \beta^2}$$

Setting $\tan \phi = \omega\beta/(k - \omega^2 m)$, we can write the steady state solution (65) as

$$x_p = \frac{f_0}{\sqrt{(k - \omega^2 m)^2 + \omega^2 \beta^2}} \cos(\omega t - \phi)$$

Since, $x_c \to 0$ as $t \to \infty$, the output $x = x_c + x_p$ will approximate a harmonic oscillation whose frequency $\omega(2\pi)$ is the same as the frequency of the input $f(t) = f_0 \cos \omega t$. The amplitude of the output is approximately equal to the coefficient

$$\frac{f_0}{\sqrt{(k - \omega^2 m)^2 + \omega^2 \beta^2}} = G(\omega) \tag{66}$$

since the contribution of the transient solution $x_c$ to the amplitude is negligible except for small values of $t$. Thus, if we regard $f_0$, $k$, $m$ and $\beta$ fixed, the amplitude in Eq. (66) depends on $\omega$, whereas the amplitude of the input has a fixed value $f_0$. It is easy to show that $G(\omega)$ takes on its maximum when

$$\omega = \sqrt{\frac{k}{m} - \frac{\beta^2}{2m^2}}$$

At this value of $\omega$, we say that the system is in damped (or practical) resonance. In designing systems it is extremely important to know when the steady state amplitude is large. Harmful and beneficial effects are possible, and even though $G(\omega)$ does not become infinite with $t$, the situation is analogous to pure resonance in an undamped system.

In Example 6.20, we found that the steady state solution of

$$\frac{32}{32}\frac{d^2x}{dt^2} + 8\frac{dx}{dt} + 36x = 72 \cos 6x$$

is given by $x_p = (3/2)\sin 6t$. The amplitude 3/2 can be obtained by setting $f_0 = 72$, $k = 36$, $\omega = 6$, $m = 1$ and $\beta = 8$ in

$$\frac{f_0}{\sqrt{(k-\omega^2 m)^2 + \omega^2\beta^2}}$$

To obtain the maximum amplitude corresponding to $F(t) = 72\cos\omega t$, we replace $\omega = 6$ by

$$\omega = \sqrt{\frac{k}{m} - \frac{\beta^2}{2m^2}} = \sqrt{\frac{36}{1} - \frac{(8)^2}{2(1)^2}} = 2$$

This yields the maximum amplitude

$$\frac{f_0}{\sqrt{(k-\omega^2 m)^2 + \omega^2\beta^2}} = \frac{9\sqrt{5}}{10} = 2.01 \text{ (approx.)}$$

## 6.6 ELECTRIC CIRCUIT

We have seen in the electric circuit of Fig. 4.13 that the current $I$ in amperes satisfies the differential equation

$$L\frac{dI}{dt} + RI = E(t)$$

Figure 6.27 contains an additional element known as *capacitor*. This type of element stores electrical energy in the circuit. The voltage drop across a capacitor is proportional to the charge $q$ (in coulombs) on the capacitor and is given by

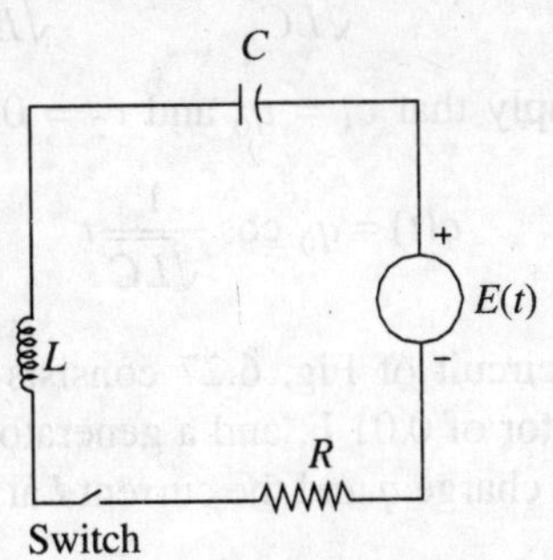


**Fig. 6.27** Circuit diagram consisting of a capacitor $C$, a resistor $R$ and an inductor $L$.

$C^{-1}q$, where $C^{-1}$ is the constant of proportionality. The constant $C$ is called the *constant of capacitance* or simply *capacitance*. Applying Kirchhoff's law to Fig. 6.27, we get the differential equation

$$L\frac{dI}{dt} + RI + \frac{1}{C}q = E(t) \tag{67}$$

The current $I$ equals the time rate of change of $q$, i.e.

$$I(t)=\frac{dq(t)}{dt} \tag{68}$$

From Eqs. (68) and (67)

$$L\frac{d^2q}{dt^2}+R\frac{dq}{dt}+\frac{1}{C}q=E(t) \tag{69}$$

Differentiating both sides of Eq. (67) with respect to $t$ and using Eq. (68), we obtain

$$L\frac{d^2I}{dt^2}+R\frac{dI}{dt}+\frac{1}{C}I=\frac{d}{dt}E(t) \tag{70}$$

We assume here that $L$ (H), $R$ ($\Omega$) and $C$ (F) are constants and $E(t)$ (V) is the impressed voltage and $t$ (s) is the time.

**EXAMPLE 6.23** A series circuit contains only a capacitor and inductor. If the capacitor has an initial charge $q_0$, determine the subsequent charge $q(t)$.

***Solution*** From Eq. (69),

$$L\frac{d^2q}{dt^2}+\frac{1}{C}q=0$$

and the given initial conditions are $q(0) = q_0$. Assume that no current flows initially; then $q'(0) = 0$, since $q'(t) = I(t)$. The general solution is, thus

$$q(t)=c_1\cos\frac{1}{\sqrt{LC}}t+c_2\sin\frac{1}{\sqrt{LC}}t$$

The initial conditions imply that $c_1 = q_0$ and $c_2 = 0$, so that

$$q(t)=q_0\cos\frac{1}{\sqrt{LC}}t$$

**EXAMPLE 6.24** The circuit of Fig. 6.27 consists of an inductor of 1 H, a resistor of 12 $\Omega$, a capacitor of 0.01 F, and a generator having voltage given by $E(t) = 24\sin 10t$. Find the charge $q$ and the current $I$ at time $t$, if $q = 0$ and $I = 0$ at $t = 0$.

***Solution*** Here, Eq. (69) takes the form

$$1\frac{d^2q}{dt^2}+12\frac{dq}{dt}+\frac{1}{0.01}q=24\sin 10t$$

or

$$\frac{d^2q}{dt^2}+12\frac{dq}{dt}+100q=24\sin 10t \tag{71}$$

The roots of the characteristic equation are $-6 \pm 8i$, and the complementary solution is

$$q_c = e^{-6t}(c_1 \cos 8t + c_2 \sin 8t)$$

Assume that a particular solution is of the form $q_p = A \cos 10t + B \sin 10t$. We find from the method of undetermined coefficients that $A = 1/5$ and $B = 0$. The complete solution of Eq. (71) is

$$q = q_c + q_p = e^{-6t}(c_1 \cos 8t + c_2 \sin 8t) - \frac{1}{5}\cos 10t$$

From the initial conditions we find that $c_1 = 1/5$ and $c_2 = 3/20$. Hence, the charge $q$ is

$$q = q(t) = \frac{e^{-6t}}{20}(4 \cos 8t + 3 \sin 8t) - \frac{1}{5} \cos 10t$$

and current $I = dq/dt$ is

$$I = I(t) = -\frac{5}{2}e^{-6t} \sin 8t + 2 \sin 10t$$

The *transient current*, given by $I_t = -(5/2)e^{-6t} \sin 8t$, is determined by the initial conditions and becomes negligible soon after $t = 0$. The *steady state current* $I_s = 2 \sin 10t$ approximates the actual current when the transient current becomes negligible (one can compare this example with Example 6.19 and the graph of Fig. 6.23).

It can be shown that the differential equation describing the torsional motion of a weight suspended from the end of an elastic shaft is

$$\tilde{I}\frac{d^2\theta}{dt^2} + c\frac{d\theta}{dt} + k\theta = T(t) \tag{72}$$

As shown in Fig. 6.28. the function $\theta(t)$ represents the amount of twist of weight at any time.

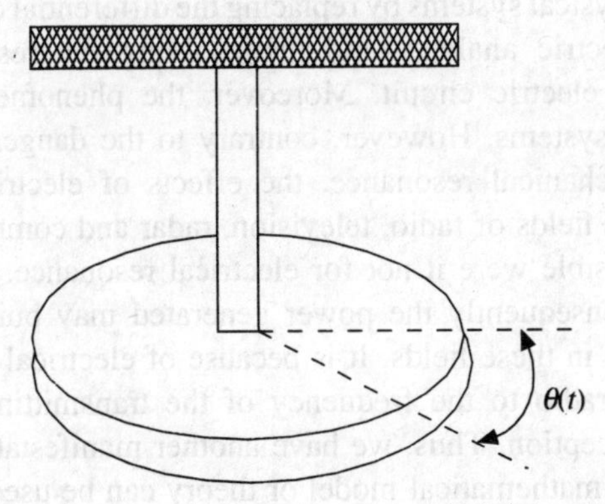


**Fig. 6.28** Torsional motion of a weight suspended from the end of an elastic shaft.

By comparing Eqs. (69), (70) and (72) with the differential equation (of the forced motion with damping)

$$m\frac{d^2x}{dt^2} + \beta\frac{dx}{dt} + kx = f(t) \tag{73}$$

we see that with the exception of terminology there is absolutely no difference between the mathematics of vibrating strings, torsional vibrations and simple series circuits. Table 6.1 gives a comparison of the analogous parts of these three kinds of systems.

**Table 6.1** Comparison of the Analogous Parts of Mechanical System, Simple Series Circuits and Torsional Vibrations

| Mechanical | Series electrical | Torsional |
|---|---|---|
| $m$ (mass) = $W/g$ | $L$ (inductance) | $\tilde{I}$ (moment of inertia) |
| $\beta$ (damping or friction constant) | $R$ (resistance) | $c$ (damping) |
| $k$ (spring constant) | $1/C$ (inverse capacitance called elastance) | $k$ (elastic shaft constant) |
| $f(t)$ (applied force) | $E(t)$ (impressed voltage) | $T(t)$ (applied torque) |
| $x$ (displacement) | $q$ (charge) | $\theta$ (amount of twist of weight) |

Because of this remarkable resemblance between the mechanical and electrical quantities, which holds in even more complicated cases, most of the statements made for mechanical systems apply to electrical systems, and vice-versa. In fact, the analogy is often made use of in industry for studying mechanical systems which may be too complicated or too expensive to build or perhaps too dangerous. An analog computer is a device that stimulates the mechanical and physical systems by replacing the differential equation governing them by their electric analogs, and then solving the resulting differential equations for the electric circuit. Moreover, the phenomenon of resonance occurs in electric systems. However, contrary to the dangerous effects which may result in mechanical resonance, the effects of electrical resonance are mainly useful. The fields of radio, television, radar and communications would virtually be impossible were it not for electrical resonance. In such instances, the current and consequently the power generated may build up to the large amounts necessary in these fields. It is because of electrical resonance that we need to tune our radio to the frequency of the transmitting radio station in order to get the reception. Thus, we have another manifestation of the manner in which the same mathematical model or theory can be used to describe more than one concrete application.

## 6.7 THE HANGING CABLE

Let a cable or rope be hung from two points $A$ and $B$ not necessarily at the same level (Fig. 6.29a). Assume that the cable is flexible so that due to the loading (which may be due to the load of the cable, to the external forces acting, or to the combination of these) it takes a shape as in the figure. Let $C$ be the lowest point on the cable, and choose $x$- and $y$-axis as in Fig. 6.29(a) so that the $y$-axis passes through point $C$.

Consider a part of the cable between $C$ and a point $P(x, y)$ on the cable. This part will be in equilibrium due to the tension at $P$ (Fig. 6.29b), the horizontal force $H$ at $C$, and the total vertical loading $W(x)$ on the portion $CP$ of the cable acting at some point $Q$ (not necessarily the centre of the arc $CP$). For equilibrium, the sum of the forces to the right equals the sum of the forces to the left, and the sum of the forces in the upward direction equals the sum of the forces in the downward direction. Thus

$$T \sin \theta = W, \qquad T \cos \theta = H \tag{74}$$

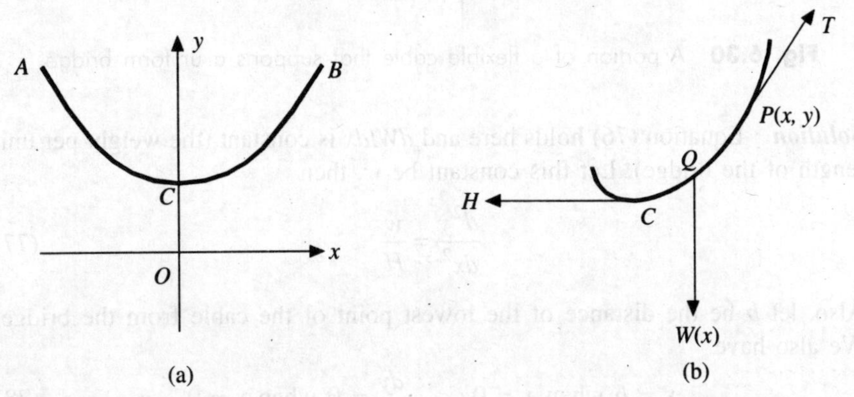


**Fig. 6.29** Schematic diagram of a hanging cable: (a) A cable hanging between two points $A$ and $B$; (b) Part of the cable between C and a point $P(x, y)$.

Since, the slope of the tangent at $P$ is $dy/dx = \tan \theta$, from Eq. (74), we have

$$\frac{dy}{dx} = \frac{W}{H} \tag{75}$$

Here, $H$ is a constant as it is the tension at the lowest point, but $W$ may depend upon $x$. Differentiating Eq. (75) with respect to $x$, we get

$$\frac{d^2 y}{dx^2} = \frac{1}{H}\frac{dW}{dx} \tag{76}$$

where $dW/dx$ represents the load per unit distance in the horizontal direction.

In the following examples, we shall see that for different loads per unit horizontal distance we obtain various differential equations that lead to the various shapes of the cables.

**EXAMPLE 6.25** A flexible cable of small (negligible) weight supports a uniform bridge (Fig. 6.30). Determine the shape of the cable. (This is the problem of determining the shape of the cable in a suspension bridge, which is of great importance in modern bridge construction.)

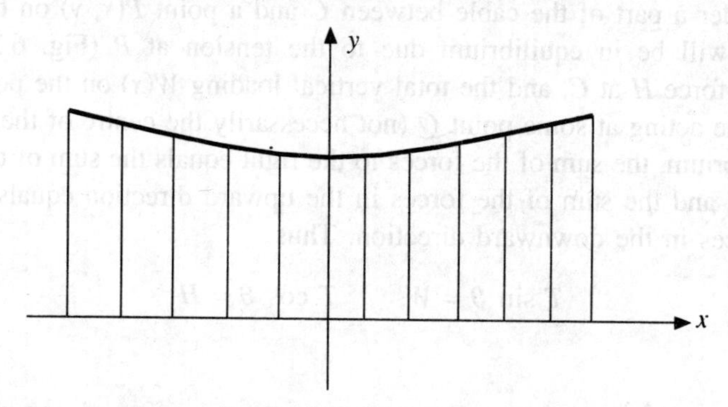


**Fig. 6.30** A portion of a flexible cable that supports a uniform bridge.

***Solution*** Equation (76) holds here and $dW/dx$ is constant (the weight per unit length of the bridge). Let this constant be $w$, then

$$\frac{d^2y}{dx^2} = \frac{w}{H} \tag{77}$$

Also, let $b$ be the distance of the lowest point of the cable from the bridge. We also have

$$y = b \text{ when } x = 0, \qquad \frac{dy}{dx} = 0 \text{ when } x = 0 \tag{78}$$

Now, integrating Eq. (77) twice and making use of the conditions (78), we get

$$y = \frac{w}{2H}x^2 + b$$

Thus, the cable assumes the shape of a parabola.

**EXAMPLE 6.26** A flexible rope of constant density hangs between two points. Determine the shape of the rope.

***Solution*** We can apply Eq. (76) here and what remains is to find $dW/dx$. For this, consider a small portion of the rope (Fig. 6.31). The weight is uniformly distributed over the arc $PQ$. If $\rho$ is the density of the rope, then $dW/ds = \rho$. Now

$$\frac{dW}{ds} = \frac{dW}{dx}\frac{dx}{ds} = \rho \quad \text{or} \quad \frac{dW}{dx} = \rho\frac{ds}{dx}$$

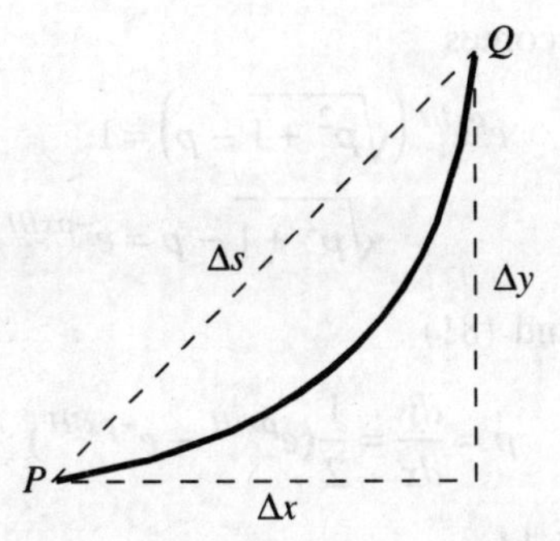


**Fig. 6.31** Schematic diagram of a portion of a flexible rope hanging between two points.

But
$$\frac{ds}{dx} = \sqrt{1 + \left(\frac{dy}{dx}\right)^2}$$

Thus
$$\frac{dW}{dx} = \rho\sqrt{1 + \left(\frac{dy}{dx}\right)^2}$$

Equation (76) now takes the form
$$\frac{d^2y}{dx^2} = \frac{\rho}{H}\sqrt{1 + \left(\frac{dy}{dx}\right)^2} \tag{79}$$

and the initial conditions are the same as those of the last example (viz., Eq. 78).

In Eq. (79), both $x$ and $y$ are missing. Putting $dy/dx = p$, we have
$$\frac{dp}{dx} = \frac{\rho}{H}\sqrt{1 + p^2}$$

Separating the variables and integrating, we get
$$\log\left(p + \sqrt{p^2 + 1}\right) = \frac{\rho x}{H} + \log c$$

or
$$p + \sqrt{p^2 + 1} = ce^{\rho x/H}$$

From Eq. (78), we have $c = 1$, so that
$$p + \sqrt{p^2 + 1} = e^{\rho x/H} \tag{80}$$

Now
$$\left(\sqrt{p^2 + 1}\right)^2 - p^2 = 1$$

or
$$\left(\sqrt{p^2 + 1} - p\right)\left(\sqrt{p^2 + 1} + p\right) = 1$$

Using Eq. (80), this becomes

$$e^{\rho x/H}\left(\sqrt{p^2+1}-p\right)=1$$

or

$$\sqrt{p^2+1}-p=e^{-\rho x/H} \tag{81}$$

Thus, from Eqs. (80) and (81)

$$p=\frac{dy}{dx}=\frac{1}{2}(e^{\rho x/H}-e^{-\rho x/H})$$

which on integration yields

$$y=\frac{H}{2\rho}(e^{\rho x/H}+e^{-\rho x/H})+c_1$$

Using the condition (78), we find $c_1 = b - H/\rho$. By suitably shifting the $x$-axis we can make the choice $b = H/\rho$; then $c_1 = 0$, and we have

$$y=\frac{H}{2\rho}(e^{\rho x/H}+e^{-\rho x/H})=\frac{H}{\rho}\cosh\frac{\rho x}{H} \tag{82}$$

The graph of Eq. (82) is called a *catenary* (derived from the Latin word *catena*, meaning *chain*). Many transmission lines, telegraph cables and cables of suspension bridges hang in the form of catenaries.

## 6.8 THE DEFLECTION OF BEAMS

The weightless beam in Fig. 6.32(a) has edges $AB$ and $CD$ horizontal and edges $AD$ and $BC$ vertical. Let $O$ be the centroid of the rectangle $ABCD$ through which the axes of $x$ and $y$ pass. The axis of $x$ is perpendicular to the plane of the rectangle and the $y$-axis is parallel to the edge $AD$. Since, the beam is supposed to

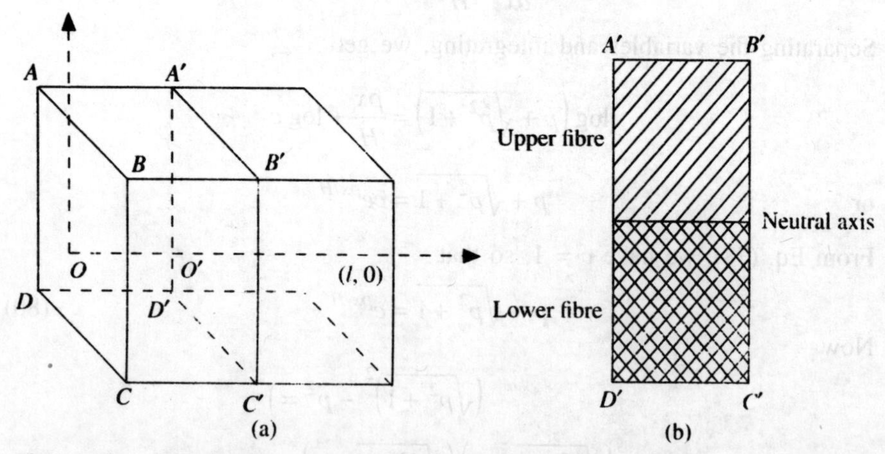


**Fig. 6.32** (a) Schematic diagram of a weightless beam having $AB$ and $CD$ as edges; (b) Longitudinal cross-section of a beam.

be weightless, the $x$-axis passes through the centroid of every cross-section of the beam parallel to the rectangle $ABCD$. For example, the centroid $O'$ of the section $A'B'C'D'$, located $x$ units from $O$, is on the $x$-axis. The same is true of centroid $(l, 0)$ of the right end of the beam, $l$ being the length of the beam.

The centroids of the cross-sections lie on the curve having equation $y = 0$, and this curve describes the shape of the beam. When the weight of the beam is taken into consideration and other forces act on the beam, the centroids of the cross-sections lie on the curve having the equation $y = f(x)$. This curve is called the *elastic curve* of the beam.

In Fig. 6.32(b), the line through $O'$ parallel to $A'B'$ is called the *neutral axis* of the cross-section. It divides the cross-section into two regions, one consisting of the upper fibres and the other the lower fibres. The forces responsible to bend the beam place one of these fibres in tension and the other in compression. The fibres on the neutral axis are neither in tension nor in compression.

The algebraic sum of the moments of the external forces and the couples acting on one side of the beam, taken about the neutral axis, is called the *bending moment* at $x$ and is denoted by $M(x)$. We also use the convention that the moment of any force or couple that is responsible to compress the upper fibre is positive and the moment of any force or couple which tends to place the upper fibre in tension is negative.

It can be shown that a differential equation of the elastic curve is

$$\frac{y''}{[1+(y')^2]^{3/2}} = \frac{M(x)}{EI} \tag{83}$$

where $E$ is Young's modulus of elasticity, and $I$ the moment of inertia of a cross-section about the neutral axis [22]. The value of $E$ depends upon the material from which the beam is made and $I$ depends upon the shape of the cross-section. We shall consider the cases where $E$ and $I$ are constants.

Equation (83) is known as the *Bernoulli-Euler equation* and this can be linearized by using the approximation $y' = 0$. This can be done because the slope $y'$ of the elastic curve is very small when a beam is bent under normal circumstances. Thus, Eq. (83) can be approximated as

$$EIY'' = M(x) \tag{84}$$

(For a proof that the solution of Eq. (84) approximate the solutions of Eq. (83), see [19]).

To solve Eq. (84), we first find $M(x)$ using either the portion of the beam to the left of the plane $x = x$ or the remaining portion to the right of the plane $x = x$. Two integrations of $M(x)$ with respect to $x$ give a solution containing two arbitrary constants. These constants are determined by two sets of initial conditions.

**EXAMPLE 6.27** Find an equation of the elastic curve for a beam $l$ ft long carrying a uniformly distributed load of $W$ 1b/ft. Assume that the beam is simply supported, that is, an upward force of $W\,(l/2)$ lb acts at each end (Fig. 6.33).

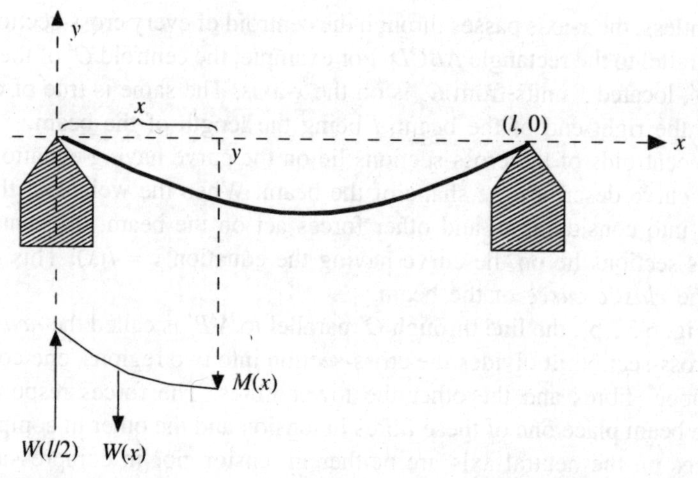


**Fig. 6.33** Simply supported beam carrying a uniformly distributed load.

***Solution*** To find $M(x)$, we shall use the forces acting on the portion of the beam to the left of $x = x$. The moment due to the left supporting force is $(Wl/2)(x)$ and is positive as this force tends to compress the upper fibre at $x = x$. To find the moment of the distributed force due to the weight of the left portion of the beam, we use the resultant of $Wx$ lb acting downward at $x/2$ ft from the origin. The moment of this force about $x = x$ is $-(Wx)(x/2)$. It is negative since this force tends to place the upper fibre at $x = x$ in tension. Thus, $M(x) = -Wx^2/2 + Wl(x/2)$. The left portion is held in equilibrium by two forces used in finding $M(x)$ in addition to an equal and opposite couple of moment $|M(x)|$ and a vertical shearing force $S$ at $x = x$ supplied by the right portion of the beam. Putting the values of $M(x)$ in Eq. (84), we get

$$EIy'' = -\frac{Wx^2}{2} + Wl\frac{x}{2}$$

Integrating twice, we have

$$EIy' = -\frac{Wx^3}{6} + Wl\frac{x^2}{4} + c_1$$

and
$$EIy = -\frac{Wx^4}{24} + \frac{Wlx^3}{12} + c_1x + c_2$$

From the initial conditions, $x = 0$, $y = 0$ and $x = l$, $y = 0$, we obtain

$$c_2 = 0, \qquad c_1 = \frac{Wl^3}{12}$$

Thus, the equation of the elastic curve is

$$y = \frac{-W}{24EI}(x^4 - 2lx^3 + 2l^3x)$$

**EXAMPLE 6.28** Find the equation of an elastic curve of a $l$ ft beam which carries a uniformly distributed load of $W$ lb/ft, and which is built in horizontally at the left end and unsupported at the right end. This type of beam, resembling a diving board, is called a *cantilever beam.*

***Solution*** To find $M(x)$, we shall use the portion of the beam to the right of $x = x$. The advantage of using the right upper portion of the beam is that it is not necessary to calculate the unknown couple and the upward supporting force exerted by the wall on the left end of the beam. The only force acting on the right portion, other than a vertical shearing $S$ at $x = x$, is the resultant of the distributed weight $W(l - x)$ acting $(l - x)/2$ ft to the right of $x = x$ (Fig. 6.34). Thus, $M(x) = -W(l - x)^2/2$. Hence, Eq. (84) becomes

$$EIy'' = \frac{-W(l-x)^2}{2}$$

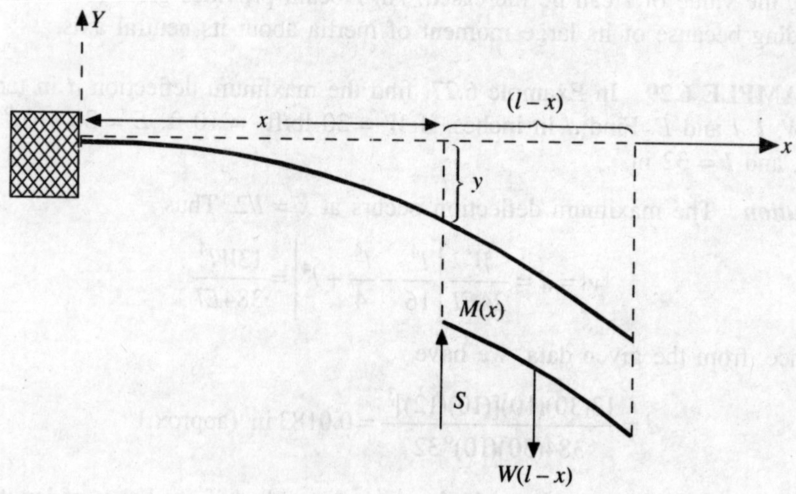


**Fig. 6.34** A cantilever beam.

Integrating, we get

$$EIy' = \frac{W(l-x)^3}{6} + c_1$$

Since, $y' = 0$ when $x = 0$, thus $c_1 = -Wl^3/6$, and

$$EIy' = \frac{W(l-x)^3}{6} - \frac{Wl^3}{6}$$

Integrating again, we get

$$EIy = \frac{-W(l-x)^4}{24} - \frac{Wl^3 x}{6} + c_2$$

Since, $y = 0$ when $x = 0$, we have $c_2 = Wl^4/24$, and

$$EIy = \frac{-W(l-x)^4}{24} - \frac{Wl^3 x}{6} + \frac{Wl^4}{24}$$

Expanding $(1 - x)^4$ and simplifying, we get

$$y = \frac{-W}{24EI}(x^4 - 4lx^3 + 6l^2x^2)$$

The maximum deflection $d$ of a beam is the maximum value assumed by $|y|$ for $0 \le x \le l$. Maximum deflection is important when the beams are designed. Since $|y|$ varies inversely with the constant $EI$, known as the *flexural rigidity*, the quantity $EI$ provides a measure of the tendency of a beam to resist bending. By choosing a less elastic material, the value of $E$ can be increased while by changing the distribution of the cross-sectional area relative to the neutral axis, the value of $I$ can be increased. An $I$ beam provides great resistance to bending because of its large moment of inertia about its neutral axis.

**EXAMPLE 6.29** In Example 6.27, find the maximum deflection $d$ in terms of $W$, $I$, $l$ and $E$. Find $d$ in inches, if $W$ = 30 lb/ft = 10 ft, $E = 30\,(10)^6$ lb/in.$^2$, and $I = 32$ in.$^4$

***Solution*** The maximum deflection occurs at $x = l/2$. Thus

$$|y| = d = \frac{W}{24EI}\left|\frac{l^4}{16} - \frac{l^4}{4} + l^4\right| = \frac{13Wl^4}{384EI}$$

Hence, from the given data, we have

$$d = \frac{13(30)(10)[(10)(12)]^3}{384(30)(10)^6 32} = 0.0183 \text{ in. (approx.)}$$

**EXAMPLE 6.30** A horizontal, simply supported, uniform beam of length $L$ and negligible weight bends under the influence of a concentrated load of $S$ pounds having a distance $L/3$ from one end. Find the equation of the elastic curve.

***Solution*** The elastic curve is shown in Fig. 6.35. The supports at $O$ and $B$ are at distances having a ratio 1:2 from the load $S$. Hence, they support loads having a ratio 2:1 so that at $O$ the amount $2S/3$ is supported while at $B$ the amount $S/3$ is supported. For the determination of the bending moment at $P$, we have the following three cases.

*Case I:* *P is left of S*, i.e. $0 \le x \le L/3$. The portion $OP$ has only one (upward) force acting at a distance $x$ from $P$. The bending moment $M(x)$ is $-2Sx/3$. Hence, Eq. (84) yields

$$EIy'' = \frac{-2Sx}{3}, \qquad 0 \le x \le L \tag{85}$$

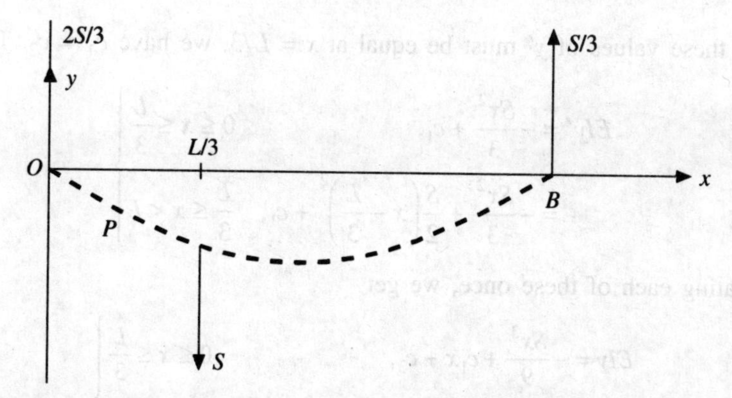


**Fig. 6.35** Schematic diagram of a horizontal, simply supported, uniform beam of length $L$.

*Case II: P is to the right of S, i.e. $L/3 \le x \le L$.* The portion $OP$ has two forces, an upward force $2S/3$, distance $x$ from $P$, and a downward force $S$, the distance $x - L/3$ from $P$. Thus, $M(x) = -2Sx/3 + S[x - (L/3)]$. Hence,

$$EIy'' = -\frac{2Sx}{3} + S\left(x - \frac{L}{3}\right), \quad \frac{L}{3} < x \le L \tag{86}$$

*Case III P is at S.* The portion $OP$ has two forces, an upward force $2S/3$, the distance $L/3$ from $P$, and a downward force $S$, the distance zero from $P$. Thus, $M(x) = -2S/3(L/3) + S(0) = -2SL/9$.

Since, this agrees with $M(x)$ given by Eqs. (85) and (86), we may combine Case III with Cases I and II, by writing the equations as

$$\left.\begin{aligned} EIy'' &= -\frac{2Sx}{3}, & 0 \le x \le \frac{L}{3} \\ &= -\frac{2Sx}{3} + S\left(x - \frac{L}{3}\right), & \frac{L}{3} \le x < L \end{aligned}\right\} \tag{87}$$

Since, each equation in (87) is of the second order, we need four conditions. Two of them are

$$y = 0 \text{ at } x = 0; \quad y = 0 \text{ at } x = L \tag{88}$$

The third condition is obtained by the fact that the two values of $y$ obtained from Eq. (87) must be equal at $x = L/3$. This is the *condition of continuity*. The fourth condition is obtained by realising that there must be a tangent at $x = L/3$. This is the condition for *continuity in the derivative*.

Integrating Eq. (87) each once, we have

$$\left.\begin{aligned} EIy' &= -\frac{Sx^2}{3} + c_1, & 0 \le x \le \frac{L}{3} \\ &= -\frac{Sx^2}{3} + \frac{S}{2}\left(x - \frac{L}{3}\right)^2 + c_2, & \frac{L}{3} \le x < L \end{aligned}\right\} \tag{89}$$

Since, these values of $y'$ must be equal at $x = L/3$, we have $c_1 = c_2$. Thus

$$\left.\begin{aligned} EIy' &= -\frac{Sx^2}{3} + c_1, & 0 \le x \le \frac{L}{3} \\ &= -\frac{Sx^2}{3} + \frac{S}{2}\left(x - \frac{L}{3}\right)^2 + c_1, & \frac{L}{3} \le x < L \end{aligned}\right\} \tag{90}$$

Integrating each of these once, we get

$$\left.\begin{aligned} EIy &= -\frac{Sx^3}{9} + c_1 x + c_3, & 0 \le x \le \frac{L}{3} \\ &= -\frac{Sx^3}{9} + \frac{S}{6}\left(x - \frac{L}{3}\right)^3 + c_1 x + c_4, & \frac{L}{3} \le x \le L \end{aligned}\right\} \tag{91}$$

Using the first and second conditions of Eq. (88) in the first and second equation of (91), respectively, and also using the condition of continuity, we find

$$c_3 = 0, \qquad c_4 = 0, \qquad c_1 = \frac{5SL^2}{81}$$

and hence

$$\left.\begin{aligned} y &= \frac{S}{81EI}(5L^2 x - 9x^3), & 0 \le x \le \frac{L}{3} \\ &= \frac{S}{81EI}\left[5L^2 x - 9x^3 + \frac{27}{2}\left(x - \frac{L}{3}\right)^3\right], & \frac{L}{3} \le x \le L \end{aligned}\right\}$$

## 6.9 COLUMNS

Beams placed vertically to support vertical loads are called *columns*. The plane of the elastic curve of a column contains the axis of each cross-section about which the moment of inertia is greatest. If we think of the column as placed horizontally without relative change of forces acting on it and with the plane of the elastic curve vertical (Fig. 6.36), we can apply Eq. (84) to obtain an approximation to the equation of the elastic curve of the column.

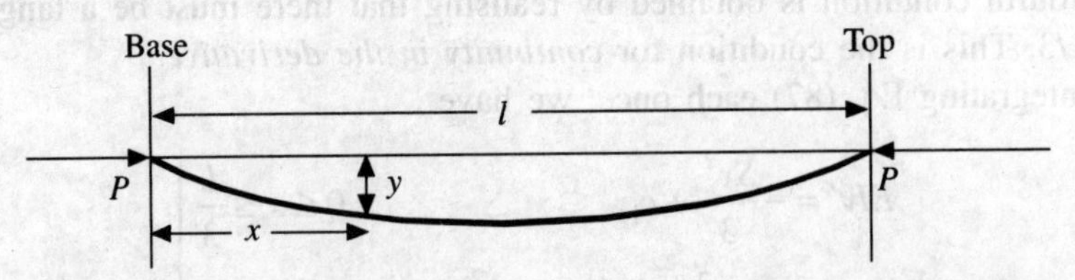


**Fig. 6.36** Schematic diagram of a column placed horizontally.

**EXAMPLE 6.31** Find the equation of the elastic curve of a column (Fig. 6.35) with both ends free to turn and find the buckling stress.*

***Solution*** Neglecting the weight of the column and applying (84), we get

$$EIy'' = -Py$$

which has the solution as

$$y = c_1 \sin ax + c_2 \cos ax, \quad a = \sqrt{\frac{P}{EI}} \tag{i}$$

The initial conditions are

$$x = 0, \quad y = 0, \quad x = l, \quad y = 0 \tag{ii}$$

Applying the first condition of (ii) to (i), we get $c_2 = 0$, and

$$y = c_1 \sin ax \tag{iii}$$

The second condition of (ii) will be satisfied if $c_1 = 0$. In this case, $y = 0$ for all values of $x$ and the column will remain straight. Also $y_1 = c_1 \sin al = 0$ if $al = \pi$; or, since $a = \sqrt{P/EI}$, $P = \pi^2 EI/l^2$. Hence, if $A$ is the area of the horizontal cross-section of the column, the stress $p$ (the force per square unit for buckling) is

$$p = \frac{P}{A} = \frac{\pi^2 EI}{Al^2} \tag{iv}$$

Note that, if $l$ is very small, then from Eq. (iv), $p$ is very large and has no design value. In fact, Eq. (iv) is useful only for a limited range of $l/(Al)^2$.

## 6.10 A PROBLEM IN CARDIOGRAPHY

The field of medicine that deals with the study of the heart is called *Cardiology.* The nature and effects of vibrations of the heart as it pumps blood through the circulatory system of the body are a great source of mathematical applications. An important aspect involves the recording of such vibrations known as *cardiography.* The instrument that records such vibrations is called an *electrocardiograph* (ECG). It translates the vibrations into electrical impulses which are then recorded.

It is interesting to translate the heart vibrations into mechanical vibrations instead of translating these vibrations into electrical impulses. This can be done in the following manner. Suppose that a person rests on a horizontal table which has springs so that it can vibrate horizontally but not vertically. Then, due to the pumping of the heart, the table will undergo small vibrations, the frequency

* If a rod or column is subjected to a sufficiently small axial force, it remains straight. If the axial forces $P$ are gradually increased, a value $P_1$ will be reached at which the rod or column suddenly begins to bend or *buckle*, and its shape varies greatly with a small increase in $P$. The critical force $P_1$, at which the rod begins to buckle, is called the *critical value* and the corresponding stress is the *buckling stress.*

and magnitude of which will depend on the various parameters associated with the heart. Thus, by investigating the motion of the table some important conclusions about the vibrations of heart can be drawn.

Let $x$ denote the horizontal displacement of some specified point of the table (as, for example, one end) from some fixed location (such as a wall). Let $M$ denote the combined mass of the person and that portion of the table which is set into motion. If we assume that there is a damping force proportional to the instantaneous velocity and a restoring force proportional to the instantaneous displacement, then the differential equation describing the motion of the table is

$$M\frac{d^2x}{dt^2}+\beta\frac{dx}{dt}+\gamma x=F \tag{92}$$

where $\beta$ and $\gamma$ are constants of proportionality and $F$ is the force on the system due to the pumping action of the heart (see also Eq. (54)). Suppose that $m$ is the mass of blood pumped out of the heart during each vibration and $y$ is the instantaneous centre of mass of this quantity of blood. Then by Newton's law, we have

$$F=m\frac{d^2y}{dt^2} \tag{93}$$

As a first approximation it may be assumed that $y$ can be expressed as a simple sinusoidal function of $t$ given by

$$y = a\sin\omega t \tag{94}$$

where $a$ and $\omega$ are constants. Equation (94) suggests that there is only one frequency associated with the vibrations of the heart, whereas evidence shows that there are many frequencies. This leads us to replace Eq. (94) by

$$y = a_1 \sin\omega t + a_2 \sin 2\omega t + a_3 \sin 3\omega t + \ldots \tag{95}$$

The series of terms on the right is called a Fourier series. The first term on the right of Eq. (95) represents a first approximation to the function, the sum of the first two terms a better approximation, and so on. Using only the first two terms of the series (95) in (93), and then putting the result into Eq. (92), we obtain

$$M\frac{d^2x}{dt^2}+\beta\frac{dx}{dt}+\gamma x=-m\omega^2(a_1\sin\omega t+a_2\sin 2\omega t) \tag{96}$$

which can be solved subject to various possible conditions.

The general solution of Eq. (96) consists of two parts:

(i) the general solution of the equation with the right side replaced by zero,

(ii) a particular solution.

The first part is the transient solution and will disappear rapidly provided $\beta > 0$. The second part will be the steady state solution in which we are interested.

This steady state solution can be easily obtained as

$$x = \frac{m\omega^2 a_1[(M\omega^2 - \gamma)\sin\omega t + \beta\omega\cos\omega t]}{(M\omega^2 - \gamma)^2 + \beta^2\omega^2}$$

$$+ \frac{4m\omega^2 a_2\,[(4M\omega^2 - \gamma)\sin(2\omega t) + 2\beta\omega\cos(2\omega t)]}{(4M\omega^2 - \gamma)^2 + 4\beta^2\omega^2}$$

A corresponding solution can be found assuming any number of terms in Eq. (95).

## 6.11 CONCENTRATION OF A SUBSTANCE INSIDE AND OUTSIDE A LIVING CELL

Let $I$ denote the region interior to a living cell and $c_i$ denote the concentration in g/cm$^3$ of a substance (solute) dissolved in the cell liquid (solvent). $CO_2$ and lactic acid are examples of such a solute. Assume that the solute is produced inside the cell at $Q$ g/cm$^3$/s, and that a steady state prevails, in which $\partial c_i/\partial t = 0$. Then $c_i$ satisfies the partial differential Eq. (15).

$$D_i\left(\frac{\partial^2 c_i}{\partial x^2} + \frac{\partial^2 c_i}{\partial y^2} + \frac{\partial^2 c_i}{\partial z^2}\right) + Q = 0 \tag{97}$$

where $D_i$ is the coefficient of diffusion for the medium interior to the cell. Let us consider a case in which $D_i$ and $Q$ are both constants, and $c_i$ depends on $x$, $y$, $z$.

Let $E$ denote the region exterior to $I$ and consider the case in which $I$ is spherical. Now introduce the spherical coordinates $(\gamma, \theta, \phi)$, such that

$$x = r\sin\phi\cos\theta, \qquad y = r\sin\phi\sin\theta, \qquad z = r\cos\phi$$

with the origin at the centre of the cell whose radius we denote by $r_0$.

In spherical coordinates

$$\left(\frac{\partial^2 c_i}{\partial x^2} + \frac{\partial^2 c_i}{\partial y^2} + \frac{\partial^2 c_i}{\partial z^2}\right)$$

takes the form

$$\frac{\partial^2 c_i}{\partial r^2} + \frac{2}{r}\frac{\partial c_i}{\partial r} + \frac{1}{r^2}\frac{\partial^2 c_i}{\partial \phi^2} + \frac{\cos\phi}{r^2\sin\phi}\frac{\partial c_i}{\partial \phi} + \frac{1}{r^2\sin^2\phi}\frac{\partial^2 c_i}{\partial \theta^2}$$

Due to the spherical symmetry of the physical situation, the partial derivatives of $c_i$ with respect to $\phi$ and $\theta$ vanish. $c_i$ depends only upon the radial distance $r$, and Eq. (97) reduces to the ordinary differential equation

$$D_i\left(\frac{d^2 c_i}{dr^2} + \frac{2}{r}\frac{dc_i}{dr}\right) + Q = 0 \tag{98}$$

The concentration $c_e$ outside the cell satisfies the equation

$$\frac{d^2 c_e}{dr^2} + \frac{2}{r}\frac{dc_e}{dr} = 0 \tag{99}$$

Here, there is no term of $Q$ as we assume that no solute is generated chemically outside the cell.

Although Eq. (98) has variable coefficients $2D_i^r{}^{-1}$, it can be solved by putting $u = rc_i$. From the relations

$$\frac{du}{dr} = r\frac{dc_i}{dr} + c_i, \qquad \frac{d^2u}{dr^2} = r\frac{d^2c_i}{dr^2} + 2\frac{dc_i}{dr}$$

and Eq. (98), we obtain

$$\frac{d^2u}{dr^2} = -\frac{Q}{D_i}r.$$

Integrating it twice with respect to $r$, we get

$$u = -\frac{Q}{6D_i}r^3 + Ar + B$$

Hence

$$c_i = -\frac{Q}{6D_i}r^2 + A + \frac{B}{r}, \qquad 0 \le r \le r_0 \tag{100}$$

Similarly, the substitution $v = rc_e$ yields

$$c_e = C + \frac{K}{r}, \qquad r_0 \le r \tag{101}$$

The constants $B$ and $C$ in Eqs. (100) and (101) are obtained by making the following assumptions. Assume that $c_i$ has a finite value at the origin and for this to hold $B$ must be zero; otherwise, $c_i$ would be undefined at origin. Next, suppose that $\lim_{r\to\infty} c_e = k$, where $k$ is the concentration that prevailed at every point exterior to the cell before the diffusion from the cell began. It is reasonable to assume that after the steady state sets in, the values of $c_e$ will still be close to $k$ at points where $r$ is large. This is close to $k$ at points where $r$ is large. However, it would not hold if the solute were being created outside the cell. Thus, $C = k$. The value $k = 0$ would prevail if there were no solute outside the cell before the outward diffusion process began. Thus, Eqs. (100) and (101) become

$$c_i = -\frac{Q}{6D_i}r^2 + A \tag{102}$$

$$c_e = k + \frac{K}{r} \tag{103}$$

respectively, and the constants $A$ and $K$ are found from the following boundary conditions at $r = r_0$. Now

$$\frac{dc_i}{dr} = \frac{D_e}{D_i}\frac{dc_e}{dr} \tag{104}$$

$$\frac{dc_i}{dr} = -\frac{h}{D_i}(c_i - c_e) \tag{105}$$

where $D_e$ is the coefficient of diffusion for the medium exterior to the cell and $h$ is the *permeability* of the cellular membrane. (For a discussion of these boundary conditions, see [18].)

From Eqs. (102) and (103), we have

$$\frac{dc_i}{dr} = -\frac{Qr}{3D_i}, \qquad \frac{dc_e}{dr} = -\frac{K}{r^2} \tag{106}$$

Using Eqs. (106) and (104) with $r = r_0$, we obtain

$$-\frac{Qr_0}{3D_i} = \frac{D_e}{D_i}\frac{-K}{r^2}$$

which gives $K = Qr_0^3/(3D_e)$. Thus

$$c_e = k + \frac{Qr_0^3}{3D_e}r^{-1} \tag{107}$$

We also have

$$A = k + \frac{Qr_0}{3h} + \frac{Qr_0^2}{6D_i} + \frac{Qr_0^2}{3D_e} \tag{108}$$

Substituting it in Eq. (100), we get

$$c_i = k + \frac{Qr_0}{3h} + \frac{Q}{6D_i}(r_0^2 - r^2) + \frac{Qr_0^2}{3D_e} \tag{109}$$

Since we have assumed above that the solute is produced inside the cell, $Q$ is positive in the development given. If the solute is consumed inside the cell, the same model also works with $Q < 0$. Oxygen and sugar are examples of substances consumed inside the cell.

## 6.12 DETECTION OF DIABETES

Diabetes mellitus is a disease of metabolism which is characterised by excess of sugar in the blood and urine. In diabetes, the body is unable to burn off all its sugar, starches and carbohydrates due to insufficient supply of insulin. Diabetes is usually diagnosed by fasting and post periodical blood test or by *glucose tolerance test* (GTT). In GTT, the patient after fasting overnight is given a large dose of glucose, and during the next three to five hours several measurements are made of the concentration of glucose in the patient's blood. In the mid-sixties, Dr. Rosevear and Dr. Molnar of the Mayo Clinic (USA) and Dr. Ackerman and Dr. Gatewood of the University of Minnesota (USA) discovered a criterion for interpreting the results of GTT. Their model is based on the following simple and well-known facts of elementary biology.

1. *Glucose is a source of energy* for all tissues and organs. For each individual there is an optimal blood glucose concentration. Any excessive deviations from this optimal concentration leads to severe pathological conditions and potential death.
2. Blood glucose levels are influenced and controlled by a variety of hormones and metabolism such as the following:

(i) *Insulin*, a hormone secreted by $\beta$ cells of the pancreas. After absorbing carbohydrates, the pancreas secretes more insulin. In addition, the glucose in the blood directly stimulates the $\beta$ cells to secrete insulin. Insulin facilitates the tissue uptake of glucose. Without sufficient insulin, the body cannot have all the energy it needs.

(ii) *Glucagon.* It is a hormone secreted by the $\alpha$ cell of the pancreas. Any excess glucose is stored in the liver in the form of glycogen and when needed this glycogen in converted back into glucose. The hormone glucagon increases the rate of breakdown of glycogen into glucose. Evidences have shown that hypoglycaemia (low blood sugar) and fasting promote the secretion of glucagon while increased blood glucose levels suppresses its secretion.

(iii) *Epinephrine* (adrenalin). This is a hormone secreted by the adrenal medulla, and is a part of the emergency mechanism to quickly increase the concentration of glucose in the blood in case of extreme hypoglycaemia. It also increases the rate of breakdown of glycogen into glucose like glucagon, but in addition, it directly inhibits glucose uptake by muscle tissues—it acts directly on the pancreas to inhibit insulin secretion. It also aids in the conversion of lactate to glucose in the liver.

(iv) *Glucocorticoids* are hormones, such as cortisol, which are secreted by the adrenal cortex. These hormones play an important role in the metabolism of carbohydrates.

(v) *Thyroxin* is a hormone secreted by the thyroid gland. It helps the liver to form glucose from noncarbohydrate sources such as glycerol, lactate and amino acids.

(vi) *Growth hormone* (somatotropin). This is a hormone secreted by the anterior pituitary gland. This hormone not only affects glucose level in a direct manner, but also tends to block insulin. These hormones decrease the sensitivity of muscle and adipose membrane to insulin, thereby reducing the effectiveness of insulin in promoting glucose uptake.

The model postulated is a simple one, requiring only a limited number of blood sample during a GTT, and it centres on two concentrations:

(a) $G$ (glucose in the blood), and
(b) $H$ (net hormonal concentration).

The latter represents the commutative effect of all the important hormones. Insulin is considered to increase $H$ while cortisone decreases $H$. Then the basic model is described by the equations

$$\frac{dG}{dt} = F_1(G, H) + J(t) \tag{110}$$

$$\frac{dH}{dt} = F_2(G, H) \tag{111}$$

where $J(t)$ is the external rate at which the blood glucose concentration is being increased.

We assume that by the time a fasting patient arrives in the hospital, $G$ and $H$ have achieved the equilibrium values $G_0$ and $H_0$ (say), i.e.

$$F_1(G_0, H_0) = 0 = F_2(G_0, H_0)$$

We are now interested in deviations of $G$ and $H$ from their equilibrium values, so we take

$$G = G_0 + g, \qquad H = H_0 + h$$

where $g$ and $h$ are small as compared to $G_0$ and $H_0$. Then

$$\frac{dg}{dt} = F_1(G_0 + g, H_0 + h) + J(t) = f_1(g, h) + J(t) \text{ (say)}$$

$$\frac{dh}{dt} = f_2(g, h) \text{ (say)}$$

We assume also that $f_1$ and $f_2$ have a linear form. Thus

$$\frac{dg}{dt} = -ag - bh + J(t) \tag{112}$$

$$\frac{dg}{dt} = -ch + eg \tag{113}$$

Now, $a > 0$, since $dg/dt < 0$ for $h = 0$ through tissue uptake of glucose, and $b > 0$, since $h > 0$ tends to decrease blood glucose levels. Also, $c > 0$ since the concentration of hormones in the blood decreases through hormone metabolism and $e > 0$ for $g > 0$ causes the endocrine glands to secrete those hormones that tend to increase $h$.

From Eq. (112)

$$h = \frac{1}{b}\left(-ag + J - \frac{dg}{dt}\right)$$

Substituting this in Eq. (113), we get

$$\frac{d^2g}{dt^2} + (a + c)\frac{dg}{dt} + (ac + be)\,g = \frac{dJ}{dt} + cJ \tag{114}$$

The right-hand side of Eq. (114) is zero except over a small interval of time. Let time $t = 0$ be the time the glucose load has been ingested so that for $t > 0$

$$\frac{d^2g}{dt^2} + (a+c)\frac{dg}{dt} + (ac + be)\,g = 0 \tag{115}$$

and can be written as

$$\frac{d^2g}{dt^2} + 2A\frac{dg}{dt} + B^2 g = 0$$

The solutions of this equation are of three different types, depending upon whether $A^2 - B^2$ is positive, negative or zero. These three types correspond to the overdamped, critically damped and underdamped case discussed earlier. We shall take here the case $A^2 - B^2 < 0$ (the other cases may be dealt in a similar manner) and the solution is

$$g = e^{-At}\,[C\cos(B_0 t) + D\sin(B_0 t)]$$

where $B_0 = B - A$. Thus, the complete solution is

$$G = G_0 + e^{-At}\,[C\cos(B_0 t) + D\sin(B_0 t)] \tag{116}$$

This contains five constants $G_0$, $A$, $B_0$, $C$, $D$. One way of finding them is as follows: The patient's blood glucose concentration before the glucose load is ingested is $G_0$. Hence, we can determine $G_0$ by measuring the patient's blood glucose concentration immediately upon his arrival at the hospital. Next, we take four additional measurements $G_1$, $G_2$, $G_3$ and $G_4$ at time $t_1$, $t_2$, $t_3$ and $t_4$. We can then determine $A$, $B_0$, $C$ and $D$ from the four equations

$$G_i = G_0 + e^{-At_i}\,[C\cos(B_0\,t_i) + D\sin(B_0\,t_i)], \qquad i = 1, 2, 3, 4$$

The second way—the better way—is to take, say $n$ measurements, and use the 'least squares' techniques to find the best fit for the parameters. However, this has to be solved on a computer.

Observation has shown that slight errors in measuring $G$ can produce large errors in the value of $A$. However, the value of the parameter $B_0$ was relatively insensitive to experimental errors in $G$. So we determine $B_0$, and use this value as a basic description for the response to a GTT. The remarkable fact is that a value for $t_0 = 2\pi/B_0$ of less than *4 hours* indicated *normality*, while appreciably *more than 4 hours* indicated *mild diabetes*.

## 6.13 CHEMICAL KINETICS

Chemical reactions are governed by the law of mass action (see Section 4.14), which states that the rate of a reaction is proportional to the active concentrations of the reactants. For example, if a molecule each of $P$ and $Q$ combine reversibly to form $R$, we write

$$P + Q \underset{k_2}{\overset{k_1}{\rightleftharpoons}} R \tag{117}$$

and if $x_1$, $x_2$, $x_3$ are the concentrations of $P$, $Q$, $R$, respectively, the law of mass action gives

$$\frac{dx_1}{dt} = \frac{dx_2}{dt} = k_2 x_3 - k_1 x_1 x_2 \tag{118}$$

$$\frac{dx_3}{dt} = k_1 x_1 x_2 - k_2 x_3 \tag{119}$$

where $k_1$ and $k_2$ are the reaction rates.

We shall study here the simpler reaction given by

$$P + P \underset{k_2}{\overset{k_1}{\rightleftharpoons}} P_2 \tag{120}$$

If $x$ is the concentration of $P$ and $y$ of $P_2$, then

$$\frac{dx}{dt} = 2k_2 y - 2k_1 x^2 \tag{121}$$

$$\frac{dy}{dt} = k_1 x^2 - k_2 y \tag{122}$$

From Eqs. (121) and (122), we have

$$\frac{d}{dt}\left(\frac{dx}{dt} + 2k_1 x^2\right) = 2k_2 k_1 x^2 - k_2\left(\frac{dx}{dt} + 2k_1 x^2\right)$$

or

$$\frac{d^2x}{dt^2} + \frac{dx}{dt}(4k_1 x + k_2) = 0 \tag{123}$$

This is a second order nonlinear differential equation that can be solved by putting $p = dx/dt$. Solving for $p$ as a function of $x$, rather than $t$, we have

$$\frac{d^2x}{dt^2} = \frac{dp}{dt} = \frac{dp}{dx}\frac{dx}{dt} = p\frac{dp}{dx}$$

so that Eq. (123) becomes

$$p\frac{dp}{dx} + p(4k_1 x + k_2) = 0 \tag{124}$$

Assume that $p \neq 0$ (if $p = 0$, $dx/dt = 0$, and we have equilibrium)

$$\frac{dp}{dx} = -4k_1 x - k_2$$

and integrate to get

$$p = \frac{dx}{dt} = -2k_1 x^2 - k_2 x + c_1 \tag{125}$$

This is a first order differential equation. Now, separate the variables, after resolving into a partial fraction and integrating the resulting expression, we get

$$\frac{1}{a}\log\left(\frac{x - a_1}{x - a_2}\right) = -t + c_2$$

where

$$a = \left(k_2^2 + 8c_1k_1\right)^{1/2}, \quad a_1 = \frac{-k_2 + a}{4k_1}, \quad a_2 = -\frac{k_2 + a}{4k_1}$$

After rearrangement, we obtain

$$x = \frac{(k_2 + a)\, c_3 e^{-at} + a - k_2}{4k_1(1 - c_3 e^{-at})} \tag{126}$$

where $c_3 = e^{ac_1}$. Substitution of (125) into (121) yields

$$2k_2 y = -k_2 x + c_1 \tag{127}$$

so that, if the initial concentrations $x_0$, $y_0$ are given, we have

$$c_1 = 2k_2 y_0 + k_2 x_0 \tag{128}$$

Also, from Eq. (126)

$$c_3 = \frac{4k_1 x_0 - a + k_2}{4k_1 x_0 + a + k_2} \tag{129}$$

We can now obtain the complete solutions for $x$ and $y$, but more important is the approximate behaviour as $t$ gets large. From (126), as $t \to \infty$, we see that

$$x \to \frac{a - k_2}{4k_1} = \frac{(k_2^2 + 8c_1k_1)^{1/2} - k_2}{4k_1} = X \text{ (say)}$$

and from Eq. (127)

$$y \to \frac{k_2 - (k_2^2 + 8c_1k_1)^{1/2}}{8k_1} + \frac{c_1}{2k_2} = Y \text{ (say)}$$

so as $t$ increases, we tend to an equilibrium solution. This is illustrated in Fig. 6.37.

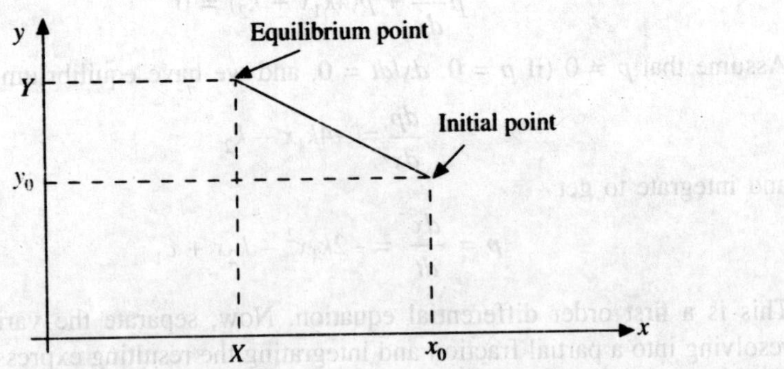


**Fig. 6.37** Equilibrium solution in chemical kinetics.

## 6.14 APPLICATIONS TO ECONOMICS

### 6.14.1 A Microeconomic Market Model

We have already considered (Section 4.17) a simple market model in which $Q_d$, the number of items of a commodity in demand per unit of time, and $Q_s$, the number of items of a commodity in supply per unit time, are linear functions of the price $P$ of a single item. We now consider a more complicated model in which $Q_d$ and $Q_s$ depend linearly not only on $P$ but also on $dP/dt$ and $d^2P/dt^2$. This model incorporates the effect of trends in price on demand and supply. Here, $dP/dt$ represents the effect of rising or falling prices while $d^2P/dt^2$ denotes the effect of increasing or decreasing price rates of change. Let

$$Q_d = a - Pb + AP' + BP'', \qquad Q_s = -c + Pd + CP' + DP''$$

where $a$, $b$, $c$, $d$ are positive constants as in the simpler model (cf. Section 4.17), and the constants $A$, $B$, $C$ and $D$ are positive, negative or zero.

For economic equilibrium to prevail, we have $Q_d = Q_s$, and this leads to the following differential equation for $P$:

$$(B - D)P'' + (A - C)P' - (b + d)P = -(a + c)$$

We consider the case $B \neq D$. Since this leads to the second order differential equation

$$P'' + \frac{A - C}{B - D}P' - \frac{b + d}{B - D}P = -\frac{a + c}{B - D} \tag{130}$$

we see by inspection that

$$P_p = \frac{a + c}{b + d} = \overline{P}$$

defines a particular solution of Eq. (130). The positive constant $\overline{P} = (a + c)/(b + d)$ is called the *equilibrium price*. The equilibrium is said to be *dynamically stable* iff $\lim_{t \to \infty} P(t) = \overline{P}$. The importance of the dynamical stability has been discussed in [4].

To determine whether a dynamically stable equilibrium prevails or not, we examine the roots of the auxiliary equation

$$m^2 + \frac{A - C}{B - D}m - \frac{b + d}{B - D} = 0$$

of Eq. (130). Since, $b + d > 0$, $m_1$ and $m_2$ are both nonzero. If $m_1$ and $m_2$ are real and unequal, the complete solution (130) is

$$P = P(t) = c_1 e^{-m_1 t} + c_2 e^{m_2 t} + \overline{P}$$

The constants $c_1$ and $c_2$ can be found from the initial values $P(0)$ and $P'(0)$. Since, $c_1$ (or $c_2$) will be zero only for very special conditions, for dynamic stability, we require that $\lim_{t\to\infty} P(t) = \overline{P}$, where $c_1$ and $c_2$ are both different from zero. It can be easily seen that

$$\lim_{t\to\infty} P(t) = \overline{P} \qquad \text{if } m_1 < 0 \text{ and } m_2 < 0 \text{ (if } m > 0,\ \lim_{t\to\infty} e^{mt} = \infty\text{)}$$

If $m_1 = m_2$, the solution of Eq. (130) is

$$P = P(t) = c_1 e^{m_1 t} + c_2 t e^{m_1 t} + \overline{P}$$

In this case, $\lim_{t\to\infty} P(t) = \overline{P}$ iff $m_1 < 0$.

Finally, if $m_1 = \alpha + i\beta$ and $m_2 = \alpha - i\beta$, then the solution of Eq. (130) is

$$P = P(t) = e^{\alpha t}(c_1 \sin \beta t + c_2 \cos \beta t) + \overline{P}$$

Here, $\lim_{t\to\infty} P(t) = \overline{P}$ iff $\alpha < 0$. (The absolute value of the harmonic factor $c_1 \sin \beta t + c_2 \cos \beta t$ cannot exceed $|c_1| + |c_2|$.)

Thus, it is possible to determine whether the solution of Eq. (130) does or does not converge to $\overline{P}$ without obtaining the complete solution.

**EXAMPLE 6.32** Given $Q_d = 8 - 2P - 3P' + 3P''$, $Q_s = -6 + 4P + 2P' + 4P''$ determine whether or not the equilibrium is dynamically stable.

***Solution*** Setting $Q_d = Q_s$, we have $P'' + 5P' + 6P = 14$, and the characteristic equation $m^2 + 5m + 6 = 0$ has the root $m_1 = -3$ and $m_2 = -2$. Since, $m_1$ and $m_2$ are both negative, the equilibrium is dynamically stable, and we have

$$\lim_{t\to\infty} P(t) = \overline{P} = \frac{a+c}{b+d} = \frac{8+6}{2+4} = \frac{7}{3}$$

**EXAMPLE 6.33** Given that $Q_d = 10 - P - P' + 2P''$, $Q_s = -5 + 4P + P' + 3P''$, find the price $P$ in terms of $t$ if $P(0) = 5$ and $P'(0) = 1$. Determine whether or not the equilibrium is dynamically stable.

***Solution*** Setting $Q_d = Q_s$, we get

$$P'' + 2P' + 5P = 15 \tag{131}$$

The roots of the auxiliary equation are $m_1 = -1 + 2i$ and $m_2 = -1 - 2i$. A particular solution of Eq. (131) is given as $P_p = 3 = \overline{P}$, and the complete solution of Eq. (131) is

$$P = P(t) = e^{-t}(c_1 \sin 2t + c_2 \cos 2t) + 3$$

Differentiate it with respect to $t$, to get

$$P'(t) = e^{-t}(2c_1 \cos 2t - 2c_2 \sin 2t) - e^{-t}(c_1 \sin 2t + c_2 \cos 2t)$$

From $P(0) = 5$ and $P'(0) = 1$, we find $5 = c_2 + 3$ and $1 = 2c_1 - c_2$, which give $c_2 = 2$ and $c_1 = 3/2$. Hence

$$P = P(t) = e^{-t}\left(\frac{3}{2}\sin 2t + 2\cos 2t\right) + 3$$

Since, $\alpha = -1 < 0$, the equilibrium is dynamically stable and $\overline{P} = \lim_{t\to\infty} P(t) = 3$.

### 6.14.2 Price and Supply Model

Another important problem in economics deals with the behaviour of the price $P$ of a commodity at some time $t$. This problem depends upon several factors. We may suppose that the commodity is such that increasing its price results in an increase of supply $S$ which, as a consequence, ultimately results in lowering of the price. Also assume that the inflation factor is given as a function of time by $F(t)$ (as prices change with time as a result of inflation), and the time rate of change in price is proportional to the difference between the supply $S$ at time $t$ and some equilibrium supply $S_0$. If $S > S_0$, the supply is too large and the price tends to decrease, while if $S < S_0$, the supply is too small and the price tends to increase so that the constant of proportionality must be negative and is denoted by $-k_1$.

We thus have

$$\frac{dP}{dt} = F(t) - k_1(S - S_0) \tag{132}$$

Now, assume that the time rate of change of supply is proportional to the difference between the price and some equilibrium price $P_0$. Then, we have

$$\frac{dS}{dt} = k_2(P - P_0) \tag{133}$$

where $k_2$ is the constant of proportionality. If $P < P_0$, the price is too low, $dS/dt$ is negative, and the supply decreases. If $P > P_0$, the price is too high, $dS/dt$ is positive, and the supply increases. Thus, $k_2 > 0$.

Solving Eq. (132) for $S$, we obtain

$$S = S_0 + \frac{1}{k_1}\left[F(t) - \frac{dP}{dt}\right] \tag{134}$$

Substituting this into Eq. (132) and assuming that $S_0$ is constant, we get

$$\frac{d^2P}{dt^2} + k_1k_2P = F(t) + k_1k_2P_0 \tag{135}$$

Now depending upon the inflation factor $F(t)$, different cases arise. We consider the special case where

$$F(t) = \alpha \tag{136}$$

with $\alpha$ as a constant. In this case, Eq. (135) has the general solution

$$P = P_0 + A\cos(\sqrt{k_1k_2})\,t + B\sin(\sqrt{k_1k_2})\,t \tag{137}$$

With this, Eq. (132) reduces to

$$S = S_0 + \frac{\alpha}{k_1} + A\sqrt{k_2/k_1}\,\sin(\sqrt{k_2/k_1})t - B\sqrt{k_2/k_1}\,\cos(\sqrt{k_2/k_1})t \tag{138}$$

Assume that at $t = 0$, $P = P_0$ and $S = S_0$, then from Eqs. (137) and (138), we find that

$$A = 0, \qquad B = \frac{\alpha}{\sqrt{k_1 k_2}}$$

so that

$$P = P_0 + \frac{\alpha}{\sqrt{k_1 k_2}} \sin(\sqrt{k_1 k_2})t \tag{139}$$

and

$$S = S_0 + \frac{\alpha}{k_1} - \frac{\alpha}{k_1} \cos(\sqrt{k_1 k_2})t \tag{140}$$

These equations show that the price and supply oscillate sinusoidally about the values $P_0$ and $S_0 + \alpha/k_1$, respectively, with period

$$T = \frac{2\pi}{\sqrt{k_1 k_2}} \tag{141}$$

If we write Eq. (140) in the form

$$S = S_0 + \frac{\alpha}{k_1} + \frac{\alpha}{k_1} \sin \sqrt{k_1 k_2}\left(t - \frac{\pi}{2\sqrt{k_1 k_2}}\right) \tag{142}$$

we see that the price leads the supply by a time equal to one quarter of the period, i.e. $1/4T$ or that the supply lags the price by $1/4T$. This means that if the maximum price occurs at some particular time when the maximum supply occurs at a time $1/4T$ later when the price has already fallen. The situation is shown graphically in Fig. 6.38, where for the purposes of comparison the same vertical axis is used for both $S$ and $P$, although the units for these variables are not same.

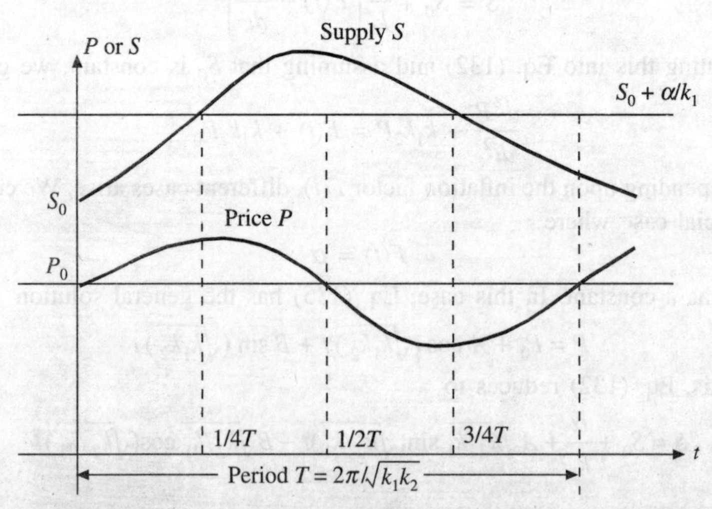


**Fig. 6.38** Schematic diagram for a price and supply model.

## EXERCISES

**1.** A 2-lb weight suspended from a spring stretches it 1.5 inches. If the weight is pulled 3 in. below equilibrium position and released, obtain the differential equation of motion and find the velocity, position of the weight, amplitude, the period and frequency of motion. Also determine the position, velocity and acceleration $\pi/64$ s after the weight is released.

**2.** A 3-lb weight on a spring stretches it 6 inches. When equilibrium is reached the weight is struck so as to give it a downward velocity of 2 ft/s. Find: (a) the velocity and position of the weight at time $t$ s after the impact, (b) the amplitude, period and frequency of the motion, and (c) the velocity and acceleration when the weight is 1 in. above the equilibrium position and moving upward.

**3.** A 4-lb weight is attached to a spring whose spring constant is 16 lb/ft. What is the period of the simple harmonic motion?

**4.** A 24-lb weight, attached to the end of a spring, stretches it 4 inches. Find the equation of motion if the weight is released from rest from a point 3 in. above the equilibrium position.

**5.** An 8-lb weight attached to a spring exhibits simple harmonic motion. Determine the equation of motion if the spring constant is 1 lb/ft and the weight is released 6 in. below the equilibrium position with a downward velocity of 3/2 ft/s. Express the solution in the form $x(t) = A \sin(\omega t + \phi)$, $A = \sqrt{a^2 + b^2}$.

**6.** A 4-lb weight suspended from a spring stretches it 3 inches. The weight is pulled 6 in. below the equilibrium position and released. Assume that the weight is acted upon by a damping force which (in pounds) is numerically equal to $2v$, where $v$ is the instantaneous velocity in ft/s.

(a) Set up a differential equation and conditions describing the motion.

(b) Determine the position of the spring at any time after the weight is released.

(c) Write the result of (b) in the form $A(t) \sin(\omega t + \phi)$.

Thus, determine the time varying amplitude, quasi-period and phase angle.

**7.** A 64-lb weight is suspended from a spring having a constant of 50 lb/ft. The weight is acted upon by a resisting force (in pounds) which is numerically equal to 12 times the instantaneous velocity. If the weight is pulled 6 in. below the equilibrium position and then released, describe the motion, giving the time-varying amplitude and the quasi-period of the motion.

**8.** A 2-lb weight on a spring stretches it to 1.5 inches. The weight is pulled 6 in. below the equilibrium position and released. Assume a damping force numerically equals to $2v$. Find the position of the weight at any time and determine whether the motion is overdamped or critically damped.

**9.** A 4-lb weight is attached to a spring whose constant is 2 lb/ft. The medium offers a resistance to the motion of the weight numerically equal to the instantaneous velocity. If the weight is released from a point 1 ft above the equilibrium position with a downward velocity of 8 ft/s, determine the time that the weight passes through the equilibrium position. Find the time for which the weight attains its extreme displacement from the equilibrium position. What is the position of the weight at this instant?

**10.** A 1-kg mass is attached to a spring whose constant is 16 N/m and the entire system is then submerged in a liquid which imparts a damping force equal to 10 $v$. Determine the equation of motion if: (a) the weight is released from rest 1 m below the equilibrium position, (b) the weight is released 1 m below the equilibrium position with an upward velocity of 12 m/s.

**11.** A force of 2-lb stretches a spring to 1 ft. A 3.2-lb weight is attached to the spring and the system is then immersed in a medium that imparts a damping force equal to 0.4 $v$.

(a) Find the equation of motion if the weight is released from rest 1 ft above the equilibrium position.

(b) Write the solution in the form $A(t) \sin (\omega t + \phi)$.

(c) Find the first time for which the weight passes through the equilibrium position heading upward.

**12.** A 10-lb weight attached to a spring stretches it 2 ft. The weight is attached to a dashpot damping device which offers a resistance equal to $\beta$ ($\beta > 0$) times the instantaneous velocity. Determine the values of the damping constant $\beta$ so that the subsequent motion is: (a) overdamped, (b) critically damped, and (c) underdamped.

**13.** A rectilinear motion is governed by $d^2x/dt^2 + \beta(dx/dt) + 64x = 0$, where $\beta > 0$. For what value of $\beta$ is the motion critically damped?

**14.** A cubical box 10 ft on a side floats in still water (density 62.5 lb/ft$^2$). It is observed that the box oscillates up and down with a period of 1/2 s. What is the weight of the box (use $g = 32$)?

**15.** A cylindrical spar buoy of diameter 1 ft and weight 100 lb floats, partially submerged, in an upright position. When it is depressed slightly from its equilibrium position and released it bobs up and down according to the equation

$$\frac{100}{g}\frac{d^2x}{dt^2} = -16\pi x - c\frac{dx}{dt}$$

Here, $c(dx/dt)$ is the frictional resistance of water. Find $c$ if the period of oscillation is 1.6 s (use $g$ = 32 ft/s$^2$).

**16.** Gravity at an internal point varies with the distance $x$ of the point from the centre of the earth. Assume that the earth is a sphere of radius 4000 miles. Show that a particle dropped from rest into a hole extending diametrically through the earth will undergo a simple harmonic motion. Find the period in hours and the speed in miles/h for $x = 0$.

**17.** The block shown in Fig. 6.39 weighs 16 lb, the spring constant $k$ = 36 lb/ft, and the resistance coefficient $\beta = 9$. If the block is pulled 4 in. to the right and released from rest, what are the velocity and displacement at time $t$?

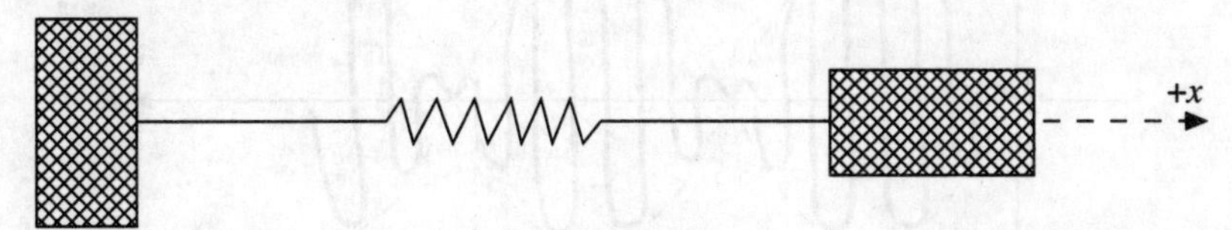


**Fig. 6.39**

**18.** The block shown in Fig. 6.39 weighs 32 lb, $k$ = 16 lb/ft, the resistance coefficient $\beta = 2$, and the block is acted on by an impressed force $F(t) = 96 \cos 4t$, for $x = 1/2$ ft and $v = 0$ ft/s when $t = 0$ s. What is the displacement and velocity at time $t$?

**19.** A vertical spring having a constant 5 lb/ft has a 16 lb weight suspended from it. An external force $F(t) = 24 \sin 10t$, $t \geq 0$ is applied. A damping force given by 4 $v$ is assumed to act. Initially the weight is at rest at its equilibrium position:

(a) Determine the position of the weight at any time.

(b) Indicate the transient and steady state solutions.

(c) Find the amplitude, period and frequency of the steady state solution.

**20.** A 16-lb weight stretches a spring 8/3 ft. Initially the weight starts from rest, 2 ft below the equilibrium position and the subsequent motion takes place in a medium that offers a damping force numerically equal to 1/2 the instantaneous velocity. Solve the equation of motion if the weight is driven by an external force $F(t) = 20 \cos 3t$.

**21.** When a mass of 2 kg is attached to a spring whose constant is 32 N/m, it comes to rest in the equilibrium position. Starting at $t = 0$, a force $F(t) = 68\, e^{-2t} \cos 4t$ is applied to the system. Solve the equation of motion in the absence of damping.

**22.** The motion of a mass on a certain vertical spring is described by

$$\frac{d^2x}{dt^2} + 100x = 36 \cos 8t, \qquad x = 0, \qquad \frac{dx}{dt} = 0 \quad \text{at} \quad t = 0$$

where $x$ is the distance of the mass from the equilibrium position, downward being taken as positive direction.

(a) Give a physical interpretation to the problem.

(b) Show that the solution can be written as $x = 2 \sin t \sin 9t$.

(c) Show that the graph of $x$ (a function of $t$) is similar to the one shown in Fig. 6.40.

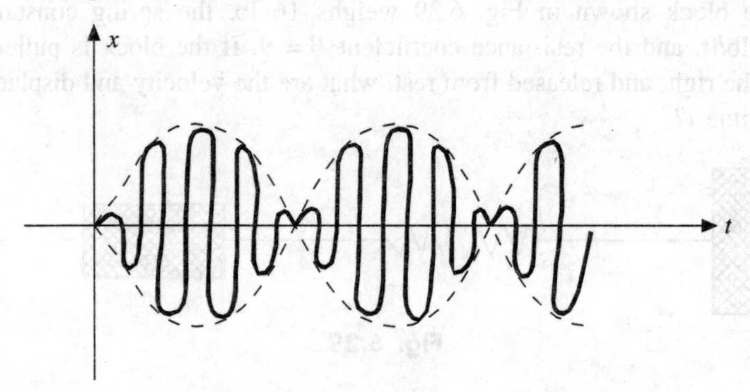


**Fig. 6.40**

**23.** The equation of forced vibration of a mass on a vertical spring is

$$m\frac{d^2x}{dt^2} + \beta\frac{dx}{dt} + kx = A\cos\omega t, \qquad t \geq 0$$

where $x$ is the displacement of the mass from the equilibrium position and $m$, $\beta$, $k$, $A$ and $\omega$ are positive constants. (a) Show that a steady state oscillation is given by

$$x = \frac{A}{\sqrt{(m\omega^2 - k)^2 + \beta^2\omega^2}}\cos(\omega t + \phi)$$

(b) Show that the maximum oscillation (resonance) will occur if $\omega$ is chosen so that

$$\omega = \sqrt{\frac{k}{m} - \frac{\beta^2}{2m^2}}$$

provided that $\beta^2 < 2$ km (c) Show that at resonance, the amplitude of the oscillation varies inversely as the damping constant.

**24.** A circular disk of mass $m$ and radius $r$ is suspended by a thin wire attached to the centre of one of its flat faces. If the disk is twisted through an angle $\theta$, torsion in the wire tends to turn the disk back in the opposite direction. The differential equation for the motion is

$$\frac{1}{2}mr^2\frac{d^2\theta}{dt^2} = -k\theta$$

where $k$ is the coefficient of torsion of the wire. Find the motion if $\theta = \theta_0$ and $d\theta/dt = v_0$ at $t = 0$.

**25.** A series LRC circuit contains $L = 1/2$ H, $R = 10\ \Omega$, $C = (1/100)$ F and $E(t) = 150$ V. Determine the charge $q(t)$ on the capacitor for $t > 0$ if $q(0) = 1$, $I(0) = 0$. What is the charge on the capacitor after a long time?

**26.** An emf of 500 V is in series with a 20 Ω resistor, an inductor of 1/0.25 H and a 0.008-F capacitor. At $t = 0$, $q$ and $I$ are zero. (a) Find $q$ and $I$ at any time $t \geq 0$. (b) Indicate the transient and steady state terms in $q$ and $I$. (c) Find the charge and current after a long time.

**27.** A 0.1 H inductor, a $(4 \times 10^{-3})$ F capacitor, and a generator having emf given by $180 \cos 40t$, $t \geq 0$, are connected in series. Find the instantaneous charge $q$ and current $I$ if $I = q = 0$ at $t = 0$.

**28.** An inductor of 0.5 H is in series with a constant 50-V emf and a capacitor 0.02 F. At $t = 0$, the charge $q$ on the capacitor is 2 coulombs and the current $I$ is zero amperes. Find $q$ and $I$ in terms of $t$.

**29.** An inductor of 0.5 H is connected in series with a resistor of 6 Ω, a capacitor of 0.02 F, a generator having alternating voltage given by $24 \sin 10t$, $t \geq 0$, and a switch $K$ (Fig. 6.41).

(a) Set up a differential equation for the charge on the capacitor.

(b) Find the charge and the current at time $t$ if the charge on the capacitor is zero when the switch is closed at $t = 0$. Compare this problem with Example 6.19 and its graph (Fig. 6.23).

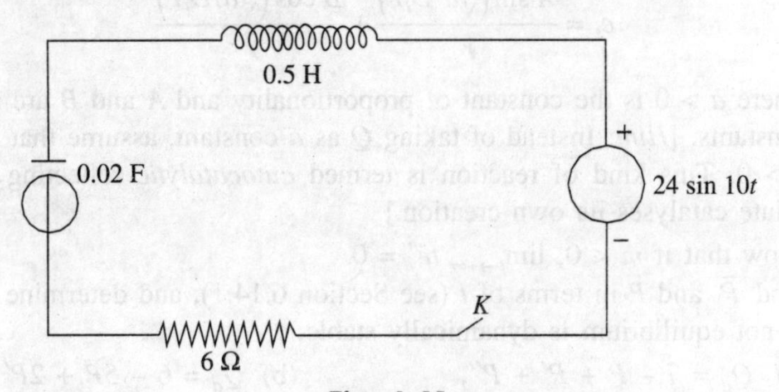


Fig. 6.41

**30.** A cable of suspension bridge has its supports at the same level, at a distance 500 ft apart. If the supports are 100 ft higher than the minimum point of the cable, use an appropriate set of axes to determine the equation of the curve in which the cable hangs, assuming the bridge is of uniform weight and that the weight of the cable is negligible. Find the slope of the cable at the support.

**31.** A cable weighs 0.15 lb/ft. It is hung from two supports which are at the same level, 100 ft apart. If the slope of the cable at one support is 12/5:

(a) Find the tension in the cable at its lowest point.

(b) Determine the equation for the curve in which the cable hungs.

**32.** A uniform cantilever beam of length $L$ and of negligible weight has a concentrated load of $S$ pounds at the free end. Find (a) the equation of the elastic curve, and (b) the maximum deflection.

**33.** A 10 ft steel cantilever beam ($E = 30 \times 10^6$ lb/in.$^2$) carries a uniformly distributed load of 1500 lb/ft. Given $I = 45$ in.$^4$, use the result given in Example 6.28 to find the maximum deflection.

**34.** A 8 ft oak beam ($E = 15 \times 10^{15}$ lb/in.$^2$) carrying a uniformly distributed load of 150 lb/ft is simply supported at both ends. Given that $I = 57$ in.$^4$, find (a) the deflection at $x = 4$ ft, and (b) the maximum deflection.

**35.** A $l$ ft beam, built in horizontally at the left end and simply supported at the right end, carries a uniformly distributed load of $W$ lb/ft. Find an equation of the elastic curve.

**36.** Use the substitution $v = rc_e$ in Eq. (99) to obtain Eq. (101).

**37.** The concentration of the solute is discontinuous at the boundary of the spherical shell. Compute $c_i - c_e$ for $r = r_0$.

**38.** A substance is produced inside a spherical living of radius $r$ at a rate proportional to the concentration $c_i$ of the dissolved substance. Use Eq. (98) to show that

$$c_i = \frac{A \sin\left(\sqrt{a/D_i r}\right)}{r} + \frac{B \cos\left(\sqrt{a/D_i r}\right)}{r}$$

where $a > 0$ is the constant of proportionality and $A$ and $B$ are arbitrary constants. [*Hint*: Instead of taking $Q$ as *a* constant, assume that $Q = ac_i$, $a > 0$. This kind of reaction is termed *autocatalytic* reflecting that the solute catalyses its own creation.]

**39.** Show that if $m < 0$, $\lim_{t \to +\infty} te^n = 0$.

**40.** Find $\overline{P}$ and $P$ in terms of $t$ (see Section 6.14.1), and determine whether or not equilibrium is dynamically stable:

(a) $Q_d = 7 - P + P' + P''$,
$Q_s = -9 + P + 4P' + 2P''$
$P(0) = 2, \quad P'(0) = -1.$

(b) $Q_d = 6 - 5P + 2P' - 4P''$,
$Q_s = -4 + 5P - 2P' - 2P''$,
$P(0) = 3, \quad P'(0) = -1.$

CHAPTER 7

# Systems of Linear Differential Equations and Their Applications

## 7.1 DEFINITIONS AND SOLUTION

Quite often we come across linear equations which contain two or more dependent variables and a single independent variable. Such differential equations constitute a system of differential equations. In this chapter, we will study the system of linear differential equations and observe how they have a wide variety of applications in physics, engineering, medicine and ecology.

**Definition 7.1** The system of $n$ equations

$$\left.\begin{aligned} \frac{dy_1}{dt} &= f_1(y_1, y_2, \ldots, y_n, t) \\ \frac{dy_2}{dt} &= f_2(y_1, y_2, \ldots, y_n, t) \\ &\vdots \\ \frac{dy_n}{dt} &= f_n(y_1, y_2, \ldots, y_n, t) \end{aligned}\right\} \tag{1}$$

where $f_1, f_2, \ldots, f_n$, are each functions of $y_1, y_2, \ldots, y_n$, $t$ defined on a common set $A$, is called *a system of n first order equations.* A *solution* of Eq. (1) is a set of functions $y_1(t)$, $y_2(t)$, $\ldots$, $y_n(t)$, each defined on a common interval $I$ contained in $A$, satisfying all equations of (1) identically.

**REMARK.** In general, it is not possible to obtain the solutions of a system of first order equations in terms of the elementary functions. Only a few very simple first order system have such solutions. Even the simple pair of equations

$$\frac{dx}{dt} = e^{t^2}, \qquad \frac{dy}{dt} = x$$

cannot be solved in terms of elementary functions.

**EXAMPLE 7.1** Solve the system of first order equations

$$\frac{dx}{dt}=\frac{t}{x^2}, \quad \frac{dy}{dt}=\frac{y}{t^2}, \qquad x \neq 0, \qquad t \neq 0. \tag{2}$$

***Solution*** Each equation of the system is of the variable separation type. From the first equation, we obtain

$$x^3 = \frac{3}{2}t^2 + c_1 \tag{3}$$

and from the second, we have

$$y = c_2 e^{-1/t} \tag{4}$$

The pair of functions given by Eqs. (3) and (4), is a solution of Eq. (2).

**EXAMPLE 7.2** Solve

$$\frac{dx}{dt} = 2e^{2t}, \qquad \frac{dy}{dx} = \frac{x-y}{t} \tag{5}$$

***Solution*** The solution of the first equation of (5) is

$$x = e^{2t} + c_1 \tag{6}$$

Now, eliminating $x$ from the second equation of (5) with the help of Eq. (6), we get

$$\frac{dy}{dt} + \frac{y}{t} = \frac{e^{2t} + c_1}{t}$$

and the solution of this equation is

$$y = \left(\frac{1}{2}e^{2t} + c_1 t + c_2\right)t^{-1} \tag{7}$$

The pair of functions defined by Eqs. (6) and (7) is a solution of (5).

## 7.2 SOLUTION OF A SYSTEM OF LINEAR EQUATIONS WITH CONSTANT COEFFICIENTS

**Definition 7.2** The system of equations

$$\begin{aligned} P_{11}(D)y_1 + P_{12}(D)y_2 + \cdots + P_{1n}(D)y_n &= Q_1(t) \\ P_{21}(D)y_1 + P_{22}(D)y_2 + \cdots + P_{2n}(D)y_n &= Q_2(t) \\ &\vdots \\ P_{n1}(D)y_1 + P_{n2}(D)y_2 + \cdots + P_{nn}(D)y_n &= Q_n(t) \end{aligned} \tag{8}$$

where $D = d/dt$ and the coefficients of $y_1, y_2, \ldots, y_n$ are polynomial operators (see Chapter 5), is called *a system of n linear differential equation.* A *solution* of Eq. (8) is a set of functions $y_1(t), y_2(t), \ldots, y_n(t)$, each defined on a common interval $I$, satisfying all equations of (8) identically. The solution is *general* if the set of functions $y_1(t)$, $y_2(t)$, ..., $y_n(t)$ contains the correct number of constants (see Theorem 7.1).

We shall solve the system (8) by means of the operators (see Chapter 5). Since, the polynomial operators obey all the rules of algebra, we shall use the method for solving system of Eq. (8) similar to that used in solving an algebraic system of simultaneous equation. However, there are two important differences between the two systems:

(i) The operator symbol $D$ is not a numerical quantity. It is a differential operator, operating on a function. Hence, the order in which the operators are written is important.

(ii) Solutions of algebraic systems do not contain arbitrary constants, while the general solution of (8) does have constants.

To decide about the number of arbitrary constants appearing in the general solution of (8), we have the following theorem (For proof, see [24]).

**Theorem 7.1** Consider the pair of equations

$$\left.\begin{aligned} p_1(D)x + p_2(D)y &= q_1(t) \\ p_3(D)x + p_4(D)y &= q_2(t) \end{aligned}\right\} \tag{9}$$

The number of arbitrary constants in the general solution $x(t)$, $y(t)$ of Eq. (9) is equal to the order of

$$p_1(D)\, p_4(D) - p_2(D)\, p_3(D) \tag{10}$$

provided that

$$p_1(D)\, p_4(D) - p_2(D)\, p_3(D) \neq 0$$

If $p_1(D)\, p_4(D) - p_2(D)\, p_3(D) = 0$, then the system (9) is called *degenerate*, and if it is nonzero, then it is *nondegenerate*.

Since, Eq. (10) has the same form as a determinant, we shall refer to it as a determinant, and write

$$\begin{vmatrix} p_1(D) & p_2(D) \\ p_3(D) & p_4(D) \end{vmatrix} = p_1(D)\, p_4(D) - p_2(D)\, p_3(D) \tag{11}$$

**EXAMPLE 7.3** Solve

$$2\frac{dx}{dt} - x + \frac{dy}{dt} + 4y = 1, \qquad \frac{dx}{dt} - \frac{dy}{dt} = t - 1 \tag{12}$$

***Solution*** Letting $D = d/dt$, we have

$$\begin{aligned} (2D - 1)x + (D + 4)y &= 1 \\ Dx - Dy &= t - 1 \end{aligned} \tag{13}$$

Multiplying the first equation by $D$, the second by $(D + 4)$, and adding them, we get

$$(3D^2 + 3D)x = 4t - 3 \tag{14}$$

The general solution of Eq. (14) is

$$x(t) = c_1 + c_2 e^{-t} + \frac{2}{3}t^2 - \frac{7}{3}t \tag{15}$$

Substituting (15) into the second equation of (13), we get after simplifying, the relation

$$Dy = -c_2 e^{-t} + \frac{t}{3} - \frac{4}{3}$$

Integrate it to get the general solution as

$$y(t) = c_2 e^{-t} + \frac{t^2}{6} - \frac{4}{3}t + c_3 \tag{16}$$

Here, the determinant of (13) is $-3D^2 - 3D$ which is of order two. Hence, by Theorem 7.1, the number of constants appearing in the general solutions must be two, but from Eqs. (15) and (16), we have three constants. To find the relation between these three constants, we use the fact that the solution of a system of equations is a set of functions, which satisfies each equation of the system identically. Now substituting, $x(t)$ and $y(t)$ from Eqs. (15) and (16) in (13), we get

$$(2D-1)\left(c_1 + c_2 e^{-t} + \frac{2}{3}t^2 - \frac{7}{3}t\right) + (D+4)\left(c_2 e^{-t} + \frac{t^2}{6} - \frac{4}{3}t + c_3\right) = 1 \tag{17}$$

Performing the indicated operations in Eq. (17), we obtain

$$c_3 = \frac{c_1 + 7}{4}$$

With this value of $c_3$, (17) will be an identity in $t$. Putting this value in Eq. (16), we have

$$y(t) = c_2 e^{-t} + \frac{t^2}{6} - \frac{4}{3}t + \frac{c_1 + 7}{4} \tag{18}$$

The pair of functions $x(t)$, $y(t)$ defined by Eqs. (15) and (18), contains only two arbitrary constants and is the general solution of (12).

**EXAMPLE 7.4** Solve

$$(D - 1)x + (D + 1)y = 0, \qquad (2D + 2)x + (2D - 2)y = t. \tag{19}$$

***Solution*** Multiplying the first equation in (19) by $(2D + 2)$ and the second by $-(D - 1)$ and then adding the two, we get

$$Dy = \frac{t}{8} - \frac{1}{8}$$

which has a solution of the form

$$y(t) = \frac{t^2}{16} - \frac{t}{8} + c_1 \tag{20}$$

Substitute it in the first equation of (19) to obtain

$$(D-1)x = -\frac{t^2}{16} + \frac{1}{8} - c_1$$

and the general solution of this equation is

$$x(t) = \frac{t^2}{16} + \frac{t}{8} + c_1 + c_2 e^t \tag{21}$$

The determinant of Eq. (19) is $-8D$ which is of order one. Hence, the pair of functions (20), (21) should have only one constant, but it has two constants. To obtain, a relationship between the two, substitute Eqs. (20) and (21) in the second equation of (19) (we have already used the first)

$$2(D+1)\left(\frac{t^2}{16} + \frac{t}{8} + c_1 + c_2 e^t\right) + 2(D-1)\left(\frac{t^2}{16} - \frac{t}{8} + c_1\right) = t$$

which simplifies to

$$2\left(\frac{t}{2} + 2c_2 e^t\right) = t, \qquad c_2 e^t = 0$$

This equation will be an identity in $t$ only when $c_2 = 0$. Putting it in Eq. (21), we get

$$x(t) = \frac{t^2}{16} + \frac{t}{8} + c_1 \tag{22}$$

Hence, by definition, Eqs. (20) and (22) constitute the general solution of (19).

**EXAMPLE 7.5** Solve

$$(D+3)x + (D+1)y = e^t, \qquad (D+1)x + (D-1)y = t \tag{23}$$

***Solution*** Multiply the first equation by $-(D-1)$ and the second by $(D+1)$, and add the two to obtain

$$x = x(t) = \frac{1}{4}(t+1) \tag{24}$$

Putting this into the first equation of (23), we get

$$(D+1)y = e^t - 1 - \frac{3}{4}t$$

which has the general solution as

$$y(t) = c_1 e^{-t} + \frac{e^t}{2} - \frac{1}{4} - \frac{3}{4}t \tag{25}$$

The determinant of Eq. (23) is of the order zero and thus, the pair of functions $x(t)$, $y(t)$ should not have any constant. Since, $x(t)$, $y(t)$ satisfy the first equation of (23), we, therefore, substitute $x(t)$, $y(t)$ in the second equation of (23). We obtain

$$(D+1)\left(\frac{t}{4}+\frac{1}{4}\right)+(D-1)\left(c_1e^{-t}+\frac{e^t}{2}-\frac{1}{4}-\frac{3}{4}t\right)=t$$

which simplifies to

$$t - 2c_1e^{-t} = t$$

This will be an identity in it only when $c_1 = 0$. Hence, the pair of functions

$$x(t)=\frac{1}{4}(t+1),\quad y(t)=\frac{e^t}{2}-\frac{1}{4}-\frac{3}{4}t$$

is a general solution of the given Eq. (23).

## 7.3 AN EQUIVALENT TRIANGULAR SYSTEM

If the number of constants appearing in the solution of the system (8) is large, then the above methods of solving (8) is very time consuming and tedious. We thus, give a method by means of which we can get the exact number of constants immediately.

Consider Eq. (9) again, i.e.

$$\left.\begin{aligned} p_1(D)x + p_2(D)y &= q_1(t) \\ p_3(D)x + p_4(D)y &= q_2(t) \end{aligned}\right\} \tag{9}$$

We obtain a new system from it as follows: We retain one of the equations in (9), let us say the first, and change the second. This is achieved by multiplying the retained equation by any arbitrary operator $k(D)$ and adding it to the second. The new system thus, takes the form

$$\left.\begin{aligned} p_1(D)x + p_2(D)y &= q_1(t) \\ [p_1(D)k(D) + p_3(D)]x + [p_2(D)\,k(D) + p_4(D)]y &= k(D)q_1(t) + q_2(t) \end{aligned}\right\} \tag{26}$$

The two systems (9) and (26) are equivalent in the sense that a pair of functions $x(t)$, $y(t)$ which satisfies the first system will also satisfy the second and vice versa.

The determinant of Eq. (9) is $p_1p_4 - p_2p_3$, while that of (26) is $p_1p_2k + p_1p_4 - p_1p_2k - p_2p_3 = p_1p_4 - p_2p_3$, showing the determinant of both systems to be same and so is the order of the determinants of both systems.

Thus, by always retaining one equation in the system and changing the other by the above procedure, we obtain a new system, equivalent to Eq. (9), in the form

$$\left.\begin{aligned} P_1(D)x &= Q_1(t) \\ P_3(D)x + P_4(D)y &= Q_2(t) \end{aligned}\right\} \tag{27}$$

Such systems, i.e. one in which a coefficient of $x$ or of $y$ is zero, is called an *equivalent triangular system.*

If the original system is already in the form

$$p_1(D)x = q_1(t), \qquad p_4(D)y = q_2(t) \tag{28a}$$

or, in the triangular form

$$p_1(D)x = q_1(t), \qquad p_3(D)x + p_4(D)y = q_2(t) \tag{28b}$$

then the pair of functions $x(t)$, $y(t)$ obtained by solving the system will contain the correct number of constants. While solving Eq. (28b), it is essential to obtain $x(t)$ first and then $y(t)$. If we solve first for $y$ by eliminating $x$, we get superfluous constants.

**EXAMPLE 7.6** Solve

$$(3D^2 + 3D)x = 4t - 3, \qquad (D - 1)x - D^2y = t^2. \tag{29}$$

***Solution*** The general solution of the first equation of (29) is

$$x(t) = c_1 + c_2e^{-t} + \frac{2}{3}t^2 - \frac{7}{3}t \tag{30}$$

Substituting it in the second equation of (29), we get

$$D^2y = -\frac{7}{3} - c_1 - 2c_2e^{-t} + \frac{11}{3}t - \frac{5}{3}t^2$$

which has a solution of the form

$$y(t) = c_4 + c_3t - \left(\frac{7}{6} + \frac{c_1}{2}\right)t^2 + \frac{11}{18}t^3 - \frac{5}{36}t^4 - 2c_2e^{-t} \tag{31}$$

The pair of functions in Eqs. (30) and (31) is the general solution of Eq. (29), and from Theorem 7.1, it contains the correct number of four arbitrary constants.

Before moving ahead, we consider one more special type, one in which a coefficient of $x$ or $y$ is constant. Hence, the system is of the form

$$\left.\begin{aligned} p_1(D)x + ky &= q_1(t) \\ p_3(D)x + p_4(D)y &= q_2(t) \end{aligned}\right\} \tag{32}$$

In this case, the equivalent triangular system is obtained in just one step. Retain the equation that contains the constant coefficient (here it is first) and change the second by multiplying the first by $-p_4(D)/k$ and adding it to the second.

**EXAMPLE 7.7** Solve

$$(3D - 1)x + 4y = t, \qquad Dx - Dy = t - 1. \tag{33}$$

***Solution*** We retain the first equation and obtain a second equation by multiplying the first by $D/4$ and adding it to the second. The equivalent triangular system thus obtained is

$$\left.\begin{aligned} (3D-1)x + 4y &= t \\ \frac{1}{4}(3D^2+3D)x &= \frac{D}{4}(t) + t - 1 = t - \frac{3}{4} \end{aligned}\right\} \tag{34}$$

The general solution of the second equation of (34) is

$$x(t) = c_1 + c_2 e^{-t} + \frac{2}{3}t^2 - \frac{7}{3}t \tag{35}$$

Substituting it into the first equation of (34), we get

$$(3D-1)\left(c_1 + c_2 e^{-t} + \frac{2}{3}t^2 - \frac{7}{3}t\right) + 4y = 1$$

which gives

$$y(t) = \frac{c_1 + 7}{4} + c_2 e^{-t} + \frac{1}{6}t^2 - \frac{4}{3}t \tag{36}$$

The determinant of Eq. (33) is $(3D - 1)(-D) - 4D = -3D^2 - 3D$, whose order is two. Thus, the pair of functions $x(t)$, $y(t)$ contains the correct number of constants and is the general solution of Eq. (33).

**EXAMPLE 7.8** Solve the system

$$(D + 4)x + Dy = 1, \qquad (D - 2)x + y = t^2. \tag{37}$$

***Solution*** We retain here the second equation of (37) and obtain a new first equation by multiplying the second by $-D$ and adding it to the first. Thus, the equivalent triangular system is

$$\left.\begin{aligned} (-D^2 + 3D + 4)x &= -2t + 1 \\ (D-2)x + y &= t^2 \end{aligned}\right\} \tag{38}$$

The general solution of the first equation in (38) is

$$x(t) = c_1 e^{4t} + c_2 e^{-t} - \frac{t}{2} + \frac{5}{8} \tag{39}$$

Substitution of (39) into the second equation of (38) yields

$$(D-2)\left(c_1 e^{4t} + c_2 e^{-t} - \frac{t}{2} + \frac{5}{8}\right) + y = t^2$$

or

$$y(t) = -2c_1 e^{4t} + 3c_2 e^{-t} + t^2 - t + \frac{7}{4} \tag{40}$$

The determinant of Eq. (37) is of order two and the pair of functions (39) and (40) contains two constants, and is thus, a general solution of Eq. (37).

We can reduce a general system to an equivalent triangular system as follows: The general system may contain polynomials of different orders. Concentrate on the equation in which the lowest order polynomial appears and retain this equation. Now multiply this equation by a suitable operator $p(D)$ and add it to the retained equation. In this way, we get a new system in which the order of the coefficient of $x$ or of $y$ is reduced. Continue this procedure until the order of the coefficient of one polynomial operator in the system is zero. When this point is reached, the system will be of the form (32). One more step will then give the required triangular system.

**EXAMPLE 7.9** Solve

$$(D + 1)x + (D + 1)y = 1, \qquad D^2x - Dy = t - 1. \tag{41}$$

***Solution*** There are three polynomial operators in Eq. (41) of the same lowest order. It is easy to retain the second equation of (41) and change the first by adding the two equations. We have

$$(D^2 + D + 1)x + y = t, \qquad D^2x - Dy = t - 1 \tag{42}$$

The system is now of the form (32). Therefore, retain the first equation and change the second by multiplying the first by $D$ and adding it to the second. We obtain the equivalent triangular system as

$$(D^2 + D + 1)x + y = t, \qquad (D^3 + 2D^2 + D)x = t \tag{43}$$

The general solution of the second equation of (43) is

$$x(t) = c_1 + c_2e^{-t} + c_3te^{-t} + \frac{t^2}{2} - 2t \tag{44}$$

Substituting this value in the first equation of (43) and simplifying, we obtain

$$y(t) = -\left[c_1 + c_2e^{-t} + c_3(-e^{-t} + te^{-t}) + \frac{t^2}{2} - 2t - 1\right] \tag{45}$$

The determinant of Eq. (41) is of the order 3. The pair of functions in Eqs. (44) and (45) also contains three arbitrary constants and thus, pair is a general solution of (41).

## 7.4 DEGENERATE CASE

We say that the system of linear differential equation (9) degenerate if its determinant

$$\begin{vmatrix} p_1(D) & p_2(D) \\ p_3(D) & p_4(D) \end{vmatrix} = 0$$

If by eliminating $x$ or $y$, the right side of the system is not zero, then there will be no solution; there will be infinitely many solutions if the right side is zero.

**EXAMPLE 7.10** Show that the system

$$Dx - Dy = t, \qquad Dx - Dy = t^2$$

is degenerate and find the number of solutions it has.

***Solution*** Here the determinant is zero and thus, the system is degenerate. When we eliminate $x$ or $y$, the right side of the given equation does not reduce to zero and thus it has no solutions.

**EXAMPLE 7.11** Solve the system

$$Dx - Dy = t, \qquad 4Dx - 4Dy = 4t.$$

***Solution*** The determinant here is zero and the system is degenerate. However, its right side reduces to zero when $x$ or $y$ is eliminated. In this case, there are infinitely many solutions of the system. For example, the pair $x = t^2/2 + c$, $y = 5c_1$ is a solution, the pair $x = t/2 + c_1$, $y = t/2 - t^2/2 + c_2$ is a solution. [In either equation, define $x(t)$ arbitrarily and solve for $y(t)$. This pair of functions will also satisfy the other equation.]

**REMARK.** Consider the system of three linear equations:

$$\left.\begin{aligned} p_1(D)x + p_2(D)y + p_3(D)z &= q_1(t) \\ p_4(D)x + p_5(D)y + p_6(D)z &= q_2(t) \\ p_7(D)x + p_8(D)y + p_9(D)z &= q_3(t) \end{aligned}\right\} \tag{46}$$

If the determinant of Eq. (46) is

$$\begin{vmatrix} p_1(D) & p_2(D) & p_3(D) \\ p_4(D) & p_5(D) & p_6(D) \\ p_7(D) & p_8(D) & p_9(D) \end{vmatrix} \neq 0$$

then the method of finding a general solution of the system of Eqs. (46) follows the same rules given while solving a two-equation system. The number of constants appearing in the general solution of (46) must be equal to the order of its determinant, provided that it is not zero. The usual way of solving (46) is first to eliminate $z$ from the three equations, so that the system is reduced to two equations. Finally, eliminate another variable (say, $y$) from these two equations to get the value of $x$. Now, this value of $x$ is to be put in either of the two equations to get the value of $y$. Now, substitute the values of $x$ and $y$ in any one of the equation of (46) to get $z$.

In the remaining part of this chapter, the application aspect has been dealt with. The system of differential equations has numerous applications in physics, chemistry, engineering, medicine, national defence, ecology and social sciences. We shall, however, present some of them here.

## 7.5 MOTION OF A PROJECTILE

Suppose that a projectile is fired from a cannon situated at the origin $O$ of an $xy$-plane, and is inclined at an angle $\theta$ with the horizontal. The initial velocity of the projectile is $u$ ft/s. Assume that there is no air resistance and a flat stationary earth. The dashed curve in Fig. 7.1 represents the path of the projectile. $OV$ represents the muzzle velocity, a vector with magnitude $u$, and a direction in the $xy$-plane making an angle $\theta$ with the $x$-axis. The components of velocity in the $x$ and $y$ directions are $u \cos \theta$ and $u \sin \theta$, respectively.

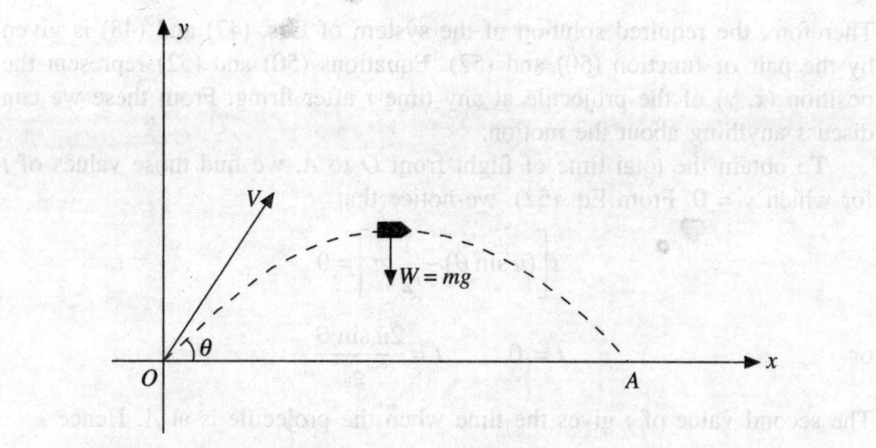


**Fig. 7.1** A projectile fired from a cannon from the origin of $xy$-plane.

Since, there is no air-resistance, the only force acting on the projectile of mass $m$ is its weight $mg$ acting vertically downward. Also, the force acting in the $x$-direction is zero, and $a_x = d^2x/dt^2$, $a_y = d^2y/dt^2$. We have, from Newton's law, the relations

$$m\frac{d^2x}{dt^2} = 0 \quad \text{or} \quad \frac{d^2x}{dt^2} = 0 \tag{47}$$

and
$$m\frac{d^2y}{dt^2} = -mg \quad \text{or} \quad \frac{d^2y}{dt^2} = -g \tag{48}$$

Moreover, the initial conditions are

$$x = 0, \quad y = 0, \quad \frac{dx}{dt} = u\cos\theta, \quad \frac{dy}{dt} = u\sin\theta \quad \text{at } t = 0 \tag{49}$$

Equations (47) and (48) together with (49) give the complete mathematical formulation of the motion of a projectile. It may be noted from Eqs. (47) and (48) that the motion does not depend on the mass, and hence on the size of the projectile, provided there is no air resistance.

Integrating Eq. (47) and using (49), we find $dx/dt = u \cos \theta$. (Thus, the horizontal component of the velocity remains constant.) Now, integrate $dx/dt = u \cos \theta$ and use Eq. (49) to find

$$x = u\cos\theta t \tag{50}$$

Similarly, integrating Eq. (48) twice and using Eq. (49) to obtain the value of the constant of integration, we get

$$\frac{dy}{dt} = -gt + u\sin\theta \tag{51}$$

$$y = (u\sin\theta)t - \frac{1}{2}gt^2 \tag{52}$$

Therefore, the required solution of the system of Eqs. (47) and (48) is given by the pair of function (50) and (52). Equations (50) and (52) represent the position $(x, y)$ of the projectile at any time $t$ after firing. From these we can discuss anything about the motion.

To obtain the total time of flight from $O$ to $A$, we find those values of $t$ for which $y = 0$. From Eq. (52), we notice that

$$t\left[(u \sin \theta) - \frac{1}{2} gt\right] = 0$$

or

$$t = 0, \qquad t = \frac{2u \sin \theta}{g}$$

The second value of $t$ gives the time when the projectile is at $A$. Hence

$$\text{Time of flight} = \frac{2u \sin \theta}{g} \tag{53}$$

To get the range (distance $OA$ along the $x$-axis), substituting the value of $t$ from Eq. (53) in (50), we have

$$\text{Range} = \frac{u^2 \sin 2\theta}{g} \tag{54}$$

The range is maximum when $2\theta = 90°$, i.e. $\theta = 45°$ and the maximum range is $u^2/g$.

Also, at the highest point $dy/dt = 0$ and from Eq. (51), we have

$$\frac{dy}{dt} = u \sin \theta - gt = 0$$

which gives

$$t = \frac{u \sin \theta}{g}$$

Putting this value of $t$ in Eq. (52), we find

$$\text{Maximum height} = \frac{u^2 \sin^2 \theta}{2g} \tag{55}$$

The path of the projectile is given by Eqs. (50) and (52), which represent the parametric equations of a parabola. By eliminating the parameter $t$ from these equations, we obtain

$$y = x \tan \theta - \frac{gx^2}{2u^2} \sec^2 \theta \tag{56}$$

which is another form for the parabola. Thus, the path of the projectile is a portion of the parabola.

When additional forces, such as frictional forces, acting on the projectile are considered, the analysis is similar except that a more complicated system of differential equation is obtained. Numerical methods of approximation are often used to integrate approximately the differential equation occurring in the model that arises. The study of ballistics discusses the effect of the Coriolis force due to the rotation of the earth, see [5].

## 7.6 CENTRAL FORCE SYSTEM, NEWTON'S LAW OF GRAVITATION: KEPLER'S LAWS OF PLANETARY MOTION

In Fig. 7.2, the particle of mass $m$ has position vector $\mathbf{R}$ and is acted upon by force $\mathbf{F}$. By Newton's law, we have

$$\mathbf{F} = m\mathbf{A} = m\frac{d\mathbf{V}}{dt} = \frac{d}{dt}(m\mathbf{V})$$

where $\mathbf{V}$ and $\mathbf{A}$ denote the velocity and acceleration of the particle at time $t$. Hence

$$\mathbf{R} \times \mathbf{F} = \mathbf{R} \times \frac{d}{dt}(m\mathbf{V})$$

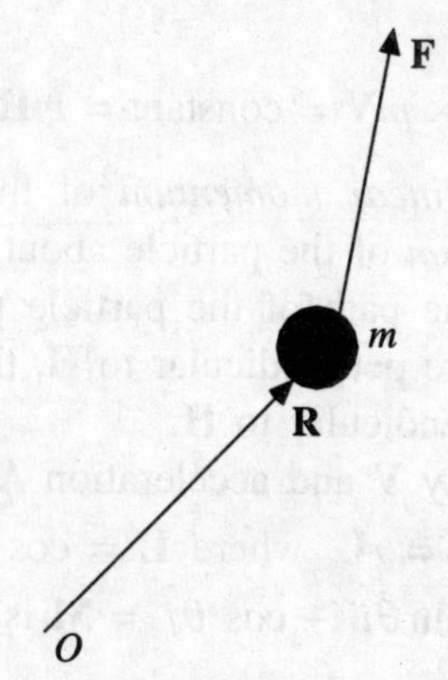


**Fig. 7.2** A force **F** is acting on a particle of mass *m*.

Since

$$\frac{d}{dt}(\mathbf{R} \times m\mathbf{V}) = \mathbf{R} \times \frac{d}{dt}(m\mathbf{V}) + \frac{d\mathbf{R}}{dt} \times m\mathbf{V}$$

$$= \mathbf{R} \times \frac{d}{dt}(m\mathbf{V}) + \mathbf{V} \times m\mathbf{V}$$

$$= \mathbf{R} \times \frac{d}{dt}(m\mathbf{V})$$

it follows that

$$\mathbf{R} \times \mathbf{F} = \frac{d}{dt}(\mathbf{R} \times m\mathbf{V}) \tag{57}$$

We now consider the case, where the force $\mathbf{F}$ always acts towards (or away from) the origin $O$ (Fig. 7.3). A force system of this type is called a *central force system.* Since, $\mathbf{R} \times \mathbf{F} = 0$, it follows that

$$\frac{d}{dt}(\mathbf{R} \times m\mathbf{V}) = 0$$

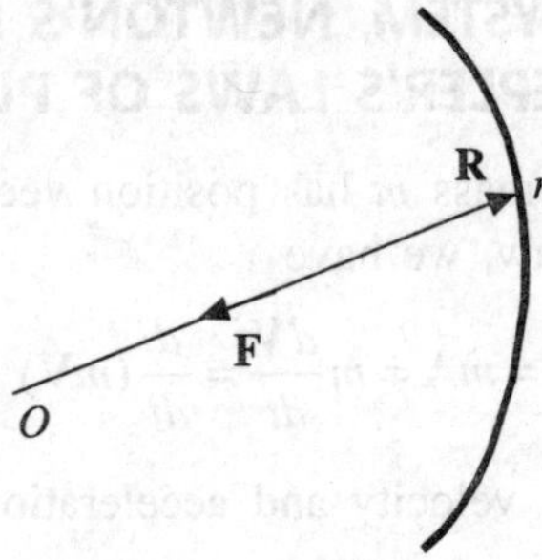

**Fig. 7.3** Force **F** is acting towards (or away from) a fixed point $O$—central force system.

from Eq. (57), and hence

$$\mathbf{R} \times m\mathbf{V} = \text{constant} = m\mathbf{H}$$

The vector $m\mathbf{V}$ is the *linear momentum* of the particle and the vector $m\mathbf{H}$, the *moment of momentum* of the particle about the origin. If $\mathbf{H} = 0$, then $\mathbf{R}$ and $\mathbf{V}$ are parallel and the path of the particle passes through the origin. Otherwise, since $\mathbf{R}$ and $\mathbf{V}$ are perpendicular to $\mathbf{H}$, the path or the orbit of the particle lies in a plane perpendicular to $\mathbf{H}$.

We now find the velocity $\mathbf{V}$ and acceleration $\mathbf{A}$ in polar coordinates. We write the position vector $\mathbf{R} = r\mathbf{L}$, where $\mathbf{L} = \cos\theta\hat{i} + \sin\theta\hat{j}$ is the *radial unit vector,* then $dL/d\theta = -\sin\theta\hat{i} + \cos\theta\hat{j} = \mathbf{M}$ is the *transverse unit vector* (Fig. 7.4) and

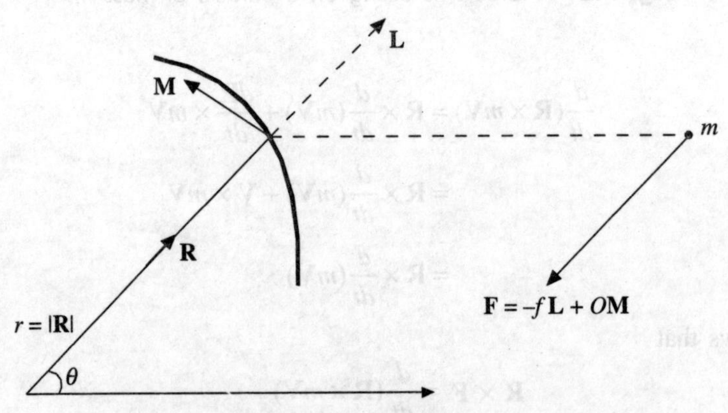


**Fig. 7.4** Radial and transverse unit vectors.

$$\mathbf{V} = \frac{d\mathbf{R}}{dt} = \frac{dr}{dt}\mathbf{L} + r\frac{d\mathbf{L}}{dt} = \frac{dr}{dt}\mathbf{L} + r\frac{d\mathbf{L}}{d\theta}\frac{d\theta}{dt} = \frac{dr}{dt}\mathbf{L} + r\frac{d\theta}{dt}\mathbf{M} \tag{58}$$

The scalar $dr/dt$ is the *radial component of* **V** and the scalar $r(d\theta/dt)$ is the *transverse component of* **V**. Differentiating Eq. (58), we get

$$\mathbf{A} = \frac{d\mathbf{V}}{dt} = \frac{dr}{dt}\frac{d\mathbf{L}}{dt} + \frac{d^2r}{dt^2}\mathbf{L} + \frac{rd\theta}{dt}\frac{d\mathbf{M}}{dt} + \left(\frac{r\,d^2\theta}{dt^2} + \frac{dr}{dt}\frac{d\theta}{dt}\right)\mathbf{M}$$

Since

$$\frac{d\mathbf{L}}{dt} = \frac{d\mathbf{L}}{d\theta}\frac{d\theta}{dt} = \frac{d\theta}{dt}\mathbf{M}$$

$$\frac{d\mathbf{M}}{dt} = \frac{d\mathbf{M}}{d\theta}\frac{d\theta}{dt} = \frac{d\theta}{dt}\mathbf{L}$$

we have

$$\mathbf{A} = \left[\frac{d^2r}{dt^2} - r\left(\frac{d\theta}{dt}\right)^2\right]\mathbf{L} + \left(r\frac{d^2\theta}{dt^2} + \frac{2dr}{dt}\frac{d\theta}{dt}\right)\mathbf{M} \tag{59}$$

In Eq. (59), the scalar coefficient of **L** is the *radial component* and the scalar coefficient of **M** is the *transverse component of acceleration* **A**. (If we take the Cartesian coordinates instead of the polar coordinates, we have the tangential and normal components of velocity and acceleration.)

Let $F$ be the magnitude of the force **F** acting on the particle. Also, let this $F$ act towards $O$. Then

$$\mathbf{F} = -f\mathbf{L} + O(\mathbf{M})$$

and Newton's second law $\mathbf{F} = m\mathbf{A}$ applied in the radial and transverse directions yields

$$\frac{d^2r}{dt^2} - r\left(\frac{d\theta}{dt}\right)^2 = -\,fm^{-1} \tag{60}$$

$$r\frac{d^2\theta}{dt^2} + 2\frac{dr}{dt}\frac{d\theta}{dt} = 0(m^{-1}) = 0 \tag{61}$$

Now, the basic problem is to find a solution of the systems (60) and (61) of differential equations involving two unknown functions $\phi$ and $\psi$, where $r = \phi(t)$ and $\theta = \psi(t)$. These functions give the parametric equation of the path of the moving particle. Another objective is to find $r$ in terms of $\theta$, i.e. to find a polar equation $r = \beta(\theta)$ of the orbit. This can be done if we have some initial condition and a knowledge of the nature of $f$ in Eq. (60).

We start from Eq. (61) which does not contain $f$ and obtain an important result as follows: As $r \neq 0$, then Eq. (61) can be written as

$$\frac{1}{r}\frac{d}{dt}\left(r^2\frac{d\theta}{dt}\right) = 0 \tag{62}$$

which gives

$$r^2 \frac{d\theta}{dt} = \text{constant} = h \tag{63}$$

We use a polar coordinate area formula to observe in Fig. 7.5 that the shaded area swept out as the particle moves from $(r_1, \theta_1)$ at time $t_1$ to $(r, \theta)$ at time $t$ is given by

$$A = \frac{1}{2} \int_{\theta_1}^{\theta} [\beta(u)]^2 \, du$$

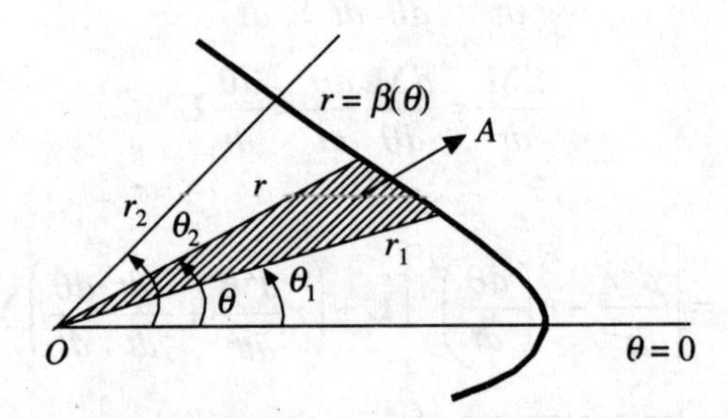


**Fig. 7.5** Area swept out by the radius vector.

Then

$$\begin{aligned}\frac{dA}{dt} &= \text{rate of change of area per unit time} \\ &= \frac{dA}{d\theta} \frac{d\theta}{dt} \\ &= \frac{1}{2}[\beta(\theta)]^2 \frac{d\theta}{dt} \\ &= \frac{1}{2} r^2 \frac{d\theta}{dt}\end{aligned}$$

and by Eq. (63), we have

$$\frac{dA}{dt} = \frac{h}{2} \tag{64}$$

The quantity $dA/dt$ is called the *areal velocity*. The area swept out as the particle moves from $(r_1, \theta_1)$ at $t_1$ to $(r_2, \theta_2)$ at $t_2$ is

$$[A]_{t_1}^{t_2} = \int_{t_1}^{t_2} \frac{h}{2} dt = \frac{h}{2}(t_2 - t_1) \tag{65}$$

which is the same for any interval of duration $t_2 - t_1$. This proves the following theorem, known as the *law of areas*, for a central force system.

**Theorem 7.2** For a particle moving in a central force system, the radius vector from the centre of the force system to the moving particle sweeps out equal area in equal interval of time.

Now, consider the motion of a planet round the sun and apply the systems (60) and (61) to this specific central force system. *Newton's law of gravitation* states that the two particles attract each other with a force whose magnitude is directly proportional to the product of their masses and inversely proportional to the distance between them. If $M$ is the mass of the sun and $m$ the mass of the planet, then $f$ in Eq. (60) is given by

$$f = G\frac{Mm}{r^2}$$

where $G$ is the universal gravitation constant. For convenience, take $k = GM$ so that $f = km/r^2$ and Eq. (60) reduces to

$$\frac{d^2r}{dt^2} - r\left(\frac{d\theta}{dt}\right)^2 = -\frac{k}{r^2} \tag{66}$$

Putting the value of $d\theta/dt$ from Eq. (63), we get

$$\frac{d^2r}{dt^2} = \frac{h^2}{r^3} - \frac{k}{r^2} \tag{67}$$

To obtain $r = \phi(t)$, let $p = dr/dt$, then

$$\frac{d^2r}{dt^2} = \frac{dp}{dt} = \frac{dr}{dt}\frac{dp}{dr} = p\frac{dp}{dt}$$

and omit this complicated calculation. To obtain $\theta = \psi(t)$, put $r = \theta(t)$ into (63) and separate the variables.

To find $r = \beta(\theta)$, we put $r = 1/u$ in Eq. (67), to obtain

$$\frac{d^2u}{d\theta^2} + u = \frac{k}{h^2} \tag{68}$$

which has the solution

$$u = A\cos\theta + B\sin\theta + \frac{k}{h^2}$$

To determine the constants $A$ and $B$, we choose the polar axis so that the moving planet is closest to the origin when $\theta = 0$.

Since, $u = 1/r$ will assume its maximum when $r$ enjoys its minimum, $du/d\theta$ will be zero and $d^2\theta/dt^2$ will be negative when $d\theta = 0$. Thus

$$\frac{du}{d\theta} = -A\sin\theta + B\cos\theta$$

$$0 = -A(0) + B(1), \qquad B = 0$$

From $d^2u/d\theta^2 = -A\cos\theta$ and $-A(1) < 0$, we conclude that $A > 0$. Hence

$$r = \frac{1}{u} = \frac{1}{(k/h^2) + A\cos\theta}$$

or
$$r = \frac{h^2/k}{1 + (Ah^2/k)\cos\theta} = \beta(\theta) \tag{69}$$

To determine $A$, let $r = r_0$ when $\theta = 0$. This gives $A = r_0^{-1} - kh^{-2}$. By assuming that $d\theta/dt = (d\theta/dt)_0$ when $r = r_0$, the constant $h$ is determined, and by Eq. (63)

$$h = r_0^2\left(\frac{d\theta}{dt}\right)_0 = r_0 v_0, \qquad v_0 = r_0\left(\frac{d\theta}{dt}\right)_0$$

Noting that Eq. (69) has the form

$$r = \frac{ep}{1 + e\cos\theta} \tag{70}$$

where $e$ and $p$ are positive, we conclude [25] that the orbit of the planet is a conic having eccentricity $e = Ah^2/k$. It is known that the planets move in closed paths and, hence, their orbits must be ellipses with the Sun at one focus. This is *Kepler's first law*.

Recurring comets, such as Halley's comet, have orbits which are elongated ellipses with eccentricities less than but close to one. The time for a comet to reappear near the earth depends upon the eccentricity of its elliptical path. For instance, Halley's comet appears once every 76 years approximately. (The last appearance of Halley's comet occurred in March 1986, it is expected to appear again in 2062.)

Objects having hyperbolic or parabolic orbits would appear only once, theoretically, never to return. Experiments with particles of atomic dimensions performed in cloud chambers show particles leaving 'fog tracks' which are parabolic or hyperbolic in shape.

*Kepler's second law* states that, as a planet moves around its orbit, the radius vector from the Sun to the planet sweeps out equal areas in equal intervals of time. This is a direct application of Theorem 7.2 to the solar system.

*Kepler's third law* states that, for each planet $T^2$ is proportional to $a^3$, where $T$ is the time required for the planet to make one complete orbit of the Sun (this is periodic time), and $a$ is the mean or average of the maximum and minimum distance of the planet from the Sun. It is easily seen that $a$ is the semi-major axis of the elliptical orbit. In fact, we have

$$T^2 = \frac{4\pi^2}{k}a^3 \tag{71}$$

where $4\pi^2/k$ is the constant of proportionality and $b^2 = h^2a/k$, with $b$ being the semi-minor axis.

Kepler's laws of planetary motion suggested the law of universal gravitation to Newton. Newton then derived Kepler's laws as we have done, from the law of gravitation. This involves a relatively simple application of the deductive mathematical analysis, whereas Kepler discovered his laws inductively from his observations during a lifetime of studies. Newton's approach is more general since it provides a model not only for planetary motion but also for satellite motion, motion of atomic particles, and so on.

In Eq. (63), we assume that $h > 0$ and that $r > 0$. This amounts to the assumption that $d\theta/dt > 0$ or that $\theta$ increases with $t$, thus making the orbit an oriented path. An example is a meteor crashing into the Sun when $h = 0$.

## 7.7 MOTION OF A PARTICLE IN THE GRAVITATIONAL FIELD OF EARTH: SATELLITE MOTION

We consider the motion of a particle under the attraction of earth and apply the system (60) and (61) to this moving particle. Let $M$ and $m$ be the masses of the earth and of the moving particle of weight $W$, respectively. Also, let $R = 3960$ miles, the mean radius of the earth. When $r = R$, then $f = W = mg$ so that $f = GMm/r^2$ becomes $mg = GMm/r^2$ giving $GM = gR^2$, and the form of $f$ in Eq. (60) is $f = mg(R/r)^2$, where $r \geq R$.

With this value of $f$, the system (60) and (61) provides a model for studying the motion of the moon, a meteorite, a rocket, a satellite, and the like.

Let us consider the possibility of launching a rocket into its orbit at a distance $r = r_0$ from the centre of the earth. It can be shown that the result does not depend upon the angle at which the satellite is launched. Hence, we assume that when $t = 0$, we have $\theta = 0$, $r = r_0$ and $\mathbf{V} = \mathbf{V}_0 = r_0(d\theta/dt)\mathbf{M} = v_0\mathbf{M}$ (see Fig. 7.6).

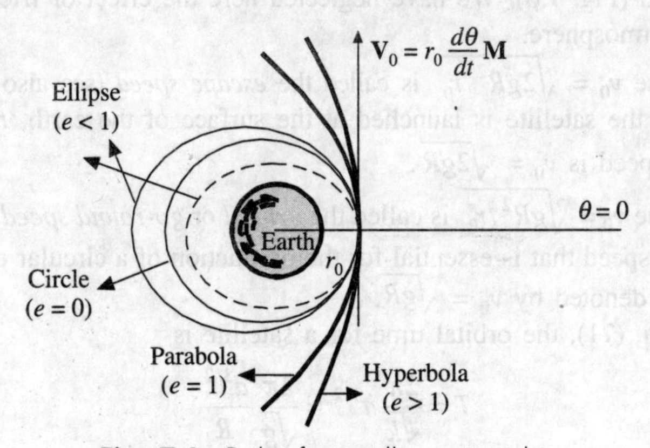


**Fig. 7.6** Path of a satellite in its orbit.

To find an equation of the orbit of the satellite, we set

$$k = gR^2, \qquad h = r_0^2\left(\frac{d\theta}{dt}\right)_0 = r_0 v_0$$

$$A = r_0^{-1} - kh^{-2} = \frac{1}{r_0} - \frac{gR^2}{r_0^2 v_0^2} = \frac{r_0 v_0^2 - gR^2}{r_0^2 v_0^2}$$

Now using Eq. (69), we get

$$r = \frac{r_0 v_0^2/(gR^2)}{1 + [(r_0 v_0^2 - gR^2)/(gR^2)]\cos\theta} = \beta(\theta) \tag{72}$$

The value of eccentricity

$$e = \frac{r_0 v_0^2 - gR^2}{gR^2} = \frac{r_0 v_0^2}{gR^2} - 1$$

depends on the initial speed $v_0$. The eccentricity will be 1 when $r_0 v_0^2 - gR^2 = gR^{-2}$ or $v_0^2 = 2gR^2/r_0$. In this case, the satellite will follow a parabolic path and leave the earth's gravitational field. When $v_0^2 > 2gR^2/r_0$, then $e > 1$, and the satellite leaves the gravitational field of the earth along a hyperbolic path. When $gR^2/r_0 < v_0^2 < 2gR^2/r_0$, then $e < 1$ and the satellite moves around the earth in an elliptical orbit.

Now, suppose that $A$ is not positive. That is, $r_0$ is not the minimum distance of the particle (of mass $m$) from the pole. However, $(dr/dt)_0$ is still zero.

Thus, when $v_0^2 = gR^2/r_0$ then $A = 0$ and $e = 0$ and the motion of the satellite is circular ($r_0$ being the radius of the circle).

If $v_0^2 > gR^2/r_0$, then the path will be an ellipse for which $r_0$ is the maximum distance from the earth, or the satellite will begin an elliptic orbit and crash into the earth (Fig. 7.6). We have neglected here the effect of friction due to the earth's atmosphere.

The value $v_0 = \sqrt{2gR^2/r_0}$ is called the *escape speed* (see also Chapter 4) at $r = r_0$. If the satellite is launched at the surface of the earth, $r_0 = R$, and the escape speed is $v_0 = \sqrt{2gR}$.

The value $v_0 = \sqrt{gR^2/r_0}$ is called the *orbital* or *go-round speed* at $r = r_0$. It is the initial speed that is essential for the production of a circular orbit, and if $r_0 = R$, it is denoted by $v_0 = \sqrt{gR}$.

From Eq. (71), the orbital time for a satellite is

$$T = \frac{2\pi}{\sqrt{k}} a^{3/2} = \frac{2\pi}{\sqrt{g}} \frac{a^{3/2}}{R}$$

where $a$ denotes the semi-major axis of the elliptic orbit. Now

$$a = \frac{1}{2}[\beta(0) + \beta(\pi)] = \frac{1}{2}\left(r_0 + \frac{r_0 v_0^2}{2gR^2 - r_0 v_0}\right) = \frac{gR^2}{2gr_0^{-1}R^2 - v_0^2}$$

Hence

$$T = \frac{2\pi}{\sqrt{g}\,R}\left(\frac{gR^2}{2gr_0^{-1}R^2 - v_0^2}\right)^{3/2}$$

or

$$T = \frac{2\pi gR^2}{(2gr_0^{-1}R^2 - v_0^2)^{3/2}} \tag{73}$$

**EXAMPLE 7.12** Find the orbital speed and the escape speed at the surface of the earth if $R = 3960$ miles and $g = 32.2/5280$ miles/s$^2$.

***Solution*** The escape speed

$$v_0 = \sqrt{2gR} = \sqrt{\frac{2(32.2)}{5280}(3960)} = (4.914)\sqrt{2} = 6.950 \text{ miles/s} \approx 25{,}019 \text{ miles/h}$$

and the orbital speed

$$v_0 = \sqrt{gR} = 4.914 \text{ miles/s}$$

**EXAMPLE 7.13** Find the orbital time for a satellite in circular orbit at the surface of the earth.

***Solution*** Setting $r_0 = R = 3960$ miles, $g = 32.2/5280$ miles/s$^2$, and $v_0 = \sqrt{gR}$ in Eq. (73), we get

$$T = 2\pi\sqrt{\frac{R}{g}} = 5063 \text{ s} \approx 84.4 \text{ min}$$

**EXAMPLE 7.14** A satellite is launched in a direction parallel to the surface of the earth with a velocity of 18,820 miles/h from an altitude of 240 miles. Determine the velocity of the satellite when it reaches its maximum altitude of 2340 miles.

***Solution*** The satellite is moving under a central force directed towards the centre of the earth. The angular momentum $H_0$ is constant. From $H_0 = rmv \sin\phi$, we have

$$rmv \sin\phi = H_0 = \text{constant}$$

which shows that $v$ is minimum at $B$ (Fig. 7.7), where both $r$ and $\sin\phi$ are maximum. Consider the conservation of angular momentum between $A$ and $B$:

$$r_A m v_A = r_B m v_B$$

we have

$$v_B = v_A \frac{r_A}{r_B} = (18{,}820)\frac{3960 + 240}{3960 + 2340} = 12{,}550 \text{ miles/h}$$

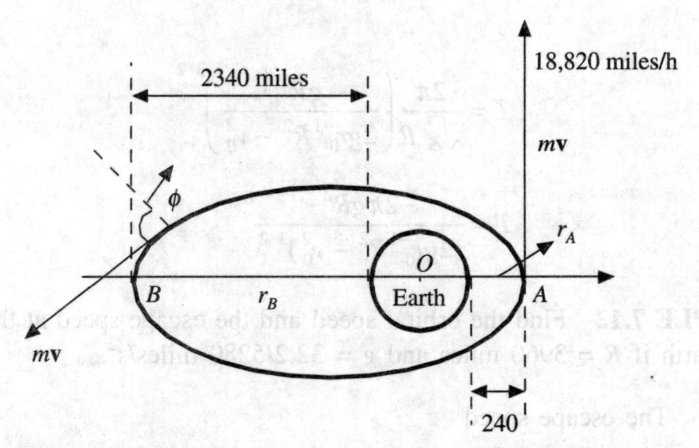


**Fig. 7.7** Diagrammatic representation of Example 7.14.

**EXAMPLE 7.15** A satellite is launched in a direction parallel to the surface of the earth with a velocity of 36,900 km/h from an altitude of 500 km (Fig. 7.8). Find: (a) the maximum altitude reached by the satellite, and (b) the periodic time of the satellite.

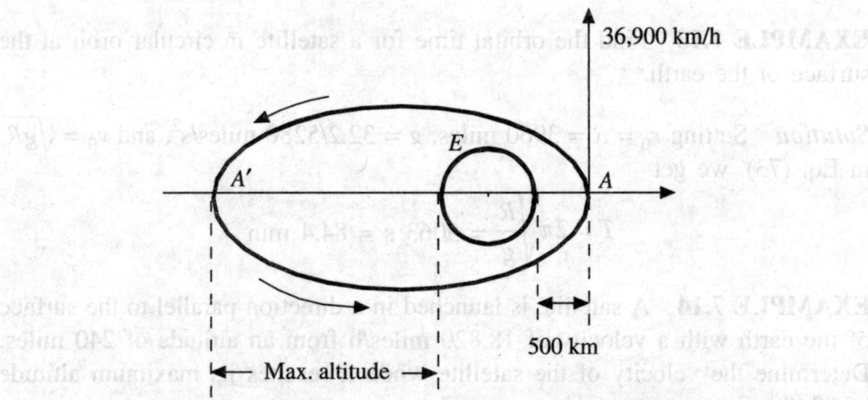


**Fig. 7.8** Diagrammatic representation of Example 7.15.

***Solution*** (a) The satellite is influenced only by the gravitational attraction of the earth after its launching and thus the motion of the satellite is given by the equation

$$\frac{1}{r} = \frac{GM}{r^2} + C \cos\theta \tag{74}$$

At the point of launching $A$, the radial component of velocity is zero, so we have, $h = r_0 v_0$. Let $R$ = 6370 km, as the radius of the earth; then

$$r_0 = 6370 + 500 \text{ km} = 6870 \text{ km} = 6.87 \times 10^6 \text{ m}$$

$$v_0 = 36{,}900 \text{ km/h} = \frac{36.9 \times 10^6 \text{ m}}{3.6 \times 10^3 \text{ s}} = 10.25 \times 10^3 \text{ m/s}$$

$$h = r_0 v_0 = (6.87 \times 10^6)\,(10.25 \times 10^3) = 70.4 \times 10^9 \text{ m}^2/\text{s}$$

$$h^2 = 4.96 \times 10^{21} \text{ m}^4/\text{s}^2$$

Since

$$GM = gR^2 = (9.81)(6.87 \times 10^6)^2 = 398 \times 10^{12} \text{ m}^3/\text{s}^2$$

$$\frac{GM}{h^2} = 80.3 \times 10^{-9} \text{ m}^{-1}$$

Substituting this value in Eq. (74), we get

$$\frac{1}{r} = 80.3 \times 10^{-9} \text{ m}^{-1} + C \cos \theta \tag{75}$$

Note that, at $A$, $\theta = 0$ and $r = r_0 = 6.87 \times 10^6$ m, so that Eq. (75) gives

$$C = 65.3 \times 10^{-9} \text{ m}^{-1}$$

At $A'$ (Fig. 7.9), the point on the orbit farthest from the earth, we have $\theta = 180°$. Using Eq. (75), we can calculate the corresponding distance $r_1$ as

$$\frac{1}{r_1} = 80.3 \times 10^{-9} + (65.3 \times 10^{-9}) \cos 180°$$

or

$$r_1 = 66.7 \times 10^6 \text{ m} = 66{,}700 \text{ km}$$

Therefore, maximum altitude = 66,700 – 6370 = 60,330 km.

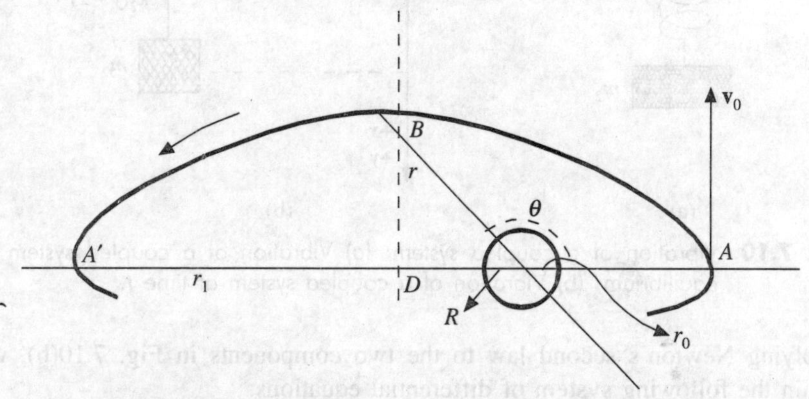


**Fig. 7.9** Schematic diagram of the path of the satellite.

(b) The perigee and apogee of the elliptic orbit are $A$ and $A'$ respectively. We thus use

$$a = AD = \frac{1}{2}(r_0 + r_1), \qquad b = BD = \sqrt{r_0 r_1}$$

and calculate the semi-major and minor axes of the orbit to get

$$a = 36.8 \times 10^6 \text{ m}, \qquad b = 21.4 \times 10^6 \text{ m}$$

Thus, the periodic time $T = 2\pi ab/h = 1171$ min. = 19 hr 31 min.

## 7.8 VIBRATION OF A COUPLED SYSTEM

The system shown in Fig. 7.10 oscillates vertically in such a way that Hooke's law holds when the springs are either in tension or compression. Since, two coordinates $x$ and $y$ are required to describe the positions of the masses $m_1$ and $m_2$ at time $t$, the system is said to have two degrees of freedom.

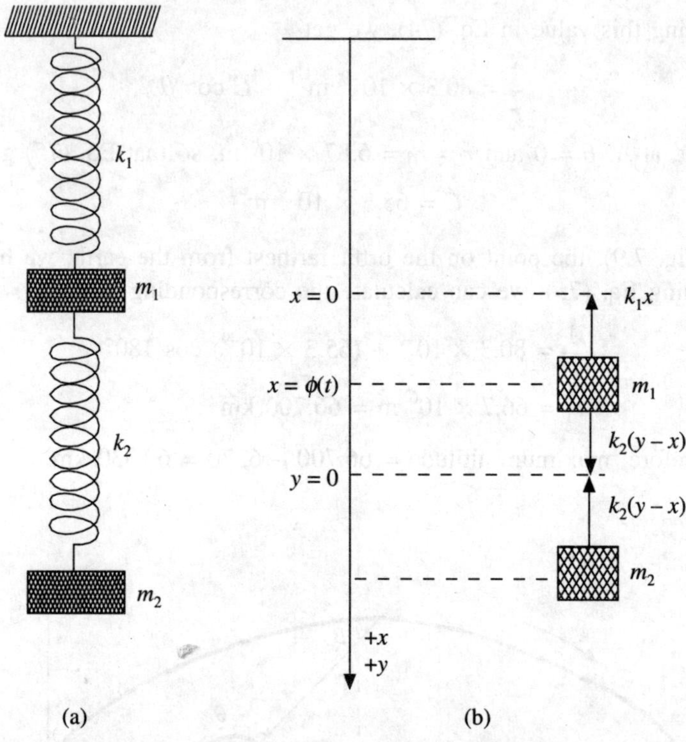


**Fig. 7.10** Vibration of a coupled system: (a) Vibration of a coupled system in equilibrium; (b) Vibration of a coupled system at time $t$.

Applying Newton's second law to the two components in Fig. 7.10(b), we obtain the following system of differential equations:

$$m_1 \frac{d^2x}{dt^2} = k_2(y - x) - k_1x \tag{76}$$

$$m_2 \frac{d^2y}{dt^2} = -k_2(y - x) \tag{77}$$

where $k_1$ and $k_2$ are the spring constants. In Fig. 7.10(b), $x > 0$ and $y - x > 0$. It can be shown that Eqs. (76) and (77) are true for all $t$, i.e., even if $x \le 0$ or $y - x \le 0$. If, for example, $x < 0$, the force having magnitude $|k_1x|$ reverses its direction. Similarly, if $y - x < 0$, the directions of the two forces, having magnitude $|k_2(y - x)|$, are reversed.

A solution of Eqs. (76) and (77) consists of a pair of functions $x = \phi(t)$ and $y = \psi(t)$ having four arbitrary constants, and the common domain of these functions is $t \geq 0$.

**EXAMPLE 7.16** Solve the system (76)–(77), with $k_1 = 2$, $k_2 = 1$, $m_1 = 3$ and $m_2 = 1$.

***Solution*** For the present problem, the system (76)–(77) takes the form

$$y = 3\frac{d^2x}{dt^2} + 3x \tag{78}$$

$$\frac{d^2y}{dt^2} = -y + x \tag{79}$$

Differentiating Eq. (78), we find

$$\frac{d^2y}{dt^2} = 3\frac{d^4x}{dt^4} + 3\frac{d^2x}{dt^2} \tag{80}$$

Substituting $y$ from Eq. (78) and $d^2y/dt^2$ from (80) into (79), we obtain

$$3\frac{d^4x}{dt^4} + 6\frac{d^2x}{dt^2} + 2x = 0 \tag{81}$$

Here, the auxiliary equation has the roots

$$m^2 = r^2 = \left(-1 \pm \frac{1}{\sqrt{3}}\right)^{1/2} \quad \text{or} \quad m = r = \pm i\omega_1, \pm i\omega_2$$

where

$$\omega_1 = -1 - \frac{1}{\sqrt{3}} = 0.650, \qquad \omega_2 = -1 + \frac{1}{\sqrt{3}} = 1.256$$

Therefore, the complete solution of Eq. (81) is

$$x = c_1 \cos \omega_1 t + c_2 \sin \omega_1 t + c_3 \cos \omega_2 t + c_4 \sin \omega_2 t \tag{82}$$

To determine $y$, we differentiate Eq. (82) twice to obtain

$$\frac{d^2x}{dt^2} = -c_1\omega_1^2 \cos \omega_1 t - c_2\omega_1^2 \sin \omega_1 t - c_3\omega_2^2 \cos \omega_2 t - c_4\omega_2^2 \sin \omega_2 t$$

Hence, Eq. (79) yields

$$y = 3c_1(1 - \omega_1^2) \cos \omega_1 t + 3c_2(1 - \omega_1^2) \sin \omega_1 t + 3c_3(1 - \omega_2^2) \cos \omega_2 t + 3c_4 (1 - \omega_2^2) \sin \omega_2 t$$

or
$$y = \sqrt{3}(c_1 \cos \omega_1 t + c_2 \sin \omega_1 t) - \sqrt{3}(c_3 \cos \omega_2 t + c_4 \sin \omega_2 t) \quad (83)$$

The pair of Eqs. (82) and (83) is the general solution of the given systems (78)–(79).

**EXAMPLE 7.17** Find the particular solution of the system (78)–(79) with the initial conditions $x = 0$, $y = 2\sqrt{3}\,\omega_1$, $dx/dt = 0$ and $dy/dt = 2\sqrt{3}\,\omega_2$ when $t = 0$.

***Solution*** From

$$\frac{dy}{dt} = -\sqrt{3}\,c_1\omega_1 \sin \omega_1 t + \sqrt{3}\,c_2\omega_1 \cos \omega_1 t + \sqrt{3}\,c_3\omega_2 \sin \omega_2 t - \sqrt{3}\,c_4\omega_2 \cos \omega_2 t$$

we obtain

$$2\sqrt{3}\,\omega_2 = \sqrt{3}\,c_2\omega_1 - \sqrt{3}\,c_4\omega_2 \quad (84)$$

and from

$$\frac{dx}{dt} = -c_1\omega_1 \sin \omega_1 t + c_2\omega_1 \cos \omega_1 t - c_3\omega_2 \sin \omega_2 t + c_4\omega_2 \cos \omega_2 t$$

we find

$$0 = c_2\omega_1 + c_4\omega_2 \quad (85)$$

Equations (84) and (85) yield $c_4 = -1$, $c_2 = 1.932$. From Eqs. (82) and (83) we find that $c_1 = 1$ and $c_3 = -1$. Therefore, the particular solution of (78)–(79) is

$$x = \cos \omega_1 t + \frac{\omega_2}{\omega_1} \sin \omega_1 t - \cos \omega_2 t - \sin \omega_2 t$$

$$y = \sqrt{3}\left( \cos \omega_1 t + \frac{\omega_2}{\omega_1} \sin \omega_1 t \right) + \sqrt{3}\,(\cos \omega_2 t + \sin \omega_2 t)$$

In general, the solution of the system (76)–(77) is of the form

$$x = A \sin (\omega_1 t + B) + C \sin (\omega_2 t + D)$$

$$y = \lambda A \sin (\omega_1 t + B) + \mu C \sin (\omega_2 t + D)$$

Thus, $x$ (as also $y$) is the superposition of the two simple harmonic motions, one with frequency $\omega_1/(2\pi)$ and the other with frequency $\omega_2/(2\pi)$. If the initial conditions are such that $C = 0$, then $x$ and $y$ describe simple harmonic motions in the ***principal*** or ***normal mode*** corresponding to the normal frequency $\omega_1/(2\pi)$. Similarly, if $A = 0$, then $x$ and $y$ describe the simple harmonic motion in another principal, or the normal mode corresponding to the normal frequency $\omega_2/(2\pi)$. The positive constants $\lambda$ and $\mu$ are the ***amplitude ratios.***

**EXAMPLE 7.18** Given the initial conditions as $x = 1$, $y = \sqrt{3}$ when $t = 0$ and when $t = \pi/(2\omega_1)$, solve the system (78)–(79).

***Solution*** Taking $t = 0$, we find from Eqs. (82) and (83) that $c_1 = 1$ and $c_3 = 0$; when $t = \pi/(2\omega_1)$, Eqs. (82) and (83) yield $c_2 = 1$ and $c_4 = 0$. Hence, the required solution is

$$x = \phi(t) = \cos \omega_1 t + \sin \omega_1 t = \sqrt{2} \sin\left(\omega_1 t + \frac{\pi}{4}\right)$$

$$y = \psi(t) = \sqrt{3} \cos \omega_1 t + \sqrt{3} \sin \omega_1 t = 6 \sin\left(\omega t + \frac{\pi}{4}\right)$$

The two masses undergo simple harmonic motion in the principal mode corresponding to the normal frequency $\omega_1/2\pi$. The amplitude ratio is $\sqrt{3}$.

More general systems having $n$ degrees of freedom can have $n$ principal modes of vibrations. While designing the structures or mechanism of, say, a building or an aeroplane, care must be taken so that resonance effects are not produced by periodic external forces having frequencies equal to or close to the various normal frequencies of the system. When the effects of frictional forces are taken into account, the vibrating systems become complicated.

Mechanical systems having more than one degree of freedom can be simulated by designing corresponding systems of coupled electrical circuits. This approach is fruitful when the complexity of the system renders a mathematical analysis difficult. For more details see [27].

## 7.9 MULTIPLE-LOOP ELECTRIC CIRCUITS

We have already applied Kirchhoff's voltage law (see Chapter 4) to the flow of electricity in a single-loop circuit. The flow of current in a multiple-loop circuit or network (Fig. 7.11) is governed by a system of differential equations.

A network consists of ***branches*** that meet at the ***branch points*** or ***nodes***. In Fig. 7.11(a), the network has nodes $B$ and $D$ and branches $DEAB$, $BD$ and $DGFB$. Let $i_1$, $i_2$ and $i_3$ represent the current in these branches. The flow of these currents is indicated by arrows in Fig. 7.11(a). This network consists of three loops—$ABDEA$, $BDGFB$ and $AEGFA$.

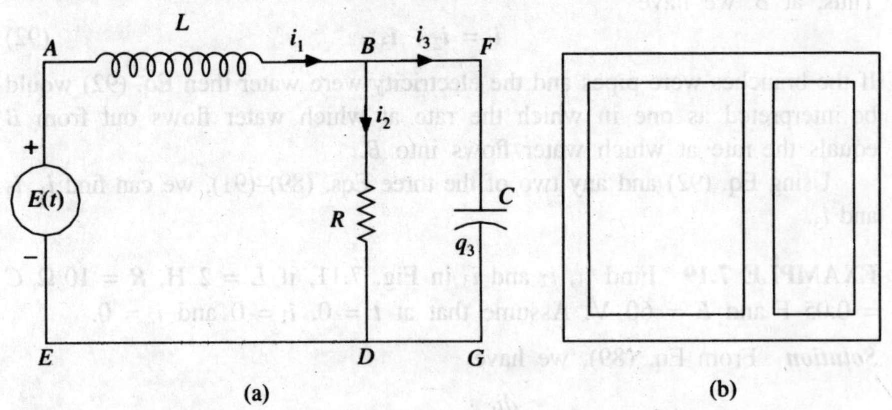


**Fig. 7.11** Multiple-loop electric circuits: (a) Flow of current in a multiple loop circuit; (b) Flow of water through pipes.

Now, apply Kirchhoff's voltage law to each loop such that a voltage drop in a branch is positive. Loop *ABDEA* yields the differential equation

$$L\frac{di_1}{dt} + Ri_2 - E = 0 \tag{86}$$

Similarly, the differential equations for the loops *BDGFB* and *AEGFA* are respectively

$$C^{-1}q_3 - Ri_2 = 0 \tag{87}$$

and

$$L\frac{di_1}{dt} + C^{-1}q_3 - E = 0 \tag{88}$$

where $i_3 = dq_3/dt$. Now, differentiating Eqs. (87) and (88) with respect to $t$ and $i_3 = dq_3/dt$, we get the following system of three differential equations:

$$L\frac{di_1}{dt} + Ri_2 = E \tag{89}$$

$$C^{-1}i_3 - R\frac{di_2}{dt} = 0 \tag{90}$$

$$L\frac{d^2i_1}{dt^2} + C^{-1}i_3 = \frac{dE}{dt} \tag{91}$$

These equations are valid even if one (or more) of the assumed directions of flow is incorrect. It is only necessary that our adopted design convention for voltage drops be applied consistently.

It can easily be shown that the three differential equations are not independent. For example, Eq. (91) can be obtained by differentiating both sides of Eq. (89) with respect to $t$ and adding the resulting differential equation and relation (90).

To determine $i_1$, $i_2$, $i_3$, we need one more equation and this is obtained by Kirchhoff's current law which states that the total current flowing into any branch point of a network equals the total current flowing out from branch. Thus, at *B*, we have

$$i_1 = i_2 + i_3 \tag{92}$$

If the branches were pipes and the electricity were water then Eq. (92) would be interpreted as one in which the rate at which water flows out from *B* equals the rate at which water flows into *B*.

Using Eq. (92) and any two of the three Eqs. (89)–(91), we can find $i_1$, $i_2$ and $i_3$.

**EXAMPLE 7.19** Find $i_1$, $i_2$ and $i_3$ in Fig. 7.11, if $L = 2$ H, $R = 10\ \Omega$, $C = 0.05$ F and $E = 60$ V. Assume that at $t = 0$, $i_1 = 0$ and $i_2 = 0$.

***Solution*** From Eq. (89), we have

$$2\frac{di_1}{dt} + 10i_2 = 60$$

and, from Eqs. (90) and (92), we get

$$20(i_1 - i_2) - 10\frac{di_2}{dt} = 0$$

In operator notations, these equations take the form

$$Di_1 + 5i_2 = 30, \qquad 2i_1 - (D + 2)\, i_2 = 0$$

By eliminating $i_2$ from these two, we get

$$[D(D + 2) + 10]i_1 = 60 \qquad \text{or} \qquad (D^2 + 2D + 10)i_1 = 60$$

The auxiliary equation has the roots $m = -1 \pm 3i$, and

$$i_1 = e^{-t}(c_1 \sin 3t + c_2 \cos 3t) + 6 \tag{93}$$

Now, at $t = 0$, $i_1 = 0$ and thus $c_2 = -6$. Also, from Eq. (92), at $t = 0$, $i_3 = 0$. Hence, from Eqs. (93), (89), (90) and (92),

$$i_1 = 6 + e^{-t}(8 \sin 3t - 6 \cos 3t)$$

$$i_2 = 6 - e^{-t}(2 \sin 3t + 6 \cos 3t)$$

$$i_3 = i_1 - i_2 = 10e^{-t} \sin 3t$$

Multiple-loop circuits are often constructed as analogs of mechanical systems having two or more degrees of freedom. Designing an appropriate network for simulating such systems often requires considerable ingenuity.

## 7.10 COMPARTMENT SYSTEMS

### 7.10.1 Mixture Problem

In Chapter 4, the mixture problem for a single compartment was done. Here, we treat the same problem for more than one compartment.

Let $X_1, X_2, \ldots, X_n$ denote a system of $n$ separate compartments and $x_i$ denote the amount of substance $S$ in compartment $X_i$ at time $t$. Assume that the concentration of $S$ is uniform throughout each compartment but varies from one compartment to another. In a *closed system*, transport of $S$ among the $n$ compartments takes place, but no amount of $S$ enters or leaves the system. In an *open system*, the input of $S$ into the system or the output of $S$ from the system or both are permitted.

The following example shows a two-compartment open system in which there is output from the system but no input to the system.

**EXAMPLE 7.20** A container $X$ contains 50-gal brine, in which 12-lb salt is dissolved, and another container $Y$ has 50-gal water containing no salt. The water without salt flows into $X$ at 8 gal/min., the mixture (well-stirred) in $X$ flows from $X$ to $Y$ at 9 gal/min., the mixture in $Y$ (well-stirred) flows back into $X$ at 1 gal/min., and out of the system at 8 gal/min. Let $x(t)$ and $y(t)$ denote the amount of salt in pounds in $X$ and $Y$ at time $t$. Find $x(t)$ and $y(t)$.

***Solution*** Salt does not flow into the two-compartment system. Since the concentration of salt in $X$ at time $t$ is $(x/50)$ lb/gal, and 9-gal solution flows from

$X$ to $Y$ per min., salt flows from $X$ to $Y$ at the rate of $9(x/50)$ lb/min. Similarly, salt flows from $Y$ to $X$ at $1(y/50)$ lb/min., and from $Y$ out of the system at $8(y/50)$ lb/ min. The transport in gal/min. of fluid is shown in Fig. 7.12(a) and that of salt in lb/min. in Fig. 7.12(b).

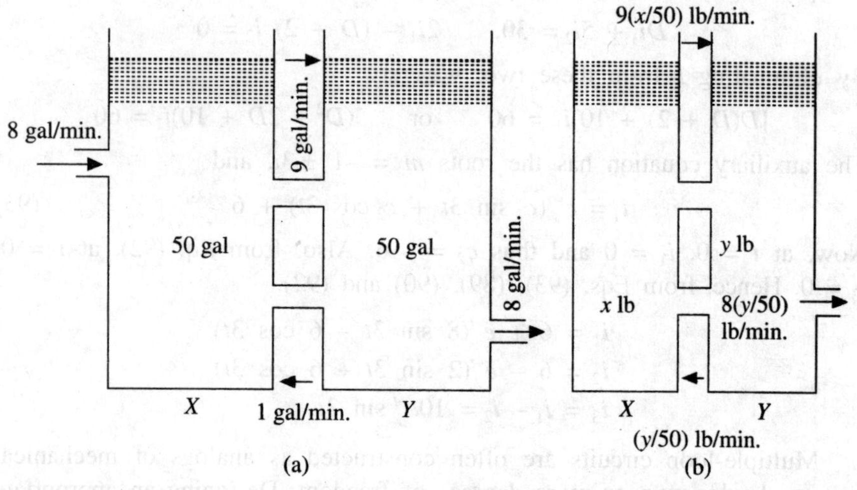


**Fig. 7.12** Two-component open system which has output from the system but no input to the system.

Applying the equation of continuity to the amount of salt in $X$ and $Y$, we obtain

$$\frac{dx}{dt} = -\frac{9}{50}x + \frac{1}{50}y$$

$$\frac{dy}{dt} = \frac{9}{50}x - \frac{1}{50}y - \frac{8}{50}y \tag{94}$$

Hence

$$\frac{d^2x}{dt^2} = -\frac{9}{50}\frac{dx}{dt} + \frac{1}{50}\frac{dy}{dt} = -\frac{9}{50}\frac{dx}{dt} + \frac{1}{50}\left(\frac{9}{50}x - \frac{9}{50}y\right)$$

$$= -\frac{9}{50}\frac{dx}{dt} + \frac{9x}{2500} - \frac{9}{50}\left(\frac{dx}{dt} + \frac{9x}{50}\right)$$

or

$$\frac{d^2x}{dt^2} + \frac{18}{50}\frac{dx}{dt} + \frac{72x}{2500} = 0$$

The auxiliary equation has the roots $m_1 = -3/25$ and $m_2 = -6/25$. Therefore

$$x(t) = c_1e^{-3t/25} + c_2e^{-6t/25}$$

$$x'(t) = \frac{dx}{dt} = -\frac{3}{25}c_1e^{-3t/25} - \frac{6}{25}c_2e^{-6t/25}$$

Using $x(0) = 12$, $y(0) = 0$ and from Eq. (94), $x'(0) = -54/25$, so that

$$12 = c_1 + c_2 \quad \text{and} \quad -\frac{54}{25} = -\frac{34}{25} - \frac{6c_2}{25}$$

which gives, $c_1 = c_2 = 6$. Therefore

$$x(t) = 6e^{-3t/25} + 6e^{-6t/25}$$

From

$$x'(t) = \frac{dx}{dt} = -\frac{18}{25}e^{-3t/25} - \frac{36}{25}e^{-6t/25}$$

and Eq. (94), we obtain

$$y(t) = 50\frac{dx}{dt} + 9x = 18e^{-3t/25} - 18e^{-6t/25}$$

**EXAMPLE 7.21** Let tank $X$ contain 100-gal of brine in which 100-lb of salt are dissolved and tank $Y$ contain 100 gal of water. Suppose water flows into tank $X$ at the rate of 2 gal/min., and the mixture flows from tank $X$ into tank $Y$ at 3 gal/ min. From $Y$ one gallon is pumped back to $X$ while 2 gal are flushed away. Find the amount of salt in both the tanks at all time $t$.

***Solution*** Let $x(t)$ and $y(t)$ denote the amount of salt in $X$ and $Y$ at time $t$ and note that the change in weight is equal to the difference between input and output. Initially, $x(0) = 100$ and $y(0) = 0$ at $t = 0$. At time $t$, let $x/100$ and $y/100$ be the amount of salt contained in each gallon of water taken from $X$ and $Y$. Three gallons are being removed from $X$ and added to $Y$, while only one of the three gallons removed from $Y$ is put in $X$. We thus have the system

$$\left.\begin{aligned} \frac{dx}{dt} &= -3\frac{x}{100} + \frac{y}{100}, \quad x(0) = 100 \\ \frac{dy}{dx} &= 3\frac{x}{100} - \frac{3y}{100}, \quad y(0) = 0 \end{aligned}\right\} \tag{95}$$

Solve the second equation for $x$ to get

$$x = y + \frac{100}{3}\frac{dy}{dt} \tag{96}$$

Differentiate Eq. (96) and equating it with Eq. (95), we have

$$-\frac{3x}{100} + \frac{y}{100} = \frac{dx}{dt} = \frac{dy}{dt} + \frac{100}{3}\frac{d^2y}{dt^2}$$

In this equation, substituting the value of $x$ from Eq. (96), we get

$$\frac{100}{3}\frac{d^2y}{dt^2} + 2\frac{dy}{dt} + \frac{2y}{100} = 0 \tag{97}$$

with $y'(0) = 3$ [from Eq. (95)]. Solving Eq. (97), we obtain

$$y(t) = 50\sqrt{3}\,[e^{[(-3+\sqrt{3})/100]t} - e^{[(-3-\sqrt{3})/100]t}]$$

and from Eq. (96)

$$x(t) = 50\,[e^{[(-3+\sqrt{3})/100]t} + e^{[(-3-\sqrt{3})/100]t}]$$

In some applications of the $n$ compartment system, $x_i$ represents the concentration of a substance $S$ in $X_i$ and $dx_i/dt$ the time rate of change of $x_i$. For example, $x_i$ may denote the concentration of a drug in a diffusion process. In cancer chemotherapy, the drug concentration varies over several compartments, which converts the injection compartment to the compartment in which malignancy occurs. The problem of drug concentration is now considered as follows.

## 7.10.2 Concentration of a Drug in a Two-compartment System

We shall determine here the concentration of some chemicals, such as a drug, in a system consisting of two compartments separated by a membrane (Fig. 7.13). The drug can pass through this membrane from compartment 1 to compartment 2, or vice versa. We also assume that the drug can escape to the external system through an opening in the second compartment.

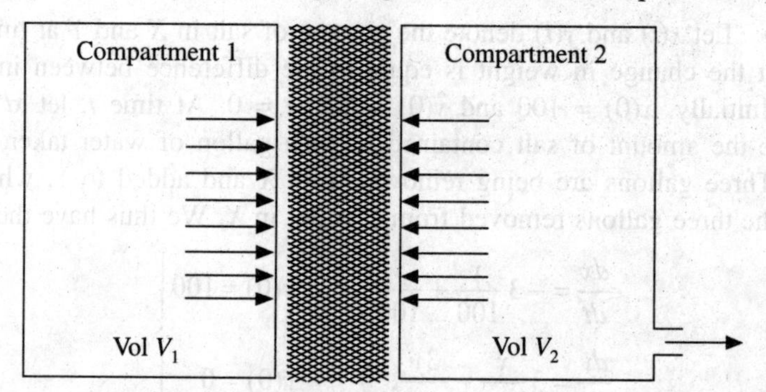


**Fig. 7.13** Concentration of chemical in a two-compartment system separated by a membrane.

Let $V_1$ and $V_2$ be the volumes of the two compartments and $A$, the cross-sectional area of the membrane. Also, let $x_1$ and $x_2$ denote the masses of the drug in compartments 1 and 2, respectively, at any time $t$. Then

Rate of change of drug mass in 1 = (rate of flow of drug mass from 2 to 1) − (rate of flow of drug mass from 1 to 2) (98)

We now find each term of Eq. (98). Thus

$$\text{Rate of change of drug mass in compartment } 1 = \frac{dx_1}{dt} \tag{99}$$

$$\text{Rate of flow of drug mass from compartment 2 to compartment } 1 = a_{21}\frac{Ax_2}{V_2} \tag{100}$$

(since, the rate of flow of the drug mass from 2 to 1 is proportional to area $A$ of the membrane and to the concentration of drug in compartment 2 given by $x_2/V_2$), where $a_{21}$ is the constant of proportionality.

By similar arguments, we have

$$\text{Rate of flow of drug mass from compartment 1 to compartment 2} = a_{12}\frac{Ax_1}{V_1} \tag{101}$$

where $a_{12}$ is the constant of proportionality.

Thus, from Eqs. (99)–(101), Eq. (98) becomes

$$\frac{dx_1}{dt} = a_{21}\frac{Ax_2}{V_2} - a_{12}\frac{Ax_1}{V_1} \tag{102}$$

The corresponding equation for compartment 2 yields

rate of change of drug mass in compartment 2
= (rate of flow of drug mass from 1 to 2
– (rate of flow of drug mass from 2 to 1)
– (rate of flow of drug mass from 2 into the external system).

A similar reasoning as above leads us to the following differential equation:

$$\frac{dx_2}{dt} = a_{12}\frac{Ax_1}{V_1} - a_{21}\frac{Ax_2}{V_2} - \frac{ax_2}{V_2} \tag{103}$$

(since the rate of flow of drug in compartment 2 is $x_2/V_2$, $a$ is the constant of proportionality.)

By the substitution

$$b_{21} = \frac{a_{21}A}{V_2}, \qquad b_{12} = \frac{a_{12}A}{V_1}, \qquad b = \frac{a}{V_2} \tag{104}$$

Equations (102) and (103) now take the form

$$\frac{dx_1}{dt} = b_{21}x_2 - b_{12}x_1, \qquad \frac{dx_2}{dt} = b_{12}x_1 - b_{21}x_2 - bx_2 \tag{105}$$

or $$(D + b_{12})x_1 - b_{21}x_2 = 0, \qquad -b_{12}x_1 + (D + b_{21} + b)\, x_2 = 0 \tag{106}$$

The simultaneous differential Eqs. (105) or (106) are to be solved under suitable conditions such as

$$x_1 = c, \qquad x_2 = d \qquad \text{at} \quad t = 0 \tag{107}$$

where $c$ and $d$ are constants and to get a solution, we proceed as follows.

Operate on the first equation of (106) with $D + b_{21} + b$ and multiply the second equation in (106) by $b_{21}$ and finally add the resulting equations to get

$$[D^2 + (b_{12} + b_{21} + b)\, D + bb_{12}]x_1 = 0 \tag{108}$$

The auxiliary equation has the roots

$$m = \frac{-(b_{12} + b_{21} + b) \pm \sqrt{(b_{12} + b_{21} + b)^2 - 4bb_{12}}}{2}$$

If we take

$$p = \frac{1}{2}(b_{12} + b_{21} + b), \qquad q = \sqrt{(b_{12} + b_{21} + b)^2 - 4bb_{12}} \tag{109}$$

then these roots are $m = -p \pm q$, and the solution is

$$x_1 = e^{-pt}(c_1 e^{qt} + c_2 e^{-qt}) \tag{110}$$

Substituting this value of $x_1$ into the first equation of (106), we get

$$x_2 = \frac{e^{-pt}}{b_{21}}[(b_{12} - p + q)\, c_1 e^{qt} + (b_{12} - p - q) c_2 e^{-qt}] \tag{111}$$

The constants $c_1$ and $c_2$ in Eqs. (110) and (111) can be obtained from Eq. (107).

The value of $q$ in Eq. (109) suggests that there are three cases which may arise corresponding to the cases where $q$ is imaginary, $q > 0$ and $q = 0$. It turns out that the first and last case cannot arise. Using $q > 0$ and $p > q$ if $b > 0$ [(as may be seen from Eq. (109)], it follows that the drug masses $x_1$ and $x_2$ approach zero steadily in a manner similar to that indicated by Fig. 6.16, corresponding to the overdamped case. We cannot have a damped oscillatory motion as in Fig. 6.21.

## 7.11 THE PROBLEM OF EPIDEMICS WITH QUARANTINE

While treating the problem of epidemics in Chapter 4, we did not consider the possibility that the students becoming infected would be removed from the community so as not to cause others to become infected. It will be interesting to consider this more realistic model of disease in which students discovered to be infected are removed or, as we often say, *quarantined*. Several circumstances may arise which are mathematically equivalent to such quarantine as follows:

(i) The individuals have an immunity and are not carriers of the disease.
(ii) Those who develop the disease do not recover.

Let $N_u$ denote the number of students who are uninfected at any time $t$ (this does not include students who may have recovered as in (i) and returned to the community); $N_i$, the number of students who are infected at time $t$ but have not yet been quarantined, and $N_q$, the number of students who are quarantined at time $t$. We thus have

$$N = N_u + N_i + N_q \tag{112}$$

where $N$ is the total number of students of the community, assumed constant.

As in Chapter 4, here also we expect that the time rate at which uninfected students become infected is proportional to the product of the number of infected and uninfected students, so that

$$\frac{dN_u}{dt} = -kN_u N_i \tag{113}$$

Also, the rate at which infected students are quarantined is proportional to the number of infected students. This leads to the relation

$$\frac{dN_q}{dt} = \lambda N_i \tag{114}$$

Using Eqs. (112)–(114), we obtain

$$\frac{dN_i}{dt} = kN_u N_i - \lambda N_i \tag{115}$$

where $k$ and $N$ are the constants of proportionality. The possible initial conditions are

$$N_u = U_0, \qquad N_i = I_0, \qquad N_q = 0 \quad \text{at} \quad t = 0 \tag{116}$$

where $U_0$ and $I_0$ are constants. From Eq. (112)

$$N = U_0 + I_0 \tag{117}$$

Dividing Eqs. (113) and (114), we get

$$\frac{dN_u}{dN_q} = -\frac{k}{\lambda} N_u \tag{118}$$

Now, separating the variable in Eq. (118) and integrating, we obtain

$$N_u = c_1 e^{-kN_q/\lambda} \tag{119}$$

where $c_1$ is an arbitrary constant. From Eq. (116), we get

$$c_1 = U_0$$

so that

$$N_u = U_0 e^{-kN_q/\lambda} \tag{120}$$

Use this in Eq. (112) to get

$$N_i = N - N_q - U_0\, e^{-kN_q/\lambda} \tag{121}$$

Thus, Eq. (114) becomes

$$\frac{dN_q}{dt} = \lambda(N - N_q - U_0\, e^{-kN_q/\lambda})$$

Separating the variables and integrating, we have

$$\int \frac{dN_q}{N - N_q - U_0 e^{-kN_q/\lambda}} = \int \lambda\, dt \tag{122}$$

The integral on the left cannot be obtained exactly. However, we can obtain a good approximation by using the first three terms of the series expansion of the exponential in Eq. (122), so that

$$N - N_q - U_0 e^{-kN_q/\lambda} = N - N_q - U_0 + \frac{kU_0 N_q}{\lambda} - \frac{k^2 U_0}{2\lambda^2} N_q^2$$

$$= I_0 - N_q + \frac{kU_0 N_q}{\lambda} - \frac{k^2 U_0 N_q^2}{2\lambda^2} \qquad \text{[from (117)]}$$

$$= I_0 + \left(\frac{kU_0}{\lambda} - 1\right) N_q - \frac{k^2 U_0 N_q^2}{2\lambda^2}$$

$$= I_0 + \alpha + (W - \alpha)^2$$

where

$$W = \frac{k\sqrt{U_0}}{\lambda\sqrt{2}} N_q \tag{123}$$

$$\alpha = \left(\frac{kU_0}{\lambda} - 1\right) \frac{\lambda\sqrt{2}}{2k\sqrt{U_0}} \tag{124}$$

Thus, Eq. (122) becomes

$$\frac{\lambda\sqrt{2}}{k\sqrt{U_0}} \int \frac{dW}{I_0 + \alpha^2 - (W - \alpha)^2} = \lambda \int dt \tag{125}$$

which on integration yields

$$\frac{\sqrt{2}}{k\sqrt{U_0}\,\sqrt{I_0 + \alpha^2}} \tanh^{-1}\left(\frac{W - \alpha}{\sqrt{I_0 + \alpha^2}}\right) = t + c_2 \tag{126}$$

where $c_2$ is the constant of integration and can be obtained by taking $N_q = 0$ or $W = 0$ at $t = 0$ in Eq. (126). Thus

$$c_2 = \frac{\sqrt{2}}{k\sqrt{U_0}\,\sqrt{I_0 + \alpha^2}} \tanh^{-1}\left(\frac{-\alpha}{\sqrt{I_0 + \alpha^2}}\right) = \frac{-\sqrt{2}}{k\sqrt{U_0}\,\sqrt{I_0 + \alpha^2}} \tanh^{-1}\left(\frac{\alpha}{\sqrt{I_0 + \alpha^2}}\right)$$

so that

$$\tanh^{-1}\left(\frac{W - \alpha}{\sqrt{I_0 + \alpha^2}}\right) = \frac{k\sqrt{U_0}\,\sqrt{I_0 + \alpha^2}\,t}{\sqrt{2}} - \tanh^{-1}\left(\frac{\alpha}{\sqrt{I_0 + \alpha^2}}\right)$$

from which

$$W = \alpha + \sqrt{I_0 + \alpha^2}\,\tanh(\beta t - \gamma)$$

where

$$\beta = \frac{k\sqrt{U_0}\,\sqrt{I_0 + \alpha^2}}{\sqrt{2}}, \qquad \gamma = \tanh^{-1}\left(\frac{a}{\sqrt{I_0 + \alpha^2}}\right)$$

Hence

$$N_q = \frac{k\sqrt{U}}{\lambda\sqrt{2}}\left[\alpha + \sqrt{I\ + \alpha}\ \ \tanh\,(\beta t - \alpha)\right]$$

$$N_u = U_0 e^{-kN_q/\lambda}$$

$$N_i = N - N_q - U_0 e^{-kN_q/\lambda} = I_0 - N_q - U_0(1 - e^{-kN_q/\lambda})$$

These equations solve the problem of epidemics completely.

## 7.12 ARMS RACE

An interesting application that leads to a system of differential equation is the study of arms race. We will present here a model for arms race. This is known as the Richardson model [17, 9].

The following assumptions are made in this model:

(i) The expenditure for armament of each country will increase at a rate which is proportional to the other country's expenditure.

(ii) The expenditure for armament of each country will decrease at a rate which is proportional to its own expenditure.

(iii) The rate of change of arms expenditure for a country has a constant component that measures the level of antagonism of that country towards the other.

(iv) The effects of all these assumptions are additive.

If $x$ and $y$ denote the expenditure incurred by two countries on armament, then under the assumptions (i)–(iv), the system of differential equations is

$$\frac{dx}{dt} = ay - px + r, \qquad \frac{dy}{dt} = bx - qy + s \tag{127}$$

The constants $a$, $b$, $p$ and $q$ are positive, but the numbers $r$ and $s$ may have any value. Positive values arise if the countries have internal attitudes of distrust of each other.

In matrix notation, see [13], the problem may be written as

$$X' = AX + B, \qquad X(0) = \begin{pmatrix} x_0 \\ y_0 \end{pmatrix} \tag{128}$$

where

$$X' = \begin{pmatrix} x' \\ y' \end{pmatrix}, \qquad X = \begin{pmatrix} x(t) \\ y(t) \end{pmatrix}, \qquad A = \begin{pmatrix} -p & a \\ b & -q \end{pmatrix}, \qquad B = \begin{pmatrix} r \\ s \end{pmatrix}$$

The nature of the solutions of the system will depend upon the eigenvalues of the matrix $A$, i.e. on the roots of the characteristic equation $|mI - A| = 0$ ($I$ is the identity matrix)

$$\begin{vmatrix} -p-m & a \\ b & -q-m \end{vmatrix} = m^2 + (p+q)\,m + (pq - ab) = 0$$

The roots are

$$\frac{-(p+q) \pm \sqrt{(p+q)^2 - 4(pq - ab)}}{2} = \frac{-(p+q) \pm \sqrt{(p-q)^2 + 4ab}}{2}$$

Since, $a$ and $b$ are positive, the eigenvalues are real and distinct. As $p > 0$, $q > 0$, it follows that if $pq - ab > 0$, then the two eigenvalues are both negative, but if $pq - ab < 0$, then the eigenvalues will have opposite signs. The presence of a positive eigenvalue is disturbing since it will lead to an exponential function that becomes unbounded as time increases, a situation that may result in a runaway arms race.

The possible consequence of this model has been illustrated in the following examples.

**EXAMPLE 7.22** Discuss the Richardson model, if the parameters in Eq. (128) have the values $a = 4$, $b = 2$, $p = 3$, $q = 1$, $r = 2$, $s = 2$, $x_0 = 4$ and $y_0 = 1$.

***Solution*** We have

$$X' = \begin{pmatrix} -3 & 4 \\ 2 & -1 \end{pmatrix} X + \begin{pmatrix} 2 \\ 2 \end{pmatrix}, \qquad X(0) = \begin{pmatrix} 4 \\ 1 \end{pmatrix}$$

The characteristic equation of the matrix equation is

$$\begin{vmatrix} -3-m & 4 \\ 2 & -1-m \end{vmatrix} = m^2 + 4m - 5 = 0$$

so that the eigenvalues are $m_1 = 1$ and $m_2 = -5$.

Now, we solve the equation $Ax = mx$ to get the eigenvectors $x$ corresponding to eigenvalues $m$. Thus, for $m_1 = 1$

$$A\begin{pmatrix} x_1 \\ x_2 \end{pmatrix} = 1\begin{pmatrix} x_1 \\ x_2 \end{pmatrix}$$

gives $x_1 = x_2$. If $x_1 = 1$, then for $m_1 = 1$ the eigenvector is

$$\begin{pmatrix} 1 \\ 1 \end{pmatrix}$$

Similarly, for $m_2 = -5$, we have $x_1 + 2x_2 = 0$. Taking $x_2 = -1$, the eigenvector is

$$\begin{pmatrix} 2 \\ -1 \end{pmatrix}$$

The general solution of the homogeneous system $X' = AX$ is

$$X(t) = c_1 \begin{pmatrix} 1 \\ 1 \end{pmatrix} e^t + c_2 \begin{pmatrix} 2 \\ -1 \end{pmatrix} e^{-5t}$$

The nonhomogeneous system $X' = AX + B$ has a solution of the form

$$\begin{pmatrix} e \\ f \end{pmatrix}$$

Substituting it into the system, we obtain

$$\begin{pmatrix} -3 & 4 \\ 2 & -1 \end{pmatrix} \begin{pmatrix} e \\ f \end{pmatrix} + \begin{pmatrix} 2 \\ 2 \end{pmatrix} = 0$$

a system with solution

$$\begin{pmatrix} -2 \\ -2 \end{pmatrix}$$

Thus, the general solution of the nonhomogeneous system is

$$X(t) = c_1 \begin{pmatrix} 1 \\ 1 \end{pmatrix} e^t + c_2 \begin{pmatrix} 2 \\ -1 \end{pmatrix} e^{-5t} + \begin{pmatrix} -2 \\ -2 \end{pmatrix}$$

while the initial condition $X(0) = \begin{pmatrix} 4 \\ 1 \end{pmatrix}$ requires that

$$\begin{pmatrix} 4 \\ 1 \end{pmatrix} = c_1 \begin{pmatrix} 1 \\ 1 \end{pmatrix} + c_2 \begin{pmatrix} 2 \\ -1 \end{pmatrix} + \begin{pmatrix} -2 \\ -2 \end{pmatrix}$$

so that $c_1 = 4$, $c_2 = 4$. Therefore, the required solution is

$$X(t) = 4 \begin{pmatrix} 1 \\ 1 \end{pmatrix} e^t + 4 \begin{pmatrix} 2 \\ -1 \end{pmatrix} e^{-5t} + \begin{pmatrix} -2 \\ -2 \end{pmatrix}$$

or $\quad x(t) = 4e^t + 8e^{-5t} - 2, \qquad y(t) = 4e^t - 2e^{-5t} - 2$

Here, we have a *runaway arms race.*

**EXAMPLE 7.23** If the parameters in Eq. (128) have the values $a = 4$, $b = 2$, $p = 3$, $q = 1$, $r = -2$, $s = -2$, $x_0 = 2$, $y_0 = 1/2$, then discuss the Richardson model.

***Solution*** The system of differential equations has the same solution as above except for the sign of the particular solution.

The general solution is

$$X(t) = c_1 \begin{pmatrix} 1 \\ 1 \end{pmatrix} e^t + c_2 \begin{pmatrix} 2 \\ -1 \end{pmatrix} e^{-5t} + \begin{pmatrix} 2 \\ 2 \end{pmatrix}$$

From the initial conditions, we have

$$\begin{pmatrix} 2 \\ \dfrac{1}{2} \end{pmatrix} = c_1 \begin{pmatrix} 1 \\ 1 \end{pmatrix} + c_2 \begin{pmatrix} 2 \\ -1 \end{pmatrix} + \begin{pmatrix} 2 \\ 2 \end{pmatrix}$$

so that $c_1 = -1$ and $c_2 = 1/2$. The solution is, therefore,

$$x(t) = -e^t + e^{-5t} + 2, \qquad y(t) = -e^t - \frac{1}{2} e^{-5t} + 2$$

and each country will eventually decrease its expenditure for arms to zero—*a condition of disarmament.*

**EXAMPLE 7.24** Solve the system (128) if the values of the parameters are $a = 3$, $b = 1$, $p = 4$, $q = 2$, $r = 6$, $s = 1$, and the initial conditions are $x(0) = 0$, $y(0) = 0$.

***Solution*** With the given values of the parameters, the system to be solved becomes

$$\begin{pmatrix} x' \\ y' \end{pmatrix} = \begin{pmatrix} -4 & 3 \\ 1 & -2 \end{pmatrix} \begin{pmatrix} x \\ y \end{pmatrix} + \begin{pmatrix} 6 \\ 1 \end{pmatrix}$$

Here, the eigenvalues are –1 and –5 and the corresponding eigenvectors are

$$\begin{pmatrix} 1 \\ 1 \end{pmatrix} \text{ and } \begin{pmatrix} 3 \\ -1 \end{pmatrix}$$

Thus, the general solution is

$$\begin{pmatrix} x \\ y \end{pmatrix} = c_1 \begin{pmatrix} 1 \\ 1 \end{pmatrix} e^{-t} + c_2 \begin{pmatrix} 3 \\ -1 \end{pmatrix} e^{-5t} + \begin{pmatrix} 3 \\ 2 \end{pmatrix}$$

From the initial conditions, we find that $c_1 = -9/4$ and $c_2 = -1/4$, so that the general solution becomes

$$\begin{pmatrix} x \\ y \end{pmatrix} = \frac{-9}{4} \begin{pmatrix} 1 \\ 1 \end{pmatrix} e^{-t} - \frac{1}{4} \begin{pmatrix} 3 \\ -1 \end{pmatrix} e^{-5t} + \begin{pmatrix} 3 \\ 2 \end{pmatrix} \tag{129}$$

which also yields

$$\begin{pmatrix} x' \\ y' \end{pmatrix} = \frac{9}{4} \begin{pmatrix} 1 \\ 1 \end{pmatrix} e^{-t} - \frac{1}{4} \begin{pmatrix} 3 \\ -1 \end{pmatrix} e^{-5t} \tag{130}$$

Equations (129) and (130) can be compared to an arms race in which each country starts with zero expenditure, but $dx/dt = 6$ and $dy/dt = 1$, both being positive quantities. Because of the negative exponents, the rate at which the expenditure is changing will tend towards zero and the arms expenditure will approach $x = 3$ and $y = 2$. There will be a *stabilised arms race.*

## 7.13 THE PREDATOR-PREY PROBLEM: A PROBLEM IN ECOLOGY

An important problem in ecology, the science which studies the interrelationships of organisms and their environment, is to investigate the question of coexistence of two species and to decide what mankind should do to preserve this ecological balance of nature.

In nature, there are many instances where one species of animal feeds on another species of animal, which in turn feeds on other things. For example, wolves in Alaska feed on caribon which in turn feed on vegetation. Sharks in the ocean feed on small fish which in turn feed on plants. The first species (wolves, sharks) is known as the *predator* and the second species as the *prey.* Theoretically, the predator can destroy all the prey so that the prey become extinct. However, if this happens, the predator will also become extinct, as it depends on the prey for its existence.

What actually happens in nature is that a cycle develops where at some time the prey may be abundant and the predator few, as indicated by the points $A_1$ and $B_1$, respectively, in Fig. 7.14. Because of the abundance of prey, the predator population grows and reduces the population of prey, leading to points $A_2$, $B_2$ of Fig. 7.14. This results in a reduction of predators and consequent increase of prey and the cycle continues.

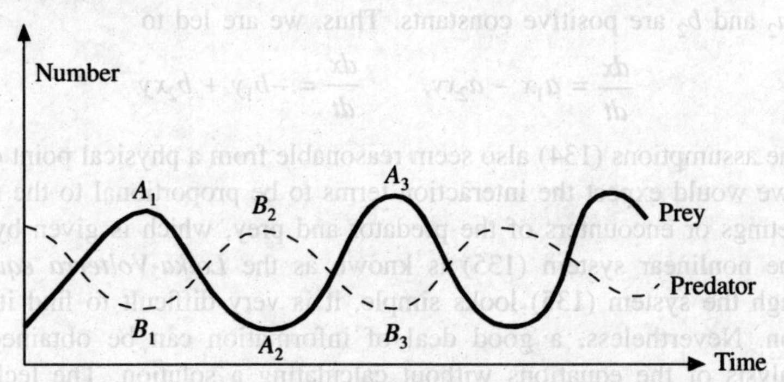


**Fig. 7.14** Predator-prey cycle.

We shall here present a mathematical formulation of this predator-prey problem (this and related problems were first investigated by Lotka and Volterra, see [14]).

Let $x$ denote the number of the prey at any time $t$ and $y$, the number of predator at any time $t$. Now, if there were no predators, then, as a first approximation, the number of prey would increase at a rate proportional to their number at any time so that

$$\frac{dx}{dt} = a_1 x \tag{131}$$

where $a_1 > 0$ is the constant of proportionality.

Similarly, if there were no prey, then, as a first approximation, the number of predators would decline at a rate proportional to their number so that

$$\frac{dy}{dt} = -b_1 y \tag{132}$$

where $b_1 > 0$ is the constant of proportionality.

As $x = k_1 e^{a_1 t}$ and $y = k_2 e^{-b_1 t}$ are the solutions of Eqs. (131) and (132), respectively, $k_1$ and $k_2$ are constants, we see that $x \to \infty$, $y \to 0$ and $t \to \infty$, which does not match with reality. To obtain a mathematical model of the actual situation, we have to modify Eqs. (131) and (132), so as to take into account the interaction of the species on each other. To do this we include in Eq. (131) an interaction term depending on $x$ and $y$, say $F_1(x, y)$, which tends to diminish the rate at which $x$ increases. Similarly, the inclusion of the interaction term, say $F_2(x, y)$, in (132) tends to increase the rate at which $y$ decreases. Thus, we have

$$\frac{dx}{dt} = a_1 x - F_1(x, y), \qquad \frac{dy}{dt} = -b_1 y + F_2(x, y) \tag{133}$$

The simplest forms of $F_1(x, y)$ and $F_2(x, y)$ are given by

$$F_1(x, y) = a_2 xy, \qquad F_2(x, y) = b_2 xy \tag{134}$$

so that they are zero if either $x = 0$ or $y = 0$, i.e. if there are no predator or prey, $a_2$ and $b_2$ are positive constants. Thus, we are led to

$$\frac{dx}{dt} = a_1 x - a_2 xy, \qquad \frac{dx}{dt} = -b_1 y + b_2 xy \tag{135}$$

The assumptions (134) also seem reasonable from a physical point of view since we would expect the interaction terms to be proportional to the number of meetings or encounters of the predator and prey, which is given by $xy$.

The nonlinear system (135) is known as the *Lotka-Volterra equations.* Although the system (135) looks simple, it is very difficult to find its exact solution. Nevertheless, a good deal of information can be obtained from an analysis of the equations without calculating a solution. The techniques described here are intended only as an indication of some of the ways by means of which one can attack a nonlinear problem (we can, however, get an approximate solution of Eq. (135) either by the series or numerical method).

To begin with, it is convenient to interpret Eq. (135) to represent the $x$ and $y$ components of the velocity of a particle moving in an $xy$-plane. The curve (known as the path, trajectory or orbit) in which the particle moves can be represented by the parametric equations

$$x = \phi_1(t, c_1, c_2), \qquad y = \phi_2(t, c_1, c_2) \tag{136}$$

where $c_1$, $c_2$ are constants, which can be determined by prescribing a point through which the particles pass at some time, say, $t = 0$. Equations (136) represent the solutions of the system (135) and also represent geometrically the family of curves or trajectories, any of which could be the path of the particle.

These ideas enable us to describe solutions of (135) by using concepts of mechanics and geometry and provide valuable insight into what is going on. The *xy*-plane in which the particle is moving is called the *phase plane* and the analysis using such interpretation is called *phase plane analysis.* The family of curves are, therefore, often referred to as *phase curves***.

The points where the particle stops moving are known as the *rest points* or *equilibrium points.* They occur where the velocity is zero, i.e. where $dx/dt = 0$ and $dy/dt = 0$. From Eq. (135), we have

$$\frac{dx}{dt} = 0, \quad \text{for } x = 0 \text{ or } y = \frac{a_1}{a_2} \qquad (137)$$

$$\frac{dy}{dt} = 0, \quad \text{for } y = 0 \text{ or } x = \frac{b_1}{b_2}$$

Thus, there are two possible equilibrium points

$$x = 0, \qquad y = 0, \qquad x = \frac{b_1}{b_2}, \qquad y = \frac{a_1}{a_2} \qquad (138)$$

It is interesting to note that these are also solutions of Eqs. (135), which are independent of time and can be considered as special cases of Eqs. (136) in which the curves degenerate into points. We shall now examine in detail the two cases in Eq. (138).

*Case I Equilibrium point* (0, 0): In this case, there is neither predator nor prey. When the particle in the phase plane interpretation is closely near this point, we can neglect the second terms $a_2xy$ and $b_2xy$ on the right-hand side of Eqs. (135) in comparison to the first term. The resulting equations are, therefore,

$$\frac{dx}{dt} = a_1 x, \qquad \frac{dy}{dt} = -b_1 y \qquad (139)$$

with the solution

$$x = k_1 e^{a_1 t}, \qquad y = k_2 e^{-b_1 t} \qquad (140)$$

Since $x$ and $y$ are non-negative, we have $k_1 \geq 0$, $k_2 \geq 0$, and the family of curves described by Eqs. (140) is shown in Fig. 7.15. The particular case $k_1 = 0$, $k_2 = 0$ corresponds to the equilibrium point (degenerate curve) $x = 0$, $y = 0$. As $t$ increases, we see from Eqs. (140) that $x$ increases while $y$ approaches zero, so that the particle moves in the direction shown in the figure. From these curves, it is evident that if we should displace the particle slightly from the equilibrium point (0, 0), it tends to move away from the point (see Fig. 7.15). It is for this reason, we call the equilibrium point an *unstable equilibrium point* (or, a point of instability). If, on the other hand, a slight displacement from the equilibrium point resulted in a tendency for the particle to return or

---

* The phase plane has been extensively used by mathematicians Poincare and Liapunov in their researches on nonlinears systems of differential equations.

move towards the equilibrium point, as in the case of a mass on a stretched spring, we would call the point a *stable equilibrium point.*

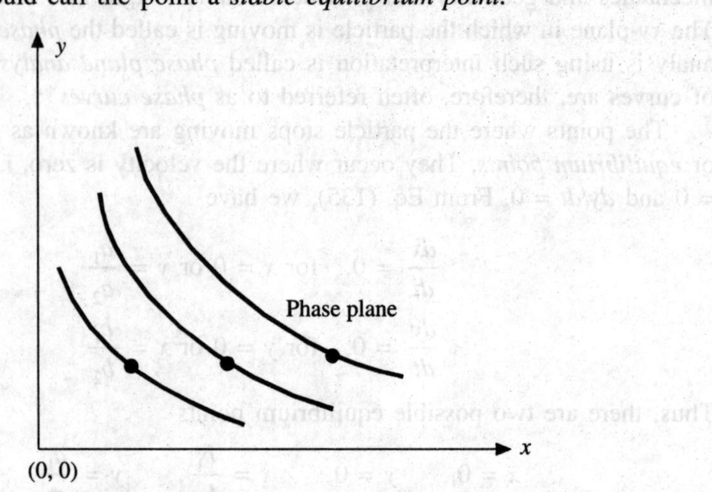


**Fig. 7.15** Unstable equilibrium.

The interpretation is that if at some time there is a small number of predators and preys, there will be a tendency for the number of preys to increase and the number of predators to decrease, as indicated in the Fig. 7.15. However, in real life situation, $x$ does not increase indefinitely since an increase in prey results in a subsequent increase in predator. This allows for the possibility that if the curves of Fig. 7.15 are extended, they will tend to rise after some point.

*Case II Equilibrium point* $(b_1/b_2, a_1/a_2)$: In this case, the predator and prey are in equilibrium state such that their numbers do not change because $x = b_1/b_2$, $y = a_1/a_2$ is a solution of Eqs. (135), which is independent of time. It is interesting to observe that what happens if there is a slight departure from this equilibrium state, which can occur if, for example, hunters destroy either the prey or the predator (or both).

For this, we consider the transformation

$$x = \frac{b_1}{b_2} + u, \qquad y = \frac{a_1}{a_2} + v \tag{141}$$

so that Eqs. (135) become

$$\frac{du}{dt} = -\frac{a_2 b_1}{b_2} v - a_2 uv, \qquad \frac{dv}{dt} = \frac{a_1 b_2}{a_2} u + b_2 uv \tag{142}$$

Now, if the particle of our phase plane interpretation is close to the point $(b_1/b_2, a_1/a_2)$ of the $xy$-plane, then $u$ and $v$ will be close to zero. In such a case, the second terms on the right of Eqs. (142) can be neglected in comparison to the first so that we obtain the linerarised system

$$\frac{du}{dt} = -\frac{a_2 b_1}{b_2} v, \qquad \frac{dv}{dt} = \frac{a_1 b_2}{a_2} u \tag{143}$$

Eliminating $v$, we obtain

$$\frac{d^2u}{dt^2} + a_1b_1u = 0 \tag{144}$$

whose solution is

$$u = c_1 \cos\left(\sqrt{a_1b_1}\right)t + c_2 \sin\left(\sqrt{a_1b_1}\right)t \tag{145}$$

so that

$$v = \frac{b_2}{a_2}\sqrt{\frac{a_1}{b_1}}\left[c_1 \sin\left(\sqrt{a_1b_1}\right)t - c_2 \cos\left(\sqrt{a_1b_1}\right)t\right] \tag{146}$$

Thus, Eqs. (141) become

$$\left.\begin{aligned} x &= \frac{b_1}{b_2} + c_1 \cos\left(\sqrt{a_1b_1}\right)t + c_2 \sin\left(\sqrt{a_1b_1}\right)t \\ y &= \frac{a_1}{a_2} + \frac{b_2}{a_2}\sqrt{\frac{a_1}{b_1}}\left[c_1 \sin\left(\sqrt{a_1b_1}\right)t - c_2 \cos\left(\sqrt{a_1b_1}\right)t\right] \end{aligned}\right\} \tag{147}$$

The parametric Eqs. (147) represent concentric ellipses having a common centre $(b_1/b_2, a_1/a_2)$ and clockwise direction as shown in Fig. 7.16. This direction can be obtained from the solution (147) by noting how the points $(x, y)$ moves as $t$ increases, or directly by Eq. (143). To see how it is obtained from (143), refer to

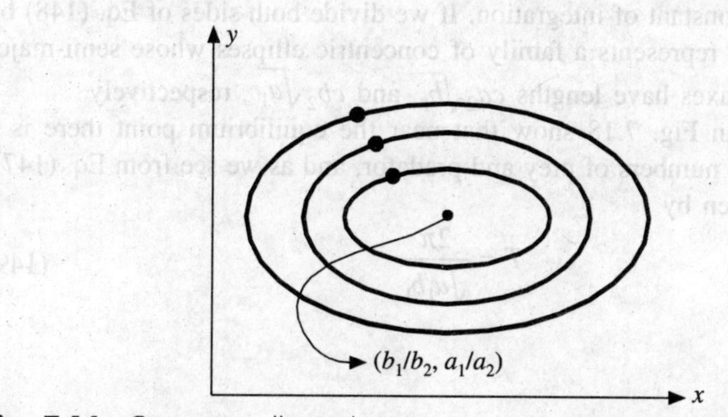


**Fig. 7.16** Concentric ellipses having a common centre and clockwise direction.

Fig. 7.17 in which a $uv$ coordinate system has been chosen with its origin at the common centre of ellipse. From Eq. (143), we see that at a point in the first quadrant where $u > 0$, $v > 0$, the components of velocity are such that $du/dt < 0$, $dv/dt > 0$, so that the velocity has the counter-clockwise direction indicated. This same direction can be confirmed by choosing points in the other quadrants. Thus, in the third quadrant, where $u < 0$, $v < 0$, we have $du/dt > 0$, $dv/dt > 0$, showing again the velocity in the counter-clockwise direction.

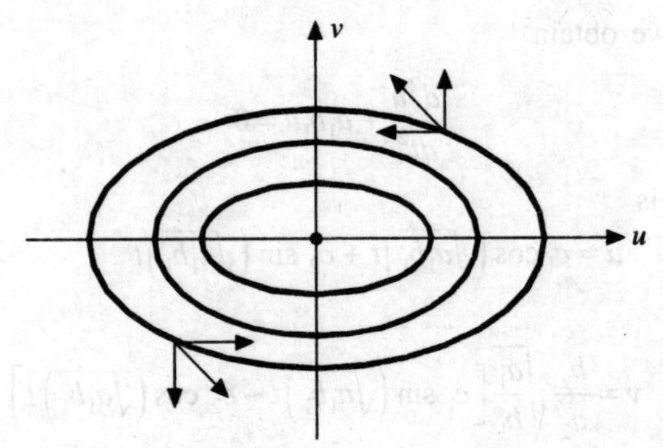


**Fig. 7.17** A uv-coordinate system for concentric ellipses to know the direction.

The fact that the phase curves are concentric ellipses can also be deduced directly from Eqs. (143) by eliminating $t$, to obtain

$$\frac{du}{dv} = -\frac{b_1 a_2^2}{a_2 b_1^2}\frac{v}{u}$$

or

$$a_1 b_2^2 u\, du + b_1 a_2^2 v\, dv = 0$$

which on integration yields

$$\frac{u^2}{b_1 a_2^2} + \frac{v^2}{a_1 b_2^2} = c^2 \tag{148}$$

where $c^2$ is the constant of integration. If we divide both sides of Eq. (148) by $c^2$, we see that it represents a family of concentric ellipses whose semi-major and semi-minor axes have lengths $ca_2\sqrt{b_1}$ and $cb_2\sqrt{a_1}$, respectively.

The curves in Fig. 7.18 show that near the equilibrium point there is a periodicity in the numbers of prey and predator, and as we see from Eq. (147), the period is given by

$$T = \frac{2\pi}{\sqrt{a_1 b_1}} \tag{149}$$

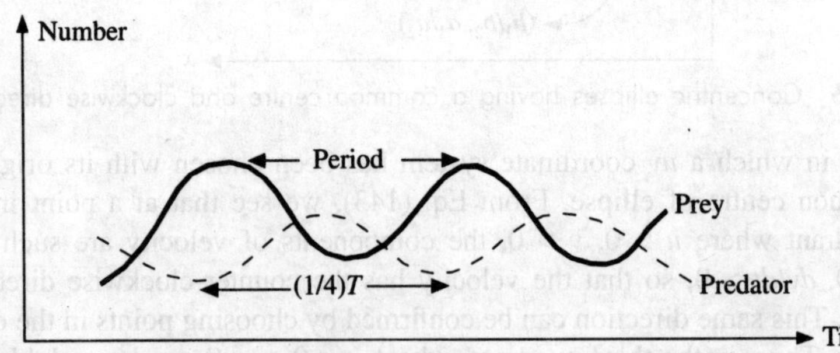


**Fig. 7.18** Periodicity in the number of prey and predator near the equilibirum point.

A question that naturally arises is whether we should consider the point $(b_1/b_2, a_1/a_2)$ a stable or unstable equilibrium point. Since, a slight displacement of our phase plane particle from the equilibrium point does not result in a movement either away or towards but only *around* the point, we might be included to call it *neither,* and may even make up a new word to describe the behaviour. (For more details of the stable and unstable equilibrium points, see [1].)

From Eqs. (147), we can obtain some more interesting information. To do so, let $\theta = \sqrt{a_1 b_1}\, t$ and note that

$$c_1 \cos \theta + c_2 \sin \theta = \sqrt{c_1^2 + c_2^2} \cos (\theta - \alpha) \tag{150}$$

where

$$\cos\alpha = \frac{c_1}{\sqrt{c_1^2 + c_2^2}}, \qquad \sin \alpha = \frac{c_2}{\sqrt{c_1^2 + c_2^2}} \tag{151}$$

Replacing $\theta$ by $\theta - (\pi/2)$ in Eq. (150), we obtain

$$c_1 \sin \theta - c_2 \cos \theta = \sqrt{c_1^2 + c_2^2} \cos\left(\theta - \alpha - \frac{\pi}{2}\right) \tag{152}$$

Thus, Eqs. (147) can be written as

$$\left.\begin{aligned} x &= \frac{b_1}{b_2} + \sqrt{c_1^2 + c_2^2} \cos \sqrt{a_1 b_1}\left(t - \frac{\alpha}{\sqrt{a_1 b_1}}\right) \\ y &= \frac{a_1}{a_2} + \frac{b_2}{a_2}\sqrt{c_1^2 + c_2^2} \cos \sqrt{a_1 b_1}\left(t - \frac{\alpha + \pi/2}{\sqrt{a_1 b_1}}\right) \end{aligned}\right\} \tag{153}$$

Equations (153) show that at time $t = \alpha/\sqrt{a_1 b_1}$, for example, the number of prey is maximum. However, the number of predator does not reach a maximum until

$$t = \frac{\alpha + (\pi/2)}{\sqrt{a_1 b_1}}$$

That is, a time $(\pi/2)\sqrt{a_1 b_1}$ or one quarter of the period (149) later. The graphs of $x$ and $y$ versus $t$ thus appear as in Fig. 7.18.

In the above analysis we examined the situation only near the equilibrium points. The question naturally arises as to the behaviour of the solutions elsewhere in the phase plane. Figures 7.15–7.18 suggest that the curves may appear as in Fig. 7.19. However, this would suggest that all the curves are in fact closed and that all the solutions of (135) are periodic. By further analysis

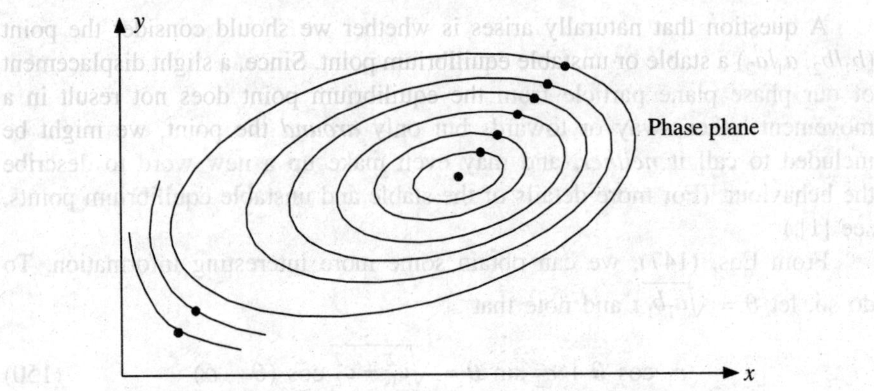


**Fig. 7.19** Concentric ellipses in the phase plane.

we can show that this is actually the case. The procedure involves elimination of $t$ between Eqs. (135). Consequently, we get

$$\frac{dy}{dx} = \frac{(-b_1 + b_2 x)y}{(a_1 - a_2 y)x} \tag{154}$$

which is a differential equation of the family of curves or trajectories in the phase plane. Separating the variables in Eq. (154), we obtain

$$\frac{b_1 - b_2 x}{x} dx + \frac{a_1 - a_2 y}{y} dy = 0$$

which on integration yields

$$b_1 \log x - b_2 x + a_1 \log y - a_2 y = \text{constant}$$

or

$$x^{b_1} y^{a_1} \exp[-(b_2 x + a_2 y)] = A \tag{155}$$

where $A$ is any arbitrary constant. These can be shown to be the closed curves, and as a consequence that solutions of Eqs. (135) are periodic. However, the period of all such curves is not as given in Eq. (149). In fact, the period is a function of $a_1$, $b_1$, $a_2$, $b_2$ which reduces to (149) near the equilibrium points. The precise period for all cases has not been found.

## 7.14 SOME FURTHER APPLICATIONS

The ideas just presented are applicable in other connections. For example, instead of having a predator and a prey, we could have a situation where a species of insects depend for their survival on certain animals or plants by actually feeding on them. The former is known as parasites and the latter as hosts. Whereas, the prey is destroyed by the predator, the parasite finds it advantageous to keep the host alive so that it may continue to feed on it. As another example, we can have more complicated situations arising when two or more species compete and may even destroy each other for the same prey.

The applications are not limited to ecology or the life sciences but can occur in other fields as well. For instance, in economics nations compete with each other for trade which in a sense is economic survival, producers compete with each other for consumers, and even consumers may compete with each other for products. In this connection, it is interesting to note similarity in appearance of the graph depicted in Fig. 6.38 involving price and supply of a commodity with the graph shown in Fig. 7.18, involving predator and prey.

The concept of phase plane and stability described in the predator-prey problem can be extended to other nonlinear systems

$$\frac{dx}{dt} = F(x, y), \qquad \frac{dy}{dt} = G(x, y)$$

## 7.15 MISCELLANEOUS SOLVED EXAMPLES

In this section, we shall work out some more problems.

**EXAMPLE 7.25** Solve the system

(i) $D^2x - 3x - 4y = 0, D^2y + x + y = 0$

(ii) $\dfrac{dy}{dx} + y - z = e^x, \dfrac{dz}{dx} + z - y = e^x$

***Solution*** (i) The given system can be expressed as follows:

$$(D^2 - 3)x - 4y = 0 \tag{156}$$

$$x + (D^2 + 1)y = 0 \tag{157}$$

Multiply Eq. (157) by $(D^2 - 3)$ and add it to Eq. (156), we get

$$(D^2 - 1)^2 y = 0$$

whose solution is

$$y = (c_1 + c_2t)e^{-t} + (c_3 + c_4t)e^t \tag{158}$$

Put this value of $y$ in Eq. (157), we get after simplification

$$x = (-2c_1 + 2c_2 - 2c_2t)e^{-t} + (-2c_3 - 2c_4 - 2c_4t)e^t \tag{159}$$

The set of Eqs. (158) and (159) represent the required solution.

(ii) Put $D = \dfrac{d}{dx}$, then the given system can be expressed as follows:

$$(D + 1)y - z = e^x \tag{160}$$

$$-y + (D + 1)z = e^x \tag{161}$$

Multiply Eq. (161) by $(D + 1)$ and add the resulting equation to Eq. (160), we get

$$(D^2 + 2D)z = 3e^x$$

whose solution is

$$z = c_1 + c_2 e^{-2x} + e^x \tag{162}$$

Put this value of $z$ in Eq. (161), we get

$$y = c_1 - c_2 e^{-2x} + e^x \tag{163}$$

Equations (162) and (163) constitute the required solution.

**EXAMPLE 7.26** Solve the system

$$(2D - 4)x + (3D + 5)y = 3t + 2$$

$$(D - 2)x + (D + 1)y = t$$

***Solution*** Multiply the second equation of the system by 2 and subtract it from the first equation of the system, we get

$$\frac{dy}{dt} + 3y = t + 2$$

which is a linear differential equation, having the solution as

$$y = \frac{1}{3}t + \frac{5}{9} + c_1 e^{-3t} \tag{164}$$

Put this value of $y$ in the second equation of the given system, we get

$$\frac{dx}{dt} - 2x = \frac{2}{3}t + 2c_1 e^{-3t} - \frac{8}{9}$$

which is again a first order linear differential equation and the solution is

$$x = \frac{5}{18} - \frac{1}{3}t - \frac{2}{5}c_1 e^{-3t} + c_2 e^2 \tag{165}$$

The determinant of the given system, from Eq. (11), is equal to $(-D^2 - D + 11)$ which is of order 2. Thus, the solution given by Eqs. (164) and (165) is the general solution as it contains the correct number of constants (which is two in this case).

**EXAMPLE 7.27** Solve the system

(i) $\omega^2 x + Dy = \sin at,\ Dx - \omega^2 y = \cos at$

(ii) $D^2 x - y = 2,\ x - D^2 y = -2$

***Solution*** (i) Operate $D$ on the second equation of the system and multiply the first equation of the system by $\omega^2$. Now, adding the two resulting equations, we get

$$D^2 x + \omega^4 x = (\omega^2 - a)\sin at$$

whose solution is

$$x = c_1 \cos \omega^2 t + c_2 \sin \omega^2 t + \frac{1}{\omega^2 + a} \sin at \tag{166}$$

Put this value of $x$ in the second equation of the system, after simplification we get

$$y = -c_1 \sin \omega^2 t + c_2 \cos \omega^2 t - \frac{1}{\omega^2 + a} \cos at \tag{167}$$

Equations (166) and (167) constitute the required solution of the given system.
(ii) Differentiate twice the second equation of the system to get $D^2 x = D^4 y$. Put this value in the first equation of the system, we get

$$(D^4 - 1)y = 2$$

which has the solution as

$$y = c_1 e^{-t} + c_2 e^t + c_3 \cos t + c_4 \sin t + 2 \tag{168}$$

Using this value of $y$ in the second equation of the system, we get

$$x = c_1 e^{-t} + c_2 e^t - c_3 \cos t - c_4 \sin t - 2 \tag{169}$$

Therefore, the required solution is given by Eqs. (168) and (169).

**EXAMPLE 7.28** Find the path of the particle if its motion is governed by the system of differential equations $\frac{dx}{dt} + \omega y = 0$, and $\frac{dy}{dt} - \omega x = 0$.

***Solution*** The given system can be written as

$$Dx + \omega y = 0 \tag{170}$$

$$Dy - \omega x = 0 \tag{171}$$

Differentiate Eq. (170) with respect to $x$, multiply Eq. (171) by $\omega$ and add the resulting equations, we get

$$(D^2 + \omega^2)x = 0$$

whose solution is

$$x = c_1 \cos \omega t + c_2 \sin \omega t \tag{172}$$

Put it in Eq. (170), we get after simplification

$$y = c_1 \sin \omega t - c_2 \cos \omega t \tag{173}$$

Squaring and adding Eqs. (172) and (173), we get

$$x^2 + y^2 = a^2, \ \ (c_1^2 + c_2^2 = a^2)$$

which shows that the path of the particle is a circle.

**EXAMPLE 7.29** Solve the system

$$Dx + y = \sin t, \; Dy + x = \cos t$$

subject to the conditions $x(0) = 2$ and $y(0) = 0$.

***Solution*** Differentiate second equation of the system and put the value of $Dx$ in the first equation of the system, we get

$$D^2 y - y = -2\sin t$$

whose solution is

$$y = c_1 e^{-t} + c_2 e^t + \sin t \tag{174}$$

From this equation, find $Dy$ and put it in the second equation of the system, we get

$$x = c_1 e^{-t} - c_2 e^t \tag{175}$$

Now, using the given conditions in Eqs. (174) and (175), we get $c_1 = 1$ and $c_2 = -1$. Thus, the required solution, from Eqs. (174) and (175), is

$$x = e^{-t} + e^t, \; y = e^{-t} - e^t + \sin t$$

**EXAMPLE 7.30** Solve the system

$$Dx = 1 - \frac{2}{t}x, \; Dy = x + y + \frac{2}{t}x - 1$$

***Solution*** First equation of the system can be expressed as

$$Dx + \frac{2}{t}x = 1$$

which is a first order linear differential equation whose solution is

$$x = \frac{1}{3}t + c_1 t^{-2} \tag{176}$$

Consider the second equation of the system, which can be written as

$$tdy = t(x + y)dt - (t - 2x)dt$$

Using the first equation of the system and separating the variables, we get

$$\frac{dx + dy}{x + y} = dt$$

which on integrating leads to

$$\log(x + y) = t + \log c_2$$

or

$$\log\left(\frac{x + y}{c_2}\right) = t$$

or $x + y = c_2e^t$. Now, using the value of $x$ from Eq. (176), this equation leads to

$$y = c_2e^t - \frac{1}{3}t - c_1t^{-2} \tag{177}$$

Therefore, the required solution is given by Eqs. (176) and (177).

**EXAMPLE 7.31** Solve the system of differential equations

$$\frac{dx}{dt} = ny - mz$$

$$\frac{dy}{dt} = lz - nx$$

$$\frac{dz}{dt} = mx - ly$$

***Solution*** Multiply the given system of equations, respectively, by $x$, $y$ and $z$ and add, we get

$$x\frac{dx}{dt} + y\frac{dy}{dt} + z\frac{dz}{dt} = 0$$

which on integration leads to

$$x^2 + y^2 + z^2 = 2a = c_1 \tag{178}$$

Now, multiply the given system of equations, respectively, by $l$, $m$ and $n$ and add, we get

$$l\frac{dx}{dt} + m\frac{dy}{dt} + n\frac{dz}{dt} = 0$$

which on integration yields

$$lx + my + nz = c_2 \tag{179}$$

Finally, multiply the given system of equations, respectively, by $lx$, $my$ and $nz$ and add, we get

$$lx\frac{dx}{dt} + my\frac{dy}{dt} + nz\frac{dz}{dt} = 0$$

Integrating this equation, we get

$$lx^2 + my^2 + nz^2 = 2b = c_3 \tag{180}$$

Therefore, the required solution is given by Eqs. (178) – (180).

**EXAMPLE 7.32** Solve the system

$$(tD + 2)x - 2y = t, \; x + (tD + 5)y = t^2$$

***Solution*** The given system can be written as

$$t\frac{dx}{dt} + 2(x - y) = t \tag{181}$$

$$t\frac{dy}{dt} + (x+5y) = t^2 \tag{182}$$

Differentiate Eq. (181) with respect to $t$ and using Eq. (182), we get

$$t\frac{d^2x}{dt^2} + 3\frac{dx}{dt} - 2[t - \frac{1}{t}(x+5y)] = 1$$

or

$$t^2\frac{d^2x}{dt^2} + 3t\frac{dx}{dt} - 2t^2 + 2x + 10y = t$$

Substituting the value of $2y$ from Eq. (181), this equation reduces to

$$t^2\frac{d^2x}{dt^2} + 8t\frac{dx}{dt} + 12x = 2t^2 + 6t \tag{183}$$

This equation can be solved by taking $t = e^z$ in Eq. (183) (cf., Section 5.7) and we have

$$(D_1^2 + 7D_1 + 12)x = 2e^{2z} + 6e^z \tag{184}$$

where $D_1 = \frac{d}{dz}$. Here

$$x_c = c_1e^{-3z} + c_2e^{-4z} = c_1t^{-3} + c_2t^{-4}$$

and

$$x_p = \frac{1}{D_1^2 + 7D_1 + 12}(2e^{2z} + 6e^z)$$

$$= \frac{1}{15}e^{2z} + \frac{3}{10}e^z$$

$$= \frac{1}{15}t^2 + \frac{3}{10}t$$

Thus, the solution of Eq. (183) is

$$x = x_c + x_p = c_1t^{-3} + c_2t^{-4} + \frac{1}{15}t^2 + \frac{3}{10}t \tag{185}$$

This equation also leads to

$$\frac{dx}{dt} = -3c_1t^{-4} - 4c_2t^{-5} + \frac{2}{15}t + \frac{3}{10} \tag{186}$$

From Eqs. (185) and (186), Eq. (181) leads to

$$y = -\frac{1}{2}c_1t^{-3} - c_2t^{-4} + \frac{2}{15}t^2 - \frac{1}{20}t \tag{187}$$

Therefore, Eqs. (185) and (187) constitute the required solution.

**EXAMPLE 7.33** Solve the system

$$t\frac{dx}{dt}+y=0,\ t\frac{dy}{dt}+x=0$$

subject to the conditions $x(1)=1$ and $y(-1)=0$.

**Solution** The given system is

$$tDx+y=0 \tag{188}$$

$$x+tDy=0 \tag{189}$$

Differentiating Eq. (188) with respect to $t$ and using Eq. (189), we get

$$t^2\frac{d^2x}{dt^2}+t\frac{dx}{dt}-x=0$$

Put $t=e^z$ in this equation, we get

$$\frac{d^2x}{dz^2}-x=0$$

which has solution as

$$x=c_1e^{-z}+c_2e^z=c_1t^{-1}+c_2t \tag{190}$$

Using this value of $x$ in Eq. (188), we get

$$y=c_1t^{-1}-c_2t \tag{191}$$

Now, using the given conditions in Eqs. (190) and (191), we get $c_1=c_2=1/2$.

Therefore, from Eqs. (190) and (191), the required solution is given by the set of equations

$$x=\frac{1}{2}\left(\frac{1}{t}+t\right),\ y=\frac{1}{2}\left(\frac{1}{t}-t\right)$$

## EXERCISES

Solve each of the following systems of equations:

1. $\frac{dx}{dt}=-x^2,\quad \frac{dy}{dt}=-y.$
2. $\frac{dx}{dt}=3e^{-t},\quad \frac{dy}{dt}=x+y.$
3. $\frac{dx}{dt}=2t,\quad \frac{dy}{dt}=3x+2t,\quad \frac{dz}{dt}=x+4y+t.$
4. $\frac{dx}{dt}=x+\sin t,\quad \frac{dy}{dt}=t-y.$
5. $\frac{dx}{dt}=3x+2e^{3t},\quad \frac{dx}{dt}+\frac{dy}{dt}-3y=\sin 2t.$
6. $3\frac{dx}{dt}+3x+2y=e^t,\quad 4x-3\frac{dy}{dt}+3y=3t.$

**7.** $\dfrac{d^2x}{dt^2} + 4x = 3\sin t, \quad \dfrac{dx}{dt} - \dfrac{d^2y}{dt^2} + y = 2\cos t.$

**8.** $\dfrac{d^2x}{dt^2} + x - \dfrac{d^2y}{dt^2} - y = -\cos 2t, \quad 2\dfrac{dx}{dt} - \dfrac{dy}{dt} - y = 0.$

**9.** $\dfrac{d^2x}{dt^2} - \dfrac{dy}{dt} = 1 - t, \quad \dfrac{dx}{dt} + \dfrac{2dy}{dt} = 4e^t + x.$

**10.** $\dfrac{dx}{dt} + \dfrac{dy}{dt} + y = t, \quad \dfrac{d^2x}{dt^2} + \dfrac{d^2y}{dt^2} + \dfrac{dy}{dt} + x + y = t^2.$

**11.** Solve the following system which is used in some problems of electron motion

$$m\frac{d^2x}{dt^2} + eH\frac{dy}{dt} = eE, \quad m\frac{d^2y}{dt^2} - eH\frac{dx}{dt} = 0$$

where $m$ and $e$ are the mass and the charge of an electron, respectively, and $E$ and $H$ are the intensities of the electric and magnetic field, respectively. Assume that initially the electron is at the origin and its velocity is zero.

Verify that each of the following systems 12–14 is degenerate. Find solutions if they exist:

**12.** $Dx + 2Dy = e^t, \quad Dx + 2Dy = t.$

**13.** $Dx - Dy = e^t, \quad 3Dx - 3Dy = 3e^t.$

**14.** $(D^2 - 1)x + (D^2 - 1)y = 0, \quad (D^2 + 4)x + (D^2 + 4)y = 0.$

**15.** Solve the following systems of equations:

(a)
$$\begin{aligned} (D-1)x &= 0, \\ -x + (D-3)y &= 0, \\ -x + y + (D-2)z &= 0. \end{aligned}$$

(b)
$$\begin{aligned} (D-1)x &= 0, \\ 3x + 2(D+1)y &= 0, \\ 2y + (2D-3)z &= 0. \end{aligned}$$

**16.** A projectile is fired from a cannon which makes an angle of 60° with the horizontal. If the muzzle velocity is 160 ft/s:

(a) Write a system of differential equations and conditions for motion.
(b) Find the position of the projectile at any time.
(c) Find the range, maximum height and the time of flight.
(d) Determine the position and velocity of the projectile after it is in flight for 2 s and 4 s.

**17.** Determine what would have been the maximum range of the 'Big Bertha' cannon of World War I, which had a muzzle velocity of 1 mile/s. (Air-resistance being neglected.) For this maximum range, what is the height reached and the total time of flight?

**18.** A stone is thrown horizontally from a cliff 256 ft high with a velocity of 50 ft/s.

(a) Find the time of flight.

(b) At what distance from the base of the cliff will the stone land?

**19.** A projectile fired from a cannon located on a horizontal plane has a range of 2000 ft and achieves a maximum height of 100 ft. Determine its muzzle velocity and the time of flight.

**20.** Show that the speed of a planet is greatest when the planet is closed to the Sun.

**21.** Find $f$ in Eq. (60) if the orbit of the particle moving in the central force system has the equation $r = e^{\theta}$.

**22.** A satellite is given an initial speed of 20,000 miles/h tangent to the surface of the earth. Show that its orbit is an ellipse. Find its maximum distance from the centre of the earth and compute its orbital time. Use $R = 3960$ miles, $g = 32.1/5280$ miles/s$^2$.

**23.** A satellite travels around the earth in a circular orbit at speed 17,000 miles/h. Find the altitude of the satellite.

**24.** A satellite situated 400 miles above the earth's surface is given an initial velocity parallel to the plane that is tangent to the earth at the point directly under the satellite. Find the orbital and escape speeds.

**25.** Show that the escape speed for a rocket fired vertically upward at the surface of the earth has a value $\sqrt{2gR} = 6.95$ miles/s.

**26.** Find the particular solution of Eqs. (78) and (79) for which $x = 1$, $y = \sqrt{3}$, $dx/dt = 0$, $dy/dt = 0$ for $t = 0$.

**27.** (a) Eliminate $y$ between Eqs. (76) and (77) to obtain a fourth order differential equation

$$\frac{d^4x}{dt^4} + \left(\frac{k_1 + k_2}{m_1} + \frac{k_2}{m_2}\right)\frac{d^2x}{dt^2} + \frac{k_1 k_2}{m_1 m_2}x = 0.$$

(b) Show that the squares of the roots of the characteristic equation of the differential equation of part (a) are negative, and hence the four roots have the form $\pm i\omega_1$, $\pm i\omega_2$, where $\omega_1 > 0$ and $\omega_2 > 0$.

(c) Show that the general solution of Eqs. (76) and (77) is given by

$$x = c_1 \cos \omega_1 t + c_2 \sin \omega_1 t + c_3 \cos \omega_2 t + c_4 \sin \omega_2 t$$

$$y = A\,(c_1 \cos \omega_1 t + c_2 \sin \omega_1 t) + B(c_3 \cos \omega_2 t + c_4 \sin \omega_2 t)$$

where

$$A = \frac{k_1 + k_2}{k_2} - \frac{m_1\omega_1^2}{k_2}, \quad B = \frac{k_1 + k_2}{k_2} - \frac{m_1\omega_1^2}{k_2}.$$

In the following problems (see Figs. 7.20 and 7.21) arrange (do not solve) a set of independent equations which are sufficient to find the unknown current in terms of arbitrary constants.

**28.**

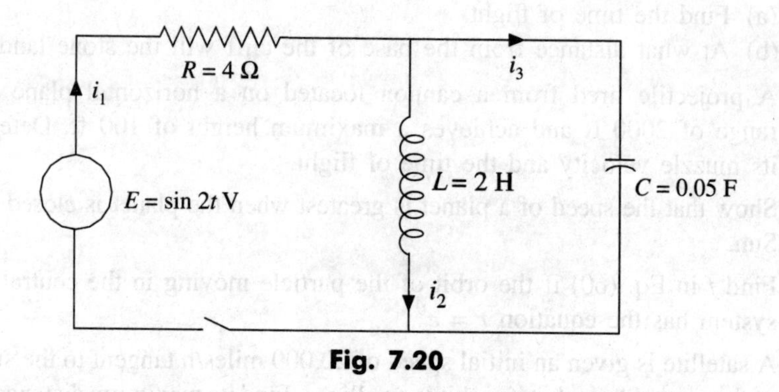


**Fig. 7.20**

**29.**

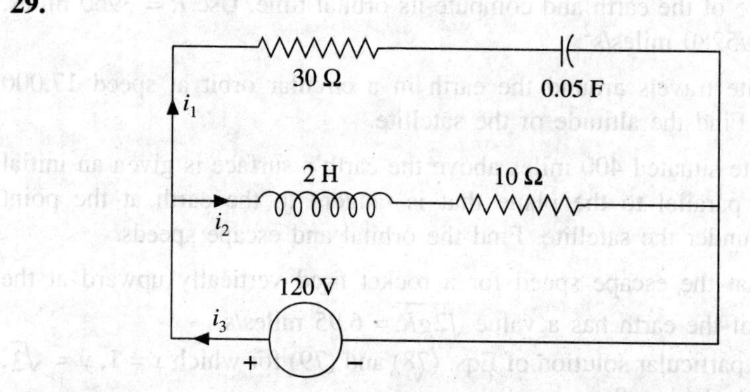


**Fig. 7.21**

**30.** In the electric network of Fig. 7.22, $E$ = 60 V. Determine the currents $I_1$ and $I_2$ as a function of time $t$, assuming that at $t$ = 0, when the key is closed, $I_1 = I_2 = 0$. Find the steady state currents.

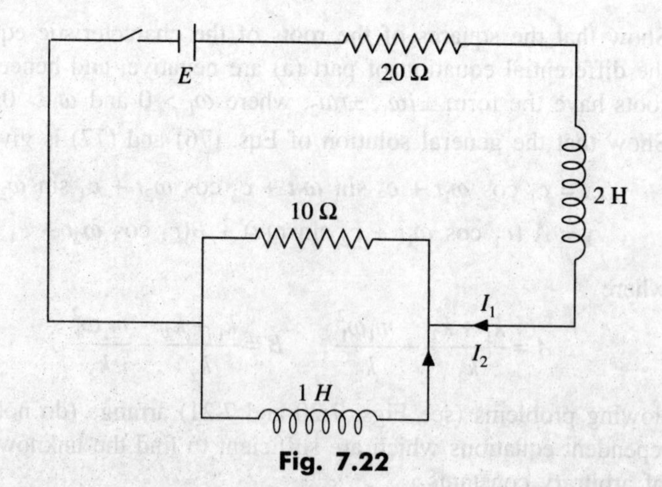


**Fig. 7.22**

**31.** Solve the above problem when $E = 150 \sin 10t$. Find the steady state currents.

**32.** In Example 7.20, find the maximum value assumed by $y(t)$ and the number of minutes before this maximum is assumed.

**33.** In Fig. 7.23, $x$, $y$ and $z$ denote the concentrations in g/cc of a substance $S$ in compartments $X$, $Y$ and $Z$. Given that $A$, $B$, $C$, $D$, $E$ and $F$ denote the rates of transport of the solvent containing $S$ in cc/min., write a system of differential equations governing the amounts $x$, $y$ and $z$ at time $t$.

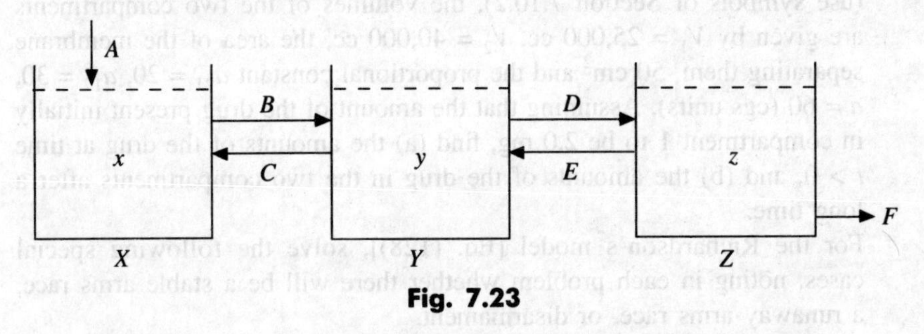


**Fig. 7.23**

**34.** In Example 7.21, when does tank $Y$ contain a maximum amount of salt? How much salt is in tank $Y$ at that time?

**35.** In an experiment of cholesterol turnover in humans, radioactive cholesterol $-4 - C^{14}$ was injected intravenously and the total plasma cholesterol and radioactivity were measured. It was discovered that the turnover of cholesterol behaves like a two-compartment system.* The compartment consisting of organs and blood has a rapid turnover, while the turnover in the other compartment is much slower. Assume that the body intakes and excretes all cholesterol through the first compartment. Let $x(t)$ and $y(t)$ denote the deviations from normal cholesterol levels in each compartment. Suppose that the daily fractional transfer coefficient from compartment $x$ is 0.134, of which 0.036 is the input to the compartment $y$, and that the transfer coefficient from $y$ is 0.02.

(a) Set up a system of differential equations of the problem.

(b) Obtain a general solution.

**36.** Suppose that in a two-compartment model we have $b = 0$, and initially the amount of the drug present in compartment 1 is $\alpha$, but the drug is not present in compartment 2.

(a) Show that the amounts of the drug present in compartments 1 and 2 at time $t$ are given by

$$x_1 = \frac{\alpha b_{21}}{b_{12} + b_{21}} + \frac{\alpha b_{12}}{b_{12} + b_{21}} e^{-(b_{12} + b_{21})t}$$

---

* D.S. Goodman and R.P. Noble: Turnover of plasma choiesterol in man. *J. Clinic. Invest.* **47** (1968) 231.

$$x_2 = \frac{\alpha b_{12}}{b_{21} + b_{12}} + \frac{\alpha b_{21}}{b_{21} + b_{12}} e^{-(b_{21}+b_{12})t}$$

(b) Show that at all time $x_1 + x_2 = \alpha$.

(c) Show that the amounts of drug present in the two compartments after a long time are given respectively by $\alpha b_{21}/(b_{12} + b_{21})$ and $\alpha b_{12}/(b_{12} + b_{21})$.

**37.** Suppose that in the two-compartment model of Problem 36 we have (use symbols of Section 7.10.2), the volumes of the two compartments are given by $V_1 = 25{,}000$ cc, $V_2 = 40{,}000$ cc, the area of the membrane separating them, 50 $cm^2$ and the proportional constant $a_{21} = 20$, $a_{12} = 30$, $a = 60$ (cgs units). Assuming that the amount of the drug present initially in compartment 1 to be 2.0 mg, find (a) the amounts of the drug at time $t > 0$, and (b) the amounts of the drug in the two compartments after a long time.

For the Richardson's model [Eq. (128)], solve the following special cases, noting in each problem whether there will be a stable arms race, a runaway arms race, or disarmament.

**38.** $a = 2$, $b = 4$, $p = 5$, $q = 3$, $r = 1$, $s = 2$, $x_0 = 8$, $y_0 = 7$.

**39.** $a = 4$, $b = 4$, $p = 2$, $q = 2$, $r = 8$, $s = 2$, $x_0 = 5$, $y_0 = 2$.

**40.** For $a = 4$, $b = 4$, $p = 2$, $q = 2$, $r = -2$, $s = -2$, show that there will be disarmament if $x_0 + y_0 < 2$ and a runaway arm race if $x_0 + y_0 > 2$.

**41.** The system of differential equations describing a particular predator-prey problem is given by

$$\frac{dx}{dt} = 10^{-2}x - 2\times 10^{-5}xy, \qquad \frac{dy}{dt} = -4\times 10^{-2}y + 10^{-5}xy$$

where $x$ and $y$ are the numbers of predators and preys, respectively, and $t$ is the time in days.

(a) Determine the number of predator and prey at equilibrium.

(b) Find the period.

(c) Write the equation of the phase curve in the neighbourhood of the equilibrium point.

**42.** In 1934, Gause gave a model for the interaction of two competing species of paramecium in the presence of limited food supply. He assumed the equation governing the population of each species to be a combination of the logistic equation (due to limited food supply) and a death rate proportional to the population of the competing species:

$$\frac{dx}{dt} = a_1x - b_1x^2 - c_1xy, \qquad \frac{dy}{dt} = a_2y - b_2y^2 - c_2xy \tag{156}$$

where the constants, $a_1$, $a_2$, $b_1$, $b_2$, $c_1$ and $c_2$ are all non-negative.

(a) Eliminate one of the variables and obtain a nonlinear second order differential equation.

(b) Find all the nonzero equilibrium points for Eq. (156).

(c) Find a solution of Eq. (156) if $a_1 = a_2 = l_1 = b_2 = c_1 = c_2 = 1$.

(d) Calculate $dy/dx$ and find a solution when $a_1 = a_2 = 0$. To what situation would this case correspond?

CHAPTER 8

# Laplace Transforms and Their Applications to Differential Equations

## 8.1 INTRODUCTION

In an attempt to solve ordinary linear differential equations in an easier manner, Oliver Heaviside devised a method of operational calculus which led to *Laplace transforms.* The methods of Laplace transforms are very simple, and they give solutions of differential equations satisfying given boundary conditions directly without the use of a general solution. Since, these particular solutions are the ones widely used in physics, mechanics, chemistry, medicine, national defence and many fields of practical research, the knowledge of Laplace transforms in recent years has become an essential part of the mathematical background required for engineers and scientists.

Although Laplace transforms have extensive applications, here we discuss only about their use in solving differential equations. (For a thorough knowledge of transforms, see [5]).

**Definition 8.1** Let $f(x)$ be a function defined for every positive $x$. Then the Laplace transform of $f(x)$, denoted by $L[f(x)]$, is defined by

$$L[]f(x) = \int_0^\infty e^{-sx} f(x)\, dx \tag{1}$$

provided that the integral exists, $s$ is a parameter which may be real or complex. $L[f(x)]$, being a function of $s$, is written as

$$L[f(x)] = \overline{f}(s)$$

which can also be written as

$$f(x) = L^{-1}[\overline{f}(s)]$$

Then $f(x)$ is called the inverse Laplace transform of $\overline{f}(s)$. The symbol $L$, which transforms $f(x)$ into $\overline{f}(s)$, is called the *Laplace transform operator.*

We give the following example so that the reader can have a familiarity with the concept.

**EXAMPLE 8.1** Find the Laplace transform of the following functions:
(i) $f(x) = 1$, (ii) $f(x) = x$, (iii) $f(x) = e^{ax}$,
(iv) $f(x) = \sin ax$, (v) $f(x) = \sinh ax$.

***Solution*** Using Eq. (1), we have

$$\text{(i) } L[f(x)] = L(1) = \int_0^\infty e^{-sx}(1)dx = \left[\frac{-e^{-sx}}{s}\right]_0^\infty = \frac{1}{s}, \qquad s > 0$$

$$\text{(ii) } L[f(x)] = L(x) = \int_0^\infty e^{-sx}x\,dx = \left[\frac{-xe^{-sx}}{s} - \frac{e^{-sx}}{s^2}\right]_0^\infty = \frac{1}{s^2}$$

$$\text{(iii) } L[f(x)] = L(e^{ax}) = \int_0^\infty e^{-sx}e^{ax}dx = \int_0^\infty e^{-(s-a)x}\,dx$$

$$= \left[\frac{-e^{-(s-a)x}}{s-a}\right]_0^\infty = \frac{1}{s-a}, \qquad s > a$$

$$\text{(iv) } L[f(x)] = L(\sin ax) = \int_0^\infty e^{-sx}\sin ax\,dx$$

$$= \left[\frac{e^{-sx}}{s^2+a^2}(-s\sin ax - a\cos ax)\right]_0^\infty = \frac{a}{s^2+a^2}$$

$$\text{(v) } L[f(x)] = L(\sinh ax) = \int_0^\infty e^{-sx}\sinh ax\,dx$$

$$= \int_0^\infty e^{-sx}\frac{e^{ax}-e^{-ax}}{2}dx$$

$$= \frac{1}{2}\left[\int_0^\infty e^{-(s-a)x}\,dx - \int_0^\infty e^{-(s+a)x}\,dx\right]$$

$$= \frac{1}{2}[L(e^{ax}) - L(e^{-ax})]$$

$$= \frac{1}{2}\left(\frac{1}{s-a} - \frac{1}{s+a}\right) = \frac{a}{s^2-a^2} \qquad s > |a|$$

## 8.2 PROPERTIES OF LAPLACE TRANSFORM

The Laplace transform satisfies the following properties:

**Property 8.1** If $a$, $b$, $c$ are constants and $f$, $g$, $h$ are functions of $x$, then

$$L[af(x) + bg(x) - ch(x)] = aL\,[f(x)] + bL[g(x)] - cL[h(x)]$$

i.e. $L$ is a linear operator.

**Property 8.2** *First shifting property* If $L[f(x)] = \bar{f}(s)$, then

$$L[e^{ax} f(x)] = \bar{f}(s-a)$$

This shifting property leads us to the following important results:

$$L(e^{ax} x^n) = \frac{n!}{(s-a)^{n+1}}$$

$$L(e^{ax} \sin bx) = \frac{b}{(s-a)^2 + b^2}$$

$$L(e^{ax} \cos bx) = \frac{s-a}{(s-a)^2 + b^2}$$

$$L(e^{ax} \sinh bx) = \frac{b}{(s-a)^2 - b^2}$$

$$L(e^{ax} \cosh bx) = \frac{s-a}{(s-a)^2 - b^2}, \text{ etc.}$$

**EXAMPLE 8.2** Find the Laplace transform of $\sin 2x \sin 3x$.

***Solution*** Since, $\sin 2x \sin 3x = (1/2)(\cos x - \cos 5x)$, we have

$$L(\sin 2x \sin 3x) = \frac{1}{2}[(L(\cos x) - L(\cos 5x)]$$

$$= \frac{1}{2}\left(\frac{s}{s^2+1} - \frac{s}{s^2+5^2}\right)$$

$$= \frac{12s}{(s^2+1)(s^2+25)}$$

**EXAMPLE 8.3** Show that

$$L(x \sin ax) = \frac{2as}{(s^2+a^2)^2}, \qquad L(x \cos ax) = \frac{s^2 - a^2}{(s^2+a^2)^2}.$$

***Solution*** Since $L(x) = 1/s^2$, we have

$$L(xe^{iax}) = \frac{1}{(s-ia)^2} = \frac{(s+ia)^2}{[(s-ia)(s+ia)]^2}$$

or
$$L[x(\cos ax + i\sin ax)] = \frac{(s^2 - a^2) + i\,2as}{(s^2 + a^2)^2}$$

Equating the real and imaginary parts, we get the required result.

The usual way of finding transforms and inverse transforms is by means of a table giving standard functions and their transforms. The transforms which are sufficient for our purpose are listed in Table 8.1.

**Table 8.1** Laplace Transforms

| $f(x)$ | $\bar{f}(s)$ | |
|---|---|---|
| $1$ | $\dfrac{1}{s}$ | $s > 0$ |
| $x^n$ | $\dfrac{n!}{s^{n+1}}$ | $n = 0, 1, 2, \ldots$ |
| $e^{ax}$ | $\dfrac{1}{s-a}$ | $s > a$ |
| $\sin ax$ | $\dfrac{a}{s^2 + a^2}$ | $s > 0$ |
| $\cos ax$ | $\dfrac{s}{s^2 + a^2}$ | $s > 0$ |
| $\sinh ax$ | $\dfrac{a}{s^2 - a^2}$ | $s > \lvert a \rvert$ |
| $\cosh ax$ | $\dfrac{s}{s^2 - a^2}$ | $s > \lvert a \rvert$ |
| $x^n e^{ax}$ | $\dfrac{n!}{(s-a)^{n+1}}$ | |
| $e^{ax} \sin bx$ | $\dfrac{b}{(s-a)^2 + b^2}$ | |
| $e^{ax} \cos bx$ | $\dfrac{s-a}{(s-a)^2 + b^2}$ | |

(*Contd.*)

| $f(x)$ | $\bar{f}(s)$ | |
|---|---|---|
| $e^{ax}\sinh bx$ | $\dfrac{b}{(s-a)^2-b^2}$ | |
| $e^{ax}\cosh bx$ | $\dfrac{s-a}{(s-a)^2-b^2}$ | |
| $\dfrac{x^{n-1}}{(n-1)!}$ | $\dfrac{1}{s^n}$ | $n = 1, 2, 3, \ldots$ |
| $\dfrac{e^{ax}x^{n-1}}{(n-1)!}$ | $\dfrac{1}{(s-a)^n}$ | $s > a$ |
| $\dfrac{e^{ax}-e^{bx}}{a-b}$ | $\dfrac{1}{(s-a)(s-b)}$ | $a \neq b$ |
| $\dfrac{ae^{ax}-be^{bx}}{a-b}$ | $\dfrac{s}{(s-a)(s-b)}$ | $a \neq b$ |
| $\dfrac{1}{2a^3}(\sin ax - ax\cos ax)$ | $\dfrac{1}{(s^2+a^2)^2}$ | |
| $\dfrac{\cos ax - \cos bx}{b^2-a^2}$ | $\dfrac{s}{(s^2+a^2)(s^2+b^2)}$ | $a^2 \neq b^2$ |
| $\dfrac{a\sin bx - b\sin ax}{ab(a^2-b^2)}$ | $\dfrac{1}{(s^2+a^2)(s^2+b^2)}$ | $a^2 \neq b^2$ |
| $\dfrac{x}{2a}\sin ax$ | $\dfrac{s}{(s^2+a^2)^2}$ | |
| $x\cos ax$ | $\dfrac{s^2-a^2}{(s^2+a^2)^2}$ | |
| $x^{k-1}$ | $\dfrac{\Gamma(k)}{s^k}$ | $k > 0$ |
| $x^{-1/2}$ | $\dfrac{\Gamma(1/2)}{s^{1/2}} = \dfrac{\sqrt{\pi}}{s^{1/2}}$ | |
| $\dfrac{1}{\sqrt{\pi x}}$ | $\dfrac{1}{s^{1/2}}$ | |
| $\dfrac{2}{\sqrt{x/\pi}}$ | $\dfrac{1}{s^{3/2}}$ | |

From the Table 8.1, it may be noted that the Laplace transform of each standard function involves a rational algebraic fraction. Hence, to find the inverse transform of a given function, we first express the given function of $s$ into a partial fraction which will then be easily recognizable as one of the standard forms of Table 8.1.

**EXAMPLE 8.4** Find the inverse transforms of

(i) $\dfrac{s^2 - 3s + 4}{s^3}$. (ii) $\dfrac{s+2}{s^2 - 4s + 13}$.

(ii) $\dfrac{2s^2 - 6s + 5}{s^3 - 6s^2 + 11s - 6}$. (iv) $\dfrac{5s + 3}{(s-1)(s^2 + 2s + 5)}$.

***Solution***

(i)
$$L^{-1}\left(\frac{s^2 - 3s + 4}{s^3}\right) = L^{-1}\left(\frac{1}{s}\right) - 3L^{-1}\left(\frac{1}{s^2}\right) + 4L^{-1}\left(\frac{1}{s^3}\right)$$

$$= 1 - 3x + \frac{4x^2}{2!}$$

$$= 1 - 3x + 2x^2$$

(ii)
$$L^{-1}\left(\frac{s+2}{s^2 - 4s + 13}\right) = L^{-1}\left[\frac{s+2}{(s-2)^2 + 9}\right]$$

$$= L^{-1}\left[\frac{s - 2 + 4}{(s-2)^2 + 3^2}\right]$$

$$= L^{-1}\left[\frac{s-2}{(s-2)^2 + 3^2}\right] + 4L^{-1}\left[\frac{1}{(s-2)^2 + 3^2}\right]$$

$$= e^{2x}\cos 3x + \frac{4}{3}e^{2x}\sin 3x$$

(iii) Here

$$\frac{2s^2 - 6s + 5}{s^3 - 6s^2 + 11s - 6} = \frac{2s^2 - 6s + 5}{(s-1)(s-2)(s-3)} = \frac{A}{s-1} + \frac{B}{s-2} + \frac{C}{s-3}$$

which gives $A = 1/2$, $B = -1$, $C = 5/2$. Thus

$$L^{-1}\left(\frac{2s^2 - 6S + 5}{s^3 - 6s^2 + 11s - 6}\right) = \frac{1}{2}L^{-1}\left(\frac{1}{s-1}\right) - L^{-1}\left(\frac{1}{s-2}\right) + \frac{5}{2}L^{-1}\left(\frac{1}{s-3}\right)$$

$$= \frac{1}{2}e^{x} - e^{2x} + \frac{5}{2}e^{3x}$$

(iv) Here

$$\frac{5s+3}{(s-1)(s^2+2s+5)} = \frac{5(1)+3}{(s-1)(1^2+2\times 1+5)} + \frac{As+B}{s^2+2s+5}$$

which gives $A = -1$ and $B = 2$. Thus

$$L^{-1}\left[\frac{5s+3}{(s-1)(s^2+2s+5)}\right] = L^{-1}\left(\frac{1}{s-1}\right) + L^{-1}\left(\frac{-s+2}{s^2+2s+5}\right)$$

$$= L^{-1}\left(\frac{1}{s-1}\right) + L^{-1}\left[\frac{-(s+1)+3}{(s+1)^2+4}\right]$$

$$= L^{-1}\left(\frac{1}{s-1}\right) - L^{-1}\left[\frac{s+1}{(s+1)^2+2^2}\right] + 3L^{-1}\left[\frac{1}{(s+1)^2+2^2}\right]$$

$$= e^x - e^{-x}\cos 2x + \frac{3}{2}e^{-x}\sin 2x$$

We shall now give some more results about the transforms in the form of theorems. These theorems will enable us to derive formulas involving transforms to solve difficult type of equations and to provide short cuts and simplifications in using transforms. The proofs of these theorems can be found in [12].

### 8.2.1 Transforms of Derivatives

**Theorem 8.1** If $f'(x)$ is a continuous function and $L[f(x)] = \bar{f}(s)$, then

$$L[f'(x)] = s\bar{f}(s) - f(0)$$

**Theorem 8.2** If $f(x)$ and its first $(n-1)$ derivatives are continuous, then

$$L[f^n(x)] = s^n\bar{f}(s) - s^{n-1}f(0) - s^{n-2}f'(0) - \cdots - f^{n-1}(0)$$

**Theorem 8.3** If $L[f(x)] = \bar{f}(s)$, then

$$L[x^n f(x)] = (-1)^n \frac{d^n}{ds^n}[\bar{f}(s)], \qquad n = 1, 2, 3, \ldots$$

### 8.2.2 Transforms of Integrals

**Theorem 8.4** If $L[f(x)] = \bar{f}(s)$, then

$$L\left[\int_0^x f(u)\,du\right] = \frac{1}{s}\bar{f}(s)$$

or
$$L^{-1}\left(\frac{1}{s}\bar{f}(s)\right)=\int_0^x f(u)\,du$$

**Theorem 8.5** If $L[f(x)] = \bar{f}(s)$, then

$$L\left[\frac{1}{x}f(x)\right]=\int_s^\infty \bar{f}(s)\,ds$$

**EXAMPLE 8.5** Find the Laplace transforms of

(i) $x \cos ax$, (ii) $x^3 e^{-3x}$

***Solution*** (i) Since, $L(\cos ax) = \dfrac{s}{s^2+a^2}$, we have

$$L(x\cos ax)=-\frac{d}{ds}\left(\frac{s}{s^2+a^2}\right)=-\frac{s^2+a^2-s\,2s}{(s^2+a^2)^2}=\frac{s^2-a^2}{(s^2+a^2)^2}$$

(ii) Again since, $L(e^{-3x}) = \dfrac{1}{s+3}$, we have

$$L(x^3e^{-3x}) = (-1)^3\frac{d^3}{ds^3}\left(\frac{1}{s+3}\right)=\frac{6}{(s+3)^4}$$

**EXAMPLE 8.6** Find the inverse transform of

$$\frac{1}{s(s^2+a^2)}.$$

***Solution*** Since

$$L^{-1}\left(\frac{1}{s^2+a^2}\right)=\frac{1}{a}\sin ax$$

Therefore, by Theorem 8.4, we have

$$L^{-1}\left[\frac{1}{s(s^2+a^2)}\right]=\int_0^x \frac{1}{a}\sin ax\,dx=\frac{1}{a^2}[-\cos ax]_0^x=\frac{1-\cos ax}{a^2}$$

**EXAMPLE 8.7** Find the Laplace transform of $(1 - e^x)/x$.

***Solution*** Since

$$L(1-e^x) = L(1) - L(e^x) = \frac{1}{s}-\frac{1}{s-1}$$

We have

$$L\left(\frac{1-e^x}{x}\right) = \int_s^\infty \left(\frac{1}{s} - \frac{1}{s-1}\right) ds$$

$$= [\log s - \log (s-1)]_s^\infty$$

$$= \log \left(\frac{s}{s-1}\right)_s^\infty$$

$$= -\log \left[\frac{1}{1-(1/s)}\right]$$

$$= \log \left(\frac{s-1}{s}\right)$$

## 8.3 UNIT STEP FUNCTIONS

Sometimes it may happen that we come across functions whose inverse Laplace transform cannot be determined by the formulas of Table 8.1. To deal with such functions, we now define the unit step function (or, *Heaviside's unit function*) as follows:

**Definition 8.2** The unit step function $u(x - a)$ is defined as

$$\left.\begin{aligned} u(x-a) &= 0 \quad \text{for} \quad x < a \\ &= 1 \quad \text{for} \quad x \geq a \end{aligned}\right\}$$

where $a > 0$ (Fig. 8.1).

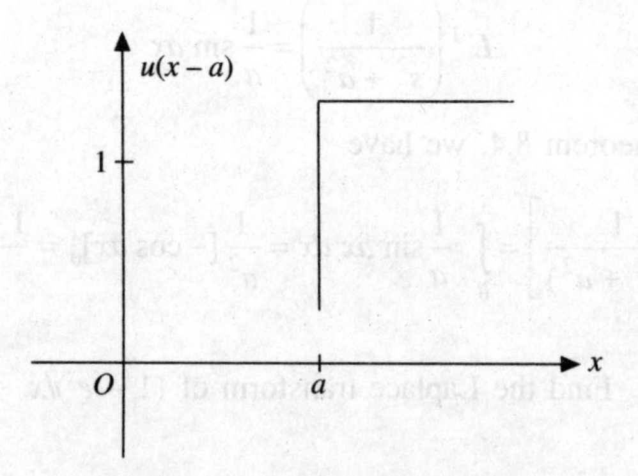


**Fig. 8.1** The unit step function.

The transform of unit function is obtained as follows:

$$L[u(x-a)] = \int_0^\infty e^{-sx} u(x-a)\, dx$$

$$= \int_0^a e^{-sx}\,(0)\, dx + \int_a^\infty e^{-sx}\,(1)\, dx$$

$$= \left(\frac{e^{-sx}}{-s}\right)_a^\infty$$

$$= \frac{e^{-as}}{s}$$

Therefore,

$$f(x)u(x-a) = \begin{cases} 0 & \text{for } x < a \\ f(x) & \text{for } x \geq a \end{cases}$$

The function $f(x - a)\, u(x - a)$ represents the graph of $f(x)$ shifted through a distance $a$ to the right.

**Property 8.3** *Second shifting property* If $L[f(x)] = \overline{f}(s)$, then

$$L[f(x - a)\, u(x - a)] = e^{-as}\, \overline{f}(s)$$

## 8.4 UNIT IMPULSE FUNCTIONS

In mechanics, there are several instances where a very large force acts for a very short time. To deal with such and other similar cases, we define the following:

**Definition 8.3** The unit impulse function $\delta(x - a)$ is defined by the relations

$$\delta(x-a) = \begin{cases} \infty & \text{for } x = a \\ 0 & \text{for } x \neq a \end{cases}$$

such that $\int_0^\infty \delta(x - a)\, dx = 1 \quad (a > 0)$.

The transform of a unit impulse function can be obtained as follows: Let $f(x)$ be a function of $x$ continuous at $x = a$; then

$$\int_0^\infty f(x)\, \delta_\varepsilon(x-a)\, dx = \int_0^{a+\varepsilon} f(x) \frac{1}{\varepsilon} dx = (a + \varepsilon - a)\, f(\eta) \frac{1}{\varepsilon} = f(\eta)$$

where $a < \eta < a + \varepsilon$. As $\varepsilon \to 0$

$$\int_0^\infty f(x)\delta(x-a) = f(a)$$

In particular, when $f(x) = e^{-sx}$

$$L[\delta(x-a)] = e^{-as}$$

**Theorem 8.6** *Convolution theorem* If

$$L^{-1}[\bar{f}(s)] = f(x) \text{ and } L^{-1}[\bar{g}(s)] = g(x)$$

then

$$L^{-1}[\bar{f}(s)\,\bar{g}(s)] = \int_0^x f(u)g(x-u)\,du$$

**Theorem 8.7** *Periodic function theorem* If $f(x)$ is a periodic function with period $T$, i.e. $f(x+T) = f(x)$, then

$$L[f(x)] = \frac{\int_0^T e^{-sx} f(x)\,dx}{1-e^{-sT}}$$

**EXAMPLE 8.8** Find the Laplace transform of the periodic function

$$\begin{aligned} f(x) &= \sin \omega x, \quad && 0 < x < \pi/\omega \\ &= 0, && \pi/\omega < x < 2\pi/\omega \end{aligned}$$

***Solution*** Since $f(x)$ is a periodic function with period $2\pi/\omega$, we have

$$L[f(x)] = \frac{1}{1-e^{-2\pi s/\omega}} \int_0^{2\pi/\omega} e^{-sx} f(x)\,dx$$

$$= \frac{1}{1-e^{-2\pi s/\omega}} \left( \int_0^{\pi/\omega} e^{-sx} \sin \omega x\,dx + \int_{\pi/\omega}^{2\pi/\omega} e^{-sx}\, 0 dx \right)$$

$$= \frac{1}{1-e^{-2\pi s/\omega}} \left[ \frac{e^{-sx}(-s \sin \omega x - \omega \cos \omega x)}{s^2+\omega^2} \right]_0^{\pi/\omega}$$

$$= \frac{\omega(1+e^{-s\pi/\omega})}{(1-e^{-2\pi s/\omega})(s^2+\omega^2)}$$

**Theorem 8.8** *Change of scale property* If $L[F(x)] = f(s)$, then

$$L[F(ax)] = \frac{1}{a} f\left(\frac{s}{a}\right).$$

## 8.5 SOLUTION OF A LINEAR DIFFERENTIAL EQUATION WITH CONSTANT COEFFICIENTS USING TRANSFORM METHODS

The procedure for finding the solution of a linear differential equation using transforms is as under:

(a) Take the Laplace transform of both sides of a given differential equation, using Theorems 8.1 and 8.2 and the given initial conditions.
(b) Transpose the terms with minus sign to the right.
(c) Divide by the coefficients of $y$, getting $y$ as a known function of $s$.
(d) Resolve the function of $s$ into partial fractions and take the inverse transform of both sides. This gives $y$ as a function of $x$ which is the desired solution satisfying the given initial conditions.

**EXAMPLE 8.9** Solve $y'' + 2y' + y = 1$ such that $y(0) = 2$ and $y'(0) = -2$.

***Solution*** Using Theorems 8.1 and 8.2 and the given initial conditions, the differential equation takes the form

$$s^2L(y) - 2s + 2 + 2sL(y) - 4 + L(y) = \frac{1}{s}$$

or

$$L(y) = \frac{2s^2 + 2s + 1}{s(s+1)^2}$$

Resolving into partial fractions, we get

$$L(y) = \frac{1}{s} + \frac{1}{s+1} - \frac{1}{(s+1)^2}$$

Now, taking the inverse transform, we obtain

$$y = L^{-1}\left(\frac{1}{s}\right) + L^{-1}\left(\frac{1}{s+1}\right) - L^{-1}\left[\frac{1}{(s+1)^2}\right] = 1 + e^{-x} - xe^{-x}$$

which is the required solution.

**EXAMPLE 8.10** Solve $y'' + 3y' + 2y = 12e^{2x}$ such that $y(0) = 1$, $y'(0) = -1$.

***Solution*** Take the Laplace transform of both sides of the given differential equation, and use the given initial conditions to obtain

$$L(y) = \frac{s^2 + 8}{(s+2)(s+1)(s-2)} = \frac{3}{s+2} - \frac{3}{s+1} + \frac{1}{s-2}$$

Now, take the inverse transform to get

$$y = 3e^{-2x} - 3e^{-x} + e^{2x}$$

as the required solution.

**EXAMPLE 8.11** Solve $\dfrac{d^2x}{dt^2} - 2\dfrac{dx}{dt} + x = e^t$ for which $x(0) = 2$, $x'(0) = -1$.

***Solution*** Under the given initial conditions, take the Laplace transform of both sides of the differential equation, then we have

$$(s^2 - 2s + 1)L(x) - 2s + 5 = \frac{1}{s-1}$$

or $$L(x) = \frac{2s^2 - 7s + 6}{(s-1)^3} = \frac{2}{s-1} - \frac{3}{(s-1)^2} + \frac{1}{(s-1)^3}$$

By taking the inverse transform, we have

$$x = 2L^{-1}\left(\frac{1}{s-1}\right) - 3L^{-1}\left[\frac{1}{(s-1)^2}\right] + L^{-1}\left[\frac{1}{(s-1)^3}\right]$$

$$= 2e^t - 3te^t + \frac{1}{2}t^2e^t$$

which is the required solution.

**EXAMPLE 8.12** Solve the system of differential equations

$$\frac{dx}{dt} + 5x - 2y = t, \qquad \frac{dy}{dt} + 2x + y = 0$$

such that $x(0) = y(0) = 0$.

***Solution*** Taking the Laplace transform of both equations using the given conditions, we get

$$(s + 5)L(x) - 2L(y) = \frac{1}{s^2} \tag{2}$$

$$2L(x) + (s + 1)L(y) = 0 \tag{3}$$

Solving Eqs. (2) and (3) for $L(x)$, we have

$$L(x) = \frac{s+1}{s^2(s+3)^2} = \frac{1}{27s} + \frac{1}{9s^2} - \frac{1}{27(s+3)} - \frac{2}{9}\frac{1}{(s+3)^2} \tag{4}$$

Substituting this value in Eq. (3), we get

$$L(y) = \frac{-2}{s^2(s+3)^2} = \frac{4}{27s} - \frac{2}{9s^2} - \frac{4}{27(s+3)} - \frac{2}{9(s+3)^2} \tag{5}$$

Now, by taking the inverse transforms of Eqs. (4) and (5), we get the required solution as

$$x = \frac{1}{27} + \frac{t}{9} - \frac{1}{27}e^{-3t} - \frac{2}{9}te^{-3t}$$

$$y = \frac{4}{27} - \frac{2}{9}t - \frac{4}{27}e^{-3t} - \frac{2}{9}te^{-3t}$$

**EXAMPLE 8.13** Solve

$$\frac{d^2x}{dt^2} + 3x - 2y = 0$$

$$\frac{d^2x}{dt^2} + \frac{d^2y}{dt^2} - 3x + 5y = 0$$

for which $x(0) = y(0) = 0$, $x'(0) = 3$, $y'(0) = 2$.

***Solution*** Taking the Laplace transform of the given system of differential equations under the given initial conditions, we obtain

$$(s^2 + 3)L(x) - 2L(y) = 3$$
$$(s^2 - 3)L(x) + (s^2 + 5)L(y) = 5$$

Solving these equations for $L(x)$ and $L(y)$, we get

$$L(x) = \frac{11}{4}\frac{1}{s^2+1} + \frac{1}{4}\frac{1}{s^2+9}$$

$$L(y) = \frac{11}{4}\frac{1}{s^2+1} - \frac{3}{4}\frac{1}{s^2+9}$$

which, by taking the inverse Laplace transform, reduce to

$$x = \frac{11}{4}\sin t + \frac{1}{12}\sin 3t$$

$$y = \frac{11}{4}\sin t - \frac{1}{4}\sin 3t$$

This is the required solution.

## 8.6 APPLICATIONS OF LAPLACE TRANSFORMS

### 8.6.1 Vibrating Motion

***Example 8.14*** An 8-lb weight is hung on the end of a vertically suspended spring, thereby stretching the spring 6 inches. The weight is raised 3 in. above its equilibrium position and released from rest at time $t = 0$. Find the displacement $x$ of the weight from its equilibrium position at time $t$. Use $g = 32$ ft/s$^2$.

***Solution*** By Hooke's law, $8 = k(1/2)$ and $k = 8$ lb/ft. By Newton's second law

$$\frac{8}{32}\frac{d^2x}{dt^2} = -16x, \qquad x(0) = -\frac{1}{4}, \qquad x'(0) = 0$$

where $x$ is measured positively downward. From $x''(t) + 64x(t) = 0$, we obtain

$$s^2L(x) + \frac{s}{4} + 64L(x) = 0$$

which gives

$$(s^2 + 64)L(x) = -\frac{s}{4} \quad \text{or} \quad L(x) = -\frac{1}{4}\frac{s}{s^2+64}$$

Thus

$$x(t) = -\frac{1}{4}\cos 8t$$

**EXAMPLE 8.15** Solve $x'' + 16x = \cos 4t$ for which $x(0) = 0$ and $x'(0) = 1$.

***Solution*** This initial value problem can describe the forced, undamped and resonant motion of a mass on a spring. The mass starts with an initial velocity of one ft/s in downward direction from the equilibrium position. Transforming the given equation, we get

$$(s^2 + 16)L(x) = 1 + \frac{s}{s^2+16}$$

or

$$L(x) = \frac{1}{s^2+16} + \frac{s}{(s^2+16)^2}$$

or

$$x(t) = \frac{1}{4}L^{-1}\left(\frac{4}{s^2+16}\right) + \frac{1}{8}L^{-1}\left[\frac{8s}{(s^2+16)^2}\right]$$

$$= \frac{1}{4}\sin 4t + \frac{1}{8}t\sin 4t$$

***Example 8.16*** A spring suspended vertically from a fixed support has spring constant $k = 50$ lb/ft. A 64 lb weight is placed on the spring and an external force $F(t) = 4\sin 4t$ is applied to the weight. Find the displacement $x$ in terms of $t$ if $x(0) = 0$ and $x'(0) = 0$. (Measure $x$ positively downward and use $g = 32$ ft/s$^2$)

***Solution*** By Newton's second law, we have

$$\frac{64}{32}x''(t) + 50x(t) = 4\sin 4t, \quad x(0) = x'(0) = 0$$

or

$$x''(t) + 25x(t) = 2\sin 4t$$

Hence

$$s^2L(x) + 25L(x) = \frac{8}{s^2+16}$$

or

$$L(x) = \frac{8}{(s^2+16)(s^2+25)} = \frac{8}{9}\left(\frac{1}{s^2+16} - \frac{1}{s^2+25}\right)$$

Taking the inverse transform, we get

$$x(t) = \frac{8}{9}\left(\frac{1}{4}\sin 4t - \frac{1}{5}\sin 5t\right)$$

**EXAMPLE 8.17** The rectilinear motion of a block is governed by the differential equation

$$x''(t) + 8x'(t) + 36x(t) = 72 \cos 6t$$

Find $x(t)$ in terms of $t$ if $x(0) = 1/2$ and $x'(0) = 0$.

***Solution*** The given equation with the initial conditions transforms as

$$s^2L(x) - \frac{s}{2} + 8sL(x) - 4 + 36L(x) = \frac{72s}{s^2+36}$$

or

$$L(x) = \frac{s+8}{2(s^2+8s+36)} + \frac{72s}{(s^2+8s+36)(s^2+36)}$$

$$= \frac{1}{2}\frac{s+4+4}{(s+4)^2+20} + \frac{9}{s^2+36} - \frac{9}{(s+4)^2+20}$$

$$= \frac{1}{2}\frac{s+4}{(s+4)^2+20} + \frac{2}{2\sqrt{5}}\frac{2\sqrt{5}}{(s+4)^2+20} + \frac{3}{2}\frac{6}{s^2+36} - \frac{9}{2\sqrt{5}}\frac{2\sqrt{5}}{(s+4)^2+20}$$

Thus

$$x(t) = \frac{e^{-4t}}{2}\cos 2\sqrt{5}t + \frac{2e^{-4t}}{2\sqrt{5}}\sin 2\sqrt{5}t + \frac{3}{2}\sin 6t - \frac{9e^{-4t}}{2\sqrt{5}}\sin 2\sqrt{5}t$$

$$= e^{-4t}\left(\frac{-7}{2\sqrt{5}}\sin 2\sqrt{5}t + \frac{1}{2}\cos 2\sqrt{5}t\right) + \frac{3}{2}\sin 6t$$

**EXAMPLE 8.18** A spring, with spring constant 0.75-lb/ft, lies on a long smooth (frictionless) table. A 6-lb weight is attached to the spring and is at rest at the equilibrium position. A 1.5-lb force is applied to the support along the line of action of the spring for 4 s and is then removed. Discuss the motion.

***Solution*** We have to solve the initial value problem

$$\frac{6}{32}x''(t) + \frac{3}{4}x(t) = H(t), \qquad x(0) = 0,\ x'(0) = 0 \tag{6}$$

where

$$H(t) = 1.5, \qquad 0 < t < 4$$
$$= 0, \qquad t > 4$$

Now, $H(t) = 1.5[1 - \alpha(t - 4)]$, where $\alpha$ is defined as

$$\alpha(t) = 0, \quad t < 0$$
$$= 1, \quad t \geq 0$$

Therefore, we write Eq. (6) as

$$x''(t) + 4x(t) = 8[1 - \alpha(t - 4)], \quad x(0) = 0, \quad x'(0) = 0 \tag{7}$$

Taking the Laplace transform, we obtain

$$s^2L(x) + 4L(x) = \frac{8}{s}(1 - e^{-4s})$$

or

$$L(x) = \frac{8(1 - e^{-4s})}{s(s^2 + 4)} = 2\left(\frac{1}{s} - \frac{s}{s^2 + 4}\right)(1 - e^{-4s})$$

The desired solution is

$$x(t) = 2(1 - \cos 2t) - 2[1 - \cos 2(t - 4)]\,\alpha(t - 4) \tag{8}$$

Solution (8) can also be written as

$$x(t) = 2(1 - \cos 2t), \quad 0 \le t \le 4 \tag{9}$$

and

$$x(t) = 2[\cos 2(t - 4) - \cos 2t], \quad t > 4 \tag{10}$$

[We also have

$$\lim_{t \to 4^-} x(t) = \lim_{t \to 4^+} x(t) = 2(1 - \cos 8) = 2.29,$$

$$\lim_{t \to 4^-} x'(t) = \lim_{t \to 4^+} x'(t) = 4 \sin 8 = 3.96.]$$

From Eq. (8) or (9), we see that in the range $0 < t < 4$, the maximum deviation of the weight from the starting point is $x = 4$ ft and occurs at $t = \pi/2 = 1.57$ s. At $t = 4$, $x = 2.29$ ft. For $t > 4$, Eq. (10) takes over and thereafter the motion is simple harmonic with a maximum $x$ of 3.03 ft. Indeed for $t > 4$

$$\max|x(t)| = 2\sqrt{(1 - \cos 8)^2 + \sin^2 8} = 2\sqrt{2}\,\sqrt{1 - \cos 8} = 0.03$$

**EXAMPLE 8.19** Solve $x'' + 16x = f(t)$, where

$$f(t) = \cos 4t, \quad 0 \le t < \pi$$
$$= 0, \quad t \ge \pi$$

and $x(0) = 0$, $x'(0) = 1$.

***Solution*** The function $f(t)$ can be interpreted as an external force which is acting on a mechanical system only for a short period of time and is then removed. Since

$$f(t) = \cos 4t - \cos (4t)F(t - \pi)$$
$$= \cos 4t - \cos [4(t - \pi)]F(t - \pi)$$

by the periodicity of cosine, $F(t)$ is the unit step function. Therefore

$$L(x'') + 16L(x) = L[f(t)]$$

$$s^2L(x) + 16L(x) = \frac{s}{s^2+16} - \frac{s}{s^2+16}e^{-\pi s} + 1$$

$$L(x) = \frac{1}{s^2+16} + \frac{s}{(s^2+16)^2} - \frac{s}{(s^2+16)^2}e^{-\pi s}$$

Thus

$$x(t) = \frac{1}{4}L^{-1}\left(\frac{4}{s^2+16}\right) + \frac{1}{8}L^{-1}\left[\frac{8s}{(s^2+16)^2}\right] - \frac{1}{8}L^{-1}\left[\frac{8s}{(s^2+16)^2}e^{-\pi s}\right]$$

$$= \frac{1}{4}\sin 4t + \frac{1}{8}t\sin 4t - \frac{1}{8}(t-\pi)\sin(4-t)\,F(t-\pi)$$

This solution can also be written as

$$x(t) = \frac{1}{4}\sin 4t + \frac{1}{8}t\sin 4t, \quad 0 \le t < \pi$$

$$= \frac{2+\pi}{8}\sin 4t, \qquad t \ge 4$$

From Fig. 8.2, we observe that the amplitudes of vibrations become steady as soon as the external force is turned off.

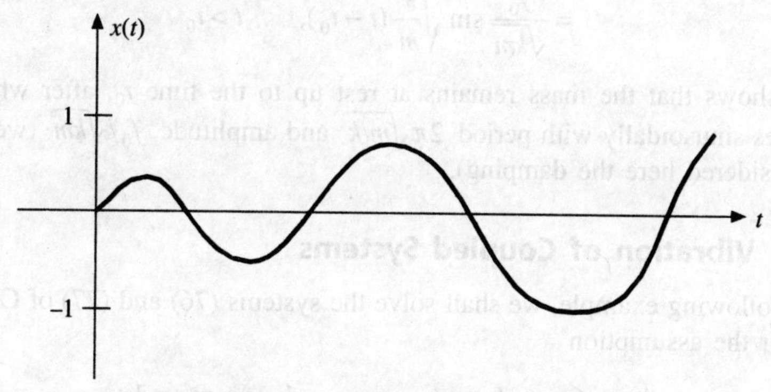


**Fig. 8.2** The displacement-time curve.

**EXAMPLE 8.20** A mass $m$ is attached to the lower end of a vertical spring of constant $k$ suspended from a fixed point. At time $t = t_0$, the mass is struck by applying a force in the upward direction lasting for a very short time. Discuss the motion.

***Solution*** Consider the vertical axis of spring as the $x$-axis and let the mass be initially at $x = 0$. Then we have

$$m\frac{d^2x}{dt^2}+kx=f_0\delta(t-t_0),\qquad x(0)=0,\quad x'(0)=0$$

Here, we have assumed that the impulse of the force applied to the mass is constant and equal to $f_0$, so that the force can be taken as $f_0\delta(t-t_0)$.

By taking the Laplace transform of the given equation alongwith the initial conditions and $L[\delta(t-t_0)]=e^{-st_0}$, we obtain

$$(ms^2+k)\,L(x)=f_0e^{-st_0}$$

$$L(x)=\frac{f_0e^{-st_0}}{ms^2+k}$$

Since

$$L^{-1}\left(\frac{f_0}{ms^2+k}\right)=\frac{f_0}{m}L^{-1}\left[\frac{1}{s^2+(k/m)}\right]=\frac{f_0}{m}\frac{\sin\sqrt{k/m}\,t}{\sqrt{k/m}}=\frac{f_0}{\sqrt{km}}\sin\sqrt{\frac{k}{m}}\,t$$

we have

$$x=L^{-1}\left(\frac{f_0e^{-st_0}}{ms^2+k}\right)=0,\qquad t<t_0$$

$$=\frac{f_0}{\sqrt{km}}\sin\sqrt{\frac{k}{m}}\,(t-t_0),\qquad t>t_0$$

which shows that the mass remains at rest up to the time $t_0$, after which it oscillates sinusoidally with period $2\pi\sqrt{m/k}$ and amplitude $f_0/\sqrt{km}$ (we have not considered here the damping).

### 8.6.2 Vibration of Coupled Systems

In the following example, we shall solve the systems (76) and (77) of Chapter 7, under the assumption

$$k_1=6,\qquad k_2=4,\qquad m_1=1,\qquad m_2=1$$

and that the masses start from their equilibrium position with opposite unit velocities.

**EXAMPLE 8.21** Solve

$$x''+10x-4y=0,\qquad -4x+y''+4y=0 \tag{11}$$

for which $x(0)=y(0)$, $x'(0)=1$ and $y'(0)=-1$.

***Solution*** The Laplace transforms of each equation are

$$s^2L(x) - sx(0) - x'(0) + 10L(x) - 4L(y) = 0$$

$$-4L(x) + s^2L(y) - sy(0) - y'(0) + 4L(y) = 0$$

or

$$\left.\begin{aligned}(s^2 + 10)\, L(x) - 4L(y) &= 1\\ -4L(x) + (s^2 + 4)\, L(y) &= 1\end{aligned}\right\} \tag{12}$$

Elimination of $L(y)$ yields

$$L(x) = \frac{s^2}{(s^2+2)(s^2+12)} = \frac{-1/5}{s^2+2} + \frac{6/5}{s^2+12}$$

Therefore

$$x(t) = -\frac{1}{5\sqrt{2}} L^{-1}\left(\frac{1}{s^2+2}\right) + \frac{6}{5\sqrt{12}} L^{-1}\left(\frac{1}{s^2+12}\right)$$

$$= -\frac{\sqrt{2}}{10} \sin \sqrt{2}\, t + \frac{\sqrt{3}}{5} \sin 2\sqrt{3}\, t \tag{13}$$

From the first equation of (12)

$$L(y) = \frac{s^2+6}{(s^2+2)(s^2+12)} = \frac{-2/5}{s^2+2} - \frac{3/5}{s^2+12}$$

Therefore

$$y(t) = \frac{-2}{5\sqrt{2}} L^{-1}\left(\frac{1}{s^2+2}\right) - \frac{3}{5\sqrt{12}} L^{-1}\left(\frac{1}{s^2+12}\right)$$

$$= -\frac{\sqrt{2}}{5} \sin \sqrt{2}\, t - \frac{\sqrt{3}}{10} \sin 2\sqrt{3}\, t \tag{14}$$

Thus, the solution of the given system (11) is given by Eqs. (13) and (14).

**EXAMPLE 8.22** In Example 7.16, the motion of two masses moving in a coupled system (Fig. 7.10) was governed by the system

$$3x''(t) + 3x(t) - y(t) = 0, \qquad y''(t) + y(t) - x(t) = 0$$

Find $x(t)$ and $y(t)$ when $x(0) = 1$, $x'(0) = 0$, $y(0) = \sqrt{3}$ and $y'(0) = 0$.

***Solution*** We have

$$3s^2L(x) - 3s + 3L(x) - L(y) = 0$$

$$s^2L(y) - \sqrt{3}\, s + L(y) - L(x) = 0$$

Elimination of $L(y)$ yields

$$L(x)=\frac{s(3s^2+3+\sqrt{3})}{3s^4+6s^2+2}=\frac{s}{s^2+1-(1/\sqrt{3})}$$

Therefore,

$$x(t)=\cos\sqrt{1-\frac{1}{\sqrt{3}}}\,t$$

Also

$$L(y)=3(s^2+1)\,L(x)-3s=\frac{3s(s^2+1)}{s^2+1-(1/\sqrt{3})}-3s=\frac{3s/\sqrt{3}}{s^2+1-(1/\sqrt{3})}$$

Thus

$$y(t)=\sqrt{3}\cos\sqrt{1-\frac{1}{\sqrt{3}}}\,t$$

### 8.6.3 Electric Circuits

**EXAMPLE 8.23** An electric circuit (Fig. 8.3) consists of a resistor of resistance $R\,\Omega$ in series with a capacitor of capacitance $C$ farads, a generator of $E$ volts, and a key. At time $t = 0$ the key is closed. Assuming that the charge on the capacitor is zero at $t = 0$, find the charge and current at any later time. Assume $R$, $C$, $E$ to be constants.

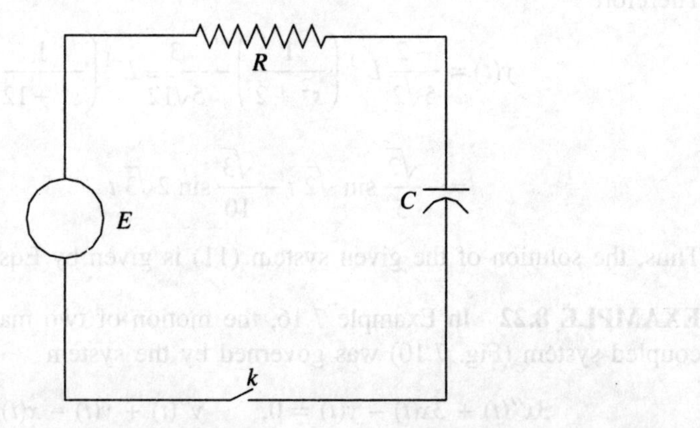


**Fig. 8.3** An electric circuit consisting of a resistor, a capacitor, a generator, and a key.

***Solution*** If $Q$ and $I = dQ/dt$ are the charge and the current at any time $t$ then by Kirchhoff's law we have

$$RI+\frac{Q}{C}=E$$

or

$$R\frac{dQ}{dt}+\frac{Q}{C}=E \quad \text{with } Q(0)=0$$

Taking the Laplace transforms of both sides, we get

$$R[sL(Q) - Q(0)] + \frac{L(Q)}{C} = \frac{E}{s}$$

or
$$L(Q)=\frac{CE}{s(RCs+1)}$$

$$=\frac{E/R}{s[s+1/(RC)]}$$

$$=\frac{E/R}{1/RC}\left[\frac{1}{s}-\frac{1}{s+1/(RC)}\right]$$

$$=CE\left[\frac{1}{s}-\frac{1}{s+1/(RC)}\right]$$

Thus,

$$Q = CE[1 - e^{-t/(RC)}], \qquad I = \frac{E}{R}e^{-t/(RC)}$$

**EXAMPLE 8.24** Solve Example 8.23 when the generator of $E$ volts is replaced by a generator with voltage given as a function of time by

$$E(t)=\begin{cases} E_0, & 0 \le t < T \\ 0, & t > R \end{cases}$$

***Solution*** Replacing $E$ by $E(t)$ in Example 8.23, we obtain

$$R\frac{dQ}{dt}+\frac{Q}{C}=E(t), \qquad Q(0)=0 \tag{15}$$

In terms of the Heaviside unit function, Eq. (15) takes the form

$$R\frac{dQ}{dt}+\frac{Q}{C}=E_0\,[1-H(t-T)] \tag{16}$$

Taking the Laplace transforms of both sides of Eqs. (15) and (16), we get

$$R[sL(Q) - Q(0)] + \frac{L(Q)}{C} = \frac{E_0(1-e^{-sT})}{s}$$

or
$$L(Q)=\frac{E_0}{R}\frac{(1-e^{-sT})}{s[s+1/(RC)]}=\frac{E_0}{Rs[s+1/(RC)]}-\frac{E_0}{Rs[s+1/(RC)]}e^{-sT}$$

$$= CE_0\left[\frac{1}{s} - \frac{1}{s + 1/(RC)}\right] - CE_0\left[\frac{1}{s} - \frac{1}{s + 1/(RC)}\right]e^{-sT}$$

If $L^{-1}[f(s)] = F(t)$, taking the inverse Laplace transform of both sides and using the result

$$L^{-1}[e^{-as}f(s)] = \begin{cases} 0, & t < a \\ F(t-a), & t > a \end{cases} = F(t-a)\,H(t-a)$$

we find

$$Q = CE_0[1 - e^{-t/(RC)}] - CE_0[1 - e^{-(t-T)/(RC)}]H(t - T)$$

$$= \begin{cases} CE_0\,[1 - e^{t/(RC)}], & t < T \\ CE_0\,[e^{-(t-T)/(RC)} - e^{-t/(RC)}], & t > T \end{cases} \tag{17}$$

For $t = T$

$$Q = CE_0[1 - e^{-T/(RC)}]$$

The problem of this example can also be treated by using the convolution theorem as follows: Let $e(s)$ be the Laplace transform of $E(t)$. Then as above, we have

$$R[sL(Q) - Q(0)] + \frac{L(Q)}{C} = e(s)$$

or

$$L(Q) = \frac{e(s)}{R[s + 1/(RC)]}$$

Now

$$L^{-1}\left[\frac{1}{R[s + 1/(RC)]}\right] = \frac{e^{-t/(RC)}}{R}, \qquad L^{-1}[e(s)] = E(t)$$

Thus, by the convolution theorem

$$Q = \frac{1}{R}\int_0^t E(u)\,e^{-(t-u)/(RC)}\,du$$

For $0 < t < T$, we have

$$Q = \frac{1}{R}\int_0^t E_0\,e^{-(t-u)/RC}\,du = CE_0\left[1 - e^{-t/(RC)}\right]$$

For $t > T$, we have

$$Q = \frac{1}{R}\int_0^T E_0\,e^{-(t-u)/(RC)}\,du = CE_0\left[e^{-(t-T)/(RC)} - e^{-t/(RC)}\right]$$

which agrees with Eq. (17).

We have already seen in Section 7.9 that an electric network having more than one loop gives rise to simultaneous differential equations. As shown in Fig. 8.4, the current $I_1(t)$ splits in the directions shown at point B (a branch point) of the network. By Kirchhoff's law

$$I_1(t) = I_2(t) + I_3(t) \tag{18}$$

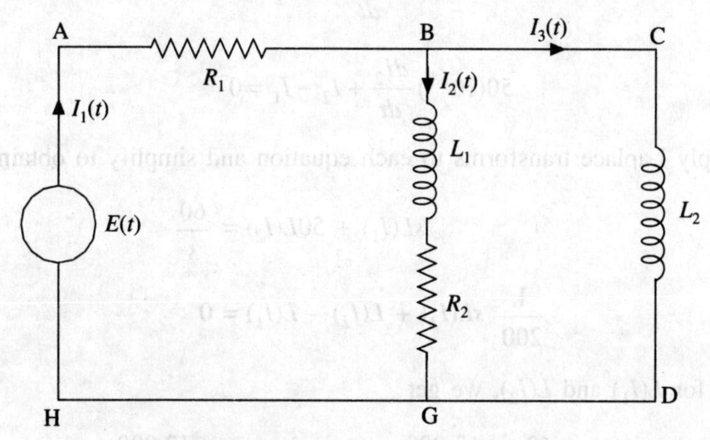


**Fig. 8.4** Multiple loop electric circuit.

We can also apply Kirchhoff's law to each loop. For loop ABGHA, we have

$$E(t) = I_1 R_1 + L_1 \frac{dI_2}{dt} + I_2 R_2 \tag{19}$$

For loop ABCDGHA, we find

$$E(t) = I_1 R_1 + L_2 \frac{dI_3}{dt} \tag{20}$$

From Eq. (18), eliminate $I_1$ in Eqs. (19) and (20) to obtain

$$\left.\begin{aligned} L_1 \frac{dI_2}{dt} + (R_1 + R_2)I_2 + R_1 I_3 &= E(t) \\ L_2 \frac{dI_3}{dt} + R_1 I_2 + R_1 I_3 &= E(t) \end{aligned}\right\} \tag{21}$$

and the initial conditions are $I_2(0) = I_3(0) = 0$.

It can also be shown that the system of differential equations describing the currents $I_1(t)$ and $I_2(t)$ in the network containing a resistor, an inductor and a capacitor (Fig. 8.4) is obtained as

$$\left.\begin{aligned} L\frac{dI_1}{dt} + RI_2 &= E(t) \\ RC\frac{dI_2}{dt} + I_2 - I_1 &= 0 \end{aligned}\right\} \tag{22}$$

**EXAMPLE 8.25** Solve the system (22) when $E = 60$ V, $L = 1$ H, $R = 50\ \Omega$, $C = 10^{-4}$ F and $I_1(0) = I_2(0) = 0$.

***Solution*** Here, Eqs. (22) take the form

$$\frac{dI_1}{dt} + 50I_2 = 60$$

$$50(10^{-4})\frac{dI_2}{dt} + I_2 - I_1 = 0$$

Now apply Laplace transforms to each equation and simplify to obtain

$$sL(I_1) + 50L(I_2) = \frac{60}{s}$$

$$\frac{1}{200}\ sL(I_2) + L(I_2) - L(I_1) = 0$$

Solving for $L(I_1)$ and $L(I_2)$, we get

$$L(I_1) = \frac{60\,s + 12{,}000}{s(s+100)^2} \quad \text{and} \quad L(I_2) = \frac{12{,}000}{s(s+100)^2}$$

which can be written as

$$L(I_1) = \frac{6/5}{s+100} - \frac{60}{(s+100)^2} + \frac{6/5}{s}$$

$$L(I_2) = \frac{6/5}{s} - \frac{6/5}{s+100} - \frac{120}{(s+100)^2}$$

Now, take the inverse Laplace transform of each of these equations to obtain

$$I_1(t) = \frac{6}{5} - \frac{6}{5}e^{-100t} - 60\,t\,e^{-100t}$$

$$I_2(t) = \frac{6}{5} - \frac{6}{5}e^{-100t} - 120t\,e^{-100t}$$

As $t \to \infty$, both $I_1(t)$ and $I_2(t)$ tend to the value $R/E = 6/5$. Moreover, since the current through the capacitor is $I_3(t) = I_1(t) - I_2(t) = 60te^{-100t}$, we find $I_3(t) \to 0$ as $t \to \infty$.

**EXAMPLE 8.26** For the network shown in Fig. 8.5, $I_1(0) = I_2(0) = I_3(0)$. Find $I_1$, $I_2$ and $I_3$ at time $t$ after the switch is closed.

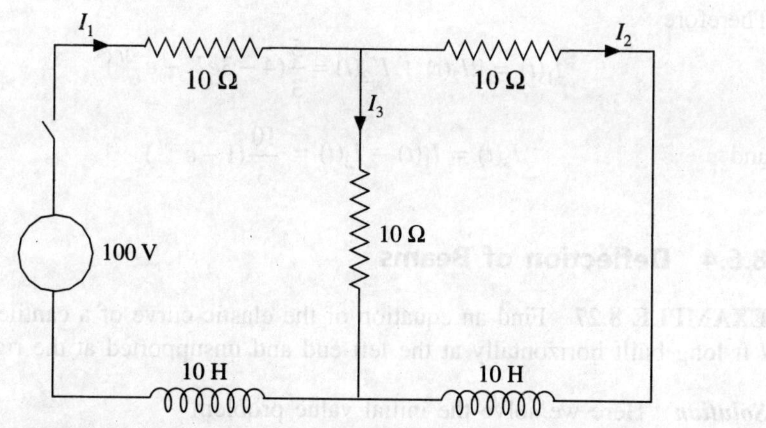


**Fig. 8.5** Diagrammatic representation of Example 8.26.

***Solution*** Applying Kirchhoff's law to the left circuit, we get

$$10I_1(t) + 10I_3(t) + 10I'_1(t) = 100$$

From the right circuit

$$10I_2(t) + 10I'_2(t) - 10I_3(t) = 0$$

Using $I_3(t) = I_1(t) - I_2(t)$, we obtain

$$2I_1(t) - I_2(t) + I'_1(t) = 10$$

$$2I_2(t) + I'_2(t) - I_1(t) = 0$$

Thus

$$2L(I_1) - L(I_2) + sL(I_1) = \frac{10}{s}$$

$$2L(I_2) + sL(I_2) - L(I_1) = 0$$

which yield

$$L(I_2) = \frac{10}{s[(s+2)^2 - 1]}$$

$$I_2(t) = 10\, e^{-2t} L^{-1}\left[\frac{1}{(s-2)\,(s^2-1)}\right]$$

$$= 10\, e^{-2t} L^{-1}\left(\frac{1}{3}\frac{1}{s-2} - \frac{1}{2}\frac{1}{s-1} + \frac{1}{6}\frac{1}{s+1}\right)$$

$$= 10e^{-2t}\left(\frac{e^{2t}}{3} - \frac{e^{t}}{2} + \frac{e^{-t}}{6}\right)$$

$$= \frac{5}{3}(2 - 3e^{-t} + e^{-3t})$$

Therefore

$$I_1(t) = 2I_2(t) + I'_2(t) = \frac{5}{3}(4 - 3e^{-t} - e^{-3t})$$

and

$$I_3(t) = I_1(t) - I_2(t) = \frac{10}{3}(1 - e^{-3t})$$

### 8.6.4 Deflection of Beams

**EXAMPLE 8.27** Find an equation of the elastic curve of a cantilever beam $l$ ft long built horizontally at the left end and unsupported at the right end.

***Solution*** Here we solve the initial value problem

$$EIY''(x) = M(x) = -\frac{W}{2}(l - x)^2 = -\frac{W}{2}(l^2 - 2lx + x^2), \qquad Y(0) = Y'(0) = 0$$

Hence

$$EIs^2 L(Y) = \frac{-W}{2}\left(\frac{l^2}{s} - \frac{2l}{s^2} + \frac{2}{s^3}\right)$$

$$L(Y) = \frac{-W}{2EI}\left(\frac{l^2}{s^3} - \frac{2l}{s^4} + \frac{2}{s^5}\right)$$

Thus

$$Y(x) = \frac{-W}{24EI}(x^4 - 4lx^3 + 6l^2x^2)$$

**EXAMPLE 8.28** Find the maximum deflection of an encastre beam $l$ ft long, carrying a uniformly distributed load $W$ lb/ft on its central half length (Fig. 8.6).

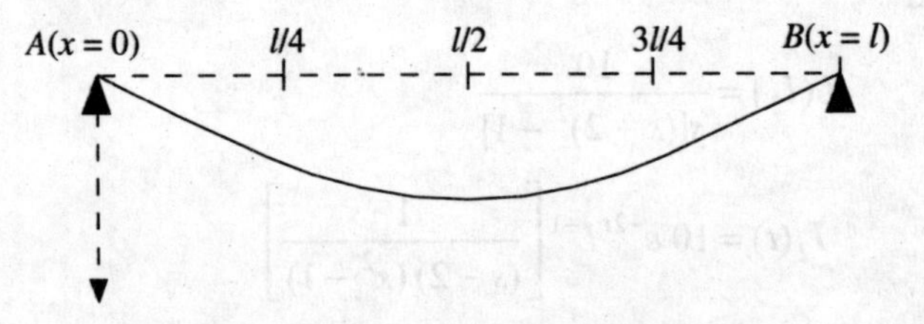


**Fig. 8.6** Deflection of an encastre beam.

***Solution*** Let the origin be at the end $A$. Then we have

$$EI\frac{d^4y}{dx^4} = W(x)$$

where

$$W(x) = W\left[u\left(x - \frac{l}{4}\right) - u\left(x - \frac{3l}{4}\right)\right]$$

Taking the Laplace transforms of both sides, we get

$$EI\,[s^4 L(y) - s^3 y(0) - s^2 y'(0) - sy''(0) - y'''(0)] = W\left(\frac{e^{-is/4}}{s} - \frac{e^{-3is/4}}{s}\right)$$

Using the conditions $y(0) = y'(0) = 0$ and taking $y''(0) = c_1$ and $y'''(0) = c_2$, we have

$$EIL(y) = W\left(\frac{e^{-is/4}}{s^5} - \frac{e^{-3is/4}}{s^5}\right) + \frac{c_1}{s^3} + \frac{c_2}{s^4}$$

and

$$EIy = \frac{W}{24}\left[\left(x - \frac{l}{4}\right)^4 u\left(x - \frac{l}{4}\right) - \left(x - \frac{3l}{4}\right)^4 u\left(x - \frac{3l}{4}\right)\right] + \frac{1}{2}c_1 x^2 + \frac{1}{6}c_2 x^3 \qquad (23)$$

For $x > 3l/4$

$$EIy = \frac{W}{24}\left[\left(x - \frac{l}{4}\right)^4 - \left(x - \frac{3l}{4}\right)^4\right] + \frac{1}{2}c_1 x^2 + \frac{1}{6}c_2 x^3$$

$$EIy' = \frac{W}{6}\left[\left(x - \frac{l}{4}\right)^3 - \left(x - \frac{3l}{4}\right)^3\right] + c_1 x + \frac{1}{2}c_2 x^2$$

Using the conditions $y(l) = 0$ and $y'(l) = 0$, we get

$$c_1 = \frac{11Wl^2}{192}, \qquad c_2 = \frac{-Wl}{4}$$

Thus, for $l/4 < x < 3l/4$, Eq. (23) becomes

$$EIy = \frac{W}{24}\left(x - \frac{1}{4}\right)^4 + \frac{11Wl^2}{384}x^2 - \frac{Wl}{24}x^3$$

Hence, maximum deflection $= y(l/2) = 13Wl^4/(6144EI)$.

**EXAMPLE 8.29** A beam is simply supported at its end $x = 0$ and is clamped at the other end $x = l$. It carries a load $W$ at $x = l/4$. Find the resulting deflection at any point.

***Solution*** The required differential equation is

$$\frac{d^4y}{dx^4} = \frac{W}{EI}\,\delta\left(x - \frac{l}{4}\right)$$

Taking the Laplace transforms of both sides and using the conditions $y(0) = 0$, $y''(0) = 0$ and $y'(0) = c_1$, $y'''(0) = c_2$, we get

$$L(y) = \frac{c_1}{s^2} + \frac{c_2}{s^4} + \frac{W}{EI}\frac{e^{-is/4}}{s^4}$$

Thus

$$y = c_1x + c_2\frac{x^3}{3!} + \frac{W}{EI}\frac{[x-(l/4)]^3}{3!}u\left(x - \frac{l}{4}\right)$$

i.e.

$$\left.\begin{aligned} y &= c_1x + \frac{1}{6}c_2x^3, \quad 0 < x < \frac{l}{4} \\ y &= c_1x + \frac{1}{6}c_2x^3 + \frac{W}{6EI}\left(x - \frac{l}{4}\right)^3, \quad \frac{l}{4} < x < l \end{aligned}\right\} \tag{24}$$

Using the conditions $y(l) = 0$, $y'(l) = 0$, we obtain

$$c_1 = \frac{45\,Wl^2}{128\,EI}, \qquad c_2 = \frac{-81W}{128EI}$$

Substituting these values in Eqs. (24), we get deflection at any point.

### 8.6.5 The Tautochrone Problem

**EXAMPLE 8.30** A wire has the shape of a curve in a vertical $xy$ plane with its lower end $O$ located at the origin (see Fig. 8.7). Find the shape which the curve must have so that a bead under the influence of gravity will slide from rest down to $O$ in a specified constant time $T$, regardless of where the bead is placed on the wire above $O$ (assume that there is no friction).

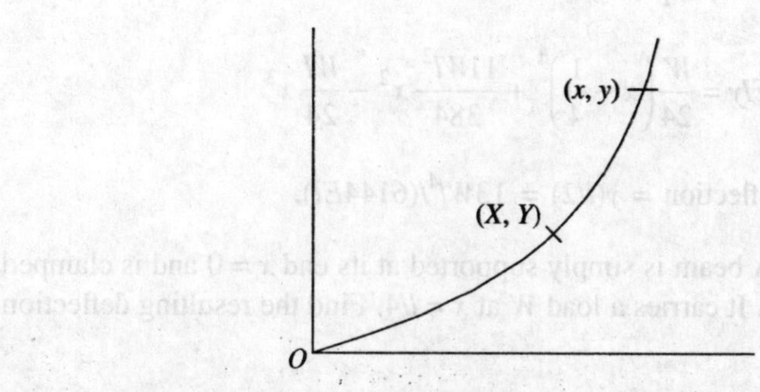


**Fig. 8.7** A bead sliding on a curve under the influence of gravity.

***Solution*** Let $(x, y)$ denote any starting point of the bead and $(X, Y)$ any point between the starting point and the point $O$. Let $\sigma$ denote the length of the wire as measured from $O$.

According to the principle of conservation of energy, we have

$$\text{P.E. at } (x, y) + \text{K.E. at } (x, y) = \text{P.E. at } (X, Y) + \text{K.E. at } (X, Y)$$

where P.E. and K.E. are the potential and kinetic energies of the bead, respectively.

If $m$ is the mass of the bead and $T$ the time of travel measured from the rest position, then

$$mgy + 0 = mgY + \frac{1}{2}m\left(\frac{d\sigma}{dt}\right)^2 \tag{25}$$

or

$$\frac{d\sigma}{dt} = \pm\sqrt{2g(y - Y)} \tag{26}$$

Since $\sigma$ decreases as $t$ increases so that $d\sigma/dt < 0$, we must choose the negative sign in Eq. (26) to obtain

$$\frac{d\sigma}{dt} = -\sqrt{2g(y - Y)} \tag{27}$$

Since $Y = y$ at $t = 0$, we have $Y = 0$ at $t = T$. From Eq. (27) on separating the variables and integrating, we obtain

$$\int_{t=0}^{T} dt = -\frac{1}{\sqrt{2g}}\int_{Y=0}^{y}\frac{d\sigma}{\sqrt{y - Y}}$$

or

$$T = \frac{1}{\sqrt{2g}}\int_{Y=0}^{y}\frac{d\sigma}{\sqrt{y - Y}} \tag{28}$$

Now, the arc length can be expressed as a function of $y$ in the form $\sigma = F(y)$ so that Eq. (28) becomes

$$T = \frac{1}{\sqrt{2g}}\int_{Y=0}^{y}\frac{F'(y)}{\sqrt{y - Y}}\,dY \tag{29}$$

We now solve Eq. (29) to obtain the required curve. Equation (29) is of the convolution type and can be written as

$$T = \frac{1}{\sqrt{2g}}\,F(y)y^{-1/2}$$

Taking the Laplace transform and using the convolution theorem, we get

$$\frac{T}{s} = \frac{1}{\sqrt{2g}}\,L[F'(y)]L(y^{-1/2}) \tag{30}$$

Now use $L(y^{-1/2}) = \Gamma(1/2)/s^{1/2} = \sqrt{\pi/s}$ and Theorem 8.1 in Eq. (30) to get

$$\frac{T}{s} = \frac{1}{\sqrt{2g}} sL[F(y)]\sqrt{\frac{\pi}{s}}$$

or
$$sL[F(y)] = T\sqrt{\frac{2g}{\pi}}\, s^{-1/2}$$

i.e.
$$L[F'(y)] = T\sqrt{\frac{2g}{\pi}}\, s^{-1/2}$$

Thus

$$F'(y) = \frac{d\sigma}{dy} = \frac{T\sqrt{2g}}{\pi\sqrt{y}} \tag{31}$$

Also,

$$\left(\frac{d\sigma}{dy}\right)^2 = \left(\frac{dx}{dy}\right)^2 + 1$$

Equation (31) thus becomes

$$\left(\frac{dx}{dy}\right)^2 + 1 = \frac{a}{y}, \quad a = \frac{2gT^2}{\pi^2}$$

or
$$\frac{dx}{dy} = \sqrt{\frac{a-y}{y}}$$

(as the slope $dy/dx$ cannot be negative). Separating the variables and integrating, we get

$$x = \int \sqrt{\frac{a-y}{y}}\, dy \tag{32}$$

Here, put $y = a \sin^2\phi$. Then

$$x = 2a\int \cos^2\phi \, d\phi = a\int (1 + \cos 2\phi)\, d\phi = \frac{a}{2}(2\phi + \sin 2\phi) + c$$

so that $y = (a/2)(1 - \cos 2\phi)$, and by substituting $2\phi = \theta$ we get $x = (a/2)(\theta + \sin\theta) + c$, $y = (a/2)(1 - \cos\theta)$. Now, for $x = 0$ when $y = 0$, we must have $c = 0$ so that the parametric equations for the required curve are given by

$$x = \frac{a}{2}(\theta + \sin\theta), \qquad y = \frac{a}{2}(1 - \cos\theta) \tag{33}$$

The curve described by Eq. (33) is a cycloid, which is generated by a fixed point on a circle of diameter $a$ as it rolls along the lower part of the line $y = a$ (see Fig. 8.8). In our case, the required shape of the wire is represented by that part of the cycloid shown heavy in Fig. 8.8. The size of the cycloid will depend on the particular value $T$.

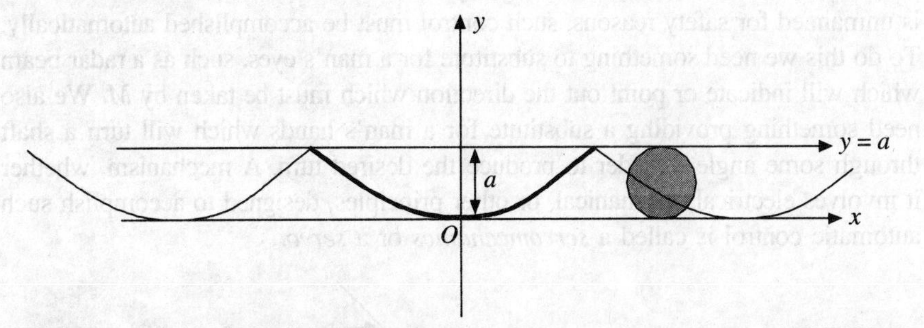


**Fig. 8.8** A cycloid generated by a fixed point on a circle of diameter $a$ as it rolls along the lower part of the line $y = a$.

The curve thus obtained is called the *tautochrone* (from the Greek *tautos,* meaning same or identical; and *chronos*, meaning time). The problem of finding the required curve known as the *tautochrone problem* (compare it with the brachistochrone problem (Example 10.16) which also involves a cycloid) was proposed near the end of the 17th century and solved in various ways by some prominent mathematicians of that time. One of them was Huygens, who employed the principle in designing a *cycloidal pendulum* for use in clocks. In this design (Fig. 8.9), the pendulum of the bob $B$ is at the end of a flexible string whose opposite end, are fixed at $O$. The pendulum is constrained by the two neighbouring arches of the cycloid, $OP$ and $OQ$, so that it swings between them. The period of oscillation is constant, and the path of the pendulum bob turns out to be a cycloid.

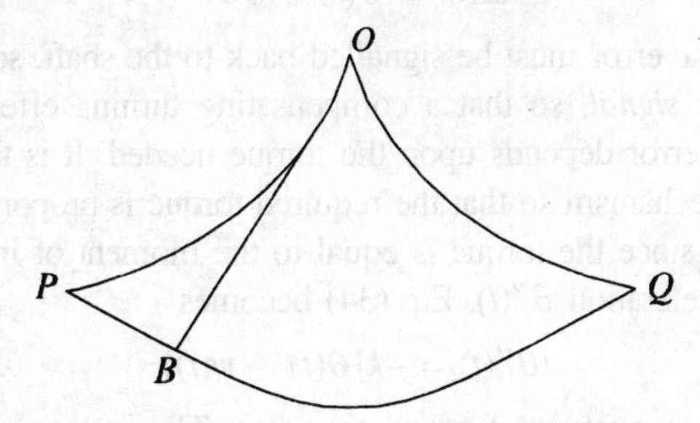


**Fig. 8.9** Schematic diagram of a cycloidal pendulum.

The tautochrone problem, although seemingly only an interesting exercise in mechanics and differential equations, actually turned out to be of greater mathematical significance because it inspired Abel in 1823 to a study of integral equations.

### 8.6.6 Theory of Automatic Control and Servomechanics

Suppose that a missile $M$ is tracking down an enemy aircraft or another missile $E$ (Fig. 8.10). If at time $t$, the enemy $E$ turns through some angle, $\psi(t)$, then $M$ must also turn through this angle if it is to catch up with $E$ and destroy it. If a man were aboard $M$, he could operate some steering mechanism to turn; but since the missile is unmanned for safety reasons, such control must be accomplished automatically. To do this we need something to substitute for a man's eyes, such as a radar beam which will indicate or point out the direction which must be taken by $M$. We also need something providing a substitute for a man's hands which will turn a shaft through some angle in order to produce the desired turn. A mechanism, whether it involves electrical, mechanical, or other principles, designed to accomplish such automatic control is called a *servomechanics* or a *servo.*

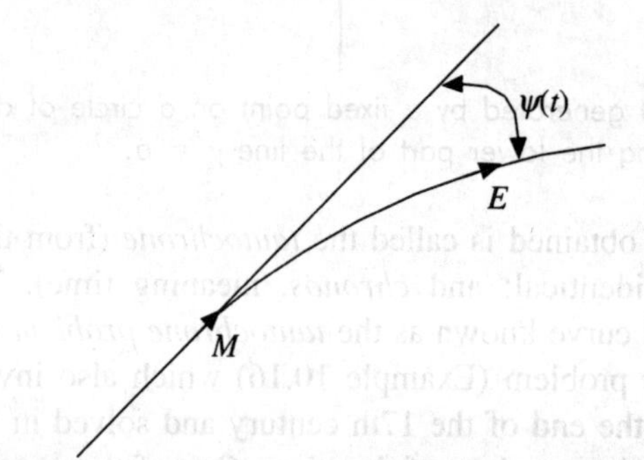


**Fig. 8.10** A missile $M$ is tracking down another missile $E$.

In the present application, let us assume that the desired angle of turn as indicated by the radar beam is $\psi(t)$. Also, let $\theta(t)$ denote the angle of turn of the shaft at time $t$. Ideally we must have $\psi(t) = \theta(t)$, but because things are happening so fast we must expect to have a discrepancy or error between the two given by

$$\text{error} = \theta(t) - \psi(t) \tag{34}$$

The existence of the error must be signalled back to the shaft, sometimes referred to as the *feedback signal,* so that a compensating turning effect or torque may be produced. The error depends upon the torque needed. It is thus reasonable to design the servomechanism so that the required torque is proportional to the error of Eq. (34). Now, since the torque is equal to the moment of inertia $I$ multiplied by the angular acceleration $\theta''(t)$, Eq. (34) becomes

$$I\theta''(t) = -k[\theta(t) - \psi(t)] \tag{35}$$

where $k > 0$ is the constant of proportionality. The negative sign before $k$ is used because, if the error is positive (i.e. if the turn is too great), then the torque must oppose it (i.e. be negative), while if the error is negative the torque must be negative. The possible initial conditions are

$$\theta(0) = 0, \qquad \theta'(0) = 0 \tag{36}$$

(Here we have neglected damping.)

Taking the Laplace transforms of Eq. (35) using Eq. (36), we have

$$Is^2 L[\theta(t)] = -k\{L[\theta(t)] - L[\psi(t)]\}$$

or
$$L[\theta(t)] = \frac{kL[\psi(t)]}{Is^2 + k} \tag{37}$$

From the convolution theorem, we have

$$\theta(t) = \sqrt{\frac{k}{l}} \int_0^1 \psi(u) \sin\left[\sqrt{\frac{k}{l}}\,(t-u)\right] du \tag{38}$$

Equation (38) determines $\theta(t)$ from $\psi(t)$. In the theory of automatic control, $\psi(t)$ and $\theta(t)$ are known as the *input* and *output*, respectively. The factor $k/(Is^2 + k)$ in Eq. (37), which serves to characterise the behaviour of the servomechanism in relating the input and output is called the *transfer* or *response function.*

Servomechanism appears in many connections in practice—for example, in the house, where a thermostat is used to regulate the temperature and on ships or planes where an automatic pilot is needed. The basic idea of a servo is shown schematically in Fig. 8.11. In the first block on the left, we have the desired state (for example, the desired position and direction in the case of a missile or the desired temperature of a room). Since the desired state is not the same as the actual state, there is an error indicated by the second block. This error is fed into an adjustor, indicated by the third block, which attempts to rectify the error (such as an adjusting torque in the case of a missile) and leads to the actual state indicated by the block on the right. This actual state is then fed back (feedback signal) to reveal the new error of departure from the desired state, and the process is repeated again and again until the desired state is achieved.

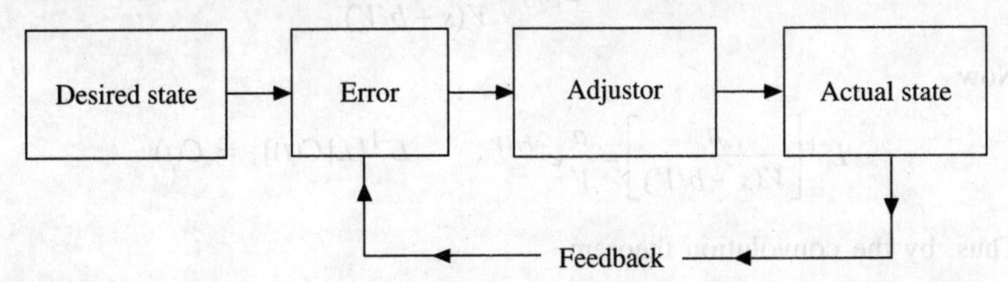


**Fig. 8.11** Block diagram illustrating the fundamental principle in servomechanics.

## 8.6.7 Absorption of Drugs in an Organ

**EXAMPLE 8.31** A liquid carries a drug into an organ of volume $V$ cm$^3$ at a rate of $a$ cm$^3$/s and leaves at a rate of $b$ cm$^3$/s where $V$, $a$ and $b$ are constants. At time $t = 0$, the concentration of the drug is zero and builds up linearly to a maximum of $k$ at $t = T$, at which time the process is stopped. What is the concentration of the drug in the organ at any time?

***Solution*** The problem is the same as that of Example 4.27 except that the concentration $C(t)$, a function of time, is given by

$$C(t) = kt, \qquad 0 \le t \le T$$
$$= 0, \qquad t > T$$

which is illustrated in Fig. 8.12. If $x$ denotes instantaneous concentration of the drug in the organ, we have

$$\frac{d}{dt}(xV) = aC(t) - bx, \qquad x(0) = 0 \tag{39}$$

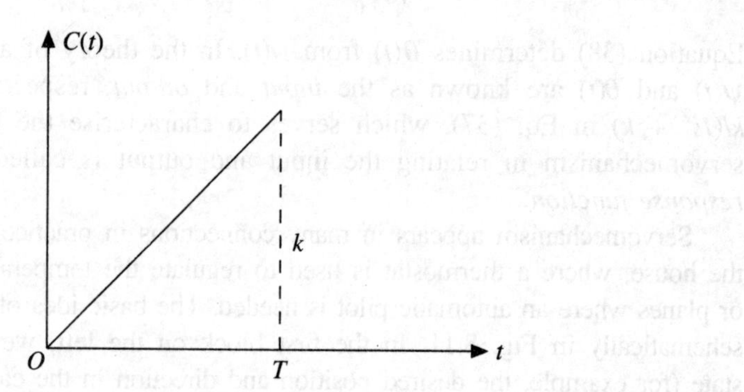


**Fig. 8.12** The concentration C of a drug in an organ at time $t$.

Taking the Laplace transform of Eq. (39), we get

$$V\{sL(x) - x(0)\} = aL[C(t)] - bL[x]$$

or
$$L[x] = \frac{aL[C(t)]}{V(s + b/V)}$$

Now

$$L^{-1}\left[\frac{a}{V(s + b/V)}\right] = \frac{a}{V}e^{-bt/V}, \qquad L^{-1}\{L[C(t)]\} = C(t)$$

Thus, by the convolution theorem

$$x = \frac{a}{V}\int_0^t C(u)\, e^{-b(t-u)/V}\, du$$

For $0 \le t \le T$, we have

$$x = \frac{a}{V}\int_0^t kue^{-b(t-u)/V}\, du = \frac{kat}{b} - \frac{Vka}{b^2}(1 - e^{-bt/V})$$

For $t > T$, we have

$$x = \frac{a}{V}\int_0^T kue^{-b(t-u)/V}\,du = \frac{Vka}{b^2}e^{-bt/V} + \left(\frac{kaT}{b} - \frac{Vka}{b^2}\right)e^{-b(t-T)V}$$

The value of $x$ for $t = T$ is found by letting $t = T$ in either of these.

From the last result we note that as $t$ increases beyond $T$, the drug generally disappears. It follows that the drug concentration in the organ will reach a maximum at some time. It can be shown that this time is $t = T$ and that this maximum (known as the *peak drug concentration*) is

$$\frac{kaT}{b} - \frac{Vka}{b^2}(1 - e^{-bT/V})$$

In practice, the peak drug concentration time will occur later than $T$ since the drug does not enter the organ instantaneously, as in this example; instead, there is a time lag.

## 8.7 MISCELLANEOUS SOLVED EXAMPLES

In this section, we shall solve a number of problems based on the different sections of this chapter.

**EXAMPLE 8.32** Find the Laplace transform of $\frac{1}{\sqrt{x}}$.

***Solution*** From the definition, we have

$$L\left(\frac{1}{\sqrt{x}}\right) = \int_0^\infty e^{-sx}x^{-1/2}dx$$

$$= \frac{2}{\sqrt{s}}\int_0^\infty e^{-t^2}\,dt \text{ (by taking } \sqrt{sx} = t)$$

$$= \frac{2}{\sqrt{s}}\left(\frac{\sqrt{\pi}}{2}\right) = \left(\frac{\pi}{s}\right)^{1/2} = \frac{\Gamma(1/2)}{\sqrt{s}}, \quad s > 0 \tag{40}$$

where $\int_0^\infty e^{-t^2}\,dt = \frac{\sqrt{\pi}}{2}$.

**EXAMPLE 8.33** Find the Laplace transform of $x^k$.

***Solution*** From the definition, we have

$$L(x^k) = \int_0^\infty e^{-sx}x^k\,dx \tag{41}$$

Put $u = sx$, then Eq. (41) reduces to

$$L(x^k) = \frac{1}{s^{k+1}} \int_0^\infty u^k e^{-u} du = \frac{1}{s^{k+1}} \Gamma(k+1) \tag{42}$$

If $k = n$, where $n$ is a positive integer, then $\Gamma(n+1) = n!$ and Eq. (42) leads to $L(x^n) = \frac{n!}{s^{n+1}}$ (cf., Table 8.1).

**EXAMPLE 8.34** Find the Laplace transform of $J_0(x)$, where $J_0(x)$ is the Bessel function of order zero and is given in Chapter 5 by Eq. (193 b) and hence obtain the Laplace transforms of $J_0(ax), xJ_0(ax)$ and $e^{-ax} J_0(ax)$.

***Solution*** Taking the Laplace transform of both sides of Eq. (193 b) Chapter 5, we get

$$L[J_0(x)] = L[1] - \frac{1}{2^2} L[x^2] + \frac{1}{2^2 \cdot 4^2} L[x^4] - \frac{1}{2^2 \cdot 4^2 \cdot 6^2} L[x^6] + \cdots$$

$$= \frac{1}{s} - \frac{1}{2^2} \frac{2!}{s^3} + \frac{1}{2^2 \cdot 4^2} \frac{4!}{s^5} - \frac{1}{2^2 \cdot 4^2 \cdot 6^2} \frac{6!}{s^7} + \cdots$$

$$= \frac{1}{s} \left[ 1 - \frac{1}{2} \left( \frac{1}{s^2} \right) + \frac{1 \cdot 3}{2 \cdot 4} \left( \frac{1}{s^4} \right) - \frac{1 \cdot 3 \cdot 5}{2 \cdot 4 \cdot 6} \left( \frac{1}{s^6} \right) + \cdots \right]$$

$$= \frac{1}{s} \left[ \left( 1 + \frac{1}{s^2} \right)^{-1/2} \right] = (1 + s^2)^{-1/2} \tag{43}$$

Now using Theorem 8.8, Eq. (43) leads to

$$L[J_0(ax)] = \frac{1}{a} \frac{1}{\left[ 1 + \left( \frac{s}{a} \right)^2 \right]^{1/2}} = (s^2 + a^2)^{-1/2} \tag{44}$$

Also, from Theorem 8.3 and Eq. (44), we have

$$L[xJ_0(ax)] = (-1)^1 \frac{d}{ds} \{L[J_0(ax)]\}$$

$$= -\frac{d}{ds} [(s^2 + a^2)^{-1/2}] = s(s^2 + a^2)^{-3/2} \tag{45}$$

Moreover, from Property 8.2 and Eq. (44), we have

$$L[e^{-ax} J_0(ax)] = \frac{1}{[(s+a)^2 + a^2]^{1/2}} = (s^2 + 2as + 2a^2)^{-1/2}$$

**EXAMPLE 8.35** Find the Laplace transform of $J_1(x)$ and $xJ_1(x)$, where $J_1(x)$ is the Bessel function of order one and is given by Eq. (193 c) (Chapter 5).

***Solution*** Taking the Laplace transform of both sides of Eq. (193c) (Chapter 5), we get

$$L[J_1(x)] = \frac{1}{2}L[x] - \frac{1}{2^2}\frac{1}{4}L[x^3] + \frac{1}{2^2\cdot 4^2}\frac{1}{6}L[x^5] + \cdots$$

$$= \frac{1}{2}\frac{1}{s^2} - \frac{1}{2^2}\frac{1}{4}\frac{3!}{s^4} + \frac{1}{2^2\cdot 4^2\cdot 6}\frac{5!}{s^6} + \cdots$$

$$= 1 - [1 - \frac{1}{2}\left(\frac{1}{s^2}\right) + \frac{1\cdot 3}{2\cdot 4}\left(\frac{1}{s^2}\right)^2 - \frac{1\cdot 3\cdot 5}{2\cdot 4\cdot 6}\left(\frac{1}{s^2}\right)^3 + \cdots]$$

$$= [1 - \left(1 + \frac{1}{s^2}\right)^{-1/2}] = 1 - \frac{s}{(1+s^2)^{1/2}} \tag{46}$$

Also, using Theorem 8.3 and Eq. (46), we get

$$L[xJ_1(x)] = -\frac{d}{ds}\left[1 - \frac{s}{(1+s^2)^{1/2}}\right] = (1+s^2)^{-3/2}$$

**EXAMPLE 8.36** Find the Laplace transform of $\sin\sqrt{x}$.

***Solution*** Since

$$\sin x = x - \frac{x^3}{3!} + \frac{x^5}{5!} - \frac{x^7}{7!} + \cdots$$

therefore

$$\sin\sqrt{x} = \sin x^{1/2} = x^{1/2} - \frac{x^{3/2}}{3!} + \frac{x^{5/2}}{5!} - \frac{x^{7/2}}{7!} + \cdots$$

Taking the Laplace transform of both sides of this equation, we get

$$L[\sin\sqrt{x}] = L[x^{1/2}] - \frac{1}{3!}L[x^{3/2}] + \frac{1}{5!}L[x^{5/2}] + \cdots$$

$$= \frac{\Gamma(3/2)}{s^{3/2}} - \frac{1}{3!}\frac{\Gamma(5/2)}{s^{5/2}} + \frac{1}{5!}\frac{\Gamma(7/2)}{s^{7/2}} + \cdots \tag{47}$$

as $L(x^n) = \dfrac{\Gamma(n+1)}{s^{n+1}}, s > 0$. Also from the properties of Gamma function that $\Gamma(n+1) = n\Gamma(n)$ and $\Gamma(1/2) = \sqrt{\pi}$, Eq. (47) after simplification reduces to

$$L[\sin\sqrt{x}] = \frac{\sqrt{\pi}}{2s^{3/2}}[1 - (\frac{1}{4s}) + \frac{1}{2!}(\frac{1}{4s})^2 - \frac{1}{3!}(\frac{1}{4s})^3 + \ldots] = \frac{\sqrt{\pi}}{2s^{3/2}}e^{-1/4s}$$

**EXAMPLE 8.37** Find the Laplace transforms of $\sinh ax\ \cos ax$ and $\sinh ax\ \sin ax$

***Solution*** We know that

$$L[\sinh ax] = \frac{a}{s^2 - a^2} = \bar{f}(s) \quad \text{(say)}$$

then from the first shifting property, we have

$$L[e^{iax} \sinh ax] = \bar{f}(s - ia)$$

$$= \frac{a}{(s - ia)^2 - a^2}$$

$$= \frac{a}{s^2 - 2a^2 - 2ias}\left(\frac{s^2 - 2a^2 + 2ias}{s^2 - 2a^2 + 2ias}\right)$$

$$= \frac{a(s^2 - 2a^2) + 2ia^2 s}{s^4 + a^4} \tag{48}$$

Since $e^{iax} = \cos ax + i \sin ax$, Eq. (48) leads to

$$L[\sinh ax(\cos ax + i \sin ax)] = \frac{a(s^2 - 2a^2) + 2ia^2 s}{s^4 + a^4}$$

which may be expressed as

$$L[\sinh ax \cos ax] + iL[\sinh ax \sin ax] = \frac{a(s^2 - 2a^2)}{s^4 + a^4} + i\frac{2a^2 s}{s^4 + a^4}$$

Equating the real and imaginary parts, we get

$$L[\sinh ax \cos ax] = \frac{a(s^2 - 2a^2)}{s^4 + a^4}$$

and $$L[\sinh ax \sin ax] = \frac{2a^2 s}{s^4 + a^4}$$

**EXAMPLE 8.38** If $L[F(x)] = \frac{1}{s} e^{-1/s}$, find $L[e^{-x} F(3x)]$.

***Solution*** Given that

$$L[F(x)] = \frac{1}{s} e^{-1/s} = \bar{f}(s)$$

From the change of scale property (Theorem 8.8), we have

$$L[F(3x)] = \frac{1}{3}\bar{f}\left(\frac{s}{3}\right) = \frac{1}{3}\left(\frac{3}{s} e^{-3/s}\right) = \frac{1}{s} e^{-3/s}$$

which from the first shifting theorem leads to

$$L[e^{-x}F(3x)]=\frac{1}{s+1}e^{-3/(s+1)}$$

**EXAMPLE 8.39** Find the Laplace transform of $x^5$ if $L(x)=\frac{1}{s^2}$.

***Solution*** Let $f(x)=x^5$, then

$$f'(x)=5x^4, f''(x)=20x^3, f'''(x)=60x^2, f^{iv}(x)=120x, f^{v}(x)=120.$$

Also, $f(0)=f'(0)=f''(0)=f'''(0)=f^{iv}(0)=0, f^{v}(0)=120$. From Theorem 8.3, we have

$$L[f^{iv}(x)]=s^4L[x^5]-s^3f(0)-s^2f'(0)-sf''(0)-f'''(0)$$

which using the above data reduces to

$$120L[x]=s^4L[x^5]$$

which leads to

$$L[x^5]=\frac{120}{s^6}$$

**EXAMPLE 8.40** Find the Laplace transform of

(i) $x^2\sin ax$ (ii) $x\cos^2 x$ (iii) $\frac{\cos ax-\cos bx}{x}$. (iv) $\frac{\sin^2 x}{x}$

***Solution*** (i) We know that

$$L[\sin ax]=\frac{a}{s^2+a^2}$$

Thus, from Theorem 8.3, we have

$$L[x^2\sin ax]=(-1)^2\frac{d^2}{ds^2}L[\sin ax]=(-1)^2\frac{d^2}{ds^2}\left(\frac{a}{s^2+a^2}\right)=\frac{2a(3s^2-a^2)}{(s^2+a^2)^3}$$

(ii) We know that

$$\cos^2 x=\frac{1}{2}(1+\cos 2x)$$

Taking the Laplace transform of both sides, we get

$$L[\cos^2 x]=\frac{1}{2}(L[1]+L[\cos 2x])=\frac{1}{2}\left(\frac{1}{s}+\frac{s}{s^2+2^2}\right)=\frac{s^2+2}{s(s^2+4)}, s>0$$

Thus, using Theorem 8.3, we have

$$L[x\cos^2 x]=-\frac{d}{ds}\left[\frac{s^2+2}{s(s^2+4)}\right]=\frac{s^4+2s^2+8}{s^2(s^2+4)^2}$$

(iii) We have

$$L[\cos ax - \cos bx] = L[\cos ax] - L[\cos bx] = \frac{s}{s^2+a^2} - \frac{s}{s^2+b^2}$$

Thus, from Theorem 8.5, we have

$$L\left[\frac{\cos ax - \cos bx}{x}\right] = \int_s^\infty \bar{f}(s)ds$$

$$= \int_s^\infty \left[\frac{s}{s^2+a^2} - \frac{s}{s^2+b^2}\right] ds$$

$$= \left[\frac{1}{2}\log(s^2+a^2) - \frac{1}{2}\log(s^2+b^2)\right]_s^\infty$$

$$= \frac{1}{2}\left[\log\frac{s^2+a^2}{s^2+b^2}\right]_s^\infty$$

$$= \frac{1}{2}\log\frac{s^2+b^2}{s^2+a^2}$$

(iv) We know that

$$\sin^2 x = \frac{1}{2}(1-\cos 2x)$$

Taking the Laplace transform of both sides of this equation, we get

$$L[\sin^2 x] = \frac{1}{2}\left[\frac{1}{s} - \frac{s}{s^2+2^2}\right]$$

Thus, using Theorem 8.5, we get

$$L\left[\frac{\sin^2 x}{x}\right] = \frac{1}{2}\int_s^\infty \left[\frac{1}{s} - \frac{s}{s^2+2^2}\right] ds$$

$$= \frac{1}{2}\left[\log s - \frac{1}{2}\log(s^2+4)\right]_s^\infty$$

$$= \frac{1}{4}\left[\log\frac{s^2}{s^2+4}\right]_s^\infty$$

$$= \frac{1}{4}\log\left(\frac{s^2+4}{s^2}\right)$$

**EXAMPLE 8.41** Find the Laplace transform of triangular wave defined by (see also Fig. 8.13)

$$f(x)=\begin{cases}\dfrac{x}{c}, & 0<x<c\\[2ex] \dfrac{2c-x}{c}, & c<x<2c\end{cases}$$

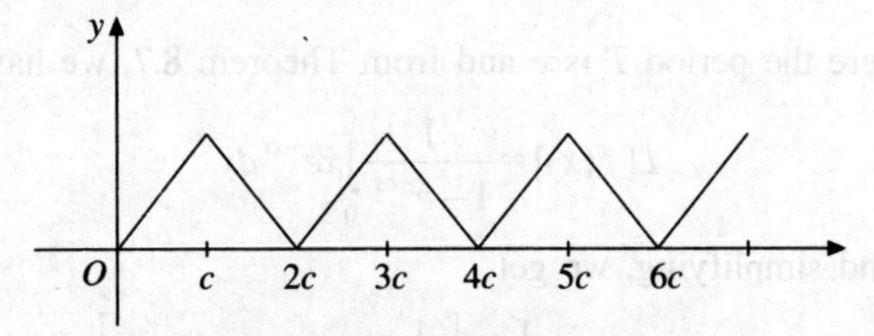


**Fig. 8.13** Triangular wave.

***Solution*** Here, the period is $2c$ and thus using Theorem 8.7, we have

$$L[f(x)]=\frac{1}{1-e^{-2cs}}\left[\int_0^c \frac{x}{c}e^{-sx}+\int_c^{2c}\left(\frac{2c-x}{c}\right)e^{-sx}\right]dx$$

which after simplification leads to

$$L[f(x)]=\frac{1}{1-e^{-2cs}}\left[\frac{1}{cs^2}(1-2e^{-cs}+e^{-2cs})\right]$$

$$=\frac{1}{cs^2(1-e^{-2cs})}(1-e^{-cs})^2$$

$$=\frac{1-e^{-cs}}{cs^2(1+e^{-cs})}$$

$$=\frac{e^{cs/2}-e^{-cs/2}}{cs^2(e^{cs/2}+e^{-cs/2})}$$

$$=\frac{1}{cs^2}\tanh\left(\frac{cs}{2}\right),\ s>0$$

**EXAMPLE 8.42** Find the Laplace transform of saw-toothed wave defined by $f(x)=x,\ 0\le x\le c$ (see also Fig. 8.14)

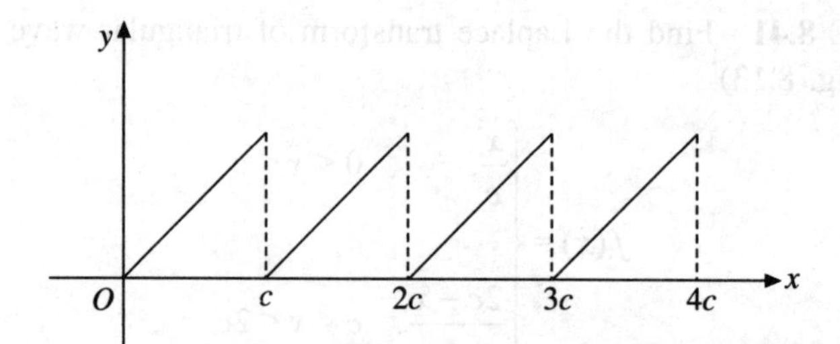


**Fig. 8.14** Saw-toothed wave.

***Solution*** Here the period $T$ is $c$ and from Theorem 8.7, we have

$$L[f(x)] = \frac{1}{1-e^{-cs}} \int_0^c x e^{-sx}\, dx$$

Integrating and simplifying, we get

$$L[f(x)] = \frac{1}{1-e^{-cs}} \left[ \frac{1}{s^2}(1-e^{-cs}) - \frac{c}{s} e^{-cs} \right]$$

$$= \frac{1}{s^2} - \frac{ce^{-cs}}{s(1-e^{-cs})}, \; s > 0$$

**EXAMPLE 8.43** Find the Laplace transform of the square wave function defined by

$$f(x) = \begin{cases} K, & 0 \le x < c \\ -K, & c \le x < 2c \end{cases}$$

where $K$ is a constant.

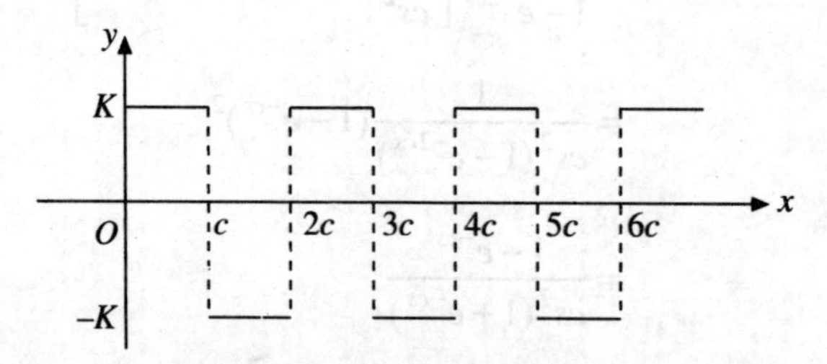


**Fig. 8.15** Square wave.

***Solution*** Here, the period $T$ is $2c$ (see Fig. 8.15) and from Theorem 8.7, we have

$$L[f(x)] = \frac{1}{1-e^{-2cs}} \left[ \int_0^c e^{-sx} K dx + \int_c^{2c} e^{-sx} (-K) dx \right]$$

Integrating and simplifying, we get

$$L[f(x)] = \frac{K}{s}\frac{e^{cs/2} - e^{-cs/2}}{e^{cs/2} + e^{-cs/2}} = \frac{K}{s}\tanh\left(\frac{cs}{2}\right), \; s > 0$$

**EXAMPLE 8.44** Find the inverse Laplace transform of the following functions

(i) $\dfrac{s^2}{(s+1)(s+2)(s+3)}$ (ii) $\dfrac{1}{s(s^2+9)}$ (iii) $\dfrac{s}{(s^2+a^2)(s^2+b^2)}$

***Solution*** (i) Here

$$\frac{s^2}{(s+1)(s+2)(s+3)} = \frac{A}{s+1} + \frac{B}{s+2} + \frac{C}{s+3} \tag{49}$$

which on solving leads to $A = 1/2$, $B = -4$, $C = 9/2$ Using these values in Eq. (49) and taking the inverse Laplace transform, we get

$$L^{-1}\left[\frac{s^2}{(s+1)(s+2)(s+3)}\right] = \frac{1}{2}L^{-1}\left[\frac{1}{s+1}\right] - 4L^{-1}\left[\frac{1}{s+2}\right] + \frac{9}{2}L^{-1}\left[\frac{1}{s+3}\right]$$

$$= \frac{1}{2}e^{-x} - 4e^{-2x} + \frac{9}{2}e^{-3x}$$

(ii) Here

$$\frac{1}{s(s^2+9)} = \frac{A}{s} + \frac{Bs+C}{s^2+9} \tag{50}$$

Solving Eq. (50), we get $A = 1/9, B = -1/9, C = 0.$ Using these values in Eq. (50) and taking the inverse Laplace transform of both sides of the resulting equation, we get

$$L^{-1}\left[\frac{1}{s(s^2+9)}\right] = \frac{1}{9}L^{-1}\left[\frac{1}{s}\right] - \frac{1}{9}L^{-1}\left[\frac{s}{s^2+9}\right] = \frac{1}{9}(1 - \cos 3x)$$

(iii) Here

$$\frac{s}{(s^2+a^2)(s^2+b^2)} = \frac{As+B}{s^2+a^2} + \frac{Cs+D}{s^2+b^2} \tag{51}$$

Solving Eq. (51), we get $A = -1/(a^2 - b^2), B = 0, C = 1/(a^2 - b^2), D = 0.$ Put these values in Eq. (51) and take the inverse Laplace transform of the resulting equation, we get

$$L^{-1}\left[\frac{s}{(s^2+a^2)(s^2+b^2)}\right] = -\frac{1}{a^2-b^2}L^{-1}\left[\frac{s}{s^2+a^2}\right] + \frac{1}{a^2-b^2}L^{-1}\left[\frac{s}{s^2+b^2}\right]$$

$$= \frac{1}{a^2-b^2}[\cos bx - \cos ax]$$

**EXAMPLE 8.45** Using the Laplace transform of derivatives, find the inverse Laplace transform of the following:

(i) $\dfrac{s}{(s^2+a^2)^2}$ (ii) $\dfrac{s+1}{(s^2+2s+2)^2}$ (iii) $\log\left(\dfrac{1+s}{s}\right)$

***Solution*** (i) Put $\bar{f}(s)=\dfrac{1}{s^2+a^2}$, then $\dfrac{d\bar{f}(s)}{ds}=-\dfrac{2s}{(s^2+a^2)^2}$ and from Theorem 8.3, we have

$$L^{-1}\left[\frac{s}{(s^2+a^2)^2}\right]=-\frac{1}{2}L^{-1}\left[\frac{d}{ds}\left(\frac{1}{s^2+a^2}\right)\right]=-\frac{1}{2}(-1)^1 xL^{-1}\left(\frac{1}{s^2+a^2}\right)$$

$$=\frac{1}{2}\frac{x}{a}\sin ax$$

(ii) Put $\bar{f}(s)=\dfrac{1}{s^2+2s+2}$, then $\dfrac{d\bar{f}(s)}{ds}=-2\left[\dfrac{s+1}{(s^2+2s+2)^2}\right]$ which leads to

$$\frac{s+1}{(s^2+2s+2)^2}=-\frac{1}{2}\frac{d\bar{f}(s)}{ds}=-\frac{1}{2}\frac{d}{ds}\left[\frac{1}{s^2+2s+2}\right] \tag{52}$$

Also, from Theorem 8.3, we have

$$L^{-1}\left[\frac{d^n}{ds^n}\bar{f}(s)\right]=(-1)^n x^n f(x)=(-1)^n x^n L^{-1}[\bar{f}(s)] \tag{53}$$

Thus, from Eqs. (52) and (53), we get

$$L^{-1}\left[\frac{s+1}{(s^2+2s+2)^2}\right]=-\frac{1}{2}L^{-1}\left[\frac{d}{ds}\left(\frac{1}{s^2+2s+2}\right)\right]$$

$$=-\frac{1}{2}(-1)^1 xL^{-1}\left[\frac{1}{s^2+2s+2}\right]$$

$$=\frac{1}{2}xL^{-1}\left[\frac{1}{(s+1)^2+1^2}\right]$$

$$=\frac{1}{2}xe^{-x}\sin x \qquad \text{(from Table 8.1)}$$

(iii) Put $\bar{f}(s)=\log\dfrac{s+1}{s}=\log(s+1)-\log s$ then $\dfrac{d\bar{f}(s)}{ds}=\dfrac{1}{s+1}-\dfrac{1}{s}$. Now taking the inverse Laplace transform of both sides, we get

$$L^{-1}\left[\frac{d\bar{f}(s)}{ds}\right]=L^{-1}\left[\frac{1}{s+1}\right]-L^{-1}\left[\frac{1}{s}\right]$$

or
$$(-1)^1 xL^{-1}[\bar{f}(s)]=e^{-x}-1$$

or
$$L^{-1}[\log\frac{s+1}{s}] = \frac{1-e^{-x}}{x}$$
(see also Example 8.7).

**EXAMPLE 8.46** Using theorems on integral transform, find the inverse Laplace transform of the following:

(i) $\dfrac{1}{s^3(s^2+1)}$, (ii) $\dfrac{s+2}{s^2(s+3)}$

***Solution*** (i) We know that

$$L^{-1}\left[\frac{1}{s^2+1}\right] = \sin x$$

Thus, from Theorem 8.4, we have

$$L^{-1}\left[\frac{1}{s(s^2+1)}\right] = \int_0^x \sin u du = [-\cos u]_0^x = 1 - \cos x$$

which leads to

$$L^{-1}\left[\frac{1}{s^2(s^2+1)}\right] = \int_0^x (1-\cos u) du = x - \sin x$$

and finally, we have

$$L^{-1}\left[\frac{1}{s^3(s^2+1)}\right] = \int_0^x (u - \sin u) du = \frac{x^2}{2} + \cos x - 1$$

(ii) The given fraction can be expressed as

$$\frac{s+2}{s^2(s+3)} = \frac{(s+3)-1}{s^2(s+3)}$$

$$= \frac{s+3}{s^2(s+3)} - \frac{1}{s^2(s+3)}$$

$$= \frac{1}{s^2} - \frac{1}{s^2(s+3)} \tag{54}$$

Taking the inverse Laplace transform of Eq. (54), we get

$$L^{-1}\left[\frac{s+2}{s^2(s+3)}\right] = L^{-1}\left[\frac{1}{s^2}\right] - L^{-1}\left[\frac{1}{s^2(s+3)}\right] \tag{55}$$

But
$$L^{-1}\left[\frac{1}{s^2}\right] = x^2 \tag{56}$$

and
$$L^{-1}\left[\frac{1}{s+3}\right] = e^{-3x}$$

Now using Theorem 8.4, we have

$$L^{-1}\left[\frac{1}{s(s+3)}\right]=\int_0^x e^{-3u}du=\frac{1}{3}(1-e^{-3x})$$

Applying Theorem 8.4 again to this equation, we get

$$L^{-1}\left[\frac{1}{s^2(s+3)}\right]=\frac{1}{3}\int_0^x(1-e^{-3u})du=\frac{1}{9}(3x+e^{-3x}-1) \tag{57}$$

Thus from Eqs. (56) and (57), Eq. (55) reduces to

$$L^{-1}\left[\frac{s+2}{s^2(s+3)}\right]=x-\frac{1}{9}(3x+e^{-3x}-1)=\frac{1}{9}(6x-e^{-3x}+1)$$

**EXAMPLE 8.47** Using convolution theorem (Theorem 8.6), find the inverse Laplace transform of the following:

(i) $\dfrac{s}{(s^2+a^2)^2}$ (ii) $\dfrac{1}{s^2(s+1)^2}$

***Solution*** (i) Here

$$\frac{s}{(s^2+a^2)^2}=\frac{s}{s^2+a^2}\cdot\frac{1}{s^2+a^2}$$

and since $L^{-1}\left[\dfrac{s}{s^2+a^2}\right]=\cos ax$ and $L^{-1}\left[\dfrac{1}{s^2+a^2}\right]=\dfrac{1}{a}\sin ax$. Therefore, using convolution Theorem 8.6, we have

$$\begin{aligned}
L^{-1}\left[\frac{s}{(s^2+a^2)^2}\right]&=L^{-1}\left[\frac{s}{s^2+a^2}\cdot\frac{1}{s^2+a^2}\right]\\
&=\int_0^x \cos au\frac{\sin a(x-u)}{a}du\\
&=\frac{1}{a}\int_0^x \cos au(\sin ax\cos au-\cos ax\sin au)du\\
&=\frac{1}{a}\int_0^x(\sin ax\cos^2 au-\cos ax\sin au\cos au)du\\
&=\frac{1}{a}\left[\sin ax\int_0^x\cos^2 au\,du-\cos ax\int_0^x\sin au\cos au\,du\right]\\
&=\frac{1}{a}\left[\sin ax\int_0^x\left(\frac{1+\cos 2au}{2}\right)du-\cos ax\int_0^x\frac{\sin 2au}{2}du\right]\\
&=\frac{1}{2a}x\sin ax
\end{aligned}$$

(after integrating and simplification)

(ii) We know that $L^{-1}\left[\dfrac{1}{s^2}\right]=x$ and $L^{-1}\left[\dfrac{1}{(s+1)^2}\right]=xe^{-x}$. Thus, by convolution theorem, we have

$$L^{-1}\left[\frac{1}{s^2(s+1)^2}\right]=L^{-1}\left[\frac{1}{(s+1)^2}\cdot\frac{1}{s^2}\right]$$

$$=\int_0^x (ue^{-u})(x-u)du$$

$$=\int_0^x (ux-u^2)e^{-u}du$$

$$=xe^{-x}+2e^{-x}+x-2$$

(after integrating by parts and simplification).

**EXAMPLE 8.48** Using the Laplace transform method, solve the differential equation $xy''+y'+4xy=0$ subject to the conditions that $y(0)=3$ and $y'(0)=0$.

***Solution*** Taking the Laplace transform of the given differential equation, we get

$$L[xy'']+L[y']+4L[xy]=0$$

Using Thorem 8.3 in the first and last terms of this equation, we get

$$(-1)^1\frac{d}{ds}(L[y''])+L[y']+4(-1)^1\frac{d}{ds}(L[y])=0$$

which from Theorem 8.2 leads to

$$-\frac{d}{ds}\{s^2L[y]-sy(0)-y'(0)\}+sL[y]-y(0)-4\frac{d}{ds}L[y]=0$$

From the given conditions this equation reduces to

$$\frac{d}{ds}L[y]+\frac{1}{2}\frac{2s}{s^2+4}L[y]=0$$

Put $L[y]=\bar{y}$ in this equation to get

$$\frac{d\bar{y}}{ds}+\frac{1}{2}\frac{2s}{s^2+4}\bar{y}=0$$

Separating the variables, this equation reduces to

$$\frac{d\bar{y}}{\bar{y}}+\frac{1}{2}\frac{2s}{s^2+4}ds=0$$

which on integration leads to

$$\log \bar{y} + \frac{1}{2}\log(s^2+4) = \log c_1$$

or

$$\log[\bar{y}(s^2+4)^{1/2}] = \log c_1$$

or

$$\bar{y} = L[y] = \frac{c_1}{(s^2+4)^{1/2}}$$

Thus

$$y = c_1 L^{-1}\left[\frac{1}{(s^2+2^2)^{1/2}}\right] = c_1 J_0(2x) \tag{58}$$

[using Eq. (44) for $J_0(2x)$].

The value of the constant $c_1$ can be obtained by using the given condition $y(0) = 3$ in Eq. (58) and we have $3 = c_1 J_0(0)$. But from Eq. (193b) (Chapter 5), $J_0(0) = 1$ and thus $c_1 = 3$. Putting it in Eq. (58), we get

$$y = 3J_0(2x)$$

as the required solution.

**EXAMPLE 8.49** Using Laplace transform method, solve

$$xy'' + (1-2x)y' - 2y = 0$$

such that $y(0) = 1$ and $y'(0) = 2$.

***Solution*** Taking the Laplace transform of the given differential equation, we have

$$L[xy''] + L[y'] - 2L[xy'] - 2L[y] = 0$$

Using Theorem 8.3 in the first and third terms of this equation, we get

$$(-1)^1 \frac{d}{ds}(L[y'']) + L[y'] - 2(-1)^1 \frac{d}{ds}(L[y']) - 2L[y] = 0$$

which on using Theorem 8.2, leads to

$$-\frac{d}{ds}\{s^2 L[y] - sy(0) - y'(0)\} + sL[y] - y(0) + 2\frac{d}{ds}\{sL[y] - y(0)\} - 2L[y] = 0$$

Applying the given conditions and taking $L[y] = \bar{y}$, this equation reduces to

$$-\frac{d}{ds}(s^2\bar{y} - s - 2) + s\bar{y} - 1 + 2\frac{d}{ds}(s\bar{y} - 1) + 2\bar{y} = 0$$

Simplifying this equation, we get

$$(-s^2 + 2s)\frac{d\bar{y}}{ds} = s\bar{y}$$

Now separating the variables and integrating, we get

$$L[y] = \bar{y} = -\frac{c_1}{s-2}$$

Thus

$$y = -c_1 L^{-1}\left[\frac{1}{s-2}\right] = -c_1 e^{2x}$$

Using the given condition $y(0) = 3$ in this equation, we get $c_1 = -1$ and therefore the required solution is

$$y = e^{2x}$$

**EXAMPLE 8.50** Using Laplace transform method, solve the differential equation

$$y'' + xy' - 2y = 6 - x$$

subject to the conditions that $y(0) = 0$ and $y'(0) = 1$.

***Solution*** Taking the Laplace transform of the given equation, we get

$$L[y''] + L[xy'] - 2L[y] = L[6] - L[x]$$

which on using Theorem 8.3 in the second term reduces to

$$L[y''] + (-1)^1 \frac{d}{ds} L[y'] - 2L[y] = L[6] - L[x]$$

Now using Theorem 8.2, this equation becomes

$$s^2 L[y] - sy(0) - y'(0) - \frac{d}{ds}\{sL[y] - y(0)\} - 2L[y] = \frac{6}{s} - \frac{1}{s^2}$$

Taking $L[y] = \bar{y}$ and applying the given conditions, this equation after simplification reduces to

$$\frac{d\bar{y}}{ds} + \left(\frac{3}{s} - s\right)\bar{y} = -\left(\frac{1}{s} + \frac{6}{s^2} - \frac{1}{s^3}\right)$$

which is a linear differential equation of order one and the solution (after simplification) is

$$\bar{y} = L[y] = \frac{6}{s^3} + \frac{1}{s^2} + \frac{c_1}{s^3} e^{s^2/2} \tag{59}$$

It may be noted that if $L[f(x)] = \bar{f}(s)$, then $\lim_{s\to\infty} \bar{f}(s) = 0$. Thus, from Eq. (59), the requirement that $\lim_{s\to\infty} \bar{y} = 0$ leads to $c_1 = 0$. Therefore, after taking the inverse Laplace transform, Eq. (59) reduces to

$$y = 3x^2 + x$$

which is the required solution.

**EXAMPLE 8.51** Solve $y'' - xy' + y = 1$ when $y(0) = 1, y'(0) = 2$.

***Solution*** Taking the Laplace transform of both sides of the given differential equation, we get

$$L[y''] - L[xy'] + L[y] = L[1]$$

which on using Theorem 8.3 in the second term, reduces to

$$L[y''] - (-1)^1 \frac{d}{ds} L[y'] + L[y] = L[1]$$

Now using Theorem 8.2, this equation leads to

$$s^2 L[y] - sy(0) - y'(0) + \frac{d}{ds}\{sL[y] - y(0)\} + L[y] = \frac{1}{s}$$

Taking $L[y] = \bar{y}$ and using the given conditions, this equation yields

$$s^2 \bar{y} - s - 2 + \frac{d}{ds}\{s\bar{y} - 1\} + \bar{y} = \frac{1}{s}$$

which on simplification reduces to

$$\frac{d\bar{y}}{ds} + \left(s + \frac{2}{s}\right)\bar{y} = \frac{2}{s} + \frac{1}{s^2} + 1 \tag{60}$$

Equation (60) is a linear differential equation of order one. Here, I.F. $= s^2 e^{s^2/2}$ and the solution is

$$\bar{y} = \frac{1}{s} + \frac{2}{s^2} + \frac{c_1}{s^2} e^{-s^2/2}$$

which can also be expressed as

$$\bar{y} = \frac{1}{s} + \frac{2}{s^2} + \frac{c_1}{s^2}\left[1 - \frac{s^2}{2} + \frac{s^4}{8} + \cdots\right]$$

or
$$\bar{y} = \frac{1}{s} + \frac{2}{s^2} + \frac{c_1}{s^2} + \frac{c_1}{s^2}\left[-\frac{s^2}{2} + \frac{s^4}{8} + \cdots\right]$$

or
$$\bar{y} = L[y] = \frac{1}{s} + \frac{1}{s^2}(2 + c_1) + c_1\left[-\frac{1}{2} + \frac{s^2}{8} + \cdots\right]$$

Taking the inverse Laplace transform of this equation, we get

$$y = 1 + (2 + c_1)x \tag{61}$$

(as $L^{-1}[s^k] = 0, k = 0, 1, 2, \ldots$).

To obtain the value of $c_1$, differentiate Eq. (61) and use the given condition $y'(0) = 2$, we get $c_1 = 0$. Thus, from Eq. (61), the required solution is

$$y = 1 + 2x$$

## EXERCISES

Find the Laplace transforms of the following functions:

**1.** $e^{2x} + 4x^3 - 2\sin 3x + 3\cos 3x.$ **2.** $\cos(ax + b).$

**3.** $\sin ax \sin bx.$ **4.** $\sin^2(3x)$

**5.** $x^3 e^{-3x}.$ **6.** $e^{-2x}\sin 4x.$

**7.** $\cosh ax \sin ax.$ **8.** $xe^{-4x}\sin ax.$

**9.** $f(x) = \sin x, \quad 0 < x < \pi$
$= 0, \quad x > \pi$

**10.** $\dfrac{\cos\sqrt{x}}{x}.$

**11.** Find $L(e^{-ax})$ and use it to find $L(\sin \omega x)$.

**12.** Show that

(a) $L[e^{-ax}J_0(bx)] = \dfrac{1}{\sqrt{s^2 - 2as + a^2 + b^2}}.$

(b) $L[J_0(a\sqrt{x})] = \dfrac{1}{s}e^{-a^2/4s}.$

Find the inverse Laplace transforms of the following:

**13.** $\dfrac{3(s^2-2)^2}{2s^5}.$ **14.** $\dfrac{s-a}{(s-a)^2+b^2}.$ **15.** $\dfrac{s^2}{s^2+4s+8}.$

**16.** $\dfrac{3s+7}{s^2-2s-3}.$ **17.** $\dfrac{1+2s}{(s+2)^2(s-1)^2}.$ **18.** $\dfrac{1}{s(s^2+4)}.$

**19.** $\dfrac{s}{(s+1)^2(s^2+1)}.$ **20.** $\dfrac{1}{(s+1)(s^2+2s+2)}.$ **21.** $\dfrac{s}{s^4+s^2+1}.$

**22.** $\dfrac{a(s^2-2a^2)}{s^4+4a^4}.$ **23.** $\dfrac{2s^2-4}{(s+1)(s-2)(s-3)}.$ **24.** $\dfrac{5s^2-15s-11}{(s+1)(s-2)^3}.$

**25.** $\dfrac{3s+1}{(s-1)(s^2+1)}.$ **26.** $\dfrac{s^2+2s+3}{(s^2+2s+2)(s^2+2s+5)}$

**27.** If $L[2\sqrt{x/\pi}] = 1/s^{3/2}$, show that $1/\sqrt{s} = L[1/\sqrt{\pi x}]$.

**28.** Use Theorem 8.3 to find the universe Laplace transform of $\dfrac{1}{(s+5)^4}.$

Using Theorem 8.4, find the inverse transforms of:

**29.** $\dfrac{1}{s^2(s+2)}.$ **30.** $\dfrac{1}{s^3(s^2+a^2)}.$ **31.** $\dfrac{s}{(s+a)^2}.$

**32.** Find $L(\cos ax)$ and deduce from it $L(\sin ax)$.

**33.** Find from definition $L(e^{ax})$ and deduce $L(xe^{ax})$.

Find the Laplace transforms of the following:

**34.** $x \sin^2(3x)$.

**35.** $x^2 \cos ax$.

**36.** $(e^{-ax} - e^{-bx})/x$.

**37.** $(\sin 2x)/x$.

**38.** Calculate the inverse transform of $\log (s + 1)/s$.

Solve the following differential equations by transform methods:

**39.** $(D^2 - 2D + 2)x = 0, \quad x = Dx = 1$ at $t = 0$.

**40.** $\dfrac{d^3y}{dt^3} + 2\dfrac{d^2y}{dt^2} - \dfrac{dy}{dt} - 2y = 0$, when $y = 1$, $dy/dt = 2 = d^2y/dt^2$ at $t = 0$.

**41.** $(D^3 + D)x = 2$, $x = 3$, $Dx = 1$, $D^2x = -2$ at $t = 0$.

**42.** $(D^2 - 3D + 2)x = 1 - e^{2t}$, $x = 1$, $Dx = 0$ at $t = 0$.

**43.** $(D^2 - 1)x = a \cosh t$, $x(0) = x'(0) = 0$.

**44.** $y'' + y' - 2y = t, \quad y(0) = 1, \quad y'(0) = 0$.

**45.** $y'' - 3y' + 2y = 4t + e^{3t}, \quad y(0) = 1$, $y'(0) = 1$.

**46.** $(D^2 + \omega^2)y = \cos \omega t$, $t > 0$, $y(0) = 0$, $y'(0) = 0$.

**47.** $\dfrac{d^2x}{dt^2} + 9x = \cos 2t$, $x(0) = 1$, $x(\pi/2) = -1$.

**48.** $\dfrac{d^2y}{dx^2} + 4\dfrac{dy}{dx} + 8y = \cos 2x$, $y(0) = 2$, $y'(0) = 1$.

**49.** $(D^2 - 2D + 2)(D^2 + 2D - 3)x = 0$, $x(0) = 0$, $x'(0) = 0$, $x''(0) = 6$, $x'''(0) = -14$.

**50.** $y'' + xy' - y = 0$, $y(0) = 0$, $y'(0) = 1$.

**51.** $xy'' + (x - 1)y' - y = 0$, $y(0) = 5$, $y'(\infty) = 0$.

**52.** $xy'' + 2xy' + 2y = 2$, $y(0) = 1$, $y'(0) =$ arbitrary.

Solve the following simultaneous equations using the transform method:

**53.** $\dfrac{dx}{dt} - y = e^t, \dfrac{dy}{dt} + x = \sin t$, $x(0) = 1$, $y(0) = 0$.

**54.** $\dfrac{dx}{dt} - \dfrac{dy}{dt} - 2x + 2y = 1 - 2t, \quad \dfrac{d^2x}{dt^2} + 2\dfrac{dy}{dt} + x = 0$

where $x = 0$, $y = 0$, $dx/dt = 0$ at $t = 0$.

**55.** $dy/dt + 2x = \sin 2t$, $dx/dt - 2y = \cos 2t$, $(t > 0)$ and $x(0) = 1$, $y(0) = 0$.

**56.** $di_1/dt - wi_2 = a\cos pt,\ di_2/dt + wi_1 = a\sin pt$, where $i_1(0) = i_2(0) = 0$.

**57.** $\dfrac{dx}{dt} = x - 4y + z, \dfrac{dy}{dt} = -2y + z, \dfrac{dz}{dt} = 4z$, where $x(0) = 0,\ y(0) = 2,\ z(0) = 3$.

Apply the convolution theorem to evaluate:

**58.** $L^{-1}\left[\dfrac{1}{s^2(s^2+a^2)}\right]$. **59.** $L^{-1}\left[\dfrac{s^2}{(s^2+4)^2}\right]$. **60.** $L^{-1}\left[\dfrac{s^2}{(s^2+a^2)(s^2+b^2)}\right]$.

**61.** $L^{-1}\left[\dfrac{1}{(s+3)(s-1)}\right]$. **62.** $L^{-1}\left[\dfrac{1}{(s+2)^2(s-2)}\right]$.

**63.** $L^{-1}\left[\dfrac{1}{(s+1)(s^2+1)}\right]$. **64.** $L^{-1}\left[\dfrac{s}{(s^2+4)^3}\right]$.

**65.** Find the Laplace transform of the *square-wave* function of period $a$ defined as

$$f(x) = 1, \quad 0 < x < a/2$$
$$= -1, \quad a/2 < x < a.$$

**66.** Find the Laplace transform of the *saw-toothed wave* of period $T$, given by:

$$f(x) = x/T \quad \text{for } 0 < x < T.$$

**67.** Find the Laplace transform of the *triangular wave* of period $2a$ given by:

$$f(x) = x, \quad 0 < x < a$$
$$= 2a - x, \quad a < x < 2a.$$

**68.** Find the Laplace transform of a half wave rectifier sine curve defined as

$$f(x) = \sin x, \quad 0 < x < \pi$$
$$= 0, \quad \pi < x < 2\pi.$$

**69.** Find the Laplace transform of the *full-wave rectifier*:

$$f(x) = E\sin \omega x,\ 0 < x < \pi/\omega, \text{ period} = \frac{\pi}{\omega}.$$

**70.** Solve

$$\frac{d^2y}{dx^2} + (1 - 2x)\frac{dy}{dx} - 2y = 0, \quad y(0) = 1, \quad y'(0) = 2.$$

**71.** Solve

$$\frac{d^2x}{dt^2} - t\left(\frac{dx}{dt}\right) + x = 1, \quad x(0) = 1,\ x'(0) = 2.$$

**72.** A 64 lb body falls from rest under the influence of gravity. The resistance due to air is $R = 8V$, where $V$ is the velocity. Find $V$ in terms of $t$ (use $g = 32\ \text{ft/s}^2$).

**73.** A weight undergoes a simple harmonic motion governed by the equation $d^2x/dt^2 + 64x = 0$. If $x(0) = 1/3$ and $x'(0) = 2$, what is $x$ in terms of $t$? Also find the amplitude, period and frequency of the motion.

**74.** A 4-lb weight stretches a spring 2 ft. The weight is released from rest 18 in. above the equilibrium position and the resulting motion takes place in a medium offering a damping force equal to 7/8 times the velocity. Use Laplace transforms to determine the equation of motion.

**75.** Obtain the equation for the forced oscillations of a mass $m$ attached to the lower end of an elastic spring whose upper end is fixed and the spring constant is $k$, when the deriving force is $F \sin at$. Solve this equation, using Laplace transforms when $a^2 \neq k/m$, given that the initial velocity and displacement (from the equilibrium position) are zero.

**76.** An impulse $I$ (kg/s) is applied to a mass $m$ attached to a spring having a spring constant $k$. The system is damped with damping constant $\mu$. Obtain the displacement and velocity of the mass, given that $x(0) = x'(0) = 0$.

**77.** A particle of mass $m$ is at rest at the origin $O$ on the $x$-axis. At $t = t_0$ it is acted upon by a force for a very short time where the impulse of the force is constant $P_0$. Solve the equation of motion.

**78.** A particle of mass $m$ can perform small oscillations about a position of equilibrium under a restoring force $mn^2$ times the displacement. It is started from rest by a constant force $F$ which acts for a time $T$ and then ceases. Show that the amplitude of the subsequent motion is $(2F/mn^2)$ $\sin(nT/2)$.

**79.** Solve Problem 35 of Chapter 6 Exercises, using the transform methods.

**80.** A cantilever beam is clamped at the end $x = 0$ and is free at the end $x = l$. It carries a uniform load of $W$ per unit length from $x = 0$ to $x = l/2$. Calculate the deflection $y$ at any point.

**81.** A voltage $Ee^{-at}$ is applied at $t = 0$ to a circuit of inductance $L$ and resistance $R$. Show by the transform method, that the current at time $t$ is $R/(R - aL)$ $(e^{-at} - e^{-Rt/L})$.

**82.** An inductor of 3 H and a 6 Ω resistor are connected in series with a generator having emf $150e^{-2t}\cos 25t$ V. Find the current $i$ in terms of $t$ if $i(0) = 20$ A.

**83.** For the network of Fig. 8.16, $i_1(0) = i_2(0) = i_3(0) = 0$. Find $i_1$, $i_2$, $i_3$ at time $t$ seconds after the switch is closed.

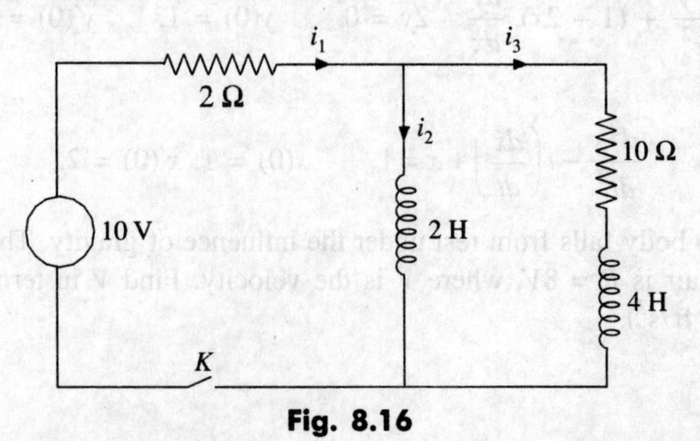


**Fig. 8.16**

**84.** Solve the equation $L(di/dt) + Ri = E(t)$ if $L = 1\,\text{H}$, $R = 10\ \Omega$, $i(0) = 0$, and

$$\begin{aligned} E(t) &= \sin t, \qquad 0 \le t < 3\pi/2 \\ &= 0 \qquad\qquad t \ge 3\pi/2. \end{aligned}$$

**85.** Determine the charge $q$ and current $i$ for a series circuit in which $L = 1$ H, $R = 20\ \Omega$, $C = 0.01$ F, $E(t) = 120 \sin 10t$, $q(0) = 0$, $i(0) = 0$. What is the steady state current?

CHAPTER 9

# Partial Differential Equations and Their Applications

## 9.1 INTRODUCTION

The notion of partial differential equations has already been introduced in Section 1.2.1. These equations play a crucial role in most sciences dealing with wave motion, for example, heat, light, electromagnetism, radar, radio, television and weather. In this chapter, we shall give a brief theory of partial differential equations and their applications.

Let $x$ and $y$ represent the independent variables and $z$ the dependent variable so that $z = f(x, y)$. We will also use the notations

$$\frac{\partial z}{\partial x} = p, \qquad \frac{\partial z}{\partial y} = q, \qquad \frac{\partial^2 z}{\partial x^2} = r, \qquad \frac{\partial^2 z}{\partial x \partial y} = s, \qquad \frac{\partial^2 z}{\partial y^2} = t$$

## 9.2 FORMATION AND SOLUTION OF PARTIAL DIFFERENTIAL EQUATIONS

In Chapter 1, we have seen how an ordinary differential equation was obtained by the elimination of arbitrary constants; but unlike ordinary differential equation, a partial differential equation can be obtained either by the elimination of arbitrary constants or by the elimination of arbitrary functions involving two or more variables. We illustrate the method with the help of the following examples:

**EXAMPLE 9.1** By eliminating the constant, obtain the partial differential equation from the relation

$$2z = \frac{x^2}{a^2} + \frac{y^2}{b^2} \tag{1}$$

***Solution*** Differentiate the given equation partially with respect to $x$ and $y$ to obtain

$$\frac{1}{a^2} = \frac{1}{x}\frac{\partial z}{\partial x} = \frac{p}{x}, \qquad \frac{1}{b^2} = \frac{1}{y}\frac{\partial z}{\partial y} = \frac{q}{y}$$

Substituting these values of $1/a^2$ and $1/b^2$ in Eq. (1), we obtain the required partial differential equation as

$$2z = px + qy$$

**EXAMPLE 9.2** Form a partial differential equation by eliminating the constants $h$ and $k$ from $(x - h)^2 + (y - k)^2 + z^2 = c^2$.

***Solution*** Differentiating the given equation partially with respect to $x$ and $y$, we obtain

$$(x - h) + zp = 0 \quad \text{and} \quad (y - k) + zq = 0$$

Substituting the values of $(x - h)$ and $(y - k)$ from these equations in the given equation, we get

$$z^2(p^2 + q^2 + 1) = c^2$$

as the desired partial differential equation.

**EXAMPLE 9.3** By eliminating the arbitrary functions, obtain the partial differential equations from (a) $z = f(x^2 + y^2)$, (b) $z = f(x + ct) + g(x - ct)$.

***Solution*** (a) Differentiate partially with respect to $x$ and $y$ to get

$$\frac{\partial z}{\partial x} = p = f'(x^2 + y^2)\, 2x$$

$$\frac{\partial z}{\partial y} = q = f'(x^2 + y^2)\, 2y$$

On dividing, we obtain $p/q = x/y$ or $yp - xq = 0$ as the required differential equation.

(b) Differentiate the given equation partially with respect to $x$ and $t$, to obtain

$$\frac{\partial z}{\partial x} = f'(x + ct) + g'(x - ct)$$

$$\frac{\partial^2 z}{\partial x^2} = f''(x + ct) + g''(x - ct)$$

$$\frac{\partial z}{\partial t} = cf'(x + ct) - cg'(x - ct)$$

$$\frac{\partial^2 z}{\partial t^2} = c^2 f''(x + ct) + c^2 g''(x - ct) = c^2 \frac{\partial^2 z}{\partial x^2}$$

Thus, the required partial differential equation is

$$\frac{\partial^2 z}{\partial t^2} = c^2 \frac{\partial^2 z}{\partial x^2}$$

**EXAMPLE 9.4** By eliminating the arbitrary function from $z = e^{ny}\phi(x - y)$, obtain a partial differential equation.

***Solution*** Differentiating with respect to $y$, we get

$$q = ne^{ny}\,\phi(x - y) - e^{ny}\,\phi'(x - y)$$

and, therefore, $q = nz - p$, i.e. $p + q - nz = 0$ as the required equation.

From these examples, it is clear that a partial differential equation can result both from the elimination of arbitrary constants and from the elimination of arbitrary functions.

A *solution* of a partial differential equation is a non-differential relation between the variables which satisfies the equation. The solution

$$f(x, y, z, a, b) = 0 \tag{2}$$

of a first order partial differential equation which contains two arbitrary constants is called a *complete integral*, and the solution obtainable from a complete integral by assigning particular values to the constants is called the *particular solution.*

If we put $b = \phi(a)$ in Eq. (2), we find the envelope of the family of surfaces $f(x, y, z, \phi(a)) = 0$ and we get a solution having one arbitrary constant, which is called the *general solution.*

The envelope of the family of surfaces (2) with parameters $a$ and $b$, if it exists, is called a *singular integral* and it differs from the particular integral in that it is not obtainable from the complete integral by assigning particular values to the constants.

## 9.3 EQUATIONS EASILY INTEGRABLE

Some of the partial differential equations can be solved by direct integration and we shall illustrate the method by the following examples. Here, the usual constant of integration consists of an arbitrary function of the variable considered constant during the integration.

**EXAMPLE 9.5** Solve $\dfrac{\partial^3 z}{\partial x^2\,\partial y} + 18\,xy^2 + \sin(2x - y) = 0$.

***Solution*** Keep $y$ fixed and integrate the given equation twice to obtain

$$\frac{\partial^2 z}{\partial x \partial y} + 9x^2y^2 - \frac{1}{2}\cos(2x - y) = f(y)$$

$$\frac{\partial z}{\partial y} + 3x^3y^2 - \frac{1}{4}\sin(2x - y) = xf(y) + g(y)$$

Now keep $x$ fixed and integrate w.r.t. $y$, we get

$$z + x^3y^3 - \frac{1}{4}\cos(x - y) = x\int f(y)\,dy + \int g(y)\,dy + h(x)$$

Denoting $\int f(y)dy = u(y)$, $\int g(y)dy = v(y)$, we can write the required solution as

$$z = \frac{1}{4}\cos(2x - y) - x^3y^3 + xu(y) + v(y) + h(x)$$

where $u$, $v$ and $h$ are arbitrary functions.

**EXAMPLE 9.6** Solve $y\dfrac{\partial^2 z}{\partial x \partial y} + \dfrac{\partial z}{\partial x} = 4xy$.

***Solution*** The given equation can be written as

$$y\frac{\partial}{\partial y}\frac{\partial z}{\partial x} + \frac{\partial z}{\partial x} = 4xy$$

or
$$y\frac{\partial p}{\partial y} + p = 4xy$$

which is a linear differential equation of the first order if $x$ is taken as a constant. Its solution is

$$py = 2xy^2 + u(x)$$

or
$$y\frac{\partial z}{\partial x} = 2xy^2 + u(x)$$

Integrating it (keeping $y$ constant), we get

$$z = x^2y + \frac{1}{y}u_1(x) + v(x)$$

where $u_1(x) = \int u(x)dx$ and $v(x)$ are arbitrary functions.

**EXAMPLE 9.7** Solve $\dfrac{\partial^2 z}{\partial x^2} + z = 0$ when $x = 0$, $z = e^y$ and $\dfrac{\partial z}{\partial x} = 1$.

***Solution*** If $z$ is a function of $x$ alone, then the solution is $z = c_1 \sin x + c_2 \cos x$, where $c_1$ and $c_2$ are constants. Since, $z$ is a function of $x$ and $y$, $c_1$ and $c_2$ are functions of $y$. Hence, the solution of the given equation is

$$z = f(y)\sin x + g(y)\cos x$$

$$\frac{\partial z}{\partial x} = f(y)\cos x - g(y)\sin x$$

when

$$x = 0, z = e^y, \qquad \text{i.e. } e^y = g(y)$$

and

$$x = 0, \frac{\partial z}{\partial x} = 1, \qquad \text{i.e. } 1 = f(y)$$

Hence, the required solution is

$$z = \sin x + e^y \cos x$$

## 9.4 LINEAR EQUATIONS OF THE FIRST ORDER

A linear partial differential equation of the first order is of the form

$$Pp + Qq = R \tag{3}$$

where $P$, $Q$ and $R$ are functions of $x$, $y$, $z$. This equation is known as *Lagrange's linear partial differential equation* and the solution is $\phi(u, v) = 0$ or $u = f(v)$.

To obtain the solution of Eq. (3), we have the following rule.

(a) Write Eq. (3) in the form

$$\frac{dx}{P} = \frac{dy}{Q} = \frac{dz}{R}$$

(b) Solve these equations by the method of Section 2.12, giving $u = a$ and $v = b$ as its solutions.

(c) Write the solution as $\phi(u, v) = 0$, or $u = f(v)$

In the same way, we can obtain the solution of the linear partial differential equation involving more than two variables.

$$P_1 \frac{\partial z}{\partial x_1} + P_2 \frac{\partial z}{\partial x_2} + \cdots + P_n \frac{\partial z}{\partial x_n} = R \tag{4}$$

First find the equations

$$\frac{dx_1}{P_1} = \frac{dx_2}{P_2} = \ldots = \frac{dx_n}{P_n} = \frac{dz}{R} \tag{5}$$

and obtain an $n$ independent solution of Eq. (5). Let these solutions be

$$u_1 = c_1, u_2 = c_2, \ldots, u_n = c_n$$

Then $\phi(u_1, u_2, \ldots, u_n) = 0$ is the solution of Eq. (4), where $\phi$ is any arbitrary function. Equation (5) is called the *subsidiary equation.*

**EXAMPLE 9.8** Solve $(y + z)p + (x + z)q = x + y$.

***Solution*** The subsidiary equation is

$$\frac{dx}{y+z} = \frac{dy}{x+z} = \frac{dz}{x+y}$$

which can be written as

$$\frac{dx+dy+dz}{2x+2y+2z}=\frac{dx-dy}{-(x-y)}=\frac{dx-dz}{-(x-z)}$$

The solutions of these equations are

$$\log(x+y+z)+2\log(x-y)=a_1$$

or
$$(x+y+z)(x-y)^2=c_1$$

and
$$\log(x-y)-\log(x-z)=a_2$$

or
$$x-y=c_2(x-z)$$

Hence, the required solution is

$$\phi\left[(x+y+z)(x-y)^2,\frac{x-y}{x-z}\right]=0$$

or
$$x-y=(x-z)\,f[(x+y+z)(x-y)^2]$$

**EXAMPLE 9.9** Solve $(mz-ny)p+(nx-lz)\,q=ly-mx$.

***Solution*** The subsidiary equation is

$$\frac{dx}{mz-ny}=\frac{dy}{nx-lz}=\frac{dz}{ly-mx}$$

Using multipliers $x$, $y$, $z$, we have

$$\text{each fraction}=\frac{xdx+ydy+zdz}{0}$$

or
$$xdx+ydy+zdz=0$$

which gives

$$x^2+y^2+z^2=a$$

Again, using multipliers $l$, $m$, $n$, we get

$$ldx+mdy+ndz=0$$

which on integration yields

$$lx+my+nz=b$$

Therefore, the required solution $a=f(b)$ is

$$x^2+y^2+z^2=f(lx+my+nz)$$

**EXAMPLE 9.10** Solve $(x^2-y^2-z^2)p+2xyq=2xz$.

***Solution*** The subsidiary equation is

$$\frac{dx}{x^2-y^2-z^2}=\frac{dy}{2xy}=\frac{dz}{2xz}$$

From the last two fractions, we have $dy/y = dz/z$ which on integration yields $y/z = a$.

Now, using multipliers $x$, $y$, $z$, we have

$$\frac{2x\,dx + 2y\,dy + 2z\,dz}{x^2 + y^2 + z^2} = \frac{dz}{z}$$

which on integration yields $(x^2 + y^2 + z^2)/z = b$. Hence, the required solution $a = f(b)$ is

$$x^2 + y^2 + z^2 = zf\left(\frac{y}{z}\right)$$

**EXAMPLE 9.11** Solve

$$(w + y + z)\frac{\partial w}{\partial x} + (w + x + z)\frac{\partial w}{\partial y} + (w + x + y)\frac{\partial w}{\partial z} = x + y + z$$

***Solution*** The subsidiary equation is

$$\frac{dw}{x + y + z} = \frac{dx}{y + z + w} = \frac{dy}{z + w + x} = \frac{dz}{w + x + y}$$

which gives

$$\frac{dw + dx + dy + dz}{3(w + x + y + z)} = \frac{dw - dx}{x - w} = \ldots$$

Integrating this equation, we get

$$(x - w)\,(w + x + y + z)^{1/3} = c_1$$

Similarly, the other two solutions are

$$(y - w)\,(w + x + y + z)^{1/3} = c_2$$

$$(z - w)\,(w + x + y + z)^{1/3} = c_3$$

Therefore, the required solution is

$$\phi\{(x - w)u,\ (y - w)u,\ (z - w)u\} = 0$$

where $u = (w + x + y + z)^{1/3}$.

## 9.5 NONLINEAR EQUATIONS OF THE FIRST ORDER

The equations which involve $p$ and $q$ other than in the first degree are called *nonlinear partial differential equations of the first order*. For such equations, the complete solution consists of only two arbitrary constants (i.e. equal to the number of independent variables involved) and the particular integral is obtained by assigning particular values to the constants.

Many equations are reducible to some standard forms. We shall now discuss these standard forms.

**Form I** $f(p, q) = 0$, i.e. *equations containing only p and q.*

A complete solution of such an equation is

$$z = ax + by + c \tag{6}$$

where $a$ and $b$ are such that

$$F(a, b) = 0 \tag{7}$$

Write Eq. (7) as $b = \phi(a)$ and use it in Eq. (6) to obtain the required solution

$$z = ax + \phi(a)\, y + c \tag{8}$$

where $a$ and $c$ are arbitrary constants.

To know the general solution, we put $c = f(a)$ in Eq. (8) so that

$$z = ax + \phi(a)y + f(a) \tag{9}$$

Now eliminate $a$ from Eq. (9), and the equation $[0 = x + \phi'(x)y + f'(a)]$ obtained by differentiating with respect to $a$.

**EXAMPLE 9.12** Solve $\sqrt{p} + \sqrt{q} = 1$.

***Solution*** The complete solution is

$$z = ax + by + c \tag{10}$$

where $\sqrt{a} + \sqrt{b} = 1$, or $b = (1 - \sqrt{a})^2$. Therefore, Eq. (10) becomes

$$z = ax + (1 - \sqrt{a})^2 y + c$$

which is the required solution.

**EXAMPLE 9.13** Solve $p^2 + q^2 = m^2$.

***Solution*** The complete solution is

$$z = ax + by + c \tag{11}$$

where $a^2 + b^2 = m^2$, or $b = \sqrt{m^2 - a^2}$.. Thus, the desired solution from Eq. (11) is

$$z = ax + \sqrt{m^2 - a^2}\, y + c \tag{12}$$

*Note* In order to get the general solution, put $c = f(a)$ in Eq. (12), so that

$$z = ax + \sqrt{m^2 - a^2}\, y + f(a)$$

Differentiating it with respect to $a$, we get

$$0 = x - \frac{a}{\sqrt{(m^2 - a^2)}}\, y + f'(a)$$

Now, from these two equations, eliminate $a$. In particular, if we choose $c$ or $f(a)$ as zero, then by the elimination of $a$, we obtain

$$z^2 = m^2(x^2 + y^2)$$

which represents a cone.

**EXAMPLE 9.14** Solve $x^2p^2 + y^2q^2 = z^2$.

***Solution*** We write the given equation as

$$\left(\frac{x}{z}\frac{\partial z}{\partial x}\right)^2 + \left(\frac{y}{z}\frac{\partial z}{\partial y}\right)^2 = 1 \tag{13}$$

so that it reduces to the standard Form I. Now set

$$\frac{dx}{x} = du, \quad \frac{dy}{y} = dv, \quad \frac{dz}{z} = dw$$

so that $u = \log x$, $v = \log y$, $w = \log z$ and Eq. (13) becomes

$$\left(\frac{\partial w}{\partial u}\right)^2 + \left(\frac{\partial w}{\partial v}\right)^2 = 1$$

i.e. $P^2 + Q^2 = 1$, where $P = \partial w/\partial u$, $Q = \partial w/\partial v$. The complete solution of this equation is

$$w = au + bv + c \tag{14}$$

where $b = \sqrt{1 - a^2}$. Thus, Eq. (14) becomes

$$w = au + \sqrt{1 - a^2}\, v + c$$

or

$$\log z = a \log x + \sqrt{1 - a^2} \log y + c$$

which is the required solution.

**Form II** $f(z, p, q) = 0$, i.e. *equations containing x and y.*

To solve such equations we use the following steps:

(i) Set $u = x + ay$ and put $p = dz/du$, $q = a(dz/du)$ in the given equation.
(ii) Solve the resulting ordinary differential equation in $z$ and $u$.
(iii) Replace $u$ by $x + ay$.

**EXAMPLE 9.15** Solve $p(1 + q) = qz$.

***Solution*** Let $u = x + ay$ so that $p = dz/du$ and $q = a(dz/du)$. With these, the given equation can be written as

$$\frac{dz}{du}\left(1 + a\frac{dz}{du}\right) = az\frac{dz}{du}$$

Separate the variables and integrate to get

$$\log (az - 1) = u + b = x + ay + b$$

as the required solution.

**EXAMPLE 9.16** Solve $z^2(p^2 + q^2 + 1) = c^2$.

***Solution*** Put $u = x + ay$, $p = dz/du$, $q = a(dz/du)$ so that the given equation takes the form

$$z^2\left[\left(\frac{dz}{du}\right)^2 + \left(a\frac{dz}{du}\right)^2 + 1\right] = c^2$$

Separating the variables, we obtain

$$\frac{zdz}{\sqrt{c^2 - z^2}} = \frac{du}{\sqrt{1+a^2}}$$

Integrating, we get

$$-\sqrt{1+a^2}\,\sqrt{c^2 - z^2} = u + b$$

Thus, the desired solution is

$$\sqrt{1+a^2}\,\sqrt{c^2 - z^2} = x + ay + b$$

**EXAMPLE 9.17** Solve $z^2(p^2x^2 + q^2) = 1$.

***Solution*** The given equation can be written as

$$z^2\left[\left(x\frac{\partial z}{\partial x}\right)^2 + \left(\frac{\partial z}{\partial y}\right)^2\right] = 1$$

Putting $X = \log x$ and $x(\partial z/\partial x) = \partial z/\partial X$ in this equation, we obtain

$$z^2\left[\left(\frac{\partial z}{\partial X}\right)^2 + \left(\frac{\partial z}{\partial y}\right)^2\right] = 1$$

which is of the standard Form II. Now, let

$$u = X + ay, \qquad \frac{\partial z}{\partial X} = \frac{\partial z}{\partial u}, \qquad \frac{\partial z}{\partial y} = a\frac{\partial z}{\partial u}$$

so that the above equation takes the form

$$z^2\left[\left(\frac{\partial z}{\partial u}\right)^2 + a^2\left(\frac{\partial z}{\partial u}\right)^2\right] = 1$$

Separating the variables and integrating, we get

$$\left(1 + \sqrt{a^2}\right)z^2 = \pm 2u + b = \pm 2(X + ay) + b$$

Therefore, the required solution is

$$\left(\sqrt{1+a^2}\right)z^2 = \pm 2(\log x + ay) + b$$

**Form III** $f(x, p) = F(y, q)$, i.e. *the equations in which the variables z does not appear and the terms containing x and p can be separated from those containing y and q.*

To obtain a solution of such an equation, we proceed as follows: Let $f(x, p) = F(y, q) = a$ and solve these for $p$ and $q$ to get

$$p = \phi(x, a), \qquad q = \psi(y, a)$$

Since

$$dz = \frac{\partial z}{\partial x}dx + \frac{\partial z}{\partial y}dy = p\,dx + q\,dy$$

we have

$$\int dz = \int (pdx + qdy)$$

or

$$z = \int \phi(x, a)dx + \int \psi(y, a)dy + b$$

which is the desired complete solution containing two constants $a$ and $b$.

**EXAMPLE 9.18** Solve $q - p + x - y = 0$.

***Solution*** The given equation can be written as

$$q - y = p - x$$

Let $p - x = q - y = a$ so that $p = x + a$ and $q = y + a$, and the complete solution is

$$z = \int (x + a)dx + \int (y + a)dy + b$$

or

$$2z = (x + a)^2 + (y + a)^2 + b$$

**EXAMPLE 9.19** Solve $p^2 + q^2 = x + y$.

***Solution*** The given equation can be written as

$$p^2 - x = y - q^2 = a$$

so that $p = \sqrt{a + x}$ and $q = \sqrt{y - a}$. Substituting these values in $dz = p\,dx + q\,dy$ and integrating, we obtain

$$z = \frac{2}{3}(a + x)^{3/2} + \frac{2}{3}(y - a)^{3/2} + b$$

which is the required complete solution.

**EXAMPLE 9.20** Solve $z^2(p^2 + q^2) = x^2 + y^2$.

***Solution*** We write the given equation as

$$\left(z\frac{\partial z}{\partial x}\right)^2 + \left(z\frac{\partial z}{\partial y}\right)^2 = x^2 + y^2 \tag{15}$$

Put $zdz = dZ$, so that

$$\frac{\partial Z}{\partial x} = \frac{\partial Z}{\partial z}\frac{\partial z}{\partial x} = z\frac{\partial z}{\partial x} = P$$

$$\frac{\partial Z}{\partial y} = \frac{\partial Z}{\partial z}\frac{\partial z}{\partial y} = z\frac{\partial z}{\partial y} = Q$$

and Eq. (15) takes the form

$$P^2 + Q^2 = x^2 + y^2 \quad \text{or} \quad P^2 - x^2 = y^2 - Q^2 = a$$

Thus, $P = \sqrt{x^2 + a}$ and $Q = \sqrt{y^2 - a}$, and $dZ = Pdx + Qdy$ which on integration gives the relation

$$Z = \frac{1}{2}x\sqrt{x^2 + a} + \frac{1}{2}a\log\left(x + \sqrt{x^2 + a}\right)$$

$$+ \frac{1}{2}y\sqrt{y^2 - a} - \frac{1}{2}a\log\left(y + \sqrt{y^2 - a}\right) + b$$

Therefore, the complete solution is

$$z^2 = x\sqrt{x^2 + a} + y\sqrt{y^2 - a} + a\log\frac{x + \sqrt{x^2 + a}}{y + \sqrt{y^2 - a}} + b$$

**Form IV** $z = px + qy + f(p, q)$, i.e. *equations analogous to Clairaut's equation* [cf. Eq. (24), Chapter 3].

The complete solution for such an equation is $z = ax + by + f(a, b)$, which is obtained by writing $a$ for $p$ and $b$ for $q$ in the given equation.

**EXAMPLE 9.21** Solve $z = px + qy + pq$.

***Solution*** Putting $a = p$ and $b = q$ in the given equation, we get the complete solution as

$$z = ax + by + ab$$

## 9.6 CHARPIT'S METHOD

We will now give a general method for obtaining the complete solution of a nonlinear partial differential equation. This method is due to Charpit and is applicable to all partial differential equations of the first order.

Consider the equation

$$F(x, y, z, p, q) = 0 \tag{16}$$

Since $z$ depends on $x$ and $y$, we have

$$dz = \frac{\partial z}{\partial x}dx + \frac{\partial z}{\partial y}dy = pdx + qdy \tag{17}$$

Now, if we can find another relation between $x$, $y$, $z$, $p$, $q$ such that

$$f(x, y, z, p, q) = 0 \tag{18}$$

then, we can solve Eqs. (16) and (18) for $p$ and $q$ and substitute them in Eq. (17). This will give the solution provided Eq. (17) is integrable.

To determine $f$, differentiate Eqs. (16) and (18) with respect to $x$ and $y$ so that

$$\frac{\partial F}{\partial x}+\frac{\partial F}{\partial z}p+\frac{\partial F}{\partial p}\frac{\partial p}{\partial x}+\frac{\partial F}{\partial q}\frac{\partial q}{\partial x}=0 \tag{19}$$

$$\frac{\partial f}{\partial x}+\frac{\partial f}{\partial z}p+\frac{\partial f}{\partial p}\frac{\partial p}{\partial x}+\frac{\partial f}{\partial q}\frac{\partial q}{\partial x}=0 \tag{20}$$

$$\frac{\partial F}{\partial y}+\frac{\partial F}{\partial z}q+\frac{\partial F}{\partial p}\frac{\partial p}{\partial y}+\frac{\partial F}{\partial q}\frac{\partial q}{\partial y}=0 \tag{21}$$

$$\frac{\partial f}{\partial y}+\frac{\partial f}{\partial z}q+\frac{\partial f}{\partial p}\frac{\partial p}{\partial y}+\frac{\partial f}{\partial q}\frac{\partial q}{\partial y}=0 \tag{22}$$

Eliminate $\partial p/\partial x$ from Eqs. (19) and (20), and $\partial q/\partial y$ from Eqs. (21) and (22) to obtain

$$\left(\frac{\partial F}{\partial x}\frac{\partial f}{\partial p}-\frac{\partial f}{\partial x}\frac{\partial F}{\partial p}\right)+\left(\frac{\partial F}{\partial z}\frac{\partial f}{\partial p}-\frac{\partial f}{\partial z}\frac{\partial F}{\partial p}\right)p+\left(\frac{\partial F}{\partial q}\frac{\partial f}{\partial p}-\frac{\partial f}{\partial q}\frac{\partial F}{\partial p}\right)\frac{\partial q}{\partial x}=0$$

$$\left(\frac{\partial F}{\partial y}\frac{\partial f}{\partial q}-\frac{\partial f}{\partial y}\frac{\partial F}{\partial q}\right)+\left(\frac{\partial F}{\partial z}\frac{\partial f}{\partial q}-\frac{\partial f}{\partial z}\frac{\partial F}{\partial q}\right)q+\left(\frac{\partial F}{\partial p}\frac{\partial f}{\partial q}-\frac{\partial f}{\partial p}\frac{\partial F}{\partial q}\right)\frac{\partial p}{\partial y}=0$$

Adding these two equations and using

$$\frac{\partial q}{\partial x}=\frac{\partial^2 z}{\partial x\,\partial y}=\frac{\partial p}{\partial y}$$

after rearrangement, we find that

$$\left(-\frac{\partial F}{\partial p}\right)\frac{\partial f}{\partial x}+\left(-\frac{\partial F}{\partial q}\right)\frac{\partial f}{\partial y}+\left(-p\frac{\partial F}{\partial p}-q\frac{\partial F}{\partial q}\right)\frac{\partial f}{\partial z}+\left(\frac{\partial F}{\partial x}+p\frac{\partial F}{\partial z}\right)\frac{\partial f}{\partial p}$$

$$+\left(\frac{\partial F}{\partial y}+q\frac{\partial F}{\partial z}\right)\frac{\partial f}{\partial q}=0 \tag{23}$$

Equation (23) is Lagrange's equation with $x$, $y$, $z$, $p$, $q$ as independent variable and $f$ as the dependent variable. Its solution will depend on the subsidiary equations

$$\frac{dx}{\frac{-\partial F}{\partial p}}=\frac{dy}{\frac{-\partial F}{\partial q}}=\frac{dz}{-p\frac{\partial F}{\partial p}-q\frac{\partial F}{\partial q}}=\frac{dp}{\frac{\partial F}{\partial x}+p\frac{\partial F}{\partial z}}=\frac{dq}{\frac{\partial F}{\partial y}+q\frac{\partial F}{\partial z}}=\frac{df}{0}$$

An integral of these equations, involving $p$ or $q$ or both, can be taken as the required relation (18), which alongwith Eq. (16) will give the values of $p$ and $q$ to make Eq. (17) integrable.

**EXAMPLE 9.22** Solve $(p^2 + q^2)y = qz$.

***Solution*** Let

$$F(x, y, z, p, q) = (p^2 + q^2)y - qz = 0 \tag{24}$$

The subsidiary equations are

$$\frac{dx}{-2py} = \frac{dy}{z - 2qy} = \frac{dz}{-qz} = \frac{dp}{-pq} = \frac{dq}{p^2}$$

The last two fractions yield $p\,dp + q\,dq = 0$, which on integration gives

$$p^2 + q^2 = c^2 \tag{25}$$

In order to solve Eqs. (24) and (25), put $p^2 + q^2 = c^2$ in Eq. (24) so that $q = c^2y/z$. Now substitute this value of $q$ in Eq. (25) to get

$$p = \frac{(c/z)}{\sqrt{z^2 - c^2y^2}}$$

Hence

$$dz = pdx + qdy = \frac{c}{z}\sqrt{(z^2 - c^2y^2)}dx + \frac{c^2y}{z}dy$$

or

$$zdz - c^2ydy = c\sqrt{z^2 - c^2y^2}\ dx$$

or

$$\frac{(1/2)d(z^2 - c^2y^2)}{\sqrt{z^2 - c^2y^2}} = c\,dx$$

Integrating, we get the required solution as

$$z^2 = (a + cx)^2 + c^2y^2$$

**EXAMPLE 9.23** Solve $p(q^2 + 1) + (b - z)q = 0$.

***Solution*** Here

$$\frac{dp}{pq} = \frac{dq}{q^2} = \frac{dz}{3pq^2 + p + (b - z)q} = \frac{dx}{q^2 + 1} = \frac{dy}{-z + b + 2pq}$$

(From the given equation, the third fraction reduces to $dz/2\,pq^2$). From the first two fractions, after integration, we get

$$q = ap$$

where $a$ is an arbitrary constant. This and the given equation determine the values of $p$ and $q$ as

$$p = \frac{\sqrt{a(z-b)-1}}{a}, \qquad q = \sqrt{a(z-b)-1}$$

Substituting these values in $dz = p\,dx + q\,dy$, we obtain

$$dz = \left(\frac{dx}{a} + dy\right)\sqrt{a(z-b)-1}$$

Separating the variables and integrating, we get the solution as

$$2\sqrt{a(z-b)-1} = x + ay + b$$

## 9.7 HOMOGENEOUS LINEAR EQUATIONS WITH CONSTANT COEFFICIENTS

An equation of the form

$$\frac{\partial^n z}{\partial x^n} + k_1 \frac{\partial^n z}{\partial x^{n-1}\partial y} + \cdots + k_n \frac{\partial^n z}{\partial y^n} = F(x, y) \tag{26}$$

where $k_1, k_2, \ldots, k_n$ are constants, is called a linear homogeneous partial differential equation of order $n$ with constant coefficients.

Equation (26) can be written as

$$(D^n + k_1 D^{n-1} D' + \cdots + k_n D'^n)z = F(x, y) \tag{27}$$

or

$$f(D, D')z = F(x, y)$$

where

$$D^n = \frac{\partial^n}{\partial x^n}, \qquad D'^n = \frac{\partial^n}{\partial y^n}$$

As in the case of ordinary linear differential equations with constant coefficients, here also the complete solution of Eq. (26) consists of the complementary function (C.F.) and particular integral (P.I.). The complementary function is the complete solution of $f(D, D')z = 0$ which contains $n$ arbitrary constants. The particular integral is the particular solution of Eq. (27).

We shall now illustrate the methods of finding the complementary function and particular integral by considering a second order partial differential equation

$$\frac{\partial^2 z}{\partial x^2} + k_1 \frac{\partial^2 z}{\partial x\,\partial y} + k_2 \frac{\partial^2 z}{\partial y^2} = 0 \tag{28}$$

which in symbolic form is

$$(D^2 + k_1 DD' + k_2 D'^2)z = 0 \tag{29}$$

The auxiliary equation is

$$D^2 + k_1 DD' + k_2 D'^2 = 0$$

and let the roots be $D/D' = m_1, m_2$. We then have

*Case I:* If the roots are real and distinct then Eq. (29) is equivalent to

$$(D - m_1D')(D - m_2D')z = 0 \tag{30}$$

which will be satisfied by the solution of $(D - m_2D')z = 0$, i.e. $p - m_2q = 0$. This is a Lagrange's linear equation and the subsidiary equations are

$$\frac{dx}{1} = \frac{dy}{-mz} = \frac{dz}{0}$$

where $y + m_2x = a$ and $z = b$. Therefore, its solution is $z = \phi(y + m_2x)$. Similarly, Eq. (30) will also be satisfied by the solution of $(D - m_1D')z = 0$, i.e. $z = f(y + m_1x)$. Hence, the complete solution of Eq. (28) is

$$z = f(y + m_1x) + \phi(y + m_2x)$$

**EXAMPLE 9.24** Solve $2\frac{\partial^2 z}{\partial x^2} + 5\frac{\partial^2 z}{\partial x \partial y} + 2\frac{\partial^2 z}{\partial y^2} = 0$.

***Solution*** The given equation can be written as

$$(2D^2 + 5DD' + 2D'^2)z = 0$$

The auxiliary equation $2m^2 + 5m + 2 = 0$ has the roots $m = -2, -1/2$. Hence, the complete solution is

$$z = f_1(y - 2x) + f_2\left(y - \frac{1}{2}x\right)$$

*Case II:* If the roots are equal (i.e. $m_1 = m_2$), then Eq. (29) is equivalent to

$$(D - m_1D')^2 z = 0 \tag{31}$$

Putting $(D - m_1D')z = u$ in (31), we have $(D - m_1D')u = 0$ which gives

$$u = \phi(y + m_1x)$$

Thus, Eq. (31) becomes

$$(D - m_1D')z = \phi(y + m_1x) \quad \text{or} \quad p - m_1q = \phi(y + m_1x)$$

This is again Lagrange's linear equation and the subsidiary equations are

$$\frac{dx}{1} = \frac{dy}{-m_2} = \frac{dz}{\phi(y + m_1x)}$$

which gives $y + m_1x = a$, and $dz = \phi(a)dx$, i.e., $z = \phi(a)x + b$.

Thus, the complete solution of Eq. (28) is

$$z - x\phi(y + m_1x) = f(y + m_1x)$$

or

$$z = f(y + m_1x) + x\phi(y + m_1x)$$

**EXAMPLE 9.25** Solve $\dfrac{\partial^2 z}{\partial x^2}+6\dfrac{\partial^2 z}{\partial x\partial y}+9\dfrac{\partial^2 z}{\partial y^2}=0$.

***Solution*** The auxiliary equation $m^2 + 6m + 9 = 0$ has the roots $m = -3$, $-3$, and the complete solution is

$$z = f_1(y - 3x) + xf_2(y - 3x)$$

Now, to find the particular integral, consider the second order partial differential equation

$$(D^2 + k_1DD' + k_2D'^2)z = F(x, y)$$

or

$$f(D, D')z = F(x, y)$$

Then

$$\text{P.I.} = \frac{1}{f(D, D')}F(x, y)$$

and we have the following cases.

*Case (a): When* $F(x, y) = e^{ax+by}$. Since

$$De^{ax+by} = ae^{ax+by}, \qquad D'e^{ax+by} = be^{ax+by}$$

we have

$$D^2e^{ax+by} = a^2e^{ax+by}, \quad D'^2e^{ax+by} = b^2e^{ax+by} \quad \text{and} \quad DD'e^{ax+by} = abe^{ax+by}$$

Thus

$$(D^2 + k_1DD' + k_2D'^2)e^{ax+by} = (a^2 + k_1ab + k_2b^2)e^{ax+by}$$

or

$$f(D, D')e^{ax+by} = f(a, b)e^{ax+by}$$

Operating both sides by $1/f(D, D')$, we get

$$\text{P.I.} = \frac{1}{f(D, D')}F(x, y) = \frac{1}{f(a, b)}e^{ax+by}$$

If $f(a, b) = 0$, then

$$\text{P.I.} = \frac{1}{f(D, D')}e^{ax+by} = \frac{1}{f(D+a, D'+b)}e^{ax+by}$$

$$= e^{ax+by}[f(D+a, D'+b)]^{-1}(1)$$

Now expand $[f(D + a, D' + b)]^{-1}$ in an infinite series, noting that $D^{-1}$ and $D'^{-1}$ mean integral with respect to $x$ and $y$, respectively, keeping the other variables constant.

*Case (b): When* $F(x, y) = \sin(mx + ny) = \cos(mx + ny)$. Here

$$\text{P.I.} = \frac{1}{f(D^2, DD', D'^2)}\sin(mx+ny)$$

$$= \frac{1}{f(-m^2, -mn, -n^2)}\sin(mx+ny)$$

and
$$\text{P.I.} = \frac{1}{f(-m^2, -mn, -n^2)} \cos(mx + ny)$$

*Case (c):* *When $F(x, y) = x^m y^n$, m, n are constants.* Here

$$\text{P.I.} = \frac{1}{f(D, D')} x^m y^n = [f(D, D')]^{-1} x^m y^n$$

and to evaluate it expand $[f(D, D')]^{-1}$ in ascending powers of $D$ or $D'$ using the bionomial theorem and then operate on $x^m y^n$ term by term.

*Case (d):* *When $F(x, y)$ is any function of x and y.* Here

$$\text{P.I.} = \frac{1}{f(D, D')} F(x, y)$$

and to evaluate it, resolve $1/f(D, D')$ into partial fractions considering $f(D, D')$ as a function of $D$ only and operate each partial fraction on $F(x, y)$ noting that

$$\frac{1}{D - mD'} F(x, y) = \int F(x, c - mx)dx$$

where $c$ is replaced by $y + mx$ after integration.

*Case (e):* *If $F(x, y) = e^{ax+by} V$, where V is a function of x and y,* then

$$P.I. = \frac{1}{f(D', D)} e^{ax+by} V = e^{ax+by} \frac{1}{f(D + a, D' + b)} V$$

*Case (f):* *When $F(x, y) = \phi(ax + by)$*

Here, we have the following two cases:

(i) Replacing $D$ by $a$ and $D'$ by $b$ in $f(D, D')$ we get $f(a, b) \neq 0$. Moreover, if $f(D, D')$ is a homogeneous function of degree $n$, then

$$P.I. = \frac{1}{f(D, D')} F(x, y) = \frac{1}{f(D, D')} \phi(ax + by) = \frac{1}{f(a, b)} \iint ... \int \phi(u)\, du^n \quad (32)$$

where $u$ is replaced by $ax + by$ after integrating the right hand side of Eq. (32) $n$ times with respect to $u$.

(ii) Replacing $D$ by a and $D'$ by $b$ in $f(D, D')$, we get $f(a, b) = 0$. In such cases, $f(D, D')$ must be factorized and in general, there are two types of factors. Let $f(D, D') = (bD - aD')^m g(D, D')$, where $g(a, b) \neq 0$. Then the particular integral can be obtained by (i) above as well as by making use of the formula

$$\frac{1}{(bD - aD')^m} \phi(ax + by) = \frac{x^m}{b^m m!} \phi(ax + by) \quad (33)$$

*Note:* The auxiliary equation of (26) is

$$m^n + k_1 m^{n-1} + \cdots + k_n = 0 \quad (34)$$

(i) If the roots of Eq. (34) are all distinct, then

$$\text{C.F.} = f_1(y + m_1x) + f_2(y + m_2x) + \cdots$$

(ii) If Eq. (34) has two equal roots, then

$$\text{C.F.} = f_1(y + m_1x) + xf_2(y + m_1x) + f_3(y + m_2x) + \cdots$$

(iii) If three roots of Eq. (34) are equal, then

$$\text{C.F.} = f_1(y + m_1x) + xf_2(y + m_1x) + x^2f_3(y + m_1x) + \cdots$$

**EXAMPLE 9.26** Solve $\dfrac{\partial^3 z}{\partial x^3} - 3\dfrac{\partial^3 z}{\partial x^2 \partial y} + 4\dfrac{\partial^3 y}{\partial y^3} = e^{x+2y}$.

***Solution*** The auxiliary equation of the given equation has the roots $m = -1, 2, 2$. Thus

$$\text{C.F.} = f_1(y - x) + f_2(y + 2x) + xf_3(y + 2x)$$

$$\text{P.I.} = \frac{1}{D^3 - 3D^2D' + 4D^3} e^{x+2y} \quad (\text{Put } D = 1,\ D' = 2)$$

$$= \frac{1}{27} e^{x+2y}$$

Therefore, the complete solution is

$$z = f_1(y - x) + f_2(y + 2x) + x f_3(y + 2x) + \frac{1}{27} e^{x+2y}$$

**EXAMPLE 9.27** Solve $\dfrac{\partial^2 z}{\partial x^2} - \dfrac{\partial^2 z}{\partial x \partial y} = \cos x \cos 2y$.

***Solution*** The roots of the auxiliary equation of the given equation is $m = 0, 1$. Therefore,

$$\text{C.F.} = f_1(y) + f_2(y + x)$$

and $$\text{P.I.} = \frac{1}{D^2 - DD'} \cos x \cos 2y$$

$$= \frac{1}{2}\frac{1}{D^2 - DD'} [\cos (x + 2y) + \cos (x - 2y)]$$

Putting $D^2 = 1$, $DD' = -2$ and $D^2 = -1$, $DD' = 2$, we get

$$\text{P.I.} = \frac{1}{2} \cos (x + 2y) - \frac{1}{6} \cos (x - 2y)$$

Thus, the complete solution is

$$z = f_1(y) + f_2(y + x) + \frac{1}{2}\cos(x + 2y) - \frac{1}{6}\cos(x - 2y)$$

**EXAMPLE 9.28** Solve $\dfrac{\partial^3 z}{\partial x^3} - 2\dfrac{\partial^3 z}{\partial x^2 \partial y} = 2e^{2x} + 3x^2 y$.

***Solution*** The auxiliary equation $m^3 - 2m^2 = 0$ has the roots $m = 0, 0, 2$. Therefore

$$\text{C.F.} = f_1(y) + xf_2(y) + f_3(y + 2x)$$

and

$$\text{P.I.} = \frac{1}{D^3 - 2D^2 D'}(2e^{2x} + 3x^2 y) = 2\frac{1}{D^3 - 2D^2 D'}e^{2x} + 3\frac{1}{D^3 - 2D^2 D'}x^2 y$$

$$= 2\frac{1}{2^3 - 2\cdot 2^2 \cdot (0)}e^{2x} + \frac{3}{D^3}\left(1 - \frac{2D'}{D}\right)^{-1} x^2 y$$

$$= \frac{1}{4}e^{2x} + \frac{3}{D^3}\left(1 + \frac{2D'}{D} + \frac{4D'^2}{D^2} + \cdots\right)x^2 y$$

$$= \frac{1}{4}e^{2x} + \frac{3}{D^3}\left(x^2 y + \frac{2}{D}x^2\right)$$

$$= \frac{1}{4}e^{2x} + \frac{3}{D^3}\left(x^2 y + \frac{2}{3}x^3\right), \qquad \left[\frac{1}{D}f(x) = \int f(x)\,dx\right]$$

$$= \frac{1}{4}e^{2x} + 3y\frac{x^5}{3\cdot 4\cdot 5} + 2\frac{x^6}{4\cdot 5\cdot 6}, \qquad \left(\frac{1}{D^3}f(x) = \int\left[\int\left(\int f(x)dx\right)dx\right]dx\right)$$

$$= \frac{e^{2x}}{4} + \frac{x^5 y}{20} + \frac{x^6}{60}$$

Thus, the complete solution is

$$z = f_1(y) + xf_2(y) + f_3(x + 2y) + \frac{e^{2x}}{4} + \frac{x^5 y}{20} + \frac{x^6}{60}$$

**EXAMPLE 9.29** Solve $\dfrac{\partial^2 z}{\partial x^2} - 4\dfrac{\partial^2 z}{\partial x \partial y} + 4\dfrac{\partial^2 z}{\partial y^2} = e^{2x+y}$

***Solution*** The roots of the auxiliary equations are $m = 2, 2$. Hence

$$\text{C.F.} = f_1(y + 2x) + xf_2(y + 2x)$$

and
$$\text{P.I.} = \frac{1}{(D - 2D')^2} e^{2x+y}$$

Now, for $D = 2$ and $D' = 1$, $(D - 2D')^2 = 0$ and the usual rule cannot be applied, and to obtain the P.I. we find the solution from $(D - 2D')u = e^{2x+y}$ and $(D - 2D')z = u = xe^{2x+y}$, so that

$$u = \int F(x, c - mx)dx = \int e^{2x+(c-2x)}dx$$

$$= xe^c = xe^{2x+y} \quad \text{as } y = c - mx = c - 2x$$

and
$$z = \int xe^{2x+(c-2x)}dx = \frac{1}{2}x^2e^c = \frac{1}{2}x^2e^{2x+y} \quad \text{as } y = c - mx = c - 2x$$

Therefore, the complete solution is

$$z = f_1(y + 2x) + xf_2(y + 2x) + \frac{1}{2}x^2e^{2x+y}$$

**EXAMPLE 9.30** Solve $\dfrac{\partial^2 z}{\partial x^2} + \dfrac{\partial^2 z}{\partial x \partial y} - 6\dfrac{\partial^2 z}{\partial y^2} = y \cos x$.

***Solution*** The roots of the auxiliary equation are $-3$ and $2$. Thus

$$\text{C.F.} = f_1(y - 3x) + f_2(y + 2x)$$

and
$$\text{P.I.} = \frac{1}{(D - 2D')(D + 3D')} y \cos x$$

which is obtained from $(D + 3D')u = y \cos x$ as

$$u = \int (c + 3x) \cos x \, dx$$

$$= (c + 3x) \sin x + 3 \cos x$$

$$= y \sin x + 3 \cos x, \qquad \text{as } y = c - mx = c + 3x$$

and from $(D - 2D')z = u = y \sin x + 3 \cos x$ as

$$z = \int [(c - 2x) \sin x + 3 \cos x]dx$$

$$= (c - 2x)(-\cos x) - (-2)(-\sin x) + 3 \sin x$$

$$= \sin x - y \cos x, \qquad \text{since } y = c - mx = c - 2x$$

Hence, the complete solution is

$$z = f_1(y - 3x) + f_2(y + 2x) + \sin x - y \cos x$$

**EXAMPLE 9.31** Solve $(D^2 + DD' - 2D'^2)z = e^x(y-1)$.

***Solution*** The auxiliary equation $m^2 + m - 2 = 0$ has roots as 1, –2. Thus

$$C.F.= f_1(y+x) + f_2(y-2x)$$

Also

$$\text{P.I.} = \frac{1}{D^2 + DD' - 2D'^2} e^x(y-1)$$

$$= e^x \frac{1}{(D+1)^2 + (D+1)D' - 2D'^2}(y-1)$$

$$= e^x[1 + (2D + D' + DD' + D^2 - 2D'^2)]^{-1}(y-1)$$

$$= e^x[1 - (2D + D' + DD' + D^2 - 2D'^2) + \text{other terms}](y-1)$$

$$= e^x(y-2)$$

Therefore, the complete solution is

$$z = f_1(y+x) + f_2(y-2x) + e^x(y-2)$$

**EXAMPLE 9.32** Solve the following partial differential equations

(i) $(D^2 + 3DD' + 2D'^2)z = x + y$

(ii) $(4D^2 - 4DD' + D'^2)z = \log(x+2y)$

(iii) $(D^2 - D'^2)z = x - y$

(iv) $(D^3 - 4D^2D' + 4DD'^2)z = \sin(2x+y)$

***Solution*** (i) Here the roots of the auxiliary equation are –1 and –2. Thus,

$$C.F.= f_1(y-x) + f_2(y-2x)$$

Also

$$P.I.= \frac{1}{D^2 + 3DD' + 2D'^2}(x+y)$$

Here, $f(D, D')$ is a homogeneous function of degree 2 and $f(a,b) \neq 0$. Thus, from Eq. (32), we have

$$\text{P.I.} = \frac{1}{1^2 + 3(1)(1) + 2(1)^2} \iint u \, du^2$$

$$= \frac{1}{6}\int \frac{u^2}{2} du = \frac{1}{36}u^3$$

$$= \frac{1}{36}(x+y)^3$$

Therefore, the complete solution is

$$z = f_1(y-x) + f_2(y-2x) + \frac{1}{36}(x+y)^3$$

(ii) The roots of the auxiliary equation are 1/2 and –1/2. Thus

$$\text{C.F.} = f_1\left(y + \frac{1}{2}x\right) + f_2\left(y - \frac{1}{2}x\right)$$

Also

$$\text{P.I.} = \frac{1}{4D^2 - 4DD' + D'^2}\log(x+2y) = \frac{1}{(2D-D')^2}\log(x+2y)$$

Here $f(a,b) = 0$ and $f(D,D')$ can be factorized and thus using Eq. (33), we get

$$\text{P.I.} = \frac{x^2}{2^2 2!}\log(x+2y) = \frac{x^2}{8}\log(x+2y)$$

Therefore, the complete solution is

$$z = f_1\left(y + \frac{1}{2}x\right) + f_2\left(y - \frac{1}{2}x\right) + \frac{x^2}{8}\log(x+2y)$$

(iii) Here the auxiliary equation has the roots 1 and –1. Thus

$$\text{C.F.} = f_1(y+x) + f_2(y-x)$$

Also

$$\text{P.I.} = -\frac{1}{D^2 - D'^2}(x-y) = \frac{1}{(D+D')(D-D')}(x-y)$$

$$= \frac{1}{D+D'}\frac{1}{1-(-1)}\int u\,du \qquad \text{[from Eq. (32)]}$$

$$= \frac{1}{2}\frac{1}{D+D'}\frac{u^2}{2} = \frac{1}{4}\frac{1}{D+D'}(x-y)^2$$

$$= \frac{1}{4}\frac{x^1}{(1)^1 1!}(x-y)^2 = \frac{1}{4}x(x-y)^2 \qquad \text{[from Eq. (33)]}$$

Therefore, the complete solution is

$$z = f_1(y+x) + f_2(y-x) + \frac{1}{4}x(x-y)^2$$

(iv) The roots of the auxiliary equation are 0, 2 and 2. Thus

$$\text{C.F.} = f_1(y) + f_2(y+2x) + xf_3(y+2x)$$

Also

$$\text{P.I.} = \frac{1}{D^3 - 4D^2D' + 4DD'^2}\sin(2x+y)$$

$$= \frac{1}{(D-2D')^2}\frac{1}{D}\sin(2x+y)$$

$$= \frac{1}{(D-2D')^2}\int \sin(2x+y)\,dx$$

$$= \frac{1}{(D-2D')^2}[\frac{-1}{2}\cos(2x+y)]$$

$$= -\frac{1}{4}x^2\cos(2x+y) \qquad \text{[using Eq. (33)]}$$

Therefore, the complete solution is

$$z = f_1(y) + f_2(y+2x) + xf_3(y+2x) - \frac{1}{4}x^2\cos(2x+y)$$

## 9.8 NONHOMOGENEOUS LINEAR PARTIAL DIFFERENTIAL EQUATIONS

If in Eq. (27), the polynomial $f(D, D')$ is not homogeneous, then Eq. (27) is called a nonhomogeneous linear partial differential equation and its complete solution consists of a complementary function and a particular integral. The methods for finding the particular integral are the same as in Section 9.7, and to obtain complementary function, we factorize $f(D, D')$ into factors of the form $D - mD' - c$. Now, to find the solution of $(D - mD' - c)z = 0$, we write it as

$$p - mq = cz \qquad (35)$$

The subsidiary equation is

$$\frac{dx}{1} = \frac{dy}{-m} = \frac{dz}{cz}$$

Its integrals are

$$y + mx = a, \qquad z = be^{cx}$$

Taking $b = \phi(a)$, we get the solution of Eq. (35) as $z = e^{cx}\phi(y + mx)$. That is, when $f(D,D')$ is resolved into non-repeated factors of the form $(lD - mD' - c)$ then the complementary function is $e^{cx/l}\phi(ly+mx)$.

The polynomial $f(D,D')$ can also be factorized in other ways. For each possible factors, different complementary functions occur and we have

(i) corresponding to a repeated factor $(lD - mD' - c)^n$,

$$\text{C.F.} = e^{cx/l}\{\phi_1(ly+mx) + x\phi_2(ly+mx) + \ldots + x^{n-1}\phi_n(ly+mx)\}$$

(ii) corresponding to a factor $(mD'+c)$, the complementary function is $e^{-cy/m}\phi(mx)$. If the factor $(mD'+c)$ is repeated $n$ times, then

$$\text{C.F.} = e^{-cy/m}\{\phi_1(mx) + x\phi_2(mx) + \cdots + x^{n-1}\phi_n(mx)\}$$

(iii) for a non-repeated factor of the form $(lD - mD')$, the complementary function is $\phi(ly+mx)$, while for a repeated factor $(lD-mD')^n$,

$$\text{C.F.} = \phi_1(ly+mx) + x\phi_2(ly+mx) + \cdots + x^{n-1}\phi_n(ly+mx)$$

(iv) corresponding to a factor $D$, the complementary function is $\phi(y)$, while for a factor of the form $D^n$,

$$\text{C.F.} = \phi_1(y) + x\phi_2(y) + \cdots + x^{n-1}\phi(y)$$

(v) for a factor of the form $D'$, the complementary function is $\phi(x)$ and corresponding to a factor $D'^n$,

$$\text{C.F.} = \phi_1(x) + y\phi_2(x) + \cdots + y^{n-1}\phi(x)$$

Moreover, if $f(D,D')$ of equation $f(D,D')z = 0$ cannot be factorized into linear factors, then the method of finding the complementary function is as follows:

Let $z = ce^{hx+ky}$ be a solution of the equation $f(D,D')z = 0$. Substituting this value of $z$ in the given partial differential equation, we get

$$cf(h,k)e^{hx+ky} = 0$$

which is zero only when $f(h,k) = 0$. Solve this equation for $h$ and $k$, then $z = ce^{hx+ky}$ will be a part of the complementary function. If $D'$ in $f(D,D')z = 0$ is of degree $n$, then the solution of $f(h,k) = 0$ will give $f_1(h), f_2(h), \ldots, f_n(h)$. Corresponding to $k = f_1(h)$, the part of the solution of $f(D,D')z = 0$ is $\sum c_1 e^{hx+f_1(h)y}$, where $\sum$ denotes the infinite series obtained by assigning $c$ and $h$ all possible values. Therefore, corresponding to all the values of $k$, the general solution is

$$z = \sum c_1 e^{hx+f_1(h)y} + \sum c_2 e^{hx+f_2(h)y} + \ldots + \sum c_n e^{hx+f_n(h)y}$$

If $k = ah + b$, i.e., $f(h)$ is linear in $h$, then the solution can take a simple form. This is true in particular when $f(D,D')$ in $f(D,D')z = 0$ is homogeneous. Equation $f(h,k) = 0$ can also be solved for $h$ in terms $k$ and in such, case, we get another form of solution.

The solutions corresponding to the various factors, when added up, give the complementary function of the nonhomogeneous linear partial differential equation.

**EXAMPLE 9.33** Solve $(D^2 + 2DD' + D'^2 - 2D - 2D')z = \sin(x + 2y)$.

***Solution*** Here, $f(D, D') = (D + D')(D + D' - 2)$. The solution corresponding to the factor $D - mD' - c$ is $z = e^{cx}\phi(y + mx)$. Therefore,

$$\text{C.F.} = \phi_1(y - x) + e^{2x}\phi_2(y - x)$$

and $$\text{P.I.} = \frac{1}{D^2 + 2DD' + D'^2 - 2D - 2D'} \sin(x + 2y)$$

$$= \frac{1}{-1 + 2(-2) + (-4) - 2D - 2D'} \sin(x + 2y)$$

$$= -\frac{1}{2(D + D') + 9} \sin(x + 2y)$$

$$= -\frac{2(D + D') - 9}{4(D^2 + 2DD' + D'^2) - 81} \sin(x + 2y)$$

$$= \frac{2(D + D') - 9}{4[-1 + 2(-2) - 4] - 81} \sin(x + 2y)$$

$$= \frac{1}{117}\{2[\cos(x + 2y) + 2\cos(x + 2y)] - 9\sin(x + 2y)\}$$

$$= \frac{1}{39}[2\cos(x + 2y) - 3\sin(x + 2y)$$

Thus, the complete solution is

$$z = \phi_1(y - x) + e^{2x}\phi_2(y - x) + \frac{1}{39}[2\cos(x + 2y) - 3\sin(x + 2y)]$$

**EXAMPLE 9.34** Solve the following partial differential equations

(i) $(DD' + D - D' - 1)z = xy$

(ii) $(D^2 - 6DD' + 9D'^2 - 4D + 12D' + 4)z = 2e^x \tan(y + 3x)$

(iii) $(D - 1)(D - D' + 1)z = e^y$

***Solution*** (i) The given equation can be expressed as

$$(D - 1)(D' + 1)z = xy$$

Here $$C.F. = e^x\phi_1(y) + e^{-x}\phi_2(y)$$

and $$\text{P.I.} = \frac{1}{(D - 1)(D' + 1)} xy$$

$$= -[(1 - D)(1 + D')]^{-1} xy$$

$$= -[(1 + D + D^2 + \cdots)(1 - D' + D'^2 + \cdots)]xy$$

$$= -[(1 + D + D^2 + \cdots)](1 - D' + D'^2 + \cdots)xy$$

$$= -[1 + D + D^2 + \cdots](xy - x)$$

$$= -xy + x - y + 1$$

Therefore, the complete solution is

$$z = e^{x}\phi_1(y) + e^{-x}\phi_2(y) - xy + x - y + 1$$

(ii) The given equation can be written as

$$(D - 3D' - 2)^2 z = 2e^{2x}\tan(y + 3x)$$

Here $$\text{C.F.} = e^{2x}\phi_1(y + 3x) + xe^{2x}\phi_2(y + 3x)$$

Also $$\text{P.I.} = \frac{1}{(D - 3D' - 2)^2} 2e^{2x}\tan(y + 3x)$$

$$= 2e^{2x}\frac{1}{[(D + 2) - 3D' - 2]^2}\tan(y + 3x)$$

$$= 2e^{2x}\frac{1}{(D - 3D')^2}\tan(y + 3x)$$

$$= x^2 e^{2x}\tan(y + 3x) \qquad \text{[from Eq. (33)]}$$

Thus, the complete solution is

$$z = e^{2x}\phi_1(y + 3x) + xe^{2x}\phi_2(y + 3x) + x^2 e^{2x}\tan(y + 3x)$$

(iii) Here

$$\text{C.F.} = e^{x}\phi_1(y) + e^{-x}\phi_2(y + x)$$

and $$\text{P.I.} = \frac{1}{(D - D' + 1)}\frac{1}{(D - 1)}e^{y}$$

$$= \frac{1}{(D - D' + 1)}\frac{1}{(0 - 1)}e^{y}$$

$$= -e^{y}\frac{1}{D - D' + 1}(1)$$

$$= -e^{y}\frac{1}{(D + 0) - (D' + 1) + 1}(1)$$

$$= -e^{y}\frac{1}{D}\left(1 - \frac{D'}{D}\right)^{-1}(1)$$

$$= -e^{y}\frac{1}{D}(1) = -xe^{y}$$

Therefore, the complete solution is

$$z = e^{x}\phi_1(y) + e^{-x}\phi_2(y + x) - xe^{y}$$

**EXAMPLE 9.35** Solve the following partial differential equations:

(i) $(D^2 - D')z = 0$

(ii) $(D^2 - DD' + 2D' - 1)z = 0$

(iii) $(D^2 - D'^2 - 1)z = 0$

***Solution*** (i) Assume that a solution of the given partial differential equation be of the form

$$z = ce^{hx+ky} \tag{36}$$

so that $\quad D'z = cke^{hx+ky},\ Dz = che^{hx+ky},\ D^2z = ch^2e^{hx+ky}$

The given equation thus reduces to

$$c(h^2 - k)e^{hx+ky} = 0$$

which leads to $h^2 - k = 0$, i.e., $h^2 = k$. With this value of $k$, Eq. (36) becomes

$$z = ce^{hx+h^2y}$$

Therefore, the general solution is

$$z = \sum ce^{hx+h^2y}$$

where $c$, and $h$, are arbitrary constants.

(ii) Here, $f(D,D') = D^2 - DD' + 2D' - 1$ can not be factorized into linear factors and thus assume that the solution of the given equation be of the form

$$z = ce^{hx+ky}$$

With this value of $z$, the given equation leads to

$$c(h^2 - hk + 2k - 1)e^{hx+ky} = 0$$

which gives $h^2 - hk + 2k - 1 = 0$. Solving this equation for $k$, we get

$$k = \frac{1-h^2}{2-h} \tag{37}$$

Therefore, the general solution of the given equation is

$$z = \sum ce^{hx+ky}$$

where $k$ is given by Eq. (37), $c$ and $h$ being the arbitrary constants.

(iii) Let the solution of the given partial differential equation be of the form

$$z = ce^{hx+ky}$$

With this value of $z$, the given equation reduces to $c(h^2 - k^2 - 1)e^{hx+ky} = 0$ which gives

$$h^2 - k^2 - 1 = 0 \tag{38}$$

If we choose $h = \sec\alpha$ and $k = \tan\alpha$, then Eq. (38) is satisfied. Therefore, the general solution is

$$z = \sum ce^{x\sec\alpha + y\tan\alpha}$$

where $c$ and $\alpha$ are arbitrary constants.

### 9.8.1 Equations Reducible to Linear Partial Differential Equations

A partial differential equation of the form

$$f(xD, yD') = F(x, y) \tag{39}$$

where $D = \dfrac{\partial}{\partial x}$ and $D' = \dfrac{\partial}{\partial y}$, having variable coefficients, can be reduced to a linear equation with constant coefficients, by making some suitable substitutions. One such substitution is

$$x = e^u \quad \text{and} \quad y = e^v \tag{40a}$$

so that

$$u = \log x \quad \text{and} \quad v = \log y \tag{40b}$$

Now, let $D_1 = \dfrac{\partial}{\partial u}$ and $D_1' = \dfrac{\partial}{\partial v}$ then (cf., Section 5.7)

$$x\frac{\partial}{\partial x} = D_1, x^2\frac{\partial^2}{\partial x^2} = D_1(D_1 - 1), \ldots$$

and

$$y\frac{\partial}{\partial y} = D_1', y^2\frac{\partial^2}{\partial y^2} = D_1'(D_1' - 1), \ldots$$

Thus, in general we have

$$x^m y^n \frac{\partial^{m+n}}{\partial x^m \partial y^n} = x^m \frac{\partial^m}{\partial x^m} y^n \frac{\partial^n}{\partial y^n}$$

$$= [D_1(D_1 - 1)\ldots(D_1 - m + 1)][D_1'(D_1' - 1)\ldots(D_1' - n + 1)]$$

Using these substitutions, the given equation reduces to linear equation with constant coefficients and can now be solved by the methods of Sections 9.7 and 9.8.

**EXAMPLE 9.36** Solve the following partial differential equations:

(i) $x^2D^2z - y^2D'^2z = xy$

(ii) $x^{-2}\dfrac{\partial^2 z}{\partial x^2} - x^{-3}\dfrac{\partial z}{\partial x} - y^{-2}\dfrac{\partial^2 z}{\partial y^2} - y^{-3}\dfrac{\partial z}{\partial y} = 0$

(iii) $x^2r - 4xypq + 4y^2t + 6yq = x^2y^4$

***Solution*** (i) Using the substitution given by Eqs. [40(a) and (b)], the given equation can now be written as

$$(D_1 - D_1')(D_1 + D_1' - 1)z = e^{u+v}$$

so that

$$\text{C.F.} = \phi_1(u+v) + e^u\phi_2(u-v)$$

$$= \phi_1(\log x + \log y) + x\phi_2(\log x - \log y)$$

$$= \phi_1[\log(xy)] + x\phi_2\left[\log\frac{x}{y}\right]$$

Also

$$\text{P.I.} = \frac{1}{(D_1 - D_1')(D_1 + D_1' - 1)}e^{u+v}$$

$$= \frac{1}{(D_1 - D_1')(1+1-1)}e^{u+v}$$

$$= e^{u+v}\frac{1}{D_1}\left(1 - \frac{D_1'}{D_1}\right)^{-1}(1)$$

$$= e^{u+v}\frac{1}{D_1}(1) = ue^{u+v} = xy\log x$$

Therefore, the required solution is

$$z = \phi_1[\log(xy)] + x\phi_2\left[\log\frac{x}{y}\right] + xy\log x$$

(ii) Put $u = \frac{1}{2}x^2, v = \frac{1}{2}y^2$ so that $du = xdx, dv = ydy$. Also

$$\frac{\partial z}{\partial u} = \frac{\partial z}{\partial x}\frac{\partial x}{\partial u} = \frac{1}{x}\frac{\partial z}{\partial x}$$

$$\frac{\partial^2 z}{\partial u^2} = \frac{\partial}{\partial u}\left(\frac{1}{x}\frac{\partial z}{\partial x}\right) = \frac{\partial}{\partial x}\left(\frac{1}{x}\frac{\partial z}{\partial x}\right)\frac{\partial x}{\partial u} = -x^{-3}\frac{\partial z}{\partial x} + x^{-2}\frac{\partial^2 z}{\partial x^2}$$

and

$$\frac{\partial^2 z}{\partial v^2} = -y^{-3}\frac{\partial z}{\partial y} + y^{-2}\frac{\partial^2 z}{\partial y^2}$$

With these substitutions, the given equation reduces to

$$\frac{\partial^2 z}{\partial u^2} - \frac{\partial^2 z}{\partial v^2} = 0$$

Therefore, the required solution is

$$z = \phi_1(v+u) + \phi_2(v-u) = \phi_1\left(\frac{1}{2}y^2 + \frac{1}{2}x^2\right) + \phi_2\left(\frac{1}{2}y^2 - \frac{1}{2}x^2\right)$$

(iii) Using Eq. (40), the given equation can be expressed as

$$(D_1 - 2D_1')(D_1 - 2D_1' - 1)z = e^{2u+4v}$$

Here

$$\begin{aligned}\text{C.F.} &= \phi_1(v+2u) + e^u\phi_2(v+2u)\\ &= \phi_1(\log y + 2\log x) + e^{\log x}\phi_2(\log y + 2\log x)\\ &= \phi_1(\log x^2 y) + x\phi_2(\log x^2 y)]\end{aligned}$$

and

$$\begin{aligned}\text{P.I.} &= \frac{1}{(D_1 - 2D_1')(D_1 - 2D_1' - 1)} e^{2u+4v}\\ &= \frac{1}{30}(e^u)^2(e^v)^4 = \frac{1}{30}x^2y^4\end{aligned}$$

Therefore, the complete solution is

$$z = \phi_1(\log x^2 y) + x\phi_2(\log x^2 y) + \frac{1}{30}x^2y^4$$

## 9.9 A GENERAL METHOD FOR SOLVING A NON-LINEAR PARTIAL DIFFERENTIAL EQUATION: MONGE'S METHOD

In Section 9.6, we have given a method for solving a non-linear partial differential equation of order one. In what follows, we shall discuss a method for finding the solution of a non-linear (quasi-linear) partial differential equation of order two.

The general partial differential equation of order two is of the form

$$F(x, y, z, p, q, r, s, t) = 0 \tag{41}$$

In most of the cases, Eq. (41) cannot be integrated exactly. Consider a second order non-linear partial differential equation of the form

$$R\frac{\partial^2 z}{\partial x^2} + S\frac{\partial^2 z}{\partial x \partial y} + T\frac{\partial^2 z}{\partial y^2} = V \tag{42a}$$

or

$$Rr + Ss + Tt = V \tag{42b}$$

where $R$, $S$, $T$ and $V$ are functions of $x$, $y$, $z$, $p$ and $q$. The method of solution of Eq. (42a)/(42b) is due to Gaspard Monge (1746 – 1818) and is known as *Monge's method.* This method involves the reduction of Eq. (42a) into an equivalent system of two equations. From these two equations, we find the values of $p$ or $q$ or both $p$ and $q$. If $p$ and $q$ are known, then the integration of $dz = pdx + pdy$ leads to the solution. If $p$ or $q$ is determined, then the solution is obtained by using the procedure of Section 9.4. However, the details of Monge's method are as follows:

The total derivatives of $p$ and $q$ are known to be as

$$p = \frac{\partial p}{\partial x}dx + \frac{\partial p}{\partial y}dy = rdx + sdy \tag{43a}$$

and

$$q = \frac{\partial q}{\partial x}dx + \frac{\partial q}{\partial y}dy = sdx + tdy \tag{43b}$$

Substituting the values of $r$ and $t$ from Eqs. (43a) and (43b), respectively, in Eq. (42b), we get

$$(Rdpdy + Tdqdx - Vdxdy) - s(Rdy^2 - Sdydx + Tdx^2) = 0 \tag{44}$$

A solution of Eq. (42b) is also a solution of Eq. (44). Since this equation holds for arbitrary values of $s$, therefore Eq. (44) is identically satisfied only when

$$Rdpdy + Tdqdx - Vdxdy = 0 \tag{45a}$$

and

$$Rdy^2 - Sdydx + Tdx^2 = 0 \tag{45b}$$

Equations (45a) and (45b) are known as *Monge's subsidiary equations,* and from these equations, we can find one or two relations between $x$, $y$, $z$, $p$ and $q$. These relations are called *intermediate integrals.* Solve these relations for $p$ and $q$ and integrate the equation $dz = pdx + qdy$ to get the solution of the given partial differential equation. Also, from Eq. (45b), we have

$$R\left(\frac{dy}{dx}\right)^2 - S\left(\frac{dy}{dx}\right) + T = 0 \tag{46}$$

which on solving leads to

$$\frac{dy}{dx} = \frac{1}{2R}(S \pm \sqrt{S^2 - 4RT}) \tag{47}$$

Equation (47) may lead to two distinct values of $dy/dx$ and in this case Eq. (45b) can be factorized into two distinct factors. Both these values of $dy/dx$ may be used to obtain the solution, or if we are using only one of the values of $dy/dx$ we then have to solve the first order partial differential equation (for example, Lagrange's equation) to get the solution.

Moreover, if in Eq. (47), $S^2 - 4RT = 0$, we then have only one value of $dy/dx$; and Eq. (45b) now becomes a perfect square. Thus, to obtain a solution, we use the methods for solving the partial differential equation of order one. It may be noted, from Eq. (46), that if $R = T = 0$, we then also have only one value of $dy/dx$.

Following examples will illustrate the method of solving partial differential equations by Monge's method.

**EXAMPLE 9.37** Solve $\dfrac{\partial^2 z}{\partial x^2} - a^2\dfrac{\partial^2 z}{\partial y^2} = 0$

***Solution*** The given equation can be written as $r - a^2 t = 0$. Comparing this equation with Eq. (42a) we see that $R = 1, S = 0, T = -a^2$ and thus subsidiary Eq. (45) leads to

$$dpdy - a^2 dqdx = 0 \tag{48}$$

$$dy^2 - a^2 dx^2 = 0 \tag{49}$$

From Eq. (49), we have

$$(dy - adx)(dy + adx) = 0$$

which leads to

$$dy - adx = 0 \tag{50a}$$

$$dy + adx = 0 \tag{50b}$$

Now, from Eqs. (50a) and (48), we have

$$dp - adq = 0 \tag{51}$$

which on integration leads to

$$p - aq = c_1 \tag{52}$$

while the integration of Eq. (50 a) leads to

$$y - ax = c_2 \tag{53}$$

Thus, from Eqs. (52) and (53), the intermediate integral is

$$p - aq = \phi_1(y - ax) \tag{54}$$

In a similar way, using Eqs. (50 b) and (48), we can obtain the second intermediate integral as

$$p + aq = \phi_2(y + ax) \tag{55}$$

Equations (54) and (55) now leads to

$$p = \frac{1}{2}[\phi_2(y + ax) + \phi_1(y - ax)]$$

$$q = \frac{1}{2a}[\phi_2(y + ax) - \phi_1(y - ax)]$$

Now, using these values in equation $dz = pdx + qdy$, we get

$$dz = \frac{1}{2}[\phi_2(y + ax) + \phi_1(y - ax)]dx + \frac{1}{2a}[\phi_2(y + ax) - \phi_1(y - ax)]dy$$

which can also be expressed as

$$dz = \frac{1}{2a}[\phi_2(y + ax)(dy + adx) - \phi_1(y - ax)(dy - adx)]$$

so that the integration of this equation yields

$$z = \frac{c_3}{2a}\phi_2(y+ax) - \frac{c_2}{2a}\phi_1(y-ax) = \psi_1(y+ax) - \psi_2(y+ax)$$

which is the required solution.

**EXAMPLE 9.38** Solve $\frac{\partial^2 z}{\partial x^2} - \sin^2 x\frac{\partial^2 z}{\partial y^2} - \cot x\frac{\partial z}{\partial x} = 0$

***Solution*** The given partial differential equation can be expressed as

$$\frac{\partial^2 z}{\partial x^2} - \sin^2 x\frac{\partial^2 z}{\partial y^2} = \cot x\frac{\partial z}{\partial x}$$

Comparing this equation with Eq. (42), we get $R = 1, S = 0, T = -\sin^2 x$, $V = -\cot x\frac{\partial z}{\partial x}$. The subsidiary eqn (45) now becomes

$$dy^2 - \sin^2 dx^2 = 0 \tag{56a}$$

and $$dpdy - \sin^2 xdqdx - p\cot xdxdy = 0 \tag{56b}$$

Equation (56a) has distinct factors as

$$dy + \sin xdx = 0 \;\; and \;\; dy - \sin xdx = 0$$

which have solutions as

$$y = -\cos x + c_1 \tag{57a}$$

and $$y = \cos x + c_2 \tag{57b}$$

We can now use both of these equations, one by one, in Eq. (56b) to get the values of $p$ and $q$ so that we can use them in equation $dz = pdx + qdy$ to obtain the required solution. Equations (57a) and (56b) lead to

$$\csc xdp - p\cot x\csc xdx - dq = 0$$

which may be expressed as

$$d(p\csc x) - dq = 0$$

Integrating this equation, we get

$$p\csc x - q = c_3 \tag{58}$$

which on using Eq. (57a) leads to

$$p\csc x - q = \psi_1(y + \cos x) \tag{59}$$

Now, using Eq. (57b) in Eq. (56a), we get

$$\csc xdp - p\cot x\csc xdx + dq = 0$$

which can be written as

$$d(p\csc x) + dq = 0$$

Integrating this equation, we get

$$p \csc x + q = c_4 \tag{60}$$

which from Eq. (57 b) becomes

$$p \csc x + q = \psi_2(y - \cos x) \tag{61}$$

Solving Eqs. (59) and (61) for $p$ and $q$, we get

$$p = \frac{1}{2}\sin x[\psi_1(y + \cos x) + \psi_2(y - \cos x)]$$

$$q = \frac{1}{2}[\psi_2(y - \cos x) + \psi_1(y + \cos x)]$$

Now substituting these values of $p$ and $q$ in equation $dz = pdx + qdy$, we get

$$dz = -\frac{1}{2}\psi_1(y + \cos x)d(y + \cos x) + \frac{1}{2}\psi_2(y - \cos x)d(y - \cos x)$$

Integrating this equation, we get

$$z = -\frac{1}{2}\psi_1(y + \cos x)(y + \cos x) + \frac{1}{2}\psi_2(y - \cos x)(y - \cos x)$$

$$= f(y + \cos x) + g(y - \cos x)$$

which is the required solution.

**EXAMPLE 9.39** Solve $q^2 r - 2pqs + p^2 t = 0$.

***Solution*** Here, $R = q^2, S = -2pq, T = p^2, V = 0$, and Monge's subsidiary equations are as follows:

$$q^2 dpdy + p^2 dqdx = 0 \tag{62}$$

$$q^2 dy^2 + 2pqdydx + p^2 dx^2 = 0 \tag{63}$$

Equation (63) may be expressed as

$$qdy + pdx = 0 \tag{64}$$

which leads to

$$dz = pdx + qdy = 0$$

so that

$$z = c_1 \tag{65}$$

Using Eq. (64) in Eq. (62), we get after simplification

$$qdp = pdq$$

Separating the variables and integrating, we get

$$\frac{p}{q} = c_2 \tag{66}$$

From Eqs. (65) and (66), we have

$$\frac{p}{q} = \phi(z)$$

which may be expressed as

$$p - q\phi(z) = 0$$

This is the Lagrange's equation and the subsidiary equation is

$$\frac{dx}{1} = \frac{dy}{-\phi(z)} = \frac{dz}{0}$$

These equations lead to

$$z = c_2 \quad \text{and} \quad y + x\phi(z) = c_3$$

Therefore, the complete solution is

$$y + x\phi(z) = \psi(z)$$

**EXAMPLE 9.40** Solve $y^2 r + 2xys + x^2 t = -(px + qy)$.

***Solution*** Here $R = y^2, S = 2xy, T = x^2, V = -(px + qy)$ and $S^2 - 4RT = 0$. Thus, from Eq. (47), we have

$$\frac{dy}{dx} = \frac{1}{2R} S = \frac{x}{y}$$

which leads to

$$ydy = xdx \tag{67}$$

Also, Monge's subsidiary Eq. (45 a) leads to

$$y^2 dpdy + x^2 dqdx + (px + qy)dydx = 0 \tag{68}$$

which, using Eq. (67) takes the form

$$d(yp) + d(xq) = 0$$

so that on integration, we have

$$yp + xq = c_1 \tag{69}$$

Also, integration of Eq. (67) leads to

$$x^2 - y^2 = c_2 \tag{70}$$

From Eqs. (69) and (70), one integral of the given partial differential equation is

$$yp + xq = f(x^2 - y^2) \tag{71}$$

This equation is of Lagrange's form and the subsidiary equation is

$$\frac{dx}{y} = \frac{dy}{x} = \frac{dz}{f(x^2 - y^2)} \tag{72}$$

First two fractions of Eq. (72) leads to $x^2 - y^2 = c_2$ [which is same as Eq. (70)]. With this equation, the last two fractions of Eq. (72) can be written as

$$\frac{dy}{\sqrt{y^2 + c_2}} = \frac{dz}{f(c_2)}$$

which leads to

$$dz - f(c_2)\frac{dy}{\sqrt{y^2 + c_2}} = 0$$

Integrating this equation, we get

$$z - f(c_2)\log[y + \sqrt{y \;\; + c_2}] = c_3$$

which, on using Eq. (70) takes the form

$$z - f(x^2 - y^2)\log(x + y) = c_3 \tag{73}$$

Therefore, from eqns (70) and (73), the required solution is

$$z - f(x^2 - y^2)\log(x + y) = g(x^2 - y^2)$$

**EXAMPLE 9.41** Solve $(q + 1)s - (p + 1)t = 0$.

***Solution*** Here, $R = 0, S = q + 1, T = -(p + 1), V = 0$ and Monge's subsidiary equations are as follows:

$$-(p + 1)dqdx = 0$$

$$-(q + 1)dxdy - (p + 1)dx^2 = 0$$

These equations lead to

$$dq = 0 \tag{74}$$

and
$$(q + 1)dy + (p + 1)dx = 0 \tag{75}$$

Integration of Eq. (74) leads to

$$q = c_1 \tag{76}$$

while using $dz = pdx + qdy$, Eq. (75) can be written as

$$dx + dy + dz = 0 \tag{77}$$

Integrating this equation, we get

$$x + y + z = c_2 \tag{78}$$

From Eqs. (76) and (78), we have

$$q = \frac{\partial z}{\partial y} = f(x + y + z)$$

Integrating this equation with respect to $y$ (keeping $x$ as constant), we get the required solution as

$$z = \psi(x + y + z) + \phi(x)$$

where $\psi$ and $\phi$ are arbitrary constants.

## 9.10 SEPARATION OF VARIABLES

Consider a partial differential equation of the form

$$A\frac{\partial^2 z}{\partial x^2}+B\frac{\partial^2 z}{\partial x \partial y}+C\frac{\partial^2 z}{\partial y^2}=F\left(x, y, z, \frac{\partial z}{\partial x}, \frac{\partial z}{\partial y}\right) \tag{79}$$

where $A$, $B$ and $C$ are continuous functions of $x$ and $y$, the derivatives are also continuous and $F$ denotes a polynomial function of $x$, $y$, $z$, $\partial z/\partial x$ and $\partial z/\partial y$. Equation (79) is called

hyperbolic, if $B^2 - 4AC > 0$,
parabolic, if $B^2 - 4AC = 0$,
elliptic, if $B^2 - 4AC < 0$.

For example, the differential equations governing the phenomenon of one- and two-dimensional heat flow are parabolic and elliptic, respectively, while the motion of a vibrating string is described by a hyperbolic partial differential equation.

However, whether an equation is parabolic, elliptic or hyperbolic depends upon the values of $A$, $B$, $C$ in Eq. (79). Thus, the equation

$$\frac{\partial^2 u}{\partial x^2}+x\frac{\partial^2 u}{\partial y^2}+\frac{\partial u}{\partial y}=0$$

is elliptic when $x > 0$ and is hyperbolic when $x < 0$.

Partial differential equations of the second order are of great importance, as we shall see in the subsequent sections. In physical problems, we usually look for a solution of a partial differential equation which satisfies certain specified conditions, known to be the boundary conditions. The differential equation, together with these boundary conditions, constitutes a *boundary value problem.* A process used in finding the solutions of boundary value problems involving partial differential equations is known as *separation of variables*, and can best be explained with the help of the following examples.

**EXAMPLE 9.42** Solve $\dfrac{\partial^2 z}{\partial x^2}+4\dfrac{\partial^2 z}{\partial y^2}=0$.

***Solution*** We assume a solution of the form

$$z = X(x)Y(y) \tag{80}$$

where $X$ is a function of $x$ and only $Y$ is a function of $y$ only. From Eq. (80), the given equation reduces to

$$X''Y + 4Y''X = 0 \tag{81}$$

where $X' = dX/dx$, $Y' = dY/dy$, etc. Equation (81) can be written as

$$\frac{X''}{X} = -4\frac{Y''}{Y}$$

Here $X''/X$, being a function of $x$, does not change when $y$ alone changes and $-4Y''/Y$ does not change when $x$ alone changes. Hence, the two ratios will be equal only when both are equal to the same constant $k$, that is, when

$$X'' = kX, \qquad -4Y'' = kY \tag{82}$$

The solutions of these ordinary differential Eq. (82) for $k > 0$ are

$$\left.\begin{aligned} X &= a_1 e^{\sqrt{k}x} + a_2 e^{-\sqrt{k}x} \\ Y &= a_3 \sin\left(\frac{1}{2}\sqrt{k}\,y\right) + a_4 \cos\left(\frac{1}{2}\sqrt{k}\,y\right) \end{aligned}\right\} \tag{83}$$

Now, Eqs. (80) and (83) satisfy the given equation for any value of $k$. Hence, in accordance with the principle of adding of solutions of linear equations, the solution of the given equation can be written as

$$Z = \sum_k X_k Y_k = \sum_k \left\{ e^{\sqrt{k}x}\left[a_k \sin\left(\frac{1}{2}\sqrt{k}\,y\right) + b_k \cos\left(\frac{1}{2}\sqrt{k}\,y\right)\right]\right.$$
$$\left. + e^{-\sqrt{k}x}\left[c_k \sin\left(\frac{1}{2}\sqrt{k}\,y\right) + d_k \cos\left(\frac{1}{2}\sqrt{k}y\right)\right]\right\} \tag{84}$$

where $k$ may take the values of any finite set of positive numbers. If $k$ takes the values $k_1, k_2, \ldots$, then Eq. (84), involving an infinite series, is a valid solution within its region of absolute convergence.

If $k$ is negative, the sine-cosine part of the solution would go with $X$ and the exponential part with $Y$. When $k = 0$, we have

$$X = c_1 x + c_2, \qquad Y = c_3 y + c_4$$

*Note:* From the above discussions, it is clear that the solution (84) is made up of three types. For any specific application, the appropriate types are to be used.

**EXAMPLE 9.33** Solve $\dfrac{\partial^2 z}{\partial x^2} - 2\dfrac{\partial z}{\partial x} + \dfrac{\partial z}{\partial y} = 0$.

***Solution*** We assume that the solution is of the form

$$z = X(x)Y(y) \tag{85}$$

where $X$ is a function of $x$ only and $Y$ is a function of $y$ only. From Eq. (85), the given equation reduces to

$$X''Y - 2X'Y + XY' = 0$$

where

$$X' = \frac{dX}{dx}, \qquad Y' = \frac{dY}{dy}, \ldots$$

Separating the variables, we get

$$\frac{X'' - 2X'}{X} = -\frac{Y'}{Y} = k$$

or $\quad X'' - 2X' - kX = 0$ and $Y' + kY = 0$

The solutions of these ordinary differential equations are

$$X = c_1 e^{[1+\sqrt{1+k}]x} + c_2 e^{[1-\sqrt{1+k}]x}, \qquad Y = c_3 e^{-ky}$$

Substituting these values in Eq. (85), we get the required solution as

$$z = \left[c_1 e^{\left(1+\sqrt{1+k}\right)x} + c_2 e^{\left(1-\sqrt{1+k}\right)x}\right] c_3 e^{-ky}$$

$$= \left[k_1 e^{\left(1+\sqrt{1+k}\right)x} + k_2 e^{\left(1-\sqrt{1+k}\right)x}\right] e^{-ky}$$

## 9.11 FOURIER SERIES*

In many problems occurring in the conduction of heat, electrodynamics and acoustics, it is necessary to represent a function in series of sine and cosine. Most of the single-valued function which occur in these fields can be expressed in the form of a series

$$\frac{a_0}{2} + a_1 \cos x + a_2 \cos 2x + \cdots + a_n \cos nx + \cdots$$
$$+ b_1 \sin x + b_2 \sin 2x + \cdots + b_n \sin nx + \cdots \qquad (86)$$

Such a series as (86) is called a *trigonometric series.*

The *Fourier series* for the function $f(x)$ in the interval $-\pi \le x \le \pi$ is defined by

$$f(x) = \frac{a_0}{2} + \sum_{n=1}^{\infty} (a_n \cos nx + b_n \sin nx) \qquad (87)$$

where

$$\left.\begin{aligned} a_0 &= \frac{1}{\pi}\int_{-\pi}^{\pi} f(x)dx \\ a_n &= \frac{1}{\pi}\int_{-\pi}^{\pi} f(x)\cos nx\, dx \\ b_n &= \frac{1}{\pi}\int_{-\pi}^{\pi} f(x)\sin nx\, dx \end{aligned}\right\} \qquad (88)$$

Fourier series also has applications in electronic circuits which are designed to handle sharply rising pulses.

---

* We shall present here only a brief account of the Fourier series which is necessary for our purpose. For more details, see [2].

**EXAMPLE 9.44** Expand $(x - x^2)$ in a Fourier series in the interval $[-\pi, \pi]$.

***Solution*** Let $f(x) = x - x^2$, so that Eqs. (87) and (88) take the form

$$x - x^2 = \frac{a_0}{2} + \sum_{n=1}^{\infty} (a_n \cos nx + b_n \sin nx) \tag{89}$$

Here

$$a_0 = \frac{1}{\pi} \int_{-\pi}^{\pi} (x - x^2) dx = -\frac{2\pi^2}{3}$$

and
$$a_n = \frac{1}{\pi} \int_{-\pi}^{\pi} (x - x^2) \cos nx \, dx \quad \text{(integrating by parts)}$$

$$= \frac{-4(-1)^n}{n^2} \quad (\cos n\pi = (-1)^n)$$

which gives

$$a_1 = \frac{4}{1^2}, \quad a_2 = \frac{-4}{2^2}, \quad a_3 = \frac{4}{3^2}, \ldots$$

Also,

$$b_n = \frac{1}{\pi} \int_{-\pi}^{\pi} (x - x^2) \sin nx \, dx = \frac{-2(-1)^n}{n}$$

which gives

$$b_1 = \frac{2}{1}, \; b_2 = \frac{-2}{2}, \; b_3 = \frac{2}{3}, \ldots$$

Substituting these values of $a$ and $b$ in Eq. (89), we obtain the Fourier series of $x - x^2$ as

$$x - x^2 = -\frac{\pi^2}{3} + 4\left(\frac{\cos x}{1^2} - \frac{\cos 2x}{2^2} + \frac{\cos 3x}{3^2} + \ldots\right)$$
$$+ 2\left(\frac{\sin x}{1} - \frac{\sin 2x}{2} + \frac{\sin 3x}{3} + \ldots\right)$$

If we take $x = 0$, then

$$\frac{\pi^2}{12} = \frac{1}{1^2} - \frac{1}{2^2} + \frac{1}{3^2} - \frac{1}{4^2} + \ldots.$$

**EXAMPLE 9.45** Find the Fourier series for the periodic function defined as

$$f(x) = \begin{cases} 0, & -\pi < x < 0 \\ x, & 0 < x < \pi \end{cases}$$

***Solution*** Here

$$a_0 = \frac{1}{\pi}\int_{-\pi}^{\pi} f(x)dx = \frac{\pi}{2}$$

$$a_n = \frac{1}{\pi}\int_{-\pi}^{\pi} f(x)\cos nx dx = \frac{1}{\pi n^2}(\cos n\pi - 1), n = 1,2,3,...$$

$$b_n = \frac{1}{\pi}\int_{-\pi}^{\pi} f(x)\sin nx dx = \frac{1}{n}\cos n\pi$$

Thus, the Fourier series representation of the given function is

$$f(x) = \frac{\pi}{4} + \sum_{n=1}^{\infty}[\frac{(-1)^n - 1}{\pi n^2}\cos nx - \frac{(-1)^n}{n}\sin nx]$$

where $\cos n\pi = (-1)^n$.

*Note:* It is not possible always to write the Fourier expansion of each function. However, one can obtain the Fourier expansion of a function $f(x)$ if it satisfies the following condition, known as the *Dirichlet condition:*

"Any function $f(x)$ can be written as a Fourier series

$$\frac{a_0}{2} + \sum_{n=1}^{\infty}(a_n \cos nx + b_n \sin nx)$$

where $a_0$, $a_n$, $b_n$ are constants, provided

(a) $f(x)$ is periodic, single valued and finite.
(b) $f(x)$ has a finite number of discontinuities in any one periodic.
(c) $f(x)$ has atmost a finite number of maximas and minimas."

In many problems, the period of function required to be expanded is not $2\pi$ but some other interval, say $2c$, and in order to apply the above discussions to functions of period $2c$, this interval must be converted to the length $2\pi$.

Consider the periodic function $f(x)$ defined in $(\alpha, \alpha + 2c)$. Make the following substitution:

$$z = \frac{\pi x}{c} \quad \text{or} \quad x = \frac{cz}{\pi} \tag{90}$$

so that when $x = \alpha$, we have $z = \alpha\pi/c = \beta$ (say), and when $x = \alpha + 2c$, we have $z = (\alpha + 2c)\pi/c = \beta + 2\pi$.

Thus, the function $f(x)$ of period $2c$ in $(\alpha, \alpha + 2c)$ is transformed into the function $f(cz/\pi) = F(z)$ of period $2\pi$ in $(\beta, \beta + 2\pi)$. Hence, the Fourier series of $F(z)$ is

$$f\left(\frac{cz}{\pi}\right) = \frac{a_0}{2} + \sum_{n=1}^{\infty}(a_n \cos nz + b_n \sin nz) \tag{91}$$

where

$$\left.\begin{aligned} a_0 &= \frac{1}{\pi}\int_{\alpha}^{\alpha+2\pi} f\left(\frac{cz}{\pi}\right) dz \\ a_n &= \frac{1}{\pi}\int_{\alpha}^{\alpha+2\pi} f\left(\frac{cz}{\pi}\right)\cos nz\, dz \\ b_n &= \frac{1}{\pi}\int_{\alpha}^{\alpha+2\pi} f\left(\frac{cz}{\pi}\right)\sin nz\, dz \end{aligned}\right\} \tag{92}$$

Now, using $z = \pi x/c$, $dz = (\pi/c)dx$ in Eqs. (90) and (91), the Fourier expansion of $f(x)$ in the interval $(\alpha, \alpha + 2c)$ is

$$f(x) = \frac{a_0}{2} + \sum_{n=1}^{\infty}\left(a_n \cos\frac{n\pi x}{c} + b_n \sin\frac{n\pi x}{c}\right)$$

where

$$\left.\begin{aligned} a_0 &= \frac{1}{c}\int_{\alpha}^{\alpha+2c} f(x)dx \\ a_n &= \frac{1}{c}\int_{\alpha}^{\alpha+2c} f(x)\cos\frac{n\pi x}{c}\, dx \\ b_n &= \frac{1}{c}\int_{\alpha}^{\alpha+2c} f(x)\sin\frac{n\pi x}{c}\, dx \end{aligned}\right\} \tag{93}$$

*Note:* Take $\alpha = 0$ in Eq. (93) to get the results for the interval $(0, 2c)$ and $\alpha = -c$ in Eq. (93) gives the results for the interval $(-c, c)$.

From this note, a periodic function $f(x)$ defined on $(-c, c)$ can be represented by the Fourier series

$$f(x) = \frac{a_0}{2} + \sum_{n=1}^{\infty}\left(a_n \cos\frac{n\pi x}{c} + b_n \sin\frac{n\pi x}{c}\right)$$

where

$$a_0 = \frac{1}{c}\int_{-c}^{c} f(x)dx$$

$$a_n = \frac{1}{c}\int_{-c}^{c} f(x)\cos\frac{n\pi x}{c}\, dx$$

$$b_n = \frac{1}{c}\int_{-c}^{c} f(x)\sin\frac{n\pi x}{c}\, dx$$

Thus, for even and odd periodic functions, respectively, we have

$$a_0 = \frac{2}{c}\int_0^c f(x)dx, \quad a_n = \frac{2}{c}\int_0^c f(x)\cos\frac{n\pi x}{c}\,dx, \quad b_n = 0 \tag{94}$$

and $$a_0 = 0, \quad a_n = 0, \quad b_n = \frac{2}{c}\int_0^c f(x)\sin\frac{n\pi x}{c}\,dx$$

That is, the Fourier expansion for an even periodic function contains only cosine terms while for the odd periodic function there are only sine terms in its Fourier expansion.

If a function $f(x)$ is defined in the interval $(0, \pi)$ or $[0, \pi]$, then the Fourier expansion of $f(x)$ in terms of sine and cosine series is called the *half range series* and

(i) For half range sine series

$$f(x) = \sum_{n=1}^{\infty} b_n \sin nx, \text{ where } b_n = \frac{2}{\pi}\int_0^{\pi} f(x)\sin nx dx$$

(ii) For half range cosine series

$$f(x) = \frac{a_0}{2} + \sum_{n=1}^{\infty} a_n \cos nx$$

where

$$a_0 = \frac{2}{\pi}\int_0^{\pi} f(x)dx, \; a_n = \frac{2}{\pi}\int_0^{\pi} f(x)\cos nx dx$$

**EXAMPLE 9.46** Find the Fourier series for $f(x) = x$ in $(-\pi, \pi)$.

***Solution*** The given function is odd, therefore

$$f(x) = \sum_{n=1}^{\infty} b_n \sin nx$$

where

$$b_n = \frac{2}{\pi}\int_0^{\pi} f(x)\sin nx\,dx = \frac{2}{\pi}\int_0^{\pi} x\sin nx\,dx = \frac{-2\cos n\pi}{n}$$

Therefore, $b_1 = 2/1$, $b_2 = -2/2$, $b_3 = 2/3$, $b_4 = -2/4$, ... Hence, the Fourier series is

$$x = 2\left(\sin x - \frac{1}{2}\sin 2x + \frac{1}{3}\sin 3x - \frac{1}{4}\sin 4x + \cdots\right)$$

The graph of $f(x) = x$ in $(-\pi, \pi)$ is shown in Fig. 9.1.

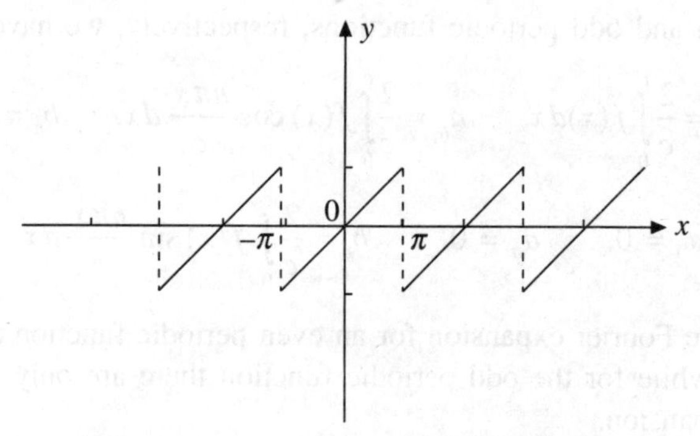


**Fig. 9.1** Graph of $f(x) = x$ in $(-\pi, \pi)$.

**EXAMPLE 9.47** Find the Fourier series for the function $f(x) = x$ defined in $(0, 2\pi)$

***Solution*** Here $f(x)$ is defined in the interval $(0, 2\pi)$ and

$$a_0 = \frac{1}{\pi}\int_0^{2\pi} f(x)dx = 2\pi,\ a_n = 0 \quad \text{and} \quad b_n = \frac{1}{\pi}\int_0^{2\pi} x\sin\frac{n\pi x}{\pi}dx = -\frac{2}{n}$$

Thus, the Fourier series of $f(x) = x$ is

$$f(x) = \pi - 2\left(\sin x + \frac{1}{2}\sin 2x + \frac{1}{3}\sin 3x + \cdots\right)$$

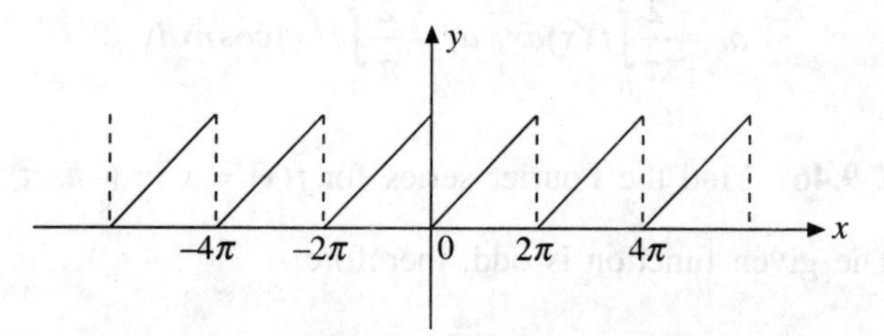


**Fig. 9.2** Saw-toothed wave.

The graph of $f(x) = x$ in the interval $(0, 2\pi)$ is shown in Fig. 9.2.

**EXAMPLE 9.48** Find the Fourier series representation of $f(x) = x^2$ defined on the interval $(-\pi, \pi)$ and obtain the value of $\pi^2/6$.

***Solution*** Here

$$a_0 = \frac{1}{\pi}\int_{-\pi}^{\pi} x^2 dx = \frac{2\pi^2}{3},\ a_n = \frac{2}{\pi}\int_0^{\pi} x^2 \cos nx dx = (-1)^n \frac{4}{n^2} \text{ and } b_n = 0$$

Thus, the Fourier series representation of the given function is

$$x^2 = \frac{\pi^2}{3} + 4\sum_{n=1}^{\infty}(-1)^n \frac{\cos nx}{n^2} \tag{95}$$

Now, put $x = \pi$ in Eq. (95), noting that $\cos n\pi = (-1)^n$, we get

$$\pi^2 = \frac{\pi^2}{3} + 4\sum_{n=1}^{\infty}\frac{1}{n^2}$$

which leads to

$$\frac{\pi^2}{6} = \sum_{n=1}^{\infty}\frac{1}{n^2} \equiv \zeta(2)$$

where $\zeta(2)$ is known as *Riemann zeta function* (in closed form).

**EXAMPLE 9.49** Find the Fourier series of the function $f(x)$ defined as

$$f(x) = \begin{cases} \sin x, & 0 \le x \le \pi \\ 0, & \pi \le x \le 2\pi \end{cases}$$

***Solution*** Here the period is $2\pi$ and

$$a_0 = \frac{1}{\pi}\int_0^{\pi} \sin x dx = \frac{2}{\pi}$$

$$a_n = \frac{1}{\pi}\int_0^{\pi} \sin x \cos nx dx = \frac{1}{\pi(n^2-1)}[(-1)^{n+1} - 1] \ \ for \ \ n \ne 1$$

and $\qquad b_n = 0$ for $n \ne 1$

While for $n = 1$, we have

$$a_1 = \frac{1}{\pi}\int_0^{\pi} \sin x \cos x dx = 0, \ b_1 = \frac{1}{\pi}\int_0^{\pi} \sin x \sin x dx = \frac{1}{2}$$

Thus, the Fourier series representation of the given function is

$$f(x) = \frac{a_0}{2} + a_1 \cos x + b_1 \sin x + \sum_{n=2}^{\infty}(a_n \cos nx + b_n \sin nx)$$

$$= \frac{1}{\pi} + \frac{1}{2}\sin x + \sum_{n=2}^{\infty}\frac{1}{n^2-1}[(-1)^{n+1} - 1]\cos nx$$

**EXAMPLE 9.50** Obtain the Fourier cosine and sine series of $f(x) = 3$ defined in the interval $0 \le x \le 5$.

***Solution*** For the Fourier cosine series, we have

$$a_0 = \frac{2}{c}\int_0^{c} f(x) dx = \frac{2}{5}\int_0^{5} 3 dx = 6$$

$$a_n = \frac{2}{c}\int_0^{c} f(x)\cos\frac{n\pi x}{c} dx = \frac{2}{5}\int_0^{5} 3\cos\frac{n\pi x}{5} dx = 0$$

and
$$f(x)=\frac{a_0}{2}+\sum_{n=1}^{\infty}a_n\cos\frac{n\pi x}{c}=3$$

Thus, the Fourier cosine series of $f(x)=3$ is the function itself. While, for the Fourier sine series, we have

$$b_n=\frac{2}{c}\int_0^c f(x)\sin\frac{n\pi x}{c}dx=\frac{2}{5}\int_0^5 3\sin\frac{n\pi x}{5}dx=\frac{6}{n\pi}[1-(-1)^n]$$

Thus, the Fourier sine series of the given function is

$$f(x)=\sum_{n=1}^{\infty}b_n\sin\frac{n\pi x}{5}=\frac{12}{\pi}\left[\sin\frac{\pi x}{5}+\frac{1}{3}\sin\frac{3\pi x}{5}+\frac{1}{5}\sin\frac{5\pi x}{5}+\cdots\right]$$

**EXAMPLE 9.51** Find the Fourier cosine series of $f(x)=\sin x$ defined in the interval $(0,\pi)$.

**Solution** Here

$$a_0=\frac{2}{c}\int_0^c f(x)dx=\frac{2}{\pi}\int_0^{\pi}\sin x dx=\frac{4}{\pi}$$

$$a_n=\frac{2}{c}\int_0^c f(x)\cos\frac{n\pi x}{c}dx=\frac{2}{\pi}\int_0^{\pi}\sin x\cos\frac{n\pi x}{\pi}dx$$

$$=\frac{2}{\pi}\int_0^{\pi}\sin x\cos nx dx$$

$$=-\frac{2(1+\cos n\pi)}{\pi(n^2-1)}\quad \text{for}\quad n\neq 1$$

For $n = 1$, $a_1=\frac{2}{\pi}\int_0^{\pi}\sin x\cos x dx=0.$

Thus, the Fourier cosine series of the given function is

$$f(x)=\frac{a_0}{2}+\sum_{n=1}^{\infty}a_n\cos\frac{n\pi x}{\pi}$$

$$=\frac{2}{\pi}+\sum_{n=2}^{\infty}\left[-\frac{2(1+\cos n\pi)}{\pi(n^2-1)}\right]\cos nx$$

$$=\frac{2}{\pi}+\sum_{n=2}^{\infty}\left[-\frac{2(1+(-1)^n)}{\pi(n^2-1)}\right]\cos nx$$

$$=\frac{2}{\pi}-\frac{4}{\pi}\left[\frac{\cos 2x}{2^2-1}+\frac{\cos 4x}{4^2-1}+\cdots\right]$$

where $\cos n\pi=(-1)^n$.

We now consider the behaviour of a Fourier series near a point of discontinuity with the help of the following example.

**EXAMPLE 9.52** Obtain the Fourier series of the function $f(x)$ defined as

$$f(x) = \begin{cases} 1, & -\pi < x < 0 \\ -1, & 0 \le x < \pi \end{cases}$$

***Solution*** The graph of the given function is shown in Fig. 9.3.

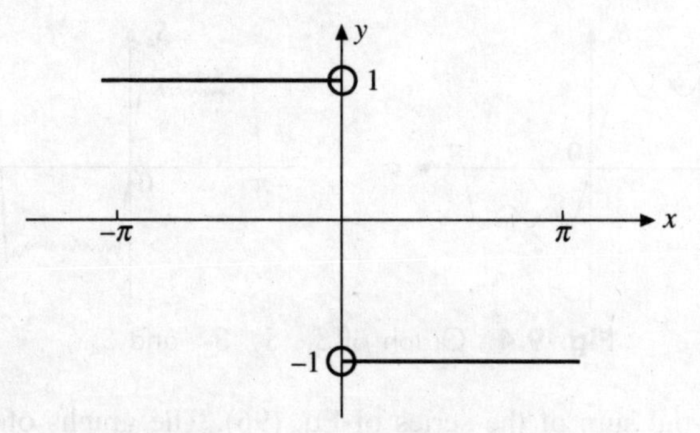


**Fig. 9.3** Graph of Example 9.52.

Here $a_0 = 0, a_n = 0$ and

$$b_n = \frac{2}{\pi}[(-1)^n - 1] = \begin{cases} 0, & \text{when } n \text{ is even} \\ -\dfrac{4}{n\pi} & \text{when } n \text{ is odd} \end{cases}$$

Thus, the Fourier series of the given function is

$$f(x) = -\frac{4}{\pi}\left[\sin x + \frac{1}{3}\sin 3x + \frac{1}{5}\sin 5x + \frac{1}{7}\sin 7x + \cdots\right] \tag{96}$$

**REMARK.** Let

$$S_1 = -\frac{4}{\pi}\sin x$$

$$S_2 = -\frac{4}{\pi}\left(\sin x + \frac{1}{3}\sin 3x\right)$$

$$S_3 = -\frac{4}{\pi}\left(\sin x + \frac{1}{3}\sin 3x + \frac{1}{5}\sin 5x\right)$$

$$\cdots$$

$$\cdots$$

$$S_{14} = -\frac{4}{\pi}\left(\sin x + \frac{1}{3}\sin 3x + \frac{1}{5}\sin 5x + \ldots + \frac{1}{27}\sin 27x\right)$$

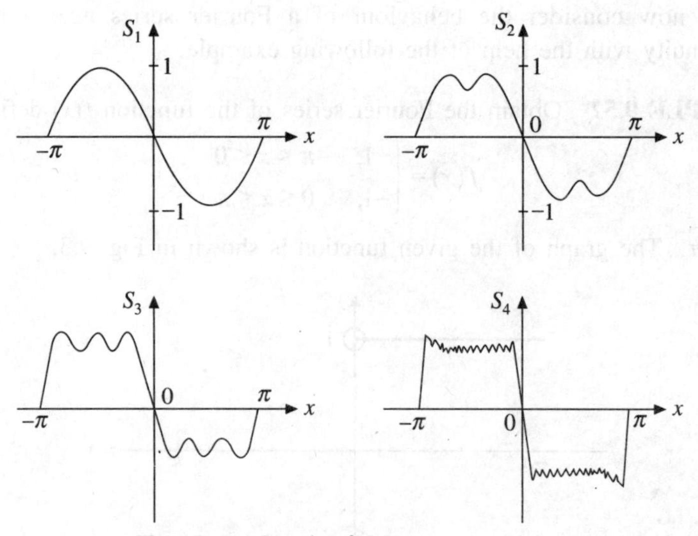


**Fig. 9.4** Graph of $S_1$, $S_2$, $S_3$, and $S_4$.

denote the partial sum of the series of Eq. (96). The graphs of $S_1$, $S_2$, $S_3$ and $S_{14}$ are shown in Fig. 9.4. By looking at the graph, it may be noted that the graph of $S_{14}$ depicts the spikes near the discontinuities at $x = -\pi$, 0 and $\pi$. Even for very large values of $n$, this oscillatory behaviour of partial sum $S_n$ about true value near a point of discontinuity does not smoothen out. This behaviour of Fourier series near a point of discontinuity is called *Gibbs phenomenon.*

## 9.12 VIBRATION OF A STRETCHED STRING—WAVE MOTION

Let $l$ be the length of a tightly stretched string whose ends $A$ and $B$ are fixed (Fig. 9.5). Also, let $T$ be the constant tension in the string which is assumed to

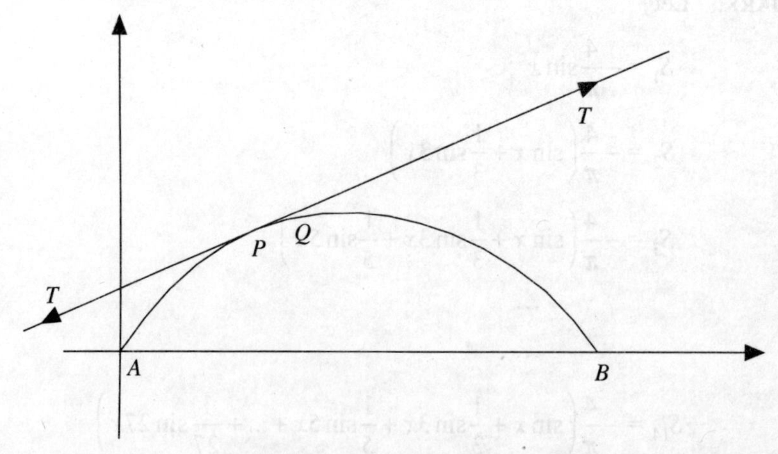


**Fig. 9.5** Schematic diagram of a tightly stretched string with fixed ends.

be large as compared to the weight of the string so that the effects of gravity are negligible. Let the string be released from rest and allowed to vibrate. Also, assume that no external forces are acting on it and that the each point of the string makes small vibrations at right angles to the equilibrium position *AB*. Under these circumstances, we shall study the subsequent motion of the string.

Consider the end *A* to be the origin, *AB*, the *x*-axis and *AY* perpendicular to it as *y*-axis so that the motion is entirely in the *xy*-plane. Let at time *t*, the position of the string be APB. Now consider the motion of the element *PQ* of the string between the points $P(x, y)$ and $Q(x + \Delta x, y + \Delta y)$, where the tangents make angles $\psi$ and $\psi + \Delta\psi$ with the *x*-axis. The movement of this element is upward with acceleration $\partial^2 y/\partial t^2$ and the vertical component of the force acting on this element is

$$T \sin (\psi + \Delta\psi) - T \sin \psi = T[\sin (\psi + \Delta\psi) - \sin \psi]$$

$$= T[\tan (\psi + \Delta\psi) - \tan \psi], \quad \text{(as } \psi \text{ is small)}$$

$$= T\left[\left(\frac{\partial y}{\partial x}\right)_{x+\Delta x} - \left(\frac{\partial y}{\partial x}\right)_{x}\right]$$

If *m* is the mass per unit length of the string, then Newton's second law gives

$$m\,\Delta x \frac{\partial^2 y}{\partial t^2} = T\left[\left(\frac{\partial y}{\partial x}\right)_{x+\Delta x} - \left(\frac{\partial y}{\partial x}\right)_{x}\right]$$

or

$$\frac{\partial^2 y}{\partial t^2} = \frac{T}{m}\left[\frac{(\partial y/\partial x)_{x+\Delta x} - (\partial y/\partial x)_{x}}{\Delta x}\right]$$

Now, take the limit as $Q \to P$, i.e. $\Delta x \to 0$ to obtain

$$\frac{\partial^2 y}{\partial t^2} = c^2 \frac{\partial^2 y}{\partial x^2} \tag{97}$$

where $c^2 = T/m$. Equation (97) is the partial differential equation describing the transverse vibrations of the string. It is a *one-dimensional wave equation*.

**REMARK.** Equation (97) also gives the vibration of a rod and in this case, $c^2 = E/\rho$, where *E* denotes the modulus of elasticity and $\rho$ the density.

To get a solution of Eq. (97), assume that the solution is of the form

$$y = X(x)T(t)$$

where *X* is a function of *x* only and *T* is a function of *t* only. Then

$$\frac{\partial^2 y}{\partial t^2} = XT'', \qquad \frac{\partial^2 y}{\partial x^2} = X''T$$

Substitution of these into Eq. (97) leads to the ordinary differential equations

$$\frac{d^2x}{dx^2} - kX = 0, \qquad \frac{d^2T}{dt^2} - kc^2T = 0$$

where $k$ is a constant. The solutions of these equations are

(i) When $k > 0$ and $= \omega^2$ (say)

$$X = c_1e^{\omega x} + c_2e^{-\omega x}, \qquad T = c_3e^{c\omega t} + c_4e^{-c\omega t}$$

(ii) When $k < 0$ and $= -\omega^2$ (say)

$$X = c_5 \cos \omega x + c_6 \sin \omega x, \qquad T = c_7 \cos (c\omega t) + c_8 \sin (c\omega t)$$

(iii) When $k = 0$,

$$X = c_9x + c_{10}, \qquad T = c_{11}t + c_{12}$$

Thus, the various possible solutions of Eq. (97) are

$$y = (c_1e^{\omega x} + c_2e^{-\omega x})\,(c_3e^{c\omega t} + c_4e^{-c\omega t})$$

$$y = (c_5 \cos \omega x + c_6 \sin \omega x)[c_7 \cos (c\omega t) + c_8 \sin (c\omega t)]$$

and $$y = (c_9x + c_{10})\,(c_{11}t + c_{12})$$

Out of these three possible solutions, we have to choose that solution which is consistent with the physical nature of the problem, and as we are working with the problem on vibrations, $y$ must be a periodic function of $x$ and $t$. Hence, the solution of Eq. (97) must involve trigonometric terms. Accordingly, the solution

$$y = (A \cos \omega x + B \sin \omega x)\,[C \cos (c\omega t) + D \sin (c\omega t)]$$

is the only suitable solution of Eq. (97).

**EXAMPLE 9.53** A string is fixed at two points $l$ apart and is stretched. The motion takes place by displacing the string in the form $y = a \sin(\pi x/l)$ from which it is released at time $t = 0$. Show that the displacement of any point at a distance $x$ from one end at time $t$ is

$$y(x, t) = a \sin\left(\frac{\pi x}{l}\right)\cos\left(\frac{\pi ct}{l}\right)$$

***Solution*** The motion of the string is given by Eq. (97). As the end points of the string are fixed, for all time $t$, we have

$$y(0, t) = 0, \qquad y(l, t) = 0 \tag{98}$$

Since, the initial transverse velocity of any point of the string is zero

$$\left(\frac{\partial y}{\partial t}\right)_{t=0} = 0 \tag{99}$$

Also,

$$y(x, 0) = a \sin\left(\frac{\pi x}{l}\right) \tag{100}$$

We will now solve Eq. (97) alongwith the boundary conditions (98)–(100). Since, the vibration of the string is periodic, the solution of Eq. (97) is given by

$$y = y(x, t) = (A \cos \omega x + B \sin \omega x)\ [C \cos (c\omega t) + D \sin (c\omega t)] \tag{101}$$

From Eq. (98), we have

$$y(0, t) = A[C \cos (c\omega t) + D \sin (c\omega t)] = 0 \tag{102}$$

and this will hold for all $t$ if $A = 0$. Hence

$$y(x, t) = B (\sin \omega x)[C \cos (c\omega t) + D \sin (c\omega t)] \tag{103}$$

and
$$\frac{\partial y}{\partial t} = B (\sin \omega x)\{C[-c\omega \sin (c\omega t)] + D[c\omega \cos (c\omega t)]\}$$

Thus, by Eq. (99), we have $(\partial y/\partial t)_{t=0} = 0 = B(\sin \omega x)(Dc\omega)$ which gives $BDc\omega = 0$. If $B = 0$, then Eq. (103) leads to $y(x, t) = 0$. This means that $D = 0$, and Eq. (103) takes the form

$$y(x, t) = BC \sin \omega x \cos (c\omega t) \tag{104}$$

From Eq. (98), $y(l, t) = 0 = BC \sin \omega l \cos (c\omega t)$, for all $t$. Since $B$ and $C$ are nonzero, we have $\sin \omega l = 0$, which gives $\omega l = n\pi$ or $\omega = n\pi/l$, where $n$ is an integer. Equation (104) thus reduces to

$$y(x, t) = BC \sin \frac{n\pi x}{l} \cos \frac{n\pi ct}{l}$$

Finally, from Eq. (100), we have

$$y(x, 0) = a \sin \frac{\pi x}{l} = BC \sin \frac{n\pi x}{l}$$

which is satisfied if $BC = a$ and $n = 1$. Therefore, the required solution is

$$y(x, t) = a \sin \frac{\pi x}{l} \cos \frac{\pi ct}{l}$$

**EXAMPLE 9.54** A string is stretched along the $x$-axis, to which it is attached at $x = 0$ and at $x = L$. Find $y$ in terms of $x$ and $t$, assuming that $y = mx(L - x)$ when $t = 0$.

***Solution*** The vibration of the string is governed by the equation

$$\frac{\partial^2 y}{\partial t^2} = c^2 \frac{\partial^2 y}{\partial x^2}$$

and the boundary conditions are

$$y(0, t) = 0, \qquad y(L, t) = 0, \qquad y(x, 0) = mx(L - x) \tag{105}$$

By the method of separation of variables, the solution of the wave equation is of the form

$$y(x, t) = (A \cos \omega x + B \sin \omega x)\ [C \cos (c\omega t) + D \sin (c\omega t)]$$

The first two conditions of Eq. (105) are satisfied if $\omega = n\pi/L$. Expanding $mx(L - x)$ in $0 < x < L$, we get

$$mx(L-x) = \frac{8L^2 m}{\pi^3}\left(\sin\frac{\pi x}{L} + \frac{1}{3^3}\sin\frac{3\pi x}{L} + \frac{1}{5^3}\sin\frac{5\pi x}{L} + \cdots\right)$$

Hence, the required solution of the wave Eq. (97) satisfying Eq. (105) is

$$y = \frac{8L^2 m}{\pi^3}\left(\frac{1}{1^3}\cos\frac{c\pi t}{L}\sin\frac{\pi x}{L} + \frac{1}{3^3}\cos\frac{3\pi ct}{L}\sin\frac{3\pi x}{L} + \cdots\right)$$

## 9.13 ONE-DIMENSIONAL HEAT FLOW

Consider a homogeneous bar of uniform cross-section $a$. Let the stream lines of the heat flow be all parallel and perpendicular to the area $a$. Take one end of the bar as the origin and the direction of flow as the positive $x$-axis. Suppose $\rho$ be the density, $s$ the specific heat, and $k$ the thermal conductivity (Fig. 9.6).

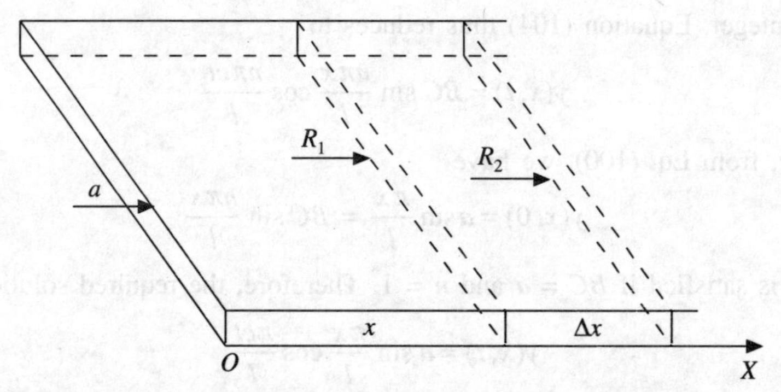


**Fig. 9.6** One-dimensional heat flow where the stream lines of the heat flow are all parallel and perpendicular to the area.

Let $u(x, t)$ be the temperature at a distance $x$ from 0. If $\Delta u$ is the change in temperature in a slab of thickness $\Delta x$ of the bar, then the quantity of heat in this slab is $s\rho a\ \Delta x \Delta u$. Hence, $s\rho a\Delta x\ (\partial u/\partial t) = R_1 - R_2$, where $R_1$ and $R_2$ represents the rate of inflow and outflow of heat, respectively. Now

$$R_1 = -ka\left(\frac{\partial u}{\partial x}\right)_x, \qquad R_2 = -ka\left(\frac{\partial u}{\partial x}\right)_{x+\Delta x}$$

Thus

$$s\rho a\,\Delta x\,\frac{\partial u}{\partial t} = -ka\left(\frac{\partial u}{\partial x}\right)_x + ka\left(\frac{\partial u}{\partial x}\right)_{x+\Delta x}$$

$$\frac{\partial u}{\partial t} = \frac{k}{s\rho}\left[\frac{(\partial u/\partial x)_{x+\Delta x} - (\partial u/\partial x)_x}{\Delta x}\right]$$

Now, taking limit as $\Delta x \to 0$, we get

$$\frac{\partial u}{\partial t} = c^2\,\frac{\partial^2 u}{\partial x^2} \tag{106}$$

where $c^2 = k/(s\rho)$ is called the *diffusivity* of the material. Equation (106) is called the *one-dimensional heat flow equation.*

To solve Eq. (106), we assume that the solution is of the form

$$u(x, t) = X(x)T(t)$$

where $X$ is a function of $x$ only and $T$ is a function of $t$ only. Substituting this into Eq. (106), we get

$$\frac{d^2X}{dx^2} - KX = 0, \qquad \frac{dT}{dt} - Kc^2T = 0 \tag{107}$$

Since we are dealing with the problem of heat condition, the solution of Eq. (107) must be a transient solution, i.e. $u$ must decrease with time. Solving ordinary differential Eq. (107) by taking $K = -\omega^2$, we get

$$u = (c_1 \cos \omega x + c_2 \sin \omega x)\; e^{-c^2\omega^2 t} \tag{108}$$

which is the only possible solution of Eq. (106).

**EXAMPLE 9.55** An insulated rod of length $L$ has its ends $A$ and $B$ maintained at 0°C and 100°C, respectively until steady-state conditions prevail. If $B$ is suddenly reduced to 0°C and maintained at 0°C, find the temperature at a distance $x$ from $A$ at time $t$.

***Solution*** We know that the solution of Eq. (106) of the heat conduction is given by Eq. (108).

Now, prior to the change in temperature at the end $B$, when $t = 0$, the heat flow was independent of time (steady state condition). When $u$ depends on $x$ only, Eq. (106) reduces to

$$\frac{\partial^2 u}{\partial x^2} = 0$$

whose general solution is

$$u = Ax + B \tag{109}$$

As $u = 0$ and $u = 100$ for $x = 0$ and $x = L$, respectively, Eq. (109) gives $B = 0$ and $A = 100/L$. Thus, the initial conditions are

$$u(x, 0) = \frac{100}{L} x \tag{110}$$

The boundary conditions for the subsequent flow are

$$u(0, t) = 0, \text{ for all } t \tag{111}$$

$$u(L, t) = 0, \text{ for all } t \tag{112}$$

We now find the solution of Eq. (106) subject to the conditions (110)–(112). The solution of Eq. (106) is

$$u(x, t) = (c_1 \cos \omega x + c_2 \sin \omega x)\, e^{-c^2\omega^2 t} \tag{113}$$

From Eq. (111), $u(0, t) = 0 = c_1 e^{-c^2\omega^2 t}$. Hence, $c_1 = 0$ and Eq. (113) reduces to

$$u(x, t) = c_2 (\sin \omega x) e^{-c^2\omega^2 t} \tag{114}$$

which, on applying Eq. (112), gives $u(L, t) = c_2 \sin \omega L e^{-c^2\omega^2 t} = 0$ and this requires $\sin \omega L = 0$, i.e. $\omega L = n\pi$, as $c_2 \neq 0$. Thus, $\omega = n\pi/L$, where $n$ is an integer. Hence, Eq. (114) becomes

$$u(x, t) = b_n \sin \frac{n\pi x}{L} e^{-n^2\pi^2c^2t/L^2}, \qquad b_n = c_2$$

Adding all such solutions, the most general solution of Eq. (106) satisfying Eq. (111) and (112) is

$$u(x, t) = \sum_{n=1}^{\infty} b_n \sin \frac{n\pi x}{L} e^{-c^2n^2\pi^2t/L^2} \tag{115}$$

Put $t = 0$ to obtain

$$u(x, 0) = \sum_{n=1}^{\infty} b_n \sin \frac{n\pi x}{L} \tag{116}$$

Now, in order that the condition (110) be satisfied, Eqs. (110) and (116) must be the same. This needs the expansion of $100x/L$ as the Fourier series in $(0, L)$. Thus

$$\frac{100\, x}{L} = \sum_{n=1}^{\infty} b_n \sin \frac{n\pi x}{L}$$

where

$$b_n = \frac{2}{L} \int_0^L \frac{100x}{L} \sin \frac{n\pi x}{L}\, dx = \frac{200}{L^2} \frac{-L^2}{n\pi} \cos n\pi = \frac{200(-1)^{n+1}}{n\pi}$$

Thus, the most general solution (115) becomes

$$u(x, t) = \frac{200}{\pi} \sum_{n=1}^{\infty} \frac{(-1)^{n+1}}{n} \sin \frac{n\pi x}{L} e^{-c^2n^2\pi^2t/L^2}$$

which is the desired result.

## 9.14 TWO-DIMENSIONAL HEAT FLOW

Consider the flow of heat in a metal plate of uniform thickness $a$. Let the density, specific heat and thermal conductivity be denoted by $\rho$, $s$ and $k$, respectively, and $XOY$ represent one face of the plate. The flow of heat is said to be two-dimensional if the temperature at any point depends only on $x$ and $y$ coordinates and time $t$, but independent of the $z$-coordinate, and in this case, the flow of heat is in the $XY$-plane and zero along the normal to the plane.

Consider a rectangular element $PQRS$ (Fig. 9.7) with sides $\Delta x$ and $\Delta y$. Then the amount of heat entering the element per second from the side $PQ = -ka\,\Delta x(\partial u/\partial y)_y$, and, the amount of heat entering the element per second from the side

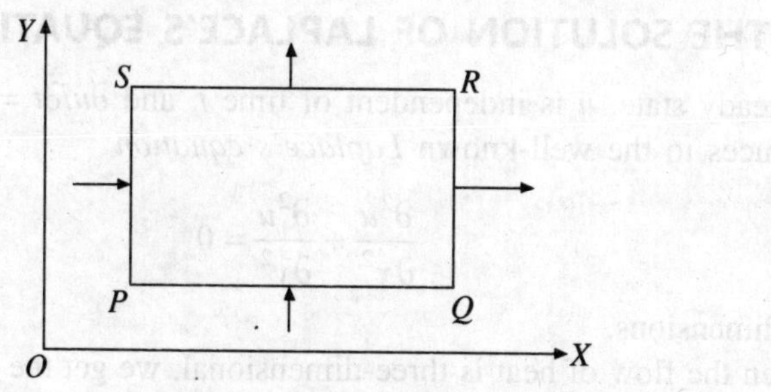


**Fig. 9.7** Flow of heat in a metal plate of uniform thickness—two-dimensional heat flow.

$PS = -ka\Delta y(\partial u/\partial x)_x$. Also, the amount of heat coming out through side $SR$ per second $= -ka\Delta x(\partial u/\partial y)_{y+\Delta y}$, and the amount of heat coming out through side $QR = -ka\Delta y(\partial u/\partial x)_{x+\Delta x}$. Thus, the total gain of heat by $PQRS$ per second is

$$-ka\,\Delta x\left(\frac{\partial u}{\partial y}\right)_y - ka\,\Delta y\left(\frac{\partial u}{\partial x}\right)_x + ka\,\Delta x\left(\frac{\partial u}{\partial y}\right)_{y+\Delta y} + ka\,\Delta y\left(\frac{\partial u}{\partial x}\right)_{x+\Delta x}$$

$$= ka\,\Delta x\Delta y\left[\frac{\left(\frac{\partial u}{\partial x}\right)_{x+\Delta x} - \left(\frac{\partial u}{\partial x}\right)_x}{\Delta x} + \frac{\left(\frac{\partial u}{\partial y}\right)_{y+\Delta y} - \left(\frac{\partial u}{\partial y}\right)_y}{\Delta y}\right] \tag{117}$$

Also

$$\text{rate of gain of heat by } PQRS = \rho\Delta x\Delta y\left(as\frac{\partial u}{\partial t}\right) \tag{118}$$

Equations (117) and (118) yield

$$ka\Delta x\Delta y\left[\frac{\left(\frac{\partial u}{\partial x}\right)_{x+\Delta x} - \left(\frac{\partial u}{\partial x}\right)_x}{\Delta x} + \frac{\left(\frac{\partial u}{\partial y}\right)_{y+\Delta y} - \left(\frac{\partial u}{\partial y}\right)_y}{\Delta y}\right] = \rho\Delta x\Delta y\left(as\frac{\partial u}{\partial t}\right)$$

which, on taking limit as $\Delta x \to 0, \Delta y \to 0$, becomes

$$k\left(\frac{\partial^2 u}{\partial x^2}+\frac{\partial^2 u}{\partial y^2}\right)=\rho s\frac{\partial u}{\partial t}$$

or
$$\frac{\partial u}{\partial t}=c^2\left(\frac{\partial^2 u}{\partial x^2}+\frac{\partial^2 u}{\partial y^2}\right) \tag{119}$$

where $c^2 = k/\rho s$ is the *diffusivity*. Equation (119) represents the distribution of temperature in the plate in the transient state.

## 9.15 THE SOLUTION OF LAPLACE'S EQUATION

In the steady state, $u$ is independent of time $t$, and $\partial u/\partial t = 0$. Equation (119) thus reduces to the well-known *Laplace's equation*

$$\frac{\partial^2 u}{\partial x^2}+\frac{\partial^2 u}{\partial y^2}=0 \tag{120}$$

in two dimensions.

When the flow of heat is three-dimensional, we get the following equation [in a similar manner as that of Eq. (115)]:

$$\frac{\partial u}{\partial t}=c^2\left(\frac{\partial^2 u}{\partial x^2}+\frac{\partial^2 u}{\partial y^2}+\frac{\partial^2 u}{\partial z^2}\right) \tag{121}$$

which in a steady state condition yields Laplace's equation

$$\frac{\partial^2 u}{\partial x^2}+\frac{\partial^2 u}{\partial y^2}+\frac{\partial^2 u}{\partial z^2}=0 \tag{122}$$

in three dimensions.

Let $u = X(x)Y(y)$ be a solution of Eq. (120). Substitute it in Eq. (120) to obtain

$$\frac{d^2X}{dx^2}Y + X\frac{d^2Y}{dy^2}=0$$

Separation of variables yields

$$\frac{1}{X}\frac{d^2X}{dx^2}=-\frac{1}{Y}\frac{d^2Y}{dy^2}=A \text{ (constant)} \tag{123}$$

The solution of the ordinary differential equation (123) depends upon the choice of the constant, and the possible solutions of Eq. (120) are (as already done):

$$u = (c_1e^{\omega x} + c_2e^{-\omega x})\,(c_3 \cos \omega y + c_4 \sin \omega y) \tag{124}$$

[if in Eq. (123) $A$ is positive and $= \omega^2$ (say)].

$$u = (c_5 \cos \omega x + c_6 \sin \omega x)\,(c_7 e^{\omega y} + c_8 e^{-\omega y}) \tag{125}$$

[if in Eq. (123) $A$ is negative and equals to $-\omega^2$ (say)].

$$u = (c_9 x + c_{10})(c_{11} y + c_{12}) \tag{126}$$

[when $A = 0$ in Eq. (123)]. Out of these solutions, we have to choose that solution which is consistent with the physical nature of the problem. The solutions of Laplace's equation are known as *harmonic functions.*

**EXAMPLE 9.56** An infinitely long plane uniform plate is bounded by two parallel edges and an end at right angles to them. The breadth is $\pi$: this end is maintained at a temperature $u_0$ at all points and other edges are at zero temperature (see Fig. 9.8). Find the temperature at any point of the plate in the steady state.

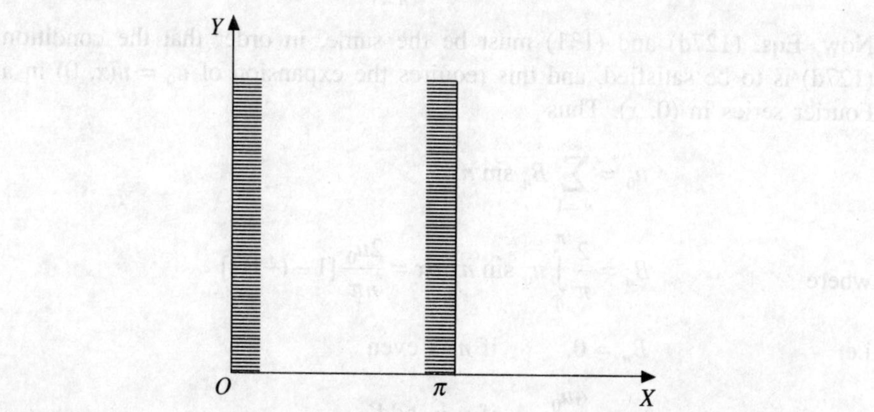


**Fig. 9.8** Schematic diagram of an infinitely long uniform plate bounded by two parallel edges and an end at right angles to them.

***Solution*** In the steady state, the temperature $u(x, y)$ at any point $P(x, y)$ satisfies the Laplace Eq. (120). The boundary conditions are

$$u(0, y) = 0, \qquad \text{for all } y \tag{127a}$$

$$u(\pi, y) = 0, \qquad \text{for all } y \tag{127b}$$

$$u(x, \infty) = 0, \qquad \text{in } 0 < x < \pi \tag{127c}$$

$$u(x, 0) = u_0, \qquad \text{in } 0 < x < \pi \tag{127d}$$

The three possible solutions of Eq. (120) are Eqs. (124)–(126). Solution (124) cannot satisfy Eq. (127a) as $u \neq 0$ for $x = 0$, for all values of $y$. The solution (126) cannot satisfy (127d). Thus, the only possible solution of Eq. (120) is of the form

$$u(x, y) = (A_1 \cos \omega x + A_2 \sin \omega x)\,(A_3 e^{\omega y} + A_4 e^{-\omega y}) \tag{128}$$

From Eq. (127a), we have $u(0, y) = A_1(A_3e^{\omega y} + A_4e^{-\omega y}) = 0$. Hence $A_1 = 0$, and Eq. (128) reduces to

$$u(x, y) = A_2 \sin \omega x \,(A_3e^{\omega y} + A_4e^{-\omega y}) \tag{129}$$

By Eq. (127b), $u(\pi, y) = A_2 \sin \omega\pi(A_3e^{\omega y} + A_4e^{-\omega y}) = 0$, which requires $\sin \omega\pi = 0$, i.e. $\omega\pi = n\pi$, as $A_2 \neq 0$. Thus $\omega = n$, an integer. Also, in order to satisfy Eq. (127c) $A_3 = 0$. Hence, Eq. (129) becomes

$$u(x, y) = B_n \sin nxe^{-ny}, \quad B_n = A_2A_4$$

Therefore, the most general solution satisfying (127a)–(127c) is of the form

$$u(x, y) = \sum_{n=1}^{\infty} B_n \sin nxe^{-ny} \tag{130}$$

Put

$$y = 0, \quad u(x, 0) = \sum_{n=1}^{\infty} B_n \sin nx \tag{131}$$

Now, Eqs. (127d) and (131) must be the same, in order that the condition (127d) is to be satisfied, and this requires the expansion of $u_0 = u(x, 0)$ in a Fourier series in $(0, x)$. Thus

$$u_0 = \sum_{n=1}^{\infty} B_n \sin nx$$

where

$$B_n = \frac{2}{\pi}\int_0^{\pi} u_0 \sin nx\, dx = \frac{2u_0}{n\pi}[1 - (-1)^n]$$

i.e.

$$B_n = 0, \qquad \text{if } n \text{ is even}$$

$$= \frac{4u_0}{n\pi}, \qquad \text{if } n \text{ is odd}$$

Hence, Eq. (130) takes the form

$$u(x, y) = \frac{4u_0}{\pi}\left(e^{-y} \sin x + \frac{1}{3}e^{-3y} \sin 3x + \frac{1}{5}e^{-5y} \sin 5x + \cdots\right)$$

**EXAMPLE 9.57** *Fourier problem.* Determine the temperature $u$ at any point $P(x, y)$ of a thin plate, $\pi$ units wide and infinitely long assuming the steady state so that $\partial^2u/\partial x^2 + \partial^2u/\partial y^2 = 0$, the short edge constantly at temperature unity and the long edge at temperature zero.

***Solution*** Take the $Y$-axis along an infinite edge and the $X$-axis along the short edge. The boundary conditions are

(a) $u = 0$, when $x = 0$.
(b) $u = 0$, when $x = \pi$.
(c) $u = 0$, when $y = \infty$.
(d) $u = 1$, when $y = 0$.

These boundary conditions are the same as those of the above example except (127d). Now take $u_0 = 1$ in Eq. (127d); then from the above example we have

$$u(x, y) = \frac{4}{\pi}\left(\frac{1}{1}e^{-y}\sin x + \frac{1}{3}e^{-3y}\sin 3x + \cdots\right)$$

which is the required temperature at any point $P(x, y)$.

## 9.16 LAPLACE'S EQUATION IN POLAR COORDINATES

Sometimes it is more convenient to write Laplace's Eq. (120) in polar coordinates. Substituting

$$x = r\cos\theta, \qquad y = r\sin\theta, \qquad r = \sqrt{x^2 + y^2}, \qquad \theta = \tan^{-1}\frac{y}{x}$$

we obtain the polar form of Laplace's Eq. (120) as

$$r^2\frac{\partial^2 u}{\partial r^2} + r\frac{\partial u}{\partial r} + \frac{\partial^2 u}{\partial \theta^2} = 0 \tag{132}$$

To obtain the solution of this equation, assume that a solution of Eq. (132) is of the form

$$u = R(r)\,\phi(\theta)$$

where $R$ is a function of $r$ only and $\phi$ is a function of $\theta$ only. Substitution of this into Eq. (132) yields

$$r^2R''\phi + rR'\phi + R\phi'' = 0$$

$$\phi(r^2R'' + rR') + R\phi'' = 0$$

Separation of variables gives

$$\frac{r^2R'' + rR'}{R} = -\frac{\phi''}{\phi} \tag{133}$$

The left-hand side of this equation is a function of $r$ only and the right-hand side is a function of $\theta$ only, and $r$ and $\theta$ are independent variables. Equation (133) is true only when each side is equal to a constant $K$ (say). Thus, Eq. (133) leads to ordinary differential equations

$$r^2\frac{d^2R}{dr^2} + r\frac{dR}{dr} - KR = 0 \tag{134}$$

and

$$\frac{d^2\phi}{d\theta^2} + K\phi = 0 \tag{135}$$

Put $r = e^z$ in Eq. (134) so that

$$\frac{d^2R}{dz^2} - KR = 0 \tag{136}$$

Thus, the three possible solutions of Eq. (132) are [depending upon the choice of $K$ in Eqs. (135) and (136)]:

$$u = (c_1 r^{\omega} + c_2 r^{-\omega})(c_3 \cos \omega\theta + c_4 \sin \omega\theta) \tag{137}$$

[if $K$ is positive and equals $\omega^2$].

$$u = [c_5 \cos (\omega \log r) + c_6 \sin (\omega \log r)]\ (c_7 e^{\omega\theta} + c_8 e^{-\omega\theta}) \tag{138}$$

[if $K$ is negative and equals $-\omega^2$].

$$u = (c_9 \log r + c_{10})(c_{11}\theta + c_{12}) \quad \text{if } K = 0 \tag{139}$$

Out of these solutions, we have to choose that solution which is consistent with the physical nature of the problem.

**EXAMPLE 9.58** The diameter of a semi-circular plate of radius $a$ is kept at 0°C and the temperature at the semi-circular boundary is $T$°C. Determine the steady state temperature in the plate.

***Solution*** Consider the pole to be the centre of the circle and the initial line to be the bounding diameter. Suppose that the steady state temperature at any point $P(r, \theta)$ is $u(r, \theta)$ (Fig. 9.9) so that $u$ satisfies Laplace's Eq. (132). The boundary conditions are

(a) $u(r, 0) = 0, \quad 0 \le r \le a$
(b) $u(r, \pi) = 0, \quad 0 \le r \le a$
(c) $u(a, \theta) = T$

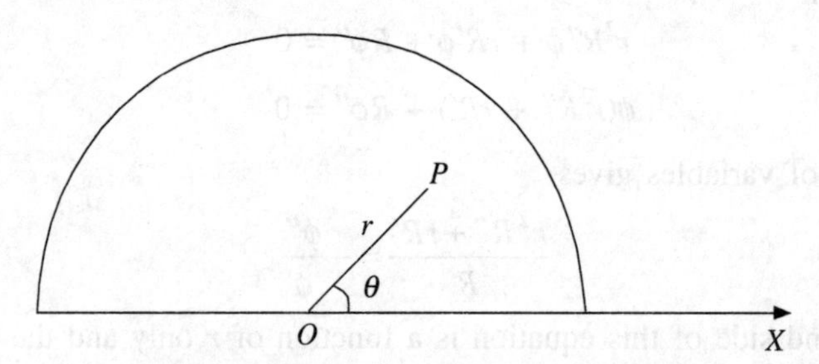


**Fig. 9.9** Semi-circular plate of radius $a$.

The three possible solutions of Eq. (132) are given by Eqs. (137)–(139). From (b) and (c), $u = 0$ when $r = 0$, i.e. $u$ must be finite at origin and solutions (138) and (139) have to be rejected. Thus, the only suitable solution is (137). From (a), we have

$$u(r, 0) = (c_1 r^{\omega} + c_2 r^{-\omega})c_3 = 0$$

Hence, $c_3 = 0$, and Eq. (137) becomes

$$u(r, \theta) = (c_1 r^{\omega} + c_2 r^{-\omega})\ c_4 \sin \omega\theta \tag{140}$$

From (b), $u(r, \pi) = (c_1 r^{\omega} + c_2 r^{-\omega})\, c_4 \sin \omega\pi = 0$, and as $c_4 \neq 0$, $\sin \omega\pi = 0$, i.e. $\omega = n$, an integer. Hence, Eq. (140) reduces to

$$u(r, \theta) = (c_1 r^n + c_2 r^{-n}) c_4 \sin n\theta \tag{141}$$

Since, $u = 0$ when $r = 0$, we have $c_2 = 0$, and Eq. (141) reduces to

$$u(r, \theta) = B_n r^n \sin n\theta, \qquad B_n = c_1 c_4$$

Thus, the most general solution of Eq. (132) for the present problem is

$$u(r, \theta) = \sum_{n=1}^{\infty} B_n r^n \sin n\theta \tag{142}$$

When $r = a$

$$u(a, \theta) = \sum_{n=1}^{\infty} B_n a^n \sin n\theta \tag{143}$$

The condition (c) above is satisfied only when Eq. (143) and (c) are the same, and this amounts to the expansion of the given function $T = u(a, \theta)$ in a Fourier series in $(0, \pi)$. Thus

$$T = \sum_{n=1}^{\infty} A_n \sin n$$

where

$$A_n = \frac{2}{\pi}\int_0^{\pi} T \sin n\theta \, d\theta = \frac{2T}{n\pi}(1 - \cos n\pi) \qquad \text{and} \quad A_n = B_n a^n$$

Therefore
$$B_n = \frac{A_n}{a^n} = \frac{2T}{n\pi a^n}(1 - \cos n\pi)$$

i.e.
$$B_n = 0 \qquad \text{if } n \text{ is even}$$

$$= \frac{4T}{n\pi a^n} \qquad \text{if } n \text{ is odd}$$

Thus, the required steady state temperature in the plate is

$$u(r, \theta) = \frac{4T}{\pi}\left[\frac{r/a}{1}\sin\theta + \frac{(r/a)^3}{3}\sin 3\theta + \frac{(r/a)^5}{5}\sin 5\theta + \cdots\right]$$

## 9.17 THE TRANSMISSION LINE

Figure 9.10 represents a long cable $PQ$ of length $l$ carrying an electric current with resistance $R$, inductance $L$, capacitance $C$ and leakage $G$ of current (or conductance to ground). Let the instantaneous voltage and current at any point $A$, distance $x$ from the sending end $P$, be $v(x, t)$ and $i(x, t)$, respectively, at time $t$. Consider a small length $AB(= \Delta x)$ of the cable.

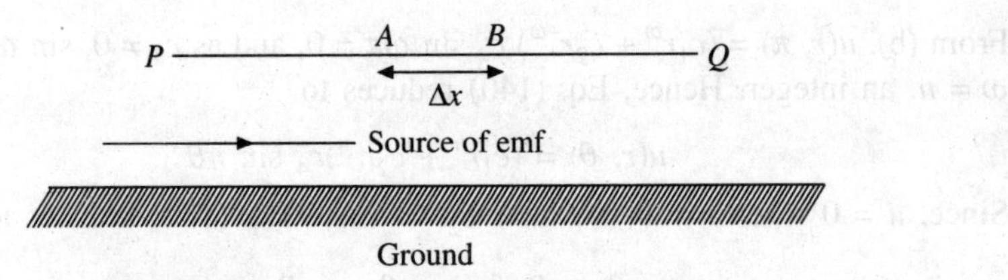


**Fig. 9.10** Flow of current in a long cable.

Since, the voltage drop across the segment $\Delta x$ = voltage drop due to resistance + voltage drop due to inductance. Therefore

$$-\Delta v = iR\Delta x + L\Delta x \frac{\partial i}{\partial t}$$

Dividing by $\Delta x$ and taking the limit as $\Delta x \to 0$, we get

$$-\frac{\partial v}{\partial x} = Ri + L\frac{\partial i}{\partial t} \tag{144}$$

Also, the current loss between $A$ and $B$ = current loss due to capacitance and leakage. Therefore,

$$-\Delta i = C\frac{\partial v}{\partial t}\Delta x + Gv\Delta x$$

which gives

$$-\frac{\partial i}{\partial x} = C\frac{\partial v}{\partial t} + Gv \tag{145}$$

Equations (144) and (145) can be written as

$$\left(R + L\frac{\partial}{\partial t}\right)i + \frac{\partial v}{\partial x} = 0 \tag{146}$$

and
$$\frac{\partial i}{\partial x} + \left(C\frac{\partial}{\partial t} + G\right)v = 0 \tag{147}$$

Now operate Eq. (146) by $\partial/\partial x$ and Eq. (147) by $\left(R + L\frac{\partial}{\partial t}\right)$ and subtract to obtain

$$\frac{\partial^2 v}{\partial x^2} = LC\frac{\partial^2 v}{\partial t^2} + (LG + RC)\frac{\partial v}{\partial t} + RGv \tag{148}$$

Similarly, operate Eq. (146) by $\left(C\frac{\partial}{\partial t} + G\right)$ and Eq. (147) by $\frac{\partial}{\partial x}$ and subtract to get

$$\frac{\partial^2 i}{\partial x^2} = LC\frac{\partial^2 i}{\partial t^2} + (LG + RC)\frac{\partial i}{\partial t} + RGi \tag{149}$$

Equations (148) and (149) are known as *telephone equations*.

Under special circumstances, these equations give rise to some important equations involved in the theory of transmissions.

(a) When $L = G = 0$ in Eqs. (148) and (149), we have

$$\frac{\partial^2 v}{\partial x^2} = RC\frac{\partial v}{\partial t} \tag{150}$$

$$\frac{\partial^2 i}{\partial x^2} = RC\frac{\partial i}{\partial t} \tag{151}$$

Equations (150) and (151) are known as the *telegraph equations.*

**Remark.** Rewrite Eq. (150) as

$$\frac{\partial v}{\partial t} = \frac{1}{RC}\frac{\partial^2 v}{\partial x^2}$$

which is similar to that of Eq. (106).

(b) When $R = G = 0$, Eqs. (148) and (149) reduce to

$$\frac{\partial^2 v}{\partial x^2} = LC\frac{\partial^2 v}{\partial t^2} \tag{152}$$

$$\frac{\partial^2 i}{\partial x^2} = LC\frac{\partial^2 i}{\partial t^2} \tag{153}$$

These equations are known as *radio equations.*

Equation (152) can be written as

$$\frac{\partial^2 v}{\partial t^2} = k^2\frac{\partial^2 v}{\partial x^2}, \qquad k^2 = \frac{1}{LC}$$

which has the general solution of the form

$$v(x, t) = \psi_1(x - kt) + \psi_2(x + kt)$$

Similarly, Eq. (153) has the general solution

$$i(x, t) = \phi_1(x - kt) + \phi_2(x + kt)$$

These solutions show that the voltage $v(x, t)$ [or current $i(x, t)$] at any point along the transmission line can be obtained by superposition of a progressive wave and a receding wave travelling with equal velocities ($k$). This corresponds to the case of oscillations of $v(x, t)$ and $i(x, t)$ at high frequencies.

(c) If $L = C = 0$, then Eq. (148) gives $\partial^2 v/\partial x^2 = RGv$, whose solution is

$$v(x) = c_1 \cosh\left(\sqrt{GR}\,x\right) + c_2 \sinh\left(\sqrt{GR}\,x\right) \tag{154}$$

Also from Eq. (144)

$$R_i = -\frac{\partial v}{\partial x} = -\sqrt{GR}\left[c_1 \sinh\left(\sqrt{GR}\,x\right) + c_2 \cosh\left(\sqrt{GR}\,x\right)\right]$$

Thus

$$i(x) = -\sqrt{G/R}\left[c_1 \sinh\left(\sqrt{GR}\,x\right) + c_2 \cosh\left(\sqrt{GR}\,x\right)\right] \tag{155}$$

If $v(0) = v_0$ and $i(0) = i_0$, then $v_0 = c_1$, and $i_0 = \sqrt{G/R}\,c_2$, Eqs. (154) and (155) reduces to

$$v(x) = v_0 \cosh Ax + i_0 B \sinh Ax \tag{156}$$

$$i(x) = i_0 \cosh Ax + \frac{v_0}{B}\sinh Ax \tag{157}$$

where $A = \sqrt{GR}$ and $B = \sqrt{R/G}$. This case corresponds to the case of a *submarine cable.*

**REMARK.** Here, we have considered only transient solutions. The steady state solutions of the transmission cable can be obtained by taking $v = Ve^{i\omega t}$ and $i = Ie^{i\omega t}$ in Eqs. (148) and (149), where $V$ and $I$ are the complex functions of $x$ only.

**EXAMPLE 9.59** A transmission cable 1000 miles long is initially under steady state conditions with potential 1300 V at the sending end ($x$ = 0) and 1200 V at the receiving end ($x$ = 1000). The terminal end of the cable is suddenly grounded, but the potential at the source is kept at 1300 V. Find the potential $v(x, t)$ when the inductance and leakage are negligible.

***Solution*** The telegraph Eq. (150) can be written as

$$\frac{\partial v}{\partial t} = \frac{1}{RC}\frac{\partial^2 v}{\partial x^2} \tag{158}$$

Now, the initial steady state voltage satisfying

$$\frac{\partial^2 v}{\partial x^2} = v_s = 0 = 1300 - \frac{x}{10} = v(x, 0) \tag{159}$$

and steady voltage (after grounding the terminal end) when steady state conditions are ultimately reached $= v'_s = 1300 - 1.3x$. Thus

$$v(x, t) = v'_s + v_t(x, t) = 1300 - 1.3x + \sum_{n=1}^{\infty} b_n \exp\left(-\frac{n^2 x^2 t}{L^2 RC}\right)\sin\frac{n\pi x}{L} \tag{160}$$

[from Eq. (115)], where $v_t(x, t)$ is the transient part and $L$ = 1000 miles. When $t$ = 0, Eqs. (159) and (160) give

$$1300 - 0.1x = v(x, 0) = 1300 - 1.3x + \sum_{n=1}^{\infty} b_n \sin \frac{n\pi x}{L}$$

i.e.
$$1.2x = \sum_{n=1}^{\infty} b_n \sin \frac{n\pi x}{L}$$

where
$$b_n = \frac{2}{L}\int_0^L 1.2x \sin \frac{nx}{L}\, dx = \frac{2400}{\pi}\frac{(-1)^{n+1}}{n}$$

Hence, the required potential voltage is

$$v(x, t) = 1300 - 1.3x + \frac{2400}{\pi}\sum_{n=1}^{\infty}\frac{(-1)^{n+1}}{n}\exp\left(-\frac{n^2\pi^2 t}{L^2 RC}\right)\sin\frac{n\pi x}{L}$$

## 9.18 NUCLEAR REACTORS

In the nuclear theory, the following partial differential equation plays a key role:

$$\frac{\partial^2 u}{\partial x^2} + \frac{\partial^2 u}{\partial y^2} + \frac{\partial^2 u}{\partial z^2} + B^2 u = 0 \tag{161}$$

where $u$ is a function of $x$, $y$ and $z$ and is known as *neutron flux*. It is the sum of the distance travelled per second per cubic centimetres at $(x, y, z)$ by bombarding neutrons. $B^2$ is a positive constant and is called *buckling*. For a certain value of $B^2$ (known as material buckling), the arrangement of the material is critical.

In this section, we shall find relations between dimensions and minimum volume of the reactors filled with material such that (161) is satisfied.

We assume that the reactor has the shape of a rectangular box having sides as $a$, $b$, $c$ (Fig. 9.11). The origin is taken as the centre of the box and

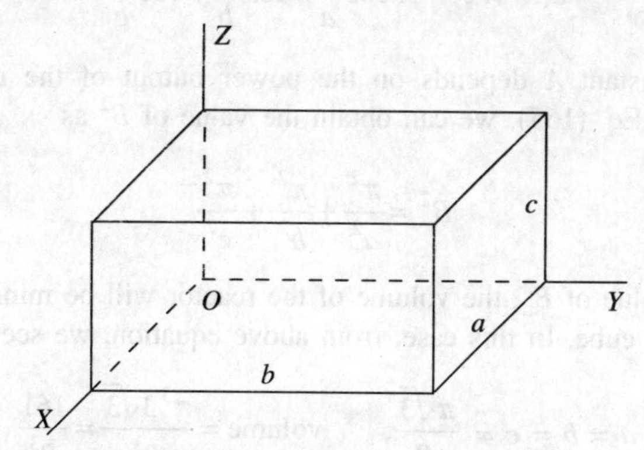


**Fig. 9.11** Rectangular shape of a nuclear reactor.

the axes are parallel to its edges. Since, outside the reactor there is no flux, the boundary conditions are

$$u = 0 \quad \text{at } x = \pm\frac{a}{2},\quad y = \pm\frac{b}{2},\quad z = \pm\frac{c}{2} \tag{162}$$

Also, assume that $u(x, y, z)$ is symmetric with respect to coordinate planes.

We now solve Eq. (161) subject to Eq. (162). Assume that the solution of Eq. (161) is of the form

$$u = X(x)\,Y(y)\,Z(z) \tag{163}$$

Substitute this value in Eq. (161) and simplify. Then we obtain

$$\frac{d^2X/dx^2}{X} + \frac{d^2Y/dy^2}{Y} + \frac{d^2Z/dz^2}{Z} + B^2 = 0 \tag{164}$$

Equate the first fraction to $-\omega^2$ and solve the resulting equation to obtain

$$X = c_1 \cos \omega x + c_2 \sin \omega x$$

Due to Eq. (163) and symmetry, the condition $u = 0$ demands that $X = 0$ when $x = \pm a/2$. Therefore, we take $c_2 = 0$ and $\omega = (2n + 1)\,(1/2)(\pi/a)$. Hence, choosing $\pi/(2a)$ for $\omega$, we have $X = c_1 \cos(\pi x/a)$. Applying the same method to each fraction, we obtain

$$X = c_1 \cos\frac{\pi}{a}x,\quad Y = d_1 \cos\frac{\pi}{b}y,\quad Z = e_1 \cos\frac{\pi}{c}z \tag{165}$$

Thus, Eq. (163) reduces to

$$u(x, y, z) = A \cos\frac{\pi}{a}x \cos\frac{\pi}{b}y \cos\frac{\pi}{c}z$$

where the constant $A$ depends on the power output of the reactor. From. Eq. (164) and Eq. (165), we can obtain the value of $B^2$ as

$$B^2 = \frac{\pi^2}{a^2} + \frac{\pi^2}{b^2} + \frac{\pi^2}{c^2}$$

For a given value of $B^2$, the volume of the reactor will be minimum if it has the shape of a cube. In this case, from above equation, we see that

$$a = b = c = \frac{\pi\sqrt{3}}{B},\qquad \text{volume} = \frac{\pi^3 3\sqrt{3}}{B^3} = \frac{161}{B^3}$$

## 9.19 MISCELLANEOUS SOLVED EXAMPLES

In this section, we shall solve a number of problems based on different sections of this chapter.

**EXAMPLE 9.60** Obtain the partial differential equation by eliminating the arbitrary constants $a$ and $b$ from the relation $z = ax^2 + by^2 + ab$.

***Solution*** Given that

$$z = ax^2 + by^2 + ab$$

Differentiating this equation partially with respect to $x$ and $y$, we get

$$\frac{\partial z}{\partial x} = p = 2ax \quad \text{and} \quad \frac{\partial z}{\partial y} = q = 2by$$

which lead to

$$a = \frac{p}{2x} \quad \text{and} \quad b = \frac{q}{2y}$$

Put these values in the given equation, we get

$$4xyz = 2px^2 y + 2qy^2 x + pq$$

which is the required partial differential equation.

**EXAMPLE 9.61** Form partial differential equation by eliminating the arbitrary constants $a$ and $\omega$ from equation $z = ae^{\omega t} \sin \omega x$.

**Solution** Given that

$$z = ae^{\omega t} \sin \omega x \tag{166}$$

Differentiating twice partially Eq. (166) with respect to $x$ and $t$, we have

$$\frac{\partial z}{\partial x} = a\omega e^{\omega t} \cos \omega x, \quad \frac{\partial z}{\partial t} = a\omega e^{\omega t} \sin \omega x$$

and

$$\frac{\partial^2 z}{\partial x^2} = -a\omega^2 e^{\omega t} \sin \omega x, \quad \frac{\partial^2 z}{\partial t^2} = a\omega^2 e^{\omega t} \sin \omega x$$

Thus, on adding these equations, we get the required partial differential equation as

$$\frac{\partial^2 z}{\partial x^2} + \frac{\partial^2 z}{\partial t^2} = 0 \tag{167}$$

It may be noted that the partial differential Eq. (167) so obtained is not unique. Also, from Eq. (166), we have

$$\frac{\partial^2 z}{\partial x \partial t} = \omega \frac{\partial z}{\partial x} \quad \text{and} \quad \frac{\partial^2 z}{\partial x^2} = -\omega \frac{\partial z}{\partial t}$$

These equations lead to

$$\frac{\partial z}{\partial t}\frac{\partial^2 z}{\partial x \partial t}+\frac{\partial z}{\partial x}\frac{\partial^2 z}{\partial x^2}=0$$

which is another partial differential equation obtained from the given equation by eliminating the arbitrary constants $a$ and $\omega$.

**EXAMPLE 9.62** Find the partial differential equation which represents the set of all right circular cones whose axes coincide with $z$-axis.

***Solution*** Let $\alpha$ be the semi-vertical angle and the coordinates of the vertex be (0, 0, $c$), then the general equation which represents the set of all right circular cones whose axes coincide with $z$-axis is given by

$$x^2+y^2=(z-c)^2\tan^2\alpha \tag{168}$$

Here, we have to eliminate the arbitrary constants $\alpha$ and $c$ to obtain a partial differential equation. Differentiating Eq. (168) partially with respect to $x$ and $y$, we get

$$2x=2(z-c)\frac{\partial z}{\partial x}\tan^2\alpha \quad \text{and} \quad 2y=2(z-c)^2\frac{\partial z}{\partial y}\tan^2\alpha$$

Now, eliminating $\tan^2\alpha$ from these two equations, we get

$$qx-py=0$$

which is the required partial differential equation.

**EXAMPLE 9.63** Find the partial differential equation which represents all planes that are at a constant distance from the origin.

***Solution*** We know that the equation which represents all planes at a constant distance $c$ from the origin is given by

$$lx+my+nz=c \tag{169}$$

such that

$$l^2+m^2+n^2=1 \tag{170}$$

Here, we have to eliminate $l$, $m$, $n$ from Eq. (169) to get the required partial differential equation. Differentiating Eq. (169) partially with respect to $x$ and $y$, we get

$$l=-n\frac{\partial z}{\partial x}=-np \quad \text{and} \quad m=-n\frac{\partial z}{\partial y}=-nq$$

Put these values in Eq. (170), we get

$$n=(p^2+q^2+1)^{-1/2}$$

and thus

$$l=-np=-p(p^2+q^2+1)^{-1/2}$$

$$m=-nq=-q(p^2+q^2+1)^{-1/2}$$

Now substituting the values of $l$, $m$, $n$ in Eq. (169), we get

$$z = px + qy + c(p^2 + q^2 + 1)^{1/2}$$

which is the required partial differential equation.

**EXAMPLE 9.64** Find the partial differential equation which represents the set of all cones having their vertex at origin.

***Solution*** We know that the equation of the cone having vertex at origin is given by

$$ax^2 + by^2 + cz^2 + 2fyz + 2gzx + 2hxy = 0 \tag{171}$$

Differentiating Eq. (171) partially with respect to $x$ and $y$, we get

$$ax + gz + hy + p(cz + fy + gx) = 0 \tag{172}$$

$$by + fz + hx + q(cz + fy + gx) = 0 \tag{173}$$

Now, multiply Eq. (172) by $x$ and Eq. (173) by $y$ and adding the resulting equations, after making use of Eq. (171) and simplification, we get

$$z = px + qy$$

which is the required partial differential equation.

**EXAMPLE 9.65** Obtain the partial differential equations by eliminating the arbitrary function from the following relations

(i) $z = yu(x) + xv(y)$ (ii) $xyz = \phi(x + y + z)$

(ii) $lx + my + nz = f(x^2 + y^2 + z^2)$ (iv) $x^2 + y^2 + z^2 = xf(y/x)$

**Solution** (i) Differentiating partially the given equation with respect to $x$ and $y$, we get

$$\frac{\partial z}{\partial x} = p = yu'(x) + v(y) \text{ and } \frac{\partial z}{\partial y} = q = u(x) + xv'(y) \tag{174}$$

so that

$$\frac{\partial}{\partial x}\left(\frac{\partial z}{\partial y}\right) = s = u'(x) + v'(y) \tag{175}$$

From Eq. (174), we have

$$u'(x) = \frac{p - v(y)}{y} \text{ and } v'(y) = \frac{q - u(x)}{x}$$

Substituting these values in Eq. (175) and using the given relation, we get

$$sxy = px + qy - z$$

which is the required partial differential equation.

(ii) Differentiating partially the given relation with respect to $x$ and $y$, we get

$$yz + xy\frac{\partial z}{\partial x} = \phi'(x + y + z)(1 + \frac{\partial z}{\partial x}) \text{ and } xz + xy\frac{\partial z}{\partial y} = \phi'(x + y + z)(1 + \frac{\partial z}{\partial y})$$

Eliminating $\phi'(x+y+z)$ from these equations, we get

$$px(y-z)+qy(z-x)=z(x-y)$$

as the required partial differential equation.

(iii) Differentiating partially the given relation with respect to $x$ and $y$, we get

$$l+np=f'(x^2+y^2+z^2)(2x+2zp) \text{ and } m+nq=f'(x^2+y^2+z^2)(2y+2zq)$$

Dividing these equations so that $f'(x^2+y^2+z^2)$ can be eliminated and after simplification, we have the required partial differential equation as

$$(mz-ny)p+(nx-lz)q=ly-mx$$

(see also Example 9.9).

(iv) Differentiating partially the given relation with respect to $x$, we get

$$2x+2zxp=f\left(\frac{y}{x}\right)+xf'\left(\frac{y}{x}\right)\left(-\frac{y}{x^2}\right)$$

Multiply this equation by $x$ to get

$$2x^2+2zxp=xf\left(\frac{y}{x}\right)-yf'\left(\frac{y}{x}\right) \tag{176}$$

Now differentiate partially the given equation with respect $y$ to get

$$2y+2zq=f'\left(\frac{y}{x}\right)$$

Multiplying this equation by $y$, we get

$$2y^2+2zqy=yf'\left(\frac{y}{x}\right) \tag{177}$$

Adding Eqs. (176) and (177) and using the given relation, we get the required partial differential equation as

$$z^2-x^2-y^2=2z(xp+yq)$$

**EXAMPLE 9.66** Solve the partial differential equation

$$y\frac{\partial^2 z}{\partial x\partial y}+\frac{\partial z}{\partial x}=\cos(x+y)+y\sin(x+y)$$

***Solution*** Integrating the given partial differential equation with respect to $x$, we get

$$y\frac{\partial z}{\partial y}+z=\sin(x+y)+y\cos(x+y)+f(y)$$

which can be expressed as

$$\frac{\partial}{\partial y}(zy)=\frac{\partial}{\partial y}[y\sin(x+y)]+f(y)$$

Integrating this equation with respect to $y$, we get

$$zy = y\sin(x+y) + \int f(y)dy + g(x)$$

or

$$zy = y\sin(x+y) + h(y) + g(x)$$

which is the required solution, where $h(y) = \int f(y)dy$.

**EXAMPLE 9.67** Solve $\frac{\partial^2 z}{\partial x^2} + \omega^2 x = 0$ under the condition that when $x = 0, \frac{\partial z}{\partial x} = \omega \sin y, \frac{\partial z}{\partial y} = 0$

***Solution*** Integrating the given equation with respect to $x$, we get

$$\frac{\partial z}{\partial x} + \frac{1}{2}\omega^2 x^2 = f(y) \tag{178}$$

which on integrating again leads to

$$z + \frac{1}{6}\omega^2 x^3 = xf(y) + g(y) \tag{179}$$

From the given conditions, Eq. (178) yields

$$\omega \sin y = f(y) \tag{180}$$

Differentiating Eq. (179) partially with respect to $y$, we get

$$\frac{\partial z}{\partial y} = xf'(y) + g'(y)$$

Using the given conditions, this equation leads to $g'(y) = 0$ and thus $g(y) = a$ (constant). Therefore, the required solution, from Eqs. (179) and (180), is

$$z + \frac{1}{6}\omega^2 x^3 = \omega x \sin y + a$$

**EXAMPLE 9.68** Solve $p = 6x - 3y$ and $q = -3x - 4y$.

***Solution*** Integration of the first equation of the given system leads to

$$z = 3x^2 - 3xy + \phi(y) \tag{181}$$

Differentiating Eq. (181) partially with respect to $y$, we get

$$\frac{\partial z}{\partial y} = -3x + \phi'(y) \tag{182}$$

Using the second equation of the given system, Eq. (182) leads to $\phi'(y) = -4y$ so that $\phi(y) = -2y^2 + c$. Thus, from Eq. (181), the required solution is

$$z = 3x(x - y) - 2y^2 + c$$

**EXAMPLE 9.69** Solve the following partial differential equations

(i) $z(y^2 p - x^2 q) = x^2 y$ (ii) $t + s + q = 0$ (iii) $xyr + x^2 s - yp = x^3 e^y$

(iv) $(x^2 - yz)p + (y^2 - zx)q = z^2 - xy$ (v) $(x + y^2)p + yq = z + x^2$

(vi) $x(y^2 - z^2)p - y(z^2 + x^2)q = z(x^2 + y^2)$

***Solution*** (i) The subsidiary equation for the given partial differential equation is

$$\frac{dx}{y^2 z} = \frac{dy}{-x^2 z} = \frac{dz}{x^2 y} \tag{183}$$

Consider the last two fractions of Eq. (183) and integrate, we have

$$z^2 + y^2 = a \tag{184}$$

where $a = 2c_1$ is an arbitrary constant. Now, consider the first two fractions of Eq. (183) and integrate so that we have

$$x^3 + y^3 = b \tag{185}$$

where $b = 3c_2$ is an arbitrary constant. Thus, from Eqs. (184) and (185), the general solution is

$$f(y^2 + z^2, x^3 + y^3) = 0$$

or

$$y^2 + z^2 = f(x^3 + y^3)$$

(ii) The given equation can be written as

$$\frac{\partial q}{\partial y} + \frac{\partial p}{\partial y} + \frac{\partial z}{\partial y} = 0$$

which on integration leads to

$$p + q = f(x) - z$$

Here the subsidiary equation is

$$dx = dy = \frac{dz}{f(x) - z}$$

First two fractions of this subsidiary equation lead to

$$x - y = c_1$$

while the first and third fractions lead to

$$\frac{dz}{dx} + z = f(x)$$

which is a first order linear differential equation having the solution as

$$ze^x - g(x) = c_2$$

where $g(x) = \int f(x)e^x\,dx$. The general solution is therefore

$$\phi[x - y, ze^x - g(x)] = 0$$

or
$$x - y = \phi[ze^x - g(x)]$$

(iii) The given partial differential equation can be written as

$$xy\frac{\partial p}{\partial x} + x^2\frac{\partial p}{\partial y} = x^3e^y + yp$$

so that the subsidiary equation is given by

$$\frac{dx}{xy} = \frac{dy}{x^2} = \frac{dp}{x^3e^y + yp} \tag{186}$$

From the first two fractions of Eq. (186), we have

$$x^2 - y^2 = c_1 \tag{187}$$

while the last two fractions of Eq. (186) lead to

$$\frac{dp}{dy} - \frac{yp}{x^2} = xe^y$$

which on using Eq. (187) leads to

$$\frac{dp}{dy} - \frac{y}{y^2 + c_1}p = (y^2 + c_1)^{1/2}e^y$$

This is a first order linear differential equation having the solution as

$$\frac{p}{(y^2 + c_1)^{1/2}} - e^y = c_2$$

Using Eq. (187), this equation simplifies to

$$\frac{p}{x} - e^y = c_2 \tag{188}$$

From Eqs. (187) and (188), the general solution is

$$\frac{p}{x} - e^y = f(x^2 - y^2)$$

which may be expressed as

$$\frac{\partial z}{\partial x} = xf(x^2 - y^2) + xe^y$$

Integrating this equation with respect to $x$, we get

$$z = \int xf(x^2 - y^2)dx + \frac{1}{2}x^2e^y + g(y)$$

Put $f(x^2 - y^2) = 2h'(x^2 - y^2)$ in this equation, we get

$$z = \int 2xh'(x^2 - y^2)dx + \frac{1}{2}x^2e^y + g(y)$$

Thus, the required solution is

$$z = h(x^2 - y^2) + \frac{1}{2}x^2 e^y + g(y)$$

(iv) Here, the subsidiary equation is

$$\frac{dx}{x^2 - yz} = \frac{dy}{y^2 - zx} = \frac{dz}{z^2 - xy}$$

which leads to

$$\frac{dx - dy}{(x-y)(x+y+z)} = \frac{dy - dz}{(y-z)(y+z+x)} = \frac{dz - dx}{(z-x)(z+x+y)} \tag{189}$$

First and second fractions of Eq. (189) lead to

$$\frac{x-y}{y-z} = c_1 \tag{190}$$

while the second and third fractions of Eq. (189) yield

$$\frac{y-z}{z-x} = c_2 \tag{191}$$

From Eqs. (190) and (191), the required solution is

$$\frac{x-y}{y-z} = f\left[\frac{y-z}{z-x}\right]$$

(v) The subsidiary equation is

$$\frac{dx}{x+y^2} = \frac{dy}{y} = \frac{dz}{z+x^2} \tag{192}$$

Consider the first two fractions of Eq. (192), we get

$$\frac{dx}{dy} - \frac{x}{y} = y$$

which is a first order linear differential equation and the solution is

$$x = y(y + c_1) \tag{193}$$

Now, consider the second and third fractions of Eq. (192) and use Eq. (193) to get

$$\frac{dy}{y} = \frac{dz}{z + y^2(y + c_1)^2}$$

which can be expressed as

$$\frac{dz}{dy} - \frac{z}{y} = y(y + c_1)^2$$

This is a first order linear differential equation and the solution is

$$z\left(\frac{1}{y}\right) = \int (y + c_1)^2\, dy + c_2 = \frac{1}{3}(y + c_1)^3 + c_2$$

which on using Eq. (193) leads to

$$\frac{3zy^2 - x^3}{3y^3} = c_2 \tag{194}$$

Therefore, from Eqs. (193) and (194), the required solution is

$$f\left[\frac{x-y^2}{y}, \frac{3zy^2 - x^3}{3y^3}\right] = 0$$

(vi) The subsidiary equation is

$$\frac{dx}{x(y^2 - z^2)} = \frac{dy}{-y(z^2 + x^2)} = \frac{dz}{z(x^2 + y^2)} \tag{195}$$

Using $x$, $y$, $z$ as multipliers, we have

$$\text{each fraction} = \frac{xdx + ydy + zdz}{0}$$

so that $xdx + ydy + zdz = 0$, which on integration leads to

$$x^2 + y^2 + z^2 = c_1 \tag{196}$$

Now using $\frac{1}{x}, -\frac{1}{y}, -\frac{1}{z}$ as multipliers in Eq. (195), we get

$$\frac{dx}{x} - \frac{dy}{y} - \frac{dz}{z} = 0$$

which on integration leads to

$$\frac{x}{yz} = c_2 \tag{197}$$

From Eqs. (196) and (197), the general solution is

$$f\left[\frac{x}{yz}, x^2 + y^2 + z^2\right] = 0$$

**EXAMPLE 9.70** Solve the following partial differential equations:

(i) $(y - x)(qy - px) = (p - q)^2$ (ii) $(1 - x^2)yp^2 + x^2 q = 0$

(iii) $q^2 y^2 = z(z - px)$ (iv) $(x^2 + y^2)(p^2 + q^2) = 1$

***Solution*** (i) Let

$$x + y = X \quad \text{and} \quad xy = Y \tag{198}$$

then
$$p = \frac{\partial z}{\partial x} = \frac{\partial z}{\partial X}\frac{\partial X}{\partial x} + \frac{\partial z}{\partial Y}\frac{\partial Y}{\partial x} = \frac{\partial z}{\partial X} + y\frac{\partial z}{\partial Y} = P + yQ \tag{199}$$

and
$$q = \frac{\partial z}{\partial y} = \frac{\partial z}{\partial X}\frac{\partial X}{\partial y} + \frac{\partial z}{\partial Y}\frac{\partial Y}{\partial y} = \frac{\partial z}{\partial X} + x\frac{\partial z}{\partial Y} = P + xQ \tag{200}$$

where $P=\frac{\partial z}{\partial X}$ and $Q=\frac{\partial z}{\partial Y}$. From Eqs. (199) and (200), we have

$$p-q=(y-x)Q \quad \text{and} \quad qy-px=(y-x)P$$

With these values, the given equation takes the form

$$P=Q^2$$

which is of the form $f(P,Q)=0$. Thus, the solution is

$$z=aX+bY+c$$

where $a=b^2$. From Eq. (198), the required solution is

$$z=b^2(x+y)+bxy+c$$

(ii) The given partial differential equation can be written as

$$\left[\frac{(1-x^2)^{1/2}}{x}\frac{\partial z}{\partial x}\right]^2+\frac{1}{y}\frac{\partial z}{\partial y}=0 \tag{201}$$

Put
$$X=(1-x^2)^{1/2},\ Y=\frac{y^2}{2} \tag{202}$$

then
$$p=\frac{\partial z}{\partial x}=\frac{\partial z}{\partial X}\frac{\partial X}{\partial x}+\frac{\partial z}{\partial Y}\frac{\partial Y}{\partial x}=-\frac{x}{(1-x^2)^{1/2}}P \tag{203}$$

and
$$q=\frac{\partial z}{\partial y}=\frac{\partial z}{\partial X}\frac{\partial X}{\partial y}+\frac{\partial z}{\partial Y}\frac{\partial Y}{\partial y}=y\frac{\partial z}{\partial Y}=yQ \tag{204}$$

where $P=\frac{\partial z}{\partial X}$ and $Q=\frac{\partial z}{\partial Y}$. From Eqs. (203) and (204), Eq. (201) reduces to

$$P^2+Q=0 \tag{205}$$

which is of the form $f(P,Q)=0$ and the solution is

$$z=aX+bY+c$$

where, from Eq. (205), $b=-a^2$. Thus, using Eq. (202), the required solution is

$$z=aX-a^2Y+c=a(1-x^2)^{1/2}-\frac{1}{2}a^2y^2+c$$

(iii) The given equation can be expressed as

$$\left(\frac{y}{z}\frac{\partial z}{\partial y}\right)^2=1-\frac{x}{z}\frac{\partial z}{\partial x} \tag{206}$$

Put
$$X=\log x,\ Y=\log y,\ Z=\log z \tag{207}$$

so that

$$\partial Z=\frac{1}{z}\partial z, \partial X=\frac{1}{x}\partial x, \partial Y=\frac{1}{y}\partial y \text{ and } \frac{\partial Z}{\partial Y}=\frac{y}{z}\frac{\partial z}{\partial y}, \frac{\partial Z}{\partial X}=\frac{x}{z}\frac{\partial z}{\partial x}$$

Equation (206) now takes the form

$$Q^2 = 1 - P$$

having the solution as

$$Z = aX + bY + c$$

where $b^2 = 1 - a$. Thus, from Eq. (207), the required solution is

$$\log z = a\log x + \sqrt{1-a}\,\log y + c$$

(iv) Consider the transformation

$$x = r\cos\theta, y = r\sin\theta, x^2 + y^2 = r^2, \theta = \tan^{-1}(y/x) \tag{208}$$

Also

$$p = \frac{\partial z}{\partial x} = \frac{\partial z}{\partial r}\frac{\partial r}{\partial x} + \frac{\partial z}{\partial \theta}\frac{\partial \theta}{\partial x} \quad \text{and} \quad q = \frac{\partial z}{\partial y} = \frac{\partial z}{\partial r}\frac{\partial r}{\partial y} + \frac{\partial z}{\partial \theta}\frac{\partial \theta}{\partial y} \tag{209}$$

From Eqs. (208) and (209), the given equation reduces to

$$\left(r\frac{\partial z}{\partial r}\right)^2 + \left(\frac{\partial z}{\partial \theta}\right)^2 = 1 \tag{210}$$

Put $R = \log r$, then Eq. (210) can be expressed as

$$\left(\frac{\partial z}{\partial R}\right)^2 + \left(\frac{\partial z}{\partial \theta}\right)^2 = 1 \tag{211}$$

that is, $P^2 + Q^2 = 1$, where $P = \dfrac{\partial z}{\partial R}$ and $Q = \dfrac{\partial z}{\partial \theta}$ and the solution is

$$z = aR + b\theta + c$$

where $a^2 + b^2 = 1$.
Thus, the required solution is

$$z = a\log r + \sqrt{1-a^2}\,\tan^{-1}(y/x) + c$$

or

$$z = \frac{1}{2}a\log(x^2 + y^2) + \sqrt{1-a^2}\,\tan^{-1}(y/x) + c$$

**EXAMPLE 9.71** Solve the following partial differential equations:

(i) $p^2z^2 + q^2 = 1$ (ii) $z(z - px) = q^2y^2$ (iii) $pq = x^3y^2z$

***Solution*** Put $p = \dfrac{dz}{du}, q = ap = a\dfrac{dz}{du}$ so that the given equation takes the form

$$z^2\left(\frac{dz}{du}\right)^2 + a^2\left(\frac{dz}{du}\right)^2 = 1$$

Separating the variables, we get

$$\sqrt{z^2 + a^2}\,dz = du$$

which on integration leads to

$$\log\left[z+\sqrt{z^2+a^2}\right]=u+c$$

Now put $u=x+ay$ to get the required solution as

$$\log[z+\sqrt{z^2+a^2}]=x+ay+c$$

(ii) Put $X=\log x, Y=\log y$ so that

$$\frac{\partial z}{\partial x}=\frac{\partial z}{\partial X}\frac{\partial X}{\partial x}+\frac{\partial z}{\partial Y}\frac{\partial Y}{\partial x}=\frac{1}{x}\frac{\partial z}{\partial X}, \frac{\partial z}{\partial y}=\frac{\partial z}{\partial X}\frac{\partial X}{\partial y}+\frac{\partial z}{\partial Y}\frac{\partial Y}{\partial y}=\frac{1}{y}\frac{\partial z}{\partial Y}$$

and the given equation can now be expressed as

$$z(z-P)=Q^2 \tag{212}$$

where $P=\dfrac{\partial z}{\partial X}$ and $Q=\dfrac{\partial z}{\partial Y}$. Now replacing $P$ by $\dfrac{dz}{du}$ and $Q$ by $a\dfrac{dz}{du}$ in Eq. (212), we get

$$a^2\left(\frac{dz}{du}\right)^2+z\frac{dz}{du}-z^2=0$$

Solving this equation for $dz/du$, we have

$$\frac{dz}{du}=\left[\frac{-1\pm(1+4a^2)^{1/2}}{2a^2}\right]z=cz \tag{213}$$

Now separating the variables in Eq. (213) and integrating, we get

$$\log z=cu+b=c(X+aY)+b=c(\log x+a\log y)+b$$

as the required solution, where $c=\dfrac{-1\pm(1+4a^2)^{1/2}}{2a^2}$.

(iii) The given equation can be written as

$$\left(\frac{1}{x^3}\frac{\partial z}{\partial x}\right)\left(\frac{1}{y^2}\frac{\partial z}{\partial x}\right)=z \tag{214}$$

Put $\dfrac{x^4}{4}=X$ and $\dfrac{y^3}{3}=Y$ so that

$$\frac{\partial z}{\partial x}=\frac{\partial z}{\partial X}\frac{\partial X}{\partial x}+\frac{\partial z}{\partial Y}\frac{\partial Y}{\partial x}=x^3\frac{\partial z}{\partial X} \text{ and } \frac{\partial z}{\partial y}=\frac{\partial z}{\partial X}\frac{\partial X}{\partial y}+\frac{\partial z}{\partial Y}\frac{\partial Y}{\partial y}=y^2\frac{\partial z}{\partial Y}$$

and Eq. (214) now reduces to

$$PQ=z \tag{215}$$

where $P=\dfrac{\partial z}{\partial X}$ and $Q=\dfrac{\partial z}{\partial Y}$. Now replacing $P$ by $\dfrac{dz}{du}$ and $Q$ by $a\dfrac{dz}{du}$ in Eq. (215), we get after simplification

$$\frac{dz}{\sqrt{z}} = \frac{1}{\sqrt{a}} du$$

Integrating this equation, we get

$$z^{1/2} = \frac{1}{2\sqrt{a}} u + b$$

Now, replacing $u$ by $X + aY$ and substituting the values of $X$ and $Y$, we get the required solution as

$$z^{1/2} = \frac{1}{2\sqrt{a}}\left(\frac{x^4}{4} + a\frac{y^3}{3}\right) + b$$

**EXAMPLE 9.72** Solve

(i) $p^2q^2 + x^2y^2 = x^2q^2(x^2 + y^2)$ (ii) $yzp^2 = q$ (iii) $z(xp - yq) = y^2 - x^2$

***Solution*** (i) The given equation can be written as

$$\frac{p^2}{x^2} - x^2 = y^2 - \frac{y^2}{q^2}$$

Let both the terms of this equation be equal to $a$ so that

$$p = x\sqrt{a^2 + x^2} \quad \text{and} \quad q = \frac{y}{\sqrt{y^2 - a^2}}$$

Now, substitute the values of $p$ and $q$ in the equation $dz = pdx + qdy$, we get the required solution as

$$z = \frac{1}{3}(x^2 + a^2)^{3/2} + (y^2 - a^2)^{1/2} + b$$

(ii) The given equation can be written as

$$y\left(z\frac{\partial z}{\partial x}\right)^2 = z\frac{\partial z}{\partial y} \tag{216}$$

Put $\frac{z^2}{2} = Z$, so that $z\frac{\partial z}{\partial x} = \frac{\partial Z}{\partial x}, z\frac{\partial z}{\partial y} = \frac{\partial Z}{\partial y}$ and Eq. (216) reduces to

$$P^2 = \frac{Q}{y}$$

where $P = \frac{\partial Z}{\partial x}$ and $Q = \frac{\partial Z}{\partial y}$. Equating both sides of this equation to a constant $a^2$, we get

$$P^2 = a^2, \frac{Q}{y} = a^2$$

Now, substitute the values of $P$ and $Q$ from these equations in equation $dZ = Pdx + Qdy$ and integrate the resulting equation, we get

$$Z = ax + \frac{1}{2}a^2y^2 + c_1$$

Thus, the complete solution is

$$z^2 = 2ax + a^2y^2 + c$$

where $c = 2c_1$.

(iii) The given equation can be written as

$$x\left(z\frac{\partial z}{\partial x}\right) - y\left(z\frac{\partial z}{\partial y}\right) = y^2 - x^2 \tag{217}$$

Put $\frac{z^2}{2} = Z$, Eq. (217) then reduces to

$$xP + x^2 = yQ + y^2$$

where $P = \frac{\partial Z}{\partial x}$ and $Q = \frac{\partial Z}{\partial y}$. Equating both sides of this equation to a constant $a$ and solving for $P$ and $Q$, we get

$$P = \frac{a}{x} - x,\ Q = \frac{a}{y} - y$$

Now, substitute these values of $P$ and $Q$ in equation $\frac{\partial z}{\partial x} = \frac{\partial z}{\partial u}\frac{\partial u}{\partial x} + \frac{\partial z}{\partial v}\frac{\partial v}{\partial x} = 2x\frac{\partial z}{\partial u}$ and $\frac{\partial z}{\partial y} = \frac{\partial z}{\partial u}\frac{\partial u}{\partial y} + \frac{\partial z}{\partial v}\frac{\partial v}{\partial y} = 2y\frac{\partial z}{\partial v}$ and integrate the resulting equation, we get

$$Z = a\log x - \frac{1}{2}x^2 + a\log y - \frac{1}{2}y^2 + c$$

which after simplification leads to the required solution as

$$z^2 = 2a\log(xy) - x^2 - y^2 + b$$

where $b = 2c$.

**EXAMPLE 9.73** Find the singular solution of the partial differential equation

$$z = px + qy + \sqrt{1 + p^2 + q^2}$$

***Solution*** The given equation is of Clairaut's form and the complete solution is

$$z = ax + by + \sqrt{1 + a^2 + b^2} \tag{218}$$

Differentiate Eq. (218) with respect to $a$ and $b$, we get

$$x + \frac{a}{\sqrt{1 + a^2 + b^2}} = 0,\ y + \frac{b}{\sqrt{1 + a^2 + b^2}} = 0 \tag{219}$$

From Eq. (219), we have

$$1 - (x^2 + y^2) = \frac{1}{1 + a^2 + b^2}$$

so that

$$\sqrt{1+a^2+b^2} = \frac{1}{\sqrt{1-x^2-y^2}}$$

Put this value in Eq. (219) so that after simplification, we get

$$a = -\frac{x}{\sqrt{1-x^2-y^2}}, b = -\frac{y}{\sqrt{1-x^2-y^2}}$$

Using these values of $a$ and $b$ in Eq. (218), we get

$$x^2+y^2+z^2=1$$

which is the required singular solution.

**EXAMPLE 9.74** Using some transformation, reduce the partial differential equation

$$2y+2zq = q(xp+yq)$$

to Clairaut's form and obtain the complete solution.

***Solution*** The given partial differential equation can be written as

$$z = x^2\left(\frac{1}{2x}\frac{\partial z}{\partial x}\right) + y^2\left(\frac{1}{2y}\frac{\partial z}{\partial y}\right) - \frac{1}{2}\left(\frac{1}{2y}\frac{\partial z}{\partial y}\right)^{-1} \tag{220}$$

where $p = \dfrac{\partial z}{\partial x}$ and $q = \dfrac{\partial z}{\partial y}$. Now use the transformation $u = x^2$ and $v = y^2$ so that

$$\frac{\partial z}{\partial x} = \frac{\partial z}{\partial u}\frac{\partial u}{\partial x} + \frac{\partial z}{\partial v}\frac{\partial v}{\partial x} = 2x\frac{\partial z}{\partial u} \text{ and } \frac{\partial z}{\partial y} = \frac{\partial z}{\partial u}\frac{\partial u}{\partial y} + \frac{\partial z}{\partial v}\frac{\partial v}{\partial y} = 2y\frac{\partial z}{\partial v}$$

Equation (220) now reduces to

$$z = Pu + Qv - \frac{1}{2Q} \tag{221}$$

which is of Clairaut's form $\left[\text{where } P = \dfrac{\partial z}{\partial u} \text{ and } Q = \dfrac{\partial z}{\partial v}\right]$ and the solution is

$$z = au + bv + -\frac{1}{2b}$$

Thus, the complete solution of the given partial differential equation is

$$z = ax^2 + by^2 - \frac{1}{2b}$$

**EXAMPLE 9.75** Find the complete solution of

$$(z-2px-2qy) = (p^2x+q^2y)^{1/2}$$

***Solution*** The given partial differential equation can be written as

$$z=\sqrt{x}\left[\frac{1}{1/(2\sqrt{x})}\frac{\partial z}{\partial x}\right]+\sqrt{y}\left[\frac{1}{1/(2\sqrt{y})}\frac{\partial z}{\partial y}\right]$$

$$+\frac{1}{2}\left[\left(\frac{1}{1/(2\sqrt{x})}\frac{\partial z}{\partial x}\right)^2+\left(\frac{1}{1/(2\sqrt{y})}\frac{\partial z}{\partial y}\right)^2\right]^{1/2}$$

Put $\sqrt{x}=u$ and $\sqrt{y}=v$ so that $2\sqrt{x}\frac{\partial z}{\partial x}=\frac{\partial z}{\partial u}$ and $2\sqrt{y}\frac{\partial z}{\partial y}=\frac{\partial z}{\partial v}$ and the above equation reduces to

$$z=Pu+Qv+\frac{1}{2}(P^2+Q^2)^{1/2},\left[\text{where } P=\frac{\partial z}{\partial u},Q=\frac{\partial z}{\partial v}\right]$$

which is of Clairaut's form and the solution is

$$z=au+bv+\frac{1}{2}\left(a^2+b^2\right)^{1/2}$$

Thus, using the values of $u$ and $v$, the required complete solution is

$$z=a\sqrt{x}+b\sqrt{y}+\frac{1}{2}(a^2+b^2)^{1/2}$$

**EXAMPLE 9.76** Solve the following partial differential equations

(i) $xq^2=px+qy+z$ (ii) $(p^2+q^2)x=pz$

(iii) $2xz-px^2-2xyq+pq=0$ (iv) $p^2x+pqy+pqx+q^2y=1$

***Solution*** (i) Let $F(x,y,z,p,q)=px+qy+z-xq^2=0$ so that the subsidiary equations are

$$\frac{dx}{x}=\frac{dy}{y-2xq}=\frac{dz}{px+q(y-2xq)}=\frac{dp}{q^2}=\frac{dq}{-2q} \tag{222}$$

Consider the first and last fractions of Eq. (222) and integrate to get

$$qx^2=a$$

Substituting this value of $q$ in the given equation, we get

$$p=\frac{1}{x^4}(a^2-axy-x^3z)$$

Now putting the values of $p$ and $q$ in equation $dz=pdx+qdy$, we get

$$dz=\frac{a}{x^3}(xdy-ydx)+\frac{a}{x^4}dx-\frac{z}{x}dx$$

which leads to

$$xdz+zdx=\frac{a}{x^2}(xdy-ydx)+a\frac{dx}{x^3}$$

Integrating this equation, we get the required solution as

$$xz = a\left(\frac{y}{x}\right) - \frac{a^2}{2x^2} + b$$

(ii) Let $F(x,y,z,p,q) = (p^2 + q^2)x - pz = 0$ so that the subsidiary equations are

$$\frac{dx}{2px - z} = \frac{dy}{2qx} = \frac{dz}{2p^2x - pz + 2q^2x} = \frac{dp}{-q^2} = \frac{dq}{pq}$$

Considering the last two fractions of this equation, we get

$$pdp + qdq = 0$$

which on integration yields

$$p^2 + q^2 = a^2$$

Putting it in the given equation, we get

$$p = \frac{a^2 x}{z}$$

Thus $$q^2 = a^2 - p^2 = \frac{a^2}{z^2}(z^2 - a^2x^2)$$

Therefore $$dz = pdx + qdy = \frac{a^2 x}{z}dx + \frac{a}{z}(z^2 - a^2x^2)^{1/2}dy$$

which on simplification reduces to

$$\frac{zdz - a^2xdx}{(z^2 - a^2x^2)^{1/2}} = ady$$

Integrating this equation, after simplification we get

$$z^2 = (ay + b)^2 + a^2x^2$$

which is the required solution (also see Example 9.22).

(iii) Let $F(x,y,z,p,q) = 2xz - px^2 - 2qxy + pq = 0$. The subsidiary equations are given by

$$\frac{dx}{x^2 - q} = \frac{dy}{2xy - p} = \frac{dz}{px^2 + 2xyq - 2pq} = \frac{dp}{2z - 2qy} = \frac{dq}{0}$$

Here, the last fraction leads to $q = a$. Using this value in the given equation, we get

$$p = \frac{2x(z - ay)}{x^2 - a}$$

Thus $$dz = pdx + qdy = \frac{2x(z - ay)}{x^2 - a}dx + ady$$

which on simplification leads to

$$\frac{dz - ady}{z - ay} = \frac{2xdx}{x^2 - a}$$

Integrating this equation, we get

$$\log(z - ay) = \log(x^2 - a) + \log b$$

Therefore, the required solution is

$$z = b(x^2 - a) + ay$$

(iv) The subsidiary equations are given by

$$\frac{dx}{-[2px + q(x+y)]} = \frac{dy}{-[p(x+y) + 2qy]} = \frac{dz}{-[2p^2x + 2pq(x+y) + 2q^2y]}$$

$$= \frac{dp}{p} = \frac{dq}{q}$$

Consider the last two fractions of this equation and integrate to get

$$p = aq$$

Using this value in the given partial differential equation, we get after simplification

$$q = \frac{1}{\sqrt{(a+1)(ax+y)}}$$

Now substituting the values of $p$ and $q$ in equation $dz = pdx + qdy$, we get

$$dz = \frac{adx + dy}{\sqrt{(a+1)(ax+y)}} = \frac{1}{\sqrt{a+1}} \frac{d(ax+y)}{\sqrt{ax+y}}$$

Integrating this equation, we get

$$z = \frac{1}{\sqrt{1+a}}(ax+y)^{1/2} + b$$

as the required solution.

**EXAMPLE 9.77** Solve the partial differential equation

$$a^2p^2z^2 + b^2q^2z^2 + c^2z^2 - d^2 = 0$$

***Solution*** Let $F(x, y, z, p, q) = a^2p^2z^2 + b^2q^2z^2 + c^2z^2 - d^2 = 0$ so that the subsidiary equations are

$$\frac{dx}{-2a^2pz^2} = \frac{dy}{-2b^2qz^2} = \frac{dz}{-p(2a^2pz^2) - q(2b^2qz^2)}$$

$$= \frac{dp}{p(2a^2p^2z + 2b^2q^2z + 2c^2z)} = \frac{dq}{q(2a^2p^2z + 2b^2q^2z + 2c^2z)}$$

Consider the last two fractions of the subsidiary equation, after integration we get $p = hq$, where $h$ is an arbitrary constant.

Using this value of $p$ in the given equation, we get after simplification

$$q = \frac{(d^2 - c^2z^2)^{1/2}}{z(a^2h^2 + b^2)^{1/2}}$$

Thus

$$p = hq = \frac{h(d^2 - c^2z^2)^{1/2}}{z(a^2h^2 + b^2)^{1/2}}$$

so that $dz = pdx + qdy$ leads to

$$dz = \frac{h(d^2 - c^2z^2)^{1/2}}{z(a^2h^2 + b^2)^{1/2}}dx + \frac{(d^2 - c^2z^2)^{1/2}}{z(a^2h^2 + b^2)^{1/2}}dy = \frac{(d^2 - c^2z^2)^{1/2}}{z(a^2h^2 + b^2)^{1/2}}[hdx + dy]$$

This equation may also be expressed as

$$\frac{z(a^2h^2 + b^2)^{1/2}}{(d^2 - c^2z^2)^{1/2}}dz = hdx + dy$$

Integrating this equation, we get the required solution as

$$(a^2h^2 + b^2)^{1/2}(d^2 - c^2z^2)^{1/2} = -c^2(hx + y) + k$$

where $k$ is an arbitrary constant.

**EXAMPLE 9.78** Solve the following partial differential equations:

(i) $(3D^2 + 10DD' + 3D'^2)z = e^{x-y}$ (ii) $(D^2 + 3DD' + 2D'^2)z = e^{x-y}$

(iii) $(4D^2 + 12DD' + 9D'^2)z = e^{3x-2y}$

***Solution*** (i) Here, the auxiliary equation $3m^2 + 10m + 3 = 0$ has roots as $-3$ and $-1/3$. Thus

$$\text{C.F.} = f_1(y - 3x) + f_2\left(y - \frac{1}{3}x\right)$$

Also

$$\text{P.I.} = \frac{1}{3D^2 + 10DD' + 3D'^2}e^{x-y}$$

$$= \frac{1}{3(1) + 10(1)(-1) + 3(-1)^2}e^{x-y}$$

$$= -\frac{1}{4}e^{x-y}$$

Therefore, the complete solution is

$$z = f_1(y - 3x) + f_2\left(y - \frac{1}{3}x\right) - \frac{1}{4}e^{x-y}$$

(ii) Here, the roots of the auxiliary equation are $-1$ and $-2$. Thus

$$\text{C.F.} = f_1(y - x) + f_2(y - 2x)$$

Also, since $F(D,D') = F(a,b) = F(1,-1) = 0$, therefore

$$\text{P.I.} = e^{x-y}F[D+a, D'+b]^{-1}(1) = e^{x-y}F[D+1, D'-1]^{-1}(1)$$

$$= -e^{x-y}\frac{1}{D}[1-(D+3D'-\frac{D'}{D}+2\frac{D'^2}{D})]^{-1}(1)$$

$$= -e^{x-y}\frac{1}{D}[1+(D+3D'-\frac{D'}{D}+2\frac{D'^2}{D})+\text{other terms}](1)$$

$$= -e^{x-y}D^{-1}(1) = -xe^{x-y}$$

Hence the complete solution is

$$z = f_1(y-x) + f_2(y-2x) - xe^{x-y}$$

(iii) Here, –3/2 and –3/2 are the roots of the auxiliary equation and we have

$$\text{C.F.} = f_1\left(y-\frac{3}{2}x\right) + xf_2\left(y-\frac{3}{2}x\right)$$

Also, $F(D,D') = 4D^2 + 12DD' + 9D'^2$ leads to $F(a,b) = F(3,-2) = 0$. Thus

$$\text{P.I.} = \frac{1}{F[D+a, D'+b]}e^{ax+by}$$

$$= e^{3x-2y}F[D+3, D'-2]^{-1}(1)$$

$$= \frac{1}{4}e^{3x-2y}D^{-2}\left[1+\left(3\frac{D'}{D}+\frac{9}{4}\frac{D'^2}{D^2}\right)\right]^{-1}(1)$$

$$= \frac{1}{4}e^{3x-2y}D^{-2}\left[1+\left(3\frac{D'}{D}+\frac{9}{4}\frac{D'^2}{D^2}\right)+\text{other terms}\right](1)$$

$$= \frac{1}{4}e^{3x-2y}D^{-2}(1) = \frac{1}{8}x^2e^{3x-2y}$$

Therefore, the complete solution is

$$z = f_1\left(y-\frac{3}{2}x\right) + xf_2\left(y-\frac{3}{2}x\right) + \frac{1}{8}x^2e^{3x-2y}$$

**EXAMPLE 9.79** Solve the following partial differential equations:

(i) $(D^2 - 2DD')z = \sin x \cos 2y$

(ii) $(D^2 - 3DD' + 2D'^2)z = e^{2x-y} + e^{x+y} + \cos(x+2y)$

***Solution*** (i) Here

$$\text{C.F.} = f_1(y) + f_2(y+2x)$$

Also

$$\text{P.I.} = \frac{1}{D^2 - 2DD'}\sin x \cos 2y = \frac{1}{2}\frac{1}{D^2 - 2DD'}(2\sin x \cos 2y)$$

$$= \frac{1}{2}\frac{1}{D^2 - 2DD'}[\sin(x+2y) + \sin(x-2y)]$$

$$= \frac{1}{2}\left[\frac{1}{D^2 - 2DD'}\sin(x+2y) + \frac{1}{D^2 - 2DD'}\sin(x-2y)\right]$$

$$= \frac{1}{6}\sin(x+2y) - \frac{1}{10}\sin(x-2y)$$

Therefore, the complete solution is

$$z = f_1(y) + f_2(y+2x) + \frac{1}{6}\sin(x+2y) - \frac{1}{10}\sin(x-2y)$$

(ii) Here

$$\text{C.F.} = f_1(y+x) + f_2(y+2x)$$

Also

$$\text{P.I.} = \frac{1}{F(D,D')}[e^{2x-y} + e^{x+y} + \cos(x+2y)]$$

Now

$$\frac{1}{F(D,D')}e^{2x-y} = \frac{1}{D^2 - 3DD' + 2D'^2}e^{2x-y} = \frac{1}{12}e^{2x-y}$$

and

$$\frac{1}{F(D,D')}e^{x+y} = \frac{1}{D^2 - 3DD' + 2D'^2}e^{x+y} = \frac{1}{0}e^{x+y}$$

Thus, for $F(x,y) = e^{x+y}$, we have

$$\text{P.I.} = \frac{1}{F(D+a, D'+b)}e^{x+y}$$

$$= \frac{1}{(D+1)^2 - 3(D+1)(D'+1) + 2(D'+1)^2}e^{x+y}$$

$$= -e^{x+y}\frac{1}{D}\left[1 - \left(D - 3D' + \frac{D'}{D} - \frac{2D'^2}{D}\right)\right]^{-1} (1)$$

$$= -e^{x+y}\frac{1}{D}[1 + (D - 3D' + \frac{D'}{D} - \frac{2D'^2}{D}) + \text{other terms}](1)$$

$$= -e^{x+y}D^{-1}(1) = -xe^{x+y}$$

Also, when $F(x,y) = \cos(x+2y)$

$$\text{P.I.} = \frac{1}{F(D,D')}\cos(x+2y) = \frac{1}{3}\cos(x+2y)$$

Therefore, the complete solution is

$$z = f_1(y+x) + f_2(y+2x) + \frac{1}{12}e^{2x-y} - xe^{x+y} + \frac{1}{3}\cos(x+2y)$$

**EXAMPLE 9.80** Solve:

(i) $(D^2 - D'^2)z = x^2 + y^2$ (ii) $(D^2 - 2DD' + D'^2)z = x\cos y$

(iii) $(D^3 - D'^3)z = x^3y^3$ (iv) $(D^2 + 6DD' + 9D'^2)z = e^{2x+y} + x + y$

***Solution*** (i) Here

$$\text{C.F.} = f_1(y+x) + f_2(y-x)$$

Also

$$\text{P.I.} = \frac{1}{D^2 - D'^2}(x^2+y^2) = \frac{1}{D^2}\left(1 - \frac{D'^2}{D^2}\right)^{-1}(x^2+y^2)$$

$$= \frac{1}{D^2}\left[1 + \frac{D'^2}{D^2} + \frac{D'^4}{D^4} + \cdots\right](x^2+y^2)$$

$$= \frac{1}{D^2}\left[(x^2+y^2) + \frac{1}{D^2}D'^2(x^2+y^2) + \cdots\right]$$

$$= \frac{1}{D^2}\left[(x^2+y^2) + \frac{1}{D^2}(2)\right]$$

$$= \frac{1}{D^2}\left[(x^2+y^2) + x^2\right]$$

$$= \frac{x^4}{6} + \frac{x^2y^2}{2}$$

Therefore, the complete solution is

$$z = f_1(y+x) + f_2(y-x) + \frac{x^4}{6} + \frac{x^2y^2}{2}$$

(ii) Here

$$\text{C.F.} = f_1(y+x) + xf_2(y+x)$$

Also

$$\text{P.I.} = \frac{1}{D^2 - 2DD' + D'^2}x\cos y = \frac{1}{D'^2}\left(1 - \frac{D}{D'}\right)^{-2}x\cos y$$

$$= \frac{1}{D'^2}\left(1 + 2\frac{D}{D'} + \frac{3D^2}{D'^2} + \cdots\right)x\cos y$$

$$= \frac{1}{D'^2}\left[x\cos y + \frac{2}{D'}\cos y\right] = \frac{1}{D'^2}(x\cos + 2\sin y)$$

$$= -x\cos y - 2\sin y$$

Therefore, the complete solution is

$$z = f_1(y+x) + xf_2(y+x) - x\cos y - 2\sin y$$

(iii) Here, the auxiliary equation $m^3 - 1 = 0$ has roots as 1, $\omega$, $\omega^2$, where $\omega$ is the cube root of unity. Thus

$$\text{C.F.} = f_1(y+x) + f_2(y+\omega x) + f_3(y+\omega^2 x)$$

Also

$$\text{P.I.} = \frac{1}{D^3 - D'^3} x^3 y^3 = \frac{1}{D^3}\left(1 - \frac{D'^3}{D^3}\right)^{-1} (x^3 y^3)$$

$$= \frac{1}{D^3}\left[1 + \frac{D'^3}{D^3} + \cdots\right](x^3 y^3)$$

$$= \frac{1}{D^3}\left[x^3 y^3 + \frac{1}{D^3} D'^3 (x^3 y^3)\right]$$

$$= \frac{1}{D^3}\left[x^3 y^3 + \frac{1}{D^3}(6x^3)\right]$$

$$= \frac{1}{D^3}\left[x^3 y^3 + \frac{x^6}{20}\right] = \frac{x^6 y^3}{120} + \frac{x^9}{10080}$$

Therefore, the complete solution is

$$z = f_1(y+x) + f_2(y+\omega x) + f_3(y+\omega^2 x) + \frac{x^6 y^3}{120} + \frac{x^9}{10080}$$

(iv) Here

$$\text{C.F.} = f_1(y-3x) + xf_2(y-3x)$$

Also

$$\text{P.I.} = \frac{1}{D^2 + 6DD' + 9D'^2} e^{2x+y} + \frac{1}{D^2 + 6DD' + 9D'^2}(x+y)$$

$$= \frac{1}{25} e^{2x+y} + \frac{1}{(D+3D')^2}(x+y)$$

$$= \frac{1}{25} e^{2x+y} + \frac{1}{D^2}\left(1 + \frac{3D'}{D}\right)^{-2}(x+y)$$

$$= \frac{1}{25} e^{2x+y} + \frac{1}{D^2}\left[1 - \frac{6D'}{D} + \frac{9D'^2}{D^2} + \cdots\right](x+y)$$

$$= \frac{1}{25} e^{2x+y} + \frac{1}{D^2}\left[(x+y) - \frac{6}{D}\frac{\partial}{\partial y}(x+y)\right]$$

$$= \frac{1}{25} e^{2x+y} + \frac{1}{D^2}\left[(x+y) - \frac{6}{D}\right]$$

$$= \frac{1}{25}e^{2x+y} + \frac{1}{D^2}[(x+y)-6x]$$

$$= \frac{1}{25}e^{2x+y} + \frac{1}{2}x^2y - \frac{5}{6}x^3$$

Therefore, the complete solution is

$$z = f_1(y-3x) + xf_2(y-3x) + \frac{1}{25}e^{2x+y} + \frac{1}{2}x^2y - \frac{5}{6}x^3$$

**EXAMPLE 9.81** Solve $(D^2 - D'^2)z = -4\pi(x^2 + y^2)$ and find a real function $z$ of $x$ and $y$ satisfying this partial differential equation and reducing to zero when $y$ is zero.

***Solution*** Here, the auxiliary equation $m^2 + 1 = 0$ has roots as $\pm i$ and thus

$$\text{C.F.} = f_1(y+ix) + f_2(y-ix)$$

Also

$$\text{P.I.} = \frac{1}{D^2 + D'^2}\{-4\pi(x^2+y^2)\}$$

$$= -4\pi\frac{1}{D^2}(1+\frac{D'^2}{D^2})^{-1}(x^2+y^2)$$

$$= -4\pi\frac{1}{D^2}\left[1 - \frac{D'^2}{D^2} + \cdots\right](x^2+y^2)$$

$$= -4\pi\frac{1}{D^2}\left[(x^2+y^2) - \frac{1}{D^2}D'^2(x^2+y^2)\right]$$

$$= -4\pi\frac{1}{D^2}\left[(x^2+y^2) - \frac{1}{D^2}(2)\right]$$

$$= -4\pi\frac{1}{D^2}[x^2+y^2-x^2] = -2\pi x^2y^2$$

The general solution is

$$z = f_1(y+ix) + f_2(y-ix) - 2\pi x^2y^2 \tag{223}$$

Moreover, since we have to obtain a real function $z(x,y)$ which satisfies the given partial differential equation and reduces to zero when $y = 0$, Eq. (223) leads to $f_1(y+ix) = f_2(y-ix) = 0$. Thus, the required solution is

$$z = -2\pi x^2y^2$$

**EXAMPLE 9.82** Solve the following partial differential equations:

(i) $(D^2 + DD' - 6D'^2)z = x^2\sin(x+y)$ (ii) $(2D^2 + 5DD' + 3D'^2)z = ye^x$

(iii) $(D^2 + 2DD' + D'^2)z = 2\cos y - x\sin y$

***Solution*** (i) Here

$$\text{C.F.} = f_1(y+2x) + f_2(y-3x)$$

Also

$$\text{P.I.} = \frac{1}{D^2 + DD' - 6D'^2} x^2 \sin(x+y)$$

$$= \frac{1}{(D+3D')(D-2D')} x^2 \sin(x+y)$$

$$= \frac{1}{(D+3D')} \int x^2 \sin(x+c-2x)dx \quad (\text{as } y = c-2x)$$

$$= \frac{1}{(D+3D')} \int x^2 \sin(c-x)dx$$

Now, integrating by parts and simplifying, we get

$$\text{P.I.} = \frac{1}{D+3D'}[(x^2-2)\cos(c-x) + 2x\sin(c-x)]$$

$$= \frac{1}{D+3D'}[(x^2-2)\cos(y+x) + 2x\sin(y+x)] \text{ (replace } c \text{ by } y+2x)$$

$$= \int[(x^2-2)\cos(c_1+4x)]dx + \int[2x\sin(c_1+4x)]dx \ (as\ y = c_1 + 3x)$$

Integrating each term of the above equation by parts, we get

$$\text{P.I.} = \frac{1}{4}(x^2-2)\sin(c_1+4x) - \frac{3}{8}x\cos(c_1+4x) + \frac{3}{32}\sin(c_1+4x)$$

Now replacing $c_1$ by $y - 3x$, after simplification we get

$$\text{P.I.} = \left[\frac{1}{4}(x^2-2) + \frac{3}{32}\right]\sin(y+x) - \frac{3}{8}x\cos(y+x)$$

Therefore, the complete solution is

$$z = f_1(y+2x) + f_2(y-3x) + \left[\frac{1}{4}(x^2-2) + \frac{3}{32}\right]\sin(y+x) - \frac{3}{8}x\cos(y+x)$$

(ii) Here

$$\text{C.F.} = f_1(y-x) + f_2\left(y - \frac{3}{2}x\right)$$

Also

$$\text{P.I.} = \frac{1}{2D^2 + 5DD' + 3D'^2} ye^x = \frac{1}{(D+D')(2D+3D')} ye^x$$

$$= \frac{1}{2(D+D')}\frac{1}{\left(D+\frac{3}{2}D'\right)}ye^x$$

$$= \frac{1}{2(D+D')}\int (c+\frac{3}{2}x)e^x dx \quad \left(\text{as } y = c+\frac{3}{2}x\right)$$

Integrating by part, after simplification we get

$$\text{P.I.} = \frac{1}{2}\frac{1}{D+D'}\left[(c+\frac{3}{2}x)e^x - \frac{3}{2}e^x\right]$$

Now replacing $c$ by $y - \frac{3}{2}x$ so that

$$\text{P.I.} = \frac{1}{2}\frac{1}{D+D'}\left[ye^x - \frac{3}{2}e^x\right] = \frac{1}{2}\int\left[(c_1+x)e^x - \frac{3}{2}e^x\right]dx$$

where $y = c_1 + x$. Integrating the above equation, we get

$$\text{P.I.} = \frac{1}{2}\left[(c_1+x)e^x - e^x - \frac{3}{2}e^x\right]$$

Finally, replacing $c_1$ by $y - x$ in the above equation, we get

$$\text{P.I.} = \frac{1}{4}(2y-5)e^x$$

Therefore, the complete solution is

$$z = f_1(y-x) + f_2\left(y - \frac{3}{2}x\right) + \frac{1}{4}(2y-5)e^x$$

(iii) Here

$$\text{C.F.} = f_1(y-x) + xf_2(y-x)$$

and $$\text{P.I.} = \frac{1}{(D+D')(D+D')}[2\cos y - x\sin y]$$

$$= \frac{1}{(D+D')}\int[2\cos(c+x) - x\sin(c+x)]dx \quad (\text{as } y = c+x)$$

Integrating, we get

$$\text{P.I.} = \frac{1}{(D+D')}[\sin(c+x) + x\cos(c+x)]$$

$$= \frac{1}{(D+D')}[\sin(y-x+x) + x\cos(y-x+x)] \quad (\text{as } c = y-x)$$

$$= \frac{1}{(D+D')}[\sin y + x\cos y]$$

$$= \int[\sin(c+x) + x\cos(c+x)]dx \quad (\text{as } y = c+x)$$

Integrating, we get

$$\text{P.I.} = x\sin(c+x)$$

Now replacing $c$ by $y - x$, we get

$$\text{P.I.} = x\sin y$$

Therefore, the required solution is

$$z = f_1(y-x) + xf_2(y-x) + x\sin y$$

**EXAMPLE 9.83** Solve the following partial differential equations

(i) $(D^3 - 3DD'^2 - 2D'^3)z = e^y(5+4x)$

(ii) $(D^2 - DD' - 2D'^2)z = 16xe^{2y}$

***Solution*** (i) Here, the roots of the auxiliary equation $m^3 - 3m - 2 = 0$ are $-1$, $-1$ and 2. Thus

$$\text{C.F.} = f_1(y-x) + xf_2(y-x) + f_3(y+2x)$$

Also

$$\text{P.I.} = \frac{1}{D^3 - 3DD'^2 - 2D'^3} e^y(5+4x)$$

$$= e^y \frac{1}{D^3 - 3D(D'+1)^2 - 2(D'+1)^3}(5+4x)$$

$$= \frac{-e^y}{2}\left[1 - \left(\frac{3DD'^2}{2} + 3DD' + \frac{3D}{2} + D'^3 + 3D'^2 + 3D' - \frac{D^3}{3}\right)\right]^{-1} (5+4x)]$$

After performing the indicated operations and simplification, we get

$$\text{P.I.} = -\frac{1}{2}e^y\left[(5+4x) - \frac{3}{2}\frac{\partial}{\partial x}(5+4x)\right] = \frac{e^y}{2}(1-4x)$$

Therefore, the complete solution is

$$z = f_1(y-x) + xf_2(y-x) + f_3(y+2x) + \frac{e^y}{2}(1-4x)$$

(ii) Here

$$\text{C.F.} = f_1(y-x) + f_2(y+2x)$$

and $$\text{P.I.} = \frac{1}{D^2 - DD' - 2D'^2} 16xe^{2y}$$

$$= 16e^{2y} \frac{1}{D^2 - D(D'+2) - 2(D'+2)^2}(x)$$

$$= -\frac{16}{2}e^{2y}\left[1 - \left(\frac{D^2}{2} - \frac{DD'}{2} - D'^2 - 4D' - 4\right)\right]^{-1} (x)$$

$$= -8e^{2y}\left[1 - \left(\frac{D^2}{2} - \frac{DD'}{2} - D'^2 - 4D' - 4\right) + \cdots\right](x)$$

$$= -8e^{2y}(x - 4x) = 24xe^{2y}$$

Therefore, the required solution is

$$z = f_1(y - x) + f_2(y + 2x) + 24xe^{2y}$$

**EXAMPLE 9.84** Solve the following partial differential equations:

(i) $(D^2 - D'^2)z = x^{-2}$  (ii) $(6D^2 + 5DD' - 6D'^2)z = 66\log(x + 3y)$

(iii) $(D^3 - 3DD'^2 + 2D'^3)z = (x + 2y)^{1/2}$

***Solution*** (i) Here

$$\text{C.F.} = f_1(y + x) + f_2(y - x)$$

Also

$$\text{P.I.} = \frac{1}{D^2 - D'^2}(x + 0 \cdot y)^{-2}$$

and $f(a, b) \neq 0$ and $f(D, D')$ is a homogeneous function of degree 2. Thus, from Eq. (32), we have

$$\text{P.I.} = \frac{1}{1 - 0}\iint u^{-2}du^2 = -\log u = -\log(x + 0 \cdot y) = -\log x$$

Therefore, the complete solution is

$$z = f_1(y + x) + f_2(y - x) - \log x$$

(ii) Here

$$\text{C.F.} = f_1\left(y + \frac{2}{3}x\right) + f_2\left(y - \frac{3}{2}x\right)$$

Also

$$\text{P.I.} = \frac{1}{6D^2 + 5DD' - 6D'^2}66\log(x + 3y)$$

and $f(a, b) \neq 0$ and $f(D, D')$ is a homogeneous function of degree 2. Thus, from Eq. (32), we have

$$\text{P.I.} = \frac{1}{6(1)^2 + 5(1)(3) - 6(3)^2}66\iint \log u\,du^2 = 2\iint \log u\,du^2$$

Integrating this equation, after simplification we get

$$\text{P.I.} = \frac{1}{2}[u^2(2\log u - 3)]$$

Now put $u = x + 3y$ in this equation, we get

$$\text{P.I.} = \frac{1}{2}[(x + 3y)^2\{2\log(x + 3y) - 3\}]$$

Therefore, the required solution is

$$z = f_1\left(y + \frac{2}{3}x\right) + f_2\left(y - \frac{3}{2}x\right) + \frac{1}{2}[(x+3y)^2\{2\log(x+3y) - 3\}]$$

(iii) The roots of the auxiliary equation $m^3 - 3m + 2 = 0$ are 1, 1 and –2. Thus

$$\text{C.F.} = f_1(y+x) + xf_2(y+x) + f_3(y-2x)$$

Also

$$\text{P.I.} = \frac{1}{D^3 - 3DD'^2 + 2D'^3}(x+2y)^{1/2}$$

Here $f(D, D')$ is a homogeneous function of degree 3 and from Eq. (32), we have

$$\text{P.I.} = \frac{1}{(1)^3 - 3(1)(2)^2 + 2(2)^3}\iiint u^{1/2}\,du^3$$

which on integration leads to

$$\text{P.I.} = \frac{8}{525}u^{7/2} = \frac{8}{525}(x+2y)^{7/2}$$

where $u = x + 2y$. Therefore, the complete solution is

$$z = f_1(y+x) + xf_2(y+x) + f_3(y-2x) + \frac{8}{525}(x+2y)^{7/2}$$

**EXAMPLE 9.85** Solve the following partial differential equations:

(i) $(D^2 + DD' + D' - 1)z = \sin(x+2y)$

(ii) $(D^2 + D'^2 - 3D + 3D')z = xy + e^{x+2y}$

(iii) $(D - 3D' - 2)^3 z = 18e^{2x}\sin(3x+y)$

(iv) $(D^3 - 3DD' + D' + 4)z = e^{2x+y}$

***Solution*** (i) The given equation can be written as

$$(D+1)(D+D'-1)z = \sin(x+2y)$$

For the factors $(D+1)$ and $(D+D'-1)$, the complementary functions, respectively, are $e^{-x}f_1(y)$ and $e^x f_2(y-x)$. Thus

$$\text{C.F.} = e^{-x}f_1(y) + e^x f_2(y-x)$$

Also

$$\text{P.I.} = \frac{1}{D^2 + DD' + D' - 1}\sin(x+2y) = \frac{1}{-1-2+D'-1}\sin(x+2y)$$

$$= \frac{1}{D'-4}\sin(x+2y) = \frac{D'+4}{D'^2 - 16}\sin(x+2y) = \frac{D'+4}{-2^2 - 16}\sin(x+2y)$$

$$= -\frac{1}{20}(D'+4)\sin(x+2y)$$

$$= -\frac{1}{10}[\cos(x+2y) + 2\sin(x+2y)]$$

Therefore the complete solution is

$$z = e^{-x} f_1(y) + e^x f_2(y-x) - \frac{1}{10}[\cos(x+2y) + 2\sin(x+2y)]$$

(ii) The given equation can be expressed as

$$(D-D')(D+D'-3)z = xy + e^{x+2y}$$

so that

$$\text{C.F.} = f_1(y+x) + e^{3x} f_2(y-x)$$

Also

$$\text{P.I.} = \frac{1}{(D-D')(D+D'-3)} xy + \frac{1}{(D-D')(D+D'-3)} e^{x+2y} = P_1 + P_2$$

Now

$$P_1 = \frac{1}{(D-D')(D+D'-3)} xy$$

$$= -\frac{1}{3D}\left(1 - \frac{D'}{D}\right)^{-1}\left(1 - \frac{D}{3} - \frac{D'}{3}\right)^{-1} xy$$

$$= -\frac{1}{3D}\left[1 + \frac{D}{3} + \frac{D'}{3} + \frac{D'}{D} + \frac{D'}{3} + \frac{2DD'}{9} + \cdots\right](xy)$$

$$= -\frac{1}{3D}\left[xy + \frac{y}{3} + \frac{x}{3} + \frac{1}{D}(x) + \frac{x}{3} + \frac{2}{9}\right]$$

$$= -\frac{1}{3D}\left[xy + \frac{y}{3} + \frac{2x}{3} + \frac{x^2}{2} + \frac{2}{9}\right]$$

$$= -\frac{1}{3}\left[\frac{x^2 y}{2} + \frac{xy}{3} + \frac{x^2}{3} + \frac{x^3}{6} + \frac{2x}{9}\right]$$

and $P_2 = \dfrac{1}{(D-D')(D+D'-3)} e^{x+2y} = \dfrac{1}{(1-2)(1+2-3)} e^{x+2y} = \dfrac{1}{0} e^{x+2y}$

Thus

$$P_2 = \frac{1}{[(D+1)-(D'+2)][(D+1)+(D'+2)-3]} e^{x+2y}$$

$$= e^{x+2y} \frac{1}{(D-D'-1)(D+D')}(1)$$

$$= e^{x+2y} \frac{1}{D}\left[(1+D')^{-1}\left(1 - \frac{D'}{D} - \frac{1}{D}\right)^{-1}\right](1)$$

$$= e^{x+2y} \frac{1}{D}(1) = xe^{x+2y}$$

Therefore

$$\text{P.I.} = P_1 + P_2 = -\frac{1}{3}\left[\frac{x^2y}{2} + \frac{xy}{3} + \frac{x^2}{3} + \frac{x^3}{6} + \frac{2x}{9}\right] + xe^{x+2y}$$

Hence, the complete solution is

$$z = f_1(y+x) + e^{3x} f_2(y-x) - \frac{1}{3}\left[\frac{x^2y}{2} + \frac{xy}{3} + \frac{x^2}{3} + \frac{x^3}{6} + \frac{2x}{9}\right] + xe^{x+2y}$$

(iii) Here

$$\text{C.F.} = e^{2x} f_1(y+3x) + xe^{2x} f_2(y+3x) + x^2 e^{2x} f_3(y+3x)$$

and

$$\text{P.I.} = \frac{1}{(D-3D'-2)^3} 18e^{2x}\sin(3x+y)$$

$$= 18e^{2x}\frac{1}{[(D+2)-3(D'+0)-2]^3}\sin(3x+y)$$

$$= 18e^{2x}\frac{1}{(D-D')^3}\sin(3x+y)$$

$$= \frac{18x^3}{1^3 3!}\sin(3x+y) \qquad \text{[from Eq. (33)]}$$

$$= 3x^3\sin(3x+y)$$

Thus, the complete solution is

$$z = e^{2x} f_1(y+3x) + xe^{2x} f_2(y+3x) + x^2 e^{2x} f_3(y+3x) + 3x^3\sin(3x+y)$$

(iv) Here $f(D,D') = D^3 - 3DD' + D' + 4$ can not be resolved into linear factors in $D$ and $D'$, and

$$\text{C.F.} = \sum c e^{hx+ky}$$

where $h^3 - 3hk + k + 4 = 0$ which leads to $k = f(h)$ and thus

$$\text{C.F.} = \sum c e^{hx+f(h)y}$$

Also

$$\text{P.I.} = \frac{1}{D^3 - 3DD' + D' + 4} e^{2x+y} = \frac{1}{2^3 - 3(2)(1) + (1) + 4} e^{2x+y} = \frac{1}{7} e^{2x+y}$$

Thus, the complete solution is given by

$$z = \sum c e^{hx+f(h)y} + \frac{1}{7} e^{2x+y}$$

**EXAMPLE 9.86** Solve the following partial differential equations:

(i) $(D^3 + 4D^2D' + 3D'^2D + D^2 - 5DD' + 2D)z = x^2 - 4xy + 2y^2$

(ii) $(D^2 - D')z = e^{x+y}$

(iii) $(D - 2D')(D^2 - D')z = e^{2x+y} + xy$

***Solution*** (i) The given equation can be expressed as

$$D(D+D'-1)(D+3D'-2)z = x^2 - 4xy + 2y^2$$

and thus

$$\text{C.F.} = f_1(y) + e^x f_2(y-x) + e^{2x} f_3(y-3x)$$

Also

$$\text{P.I.} = \frac{1}{D(D+D'-1)(D+3D'-2)}(x^2 - 4xy + 2y^2)$$

$$= \frac{1}{2D}[1-(D+D')]^{-1}\left[1-\left(\frac{D+3D'}{2}\right)\right]^{-1}(x^2-4xy+2y^2)$$

$$= \frac{1}{2D}[1+(D+D')+(D+D')^2+\cdots]$$

$$\times\left[1+\left(\frac{D+3D'}{2}\right)+\left(\frac{D+3D'}{2}\right)^2+\cdots\right](x^2-4xy+2y^2)$$

$$= \frac{1}{2D}\left[1+\frac{3D}{2}+\frac{5D'}{2}+\frac{7D^2}{4}+\frac{19D'^2}{4}+\frac{11DD'}{2}+\cdots\right](x^2-4xy+2y^2)$$

$$= \frac{1}{2D}\left[x^2-4xy+2y^2-7x+4y+\frac{1}{2}\right]$$

$$= \frac{1}{2}\left[\frac{x^3}{3}-2x^2y+2xy^2-\frac{7x^2}{2}+4xy+\frac{x}{2}\right]$$

Therefore, the complete solution is

$$z = f_1(y)+e^x f_2(y-x)+e^{2x}f_3(y-3x)+\frac{1}{2}\left[\frac{x^3}{3}-2x^2y+2xy^2-\frac{7x^2}{2}+4xy+\frac{x}{2}\right]$$

(ii) Here

$$\text{C.F.} = \sum ce^{hx+ky}$$

where $h^2 - k = 0$ which leads to $h^2 = k$ so that

$$\text{C.F.} = \sum ce^{h(x+hy)}$$

Also

$$\text{P.I.} = \frac{1}{D^2-D'}e^{x+y} = e^{x+y}\frac{1}{(D+1)^2-(D'+1)}(1)$$

$$= e^{x+y}\frac{1}{D^2-2D-D'}(1) = -e^{x+y}\frac{1}{D'}\left[1-\left(\frac{D^2+2D}{D'}\right)\right]^{-1}(1)$$

$$= -e^{x+y}\frac{1}{D'}\left[1+\left(\frac{D^2+2D}{D'}\right)+\cdots\right](1)$$

$$= -e^{x+y}\frac{1}{D'}(1) = -ye^{x+y}$$

Therefore, the required solution is:

$$z = \sum ce^{h(x+hy)} - ye^{x+y}$$

where $c$ and $h$ are arbitary constants.

(iii) Here, corresponding to the linear factor $(D-2D')$, the complementary function is $f(y+2x)$. While $(D^2-D')$ cannot be resolved into linear factors in $D$ and $D'$ and for this factor $(D^2-D')$, the complementary function is $\sum ce^{hx+ky}$, where $h$ and $k$ are related through the equation $h^2-k=0$, so that $h^2=k$. Thus, for the given equation,

$$\text{C.F.} = f(y+2x) + \sum ce^{h(x+hy)}$$

Also

$$\text{P.I.} = \frac{1}{(D-2D')(D^2-D')}e^{2x+y} + \frac{1}{(D-2D')(D^2-D')}xy = P_1 + P_2$$

Now

$$P_1 = \frac{1}{(D-2D')(D^2-D')}e^{2x+y} = \frac{1}{(D-2D')(2^2-1)}e^{2x+y}$$

$$= \frac{1}{3}e^{2x+y}\frac{1}{D-2D'}(1) = \frac{1}{3}e^{2x+y}\frac{1}{(D+2)-2(D'+1)}(1)$$

$$= \frac{1}{3}e^{2x+y}\frac{1}{D}\left[1-\frac{2D'}{D}\right]^{-1}(1) = \frac{x}{3}e^{2x+y}$$

and $$P_2 = \frac{1}{(D-2D')(D^2-D')}xy = \frac{1}{(-2D')\left(1-\dfrac{D}{2D'}\right)(-D')\left(1-\dfrac{D^2}{D'}\right)}xy$$

$$= \frac{1}{2D'^2}\left[\left(1-\frac{D}{2D'}\right)^{-1}\left(1-\frac{D^2}{D'}\right)^{-1}\right]xy$$

$$= \frac{1}{2D'^2}\left[\left(1+\frac{D}{2D'}+\cdots\right)\left(1+\frac{D^2}{D'}+\cdots\right)\right]xy$$

$$= \frac{1}{2D'^2}\left[1+\frac{D}{2D'}+\cdots\right]xy = \frac{1}{2D'^2}\left[xy+\frac{D}{2D'}(xy)\right]$$

$$= \frac{1}{2D'^2}\left[xy+\frac{1}{2D'}(y)\right] = \frac{1}{2D'^2}\left[xy+\frac{1}{2}\frac{y^2}{2}\right] = \frac{1}{2}\left[\frac{1}{6}xy^3+\frac{1}{48}y^4\right]$$

Therefore, the required solution is

$$z = f(y+2x) + \sum ce^{h(x+hy)} + \frac{1}{2}\left[\frac{1}{6}xy^3+\frac{1}{48}y^4\right]$$

**EXAMPLE 9.87** Solve the following partial differential equations:

(i) $(x^2D^2 + 2xyDD' - xD)z = x^3y^{-2}$

(ii) $(y^2D'^2 - yD')z = xy^2$

(iii) $(x^2D^2 - y^2D'^2 + xD - yD')z = \log x$

(iv) $(x^2D^2 - 4xyDD' + 4y^2D'^2 + 4yD' + xD)z = x^2y$

***Solution*** (i) Let $x = e^u$ and $y = e^v$ so that $u = \log x$ and $v = \log y$. Also, let $D = \frac{\partial}{\partial x}, D' = \frac{\partial}{\partial y}, D_1 = \frac{\partial}{\partial u}, D_1' = \frac{\partial}{\partial v}$, then the given equation reduces to

$$D_1(D_1 + 2D_1' - 2)z = e^{3u-2v}$$

Here

$$\text{C.F.} = f_1(v) + e^{2u} f_2(v - 2u)$$
$$= f_1(\log y) + x^2 f_2(\log y - 2\log x)$$
$$= f_1(\log y) + x^2 f_2\left(\log \frac{y}{x^2}\right)$$

and $$\text{P.I.} = \frac{1}{D_1(D_1 + 2D_1' - 2)} e^{3u-2v} = \frac{1}{3(3-4-2)}(e^u)^3(e^v)^{-2} = -\frac{1}{9}x^3y^{-2}$$

Therefore, the complete solution is

$$z = f_1(\log y) + x^2 f_2\left(\log \frac{y}{x^2}\right) - \frac{1}{9}x^3y^{-2}$$

(ii) The given equation can be written as

$$D_1'(D_1' - 2)z = e^{u+2v}$$

where $D_1 = \frac{\partial}{\partial u}, D_1' = \frac{\partial}{\partial v}, D = \frac{\partial}{\partial x}, D' = \frac{\partial}{\partial y}, x = e^u, y = e^v$. Here

$$\text{C.F.} = f_1(u) + e^{2v} f_2(u) = f_1(\log x) + y^2 f_2(\log x)$$

and $$\text{P.I.} = \frac{1}{(D_1' - 2)} \frac{1}{D_1'} e^{u+2v} = \frac{1}{(D_1' - 2)} \frac{1}{2} e^{u+2v}$$

$$= \frac{1}{2} e^{u+2v} \frac{1}{D_1' + 2 - 2}(1) = \frac{1}{2} e^{u+2v} \frac{1}{D_1'}(1)$$

$$= \frac{1}{2} v e^{u+2v} = \frac{1}{2} \log y (xy^2)$$

Therefore, the required solution is

$$z = f_1(\log x) + y^2 f_2(\log x) + \frac{1}{2} xy^2 \log y$$

(iii) Let $x=e^u, y=e^v, D=\frac{\partial}{\partial x}, D'=\frac{\partial}{\partial y}, D_1=\frac{\partial}{\partial u}, D_1'=\frac{\partial}{\partial v}$, so that the given equation reduces to

$$(D_1^2 - D_1'^2)z = u$$

Here $\quad$ C.F. $= f_1(v-u) + f_2(v+u) = f_1(\log y - \log x) + f_2(\log x + \log y)$

and

$$\text{P.I.} = \frac{1}{D_1^2 - D_1'^2}u = \frac{1}{D_1^2}\left(1 - \frac{D_1'^2}{D_1^2}\right)^{-1} u$$

$$= \frac{1}{D_1^2}u = \frac{1}{6}u^3 = \frac{1}{6}(\log x^3)$$

Therefore, the required solution is

$$z = f_1 \log\left(\frac{y}{x}\right) + f_2 \log(xy) + \frac{1}{6}(\log x^3)$$

(iv) The given partial differential equation can be written as

$$(D_1 - 2D_1')^2 z = e^{2u+v}$$

where $x=e^u, y=e^v$ and the operators $D_1, D_1'$, etc. have their usual meanings. Here

$$\text{C.F.} = f_1(v+2u) + uf_2(v+2u)$$

$$= f_1(\log y + 2\log x) + \log x f_2(\log y + 2\log x)$$

$$= f_1(\log y + \log x^2) + \log x f_2(\log y + \log x^2)$$

$$= f_1[\log(yx^2)] + \log x f_2[\log(yx^2)]$$

While using Eq. (33), we have

$$\text{P.I.} = \frac{1}{(D_1 - 2D_1')^2}e^{2u+v} = \frac{u^2}{2!}e^{2u+v} = \frac{1}{2}x^2 y(\log x)^2$$

Therefore, the required solution is

$$z = f_1[\log(yx^2)] + \log x f_2[\log(yx^2)] + \frac{1}{2}x^2 y(\log x)^2$$

**EXAMPLE 9.88** The points of trisection of a string are pulled aside through a distance $a$ on opposite sides of the position of equilibrium and then released from rest. Find the displacement of the string at any time $t$.

***Solution*** Let $OR = l$ be the length of the string which is trisected at the points $P$ and $Q$ (Fig 9.12).

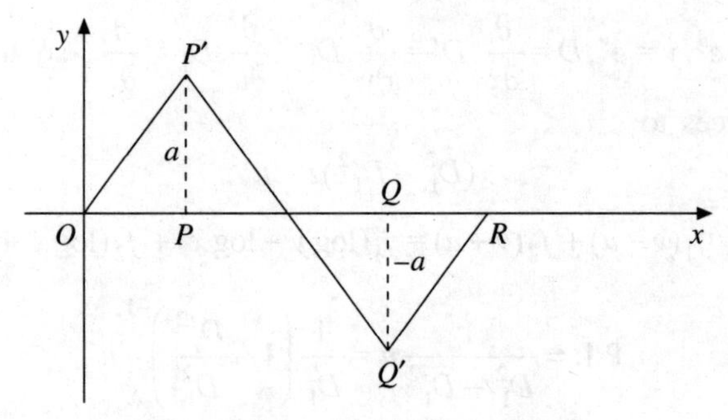


**Fig. 9.12** Graph of Ex. 9.88.

We know that the motion of the string is governed by the equation

$$\frac{\partial^2 y}{\partial t^2} = c^2 \frac{\partial^2 y}{\partial x^2}$$

whose solution is

$$y = y(x,t) = (A\cos\omega x + B\sin\omega x)(C\cos c\omega t + D\sin c\omega t) \tag{224}$$

The given conditions are

$$y(0,t) = 0 \tag{i}$$

$$y(l,t) = 0 \tag{ii}$$

$$\left(\frac{\partial y}{\partial t}\right)_{t=0} = 0 \tag{iii}$$

From condition (i), Eq. (224) reduces to

$$y(0,t) = A(C\cos c\omega t + D\sin c\omega t) = 0 \tag{225}$$

which holds for all $t$ when $A = 0$ and Eq. (224) now becomes

$$y(x,t) = B\sin\omega x(C\cos c\omega t + D\sin c\omega t) \tag{226}$$

From condition (ii), Eq. (226) leads to

$$y(l,t) = B\sin\omega l(C\cos c\omega t + D\sin c\omega t) = 0$$

If $B \neq 0$, then this equation gives $\sin\omega l = 0$ so that $\omega = n\pi / l$. Equation (226) can now be expressed as

$$y(x,t) = B\sin\frac{n\pi}{l}x\left(C\cos\frac{cn\pi}{l}t + D\sin\frac{cn\pi}{l}t\right) \tag{227}$$

Also, from Eq. (226), we have

$$\frac{\partial y}{\partial t} = B\sin\omega x[-C(c\omega)\sin c\omega t + Dc\omega\cos c\omega t]$$

which on using condition (iii) leads to

$$\left(\frac{\partial y}{\partial t}\right)_{t=0} = 0 = B\sin\omega x(Dc\omega)$$

This equation gives $D = 0$ and Eq. (226) now reduces to

$$y(x,t) = BC\sin\omega x\cos c\omega t$$

where $\omega = \frac{n\pi}{l}$. Thus, the general solution of the equation describing the motion of the string is given by

$$y(x,t) = \sum_{n=1}^{\infty} b_n \sin\frac{n\pi}{l}x\cos\frac{cn\pi}{l}t \tag{228}$$

We shall now find the equations of the lines OP′, P′Q′ and Q′R to obtain the condition for $y(x,0)$. The equation for OP′ is

$$y = \frac{ax}{l/3} = \frac{3a}{l}x$$

equation of P′Q′ is

$$y = \frac{3a}{l}(l-2x)$$

and the equation of Q′R is

$$y = \frac{3a}{l}(x-l)$$

We thus have

$$y(x,0) = \begin{cases} \frac{3a}{l}x, & 0 \le x \le \frac{l}{3} \\ \frac{3a}{l}(l-2x), & \frac{l}{3} \le x \le \frac{2l}{3} \\ \frac{3a}{l}(x-l), & \frac{2l}{3} \le x \le l \end{cases} \tag{iv}$$

Using condition (iv) in Eq. (228), we have

$$y(x,0) = \sum_{n=1}^{\infty} b_n \sin\frac{n\pi}{l}x$$

where

$$b_n = \frac{2}{l}\int_0^l y(x,0)\sin\frac{n\pi}{l}x\,dx$$

$$= \frac{2}{l}\left[\int_0^{l/3}\frac{3a}{l}x\sin\frac{n\pi x}{l}dx + \int_{l/3}^{2l/3}\frac{3a}{l}(l-2x)\sin\frac{n\pi x}{l}dx\right.$$

$$\left. + \int_{2l/3}^{l}\frac{3a}{l}(x-l)\sin\frac{n\pi x}{l}dx\right]$$

which, after simplifications, leads to

$$b_n = \begin{cases} 0, & \text{when } n \text{ is odd} \\ \dfrac{36a}{n^2\pi^2}\sin\dfrac{n\pi}{3}, & \text{when } n \text{ is even} \end{cases}$$

Thus, Eq. (228) becomes

$$y(x,t) = \sum_{n=even} \frac{36a}{n^2\pi^2}\sin\frac{n\pi}{3}\left[\sin\frac{n\pi x}{l}\cos\frac{cn\pi t}{l}\right]$$

This equation may also be expressed as

$$y(x,t) = \frac{36a}{2^2\pi^2}\sin\frac{2\pi}{3}\sin\frac{2\pi x}{l}\cos\frac{2\pi ct}{l} + \frac{36a}{4^2\pi^2}\sin\frac{4\pi}{3}\sin\frac{4\pi x}{l}\cos\frac{4\pi ct}{l} + \cdots$$

which is the required solution.

**EXAMPLE 9.89** A tightly stretched string with fixed end points $x = 0$ and $x = l$ is initially at rest in its equilibrium position. If the string is set in motion by giving to each of its points an initial velocity $\lambda x(l-x)$, find the displacement of the string at any distance $x$ from one end point at any time $t$.

***Solution*** The motion of the string is governed by the partial differential equation

$$\frac{\partial^2 y}{\partial t^2} = c^2\frac{\partial^2 y}{\partial x^2} \tag{229}$$

Given that the end points of the string are fixed, for all time $t$ we have

$$y(0,t) = 0, \; y(l,t) = 0 \tag{i}$$

Also

$$y(x,0) = 0 \tag{ii}$$

$$\left(\frac{\partial y}{\partial t}\right)_{t=0} = \lambda x(l-x) \tag{iii}$$

Here we have to solve Eq. (229) under the conditions (i) – (iii). We know that the solution of Eq. (229) is given by

$$y = y(x,t) = (A\cos\omega x + B\sin\omega x)(C\cos c\omega t + D\sin c\omega t) \tag{230}$$

From condition (i), Eq. (230) leads to

$$y(0,t) = A(C\cos c\omega t + D\sin c\omega t) = 0 \tag{231}$$

which holds for all $t$ if $A = 0$ and Eq. (230) now reduces to

$$y(x,t) = B\sin\omega x(C\cos c\omega t + D\sin c\omega t) \tag{232}$$

Using condition (ii) in Eq. (232), we get

$$y(x,0) = BC\sin\omega x = 0$$

which leads to $C = 0$ (if $B = 0$, then Eq. (232) is meaningless) and Eq. (232) now becomes

$$y(x,t) = BD\sin\omega x \sin c\omega t \tag{233}$$

Using the second equation of condition (i) in Eq. (233), we get

$$y(l,t) = 0 = BD\sin\omega l \sin c\omega t$$

Since $B$ and $D$ are non-zero, we have $\sin\omega l = 0$ which gives $\omega l = n\pi$ or $\omega = n\pi / l$, where $n$ is an integer. Equation (233) now takes the form

$$y(x,t) = BD\sin\frac{n\pi x}{l}\sin\frac{n\pi ct}{l}$$

and the general solution is

$$y(x,t) = \sum_{n=1}^{\infty} b_n \sin\frac{n\pi x}{l}\sin\frac{n\pi ct}{l} \tag{234}$$

Differentiating Eq. (234) partially with respect to $t$, we get

$$\frac{\partial y}{\partial t} = \sum_{n=1}^{\infty} b_n \sin\frac{n\pi x}{l}\left[\frac{n\pi c}{l}\cos\frac{n\pi ct}{l}\right]$$

For $t = 0$, this equation leads to

$$\left(\frac{\partial y}{\partial t}\right)_{t=0} = \sum_{n=1}^{\infty} b_n \left(\frac{n\pi c}{l}\right)\sin\frac{n\pi x}{l} \tag{235}$$

Now, in order that condition (iii) is satisfied, condition (iii) and Eq. (235) must be the same; and this requires the expansion of $\lambda x(l-x)$ as the Fourier series in (0, $l$). Thus

$$\lambda x(l-x) = \sum_{n=1}^{\infty} b_n \left(\frac{n\pi c}{l}\right)\sin\frac{n\pi x}{l}$$

where

$$b_n = \frac{2}{l}\int_0^l \lambda x(l-x)\sin\frac{n\pi x}{l}\,dx$$

which, after simplification, can be expressed as

$$b_n = \begin{cases} \dfrac{8\lambda l^3}{cn^4\pi^4}, & \text{when } n \text{ is odd} \\ 0, & \text{when } n \text{ is even} \end{cases}$$

Therefore, the required solution, from Eq. (234) is given by

$$y(x,t) = \frac{8\lambda l^3}{c\pi^4}\left[\sum_{n=\text{odd}} \frac{1}{n^4}\sin\frac{n\pi x}{l}\sin\frac{n\pi ct}{l}\right]$$

$$= \frac{8\lambda l^3}{c\pi^4}\left[\frac{1}{1^4}\sin\frac{\pi x}{l}\sin\frac{\pi ct}{l} + \frac{1}{3^4}\sin\frac{3\pi x}{l}\sin\frac{3\pi ct}{l} + \cdots\right]$$

**EXAMPLE 9.90** A homogeneous rod of conducting material of length $l$ has its ends kept at zero temperature and the initial temperature is

$$u(x,0)=\begin{cases} x, & \text{if } 0\le x\le \dfrac{l}{2} \\ l-x, & \text{if } \dfrac{l}{2}\le x\le l \end{cases}$$

Find the temperature $u(x,t)$ at any time $t$.

***Solution*** Here we have to solve the partial differential equation

$$\frac{\partial u}{\partial t}=c^2\frac{\partial^2 u}{\partial x^2} \tag{236}$$

under the conditions

$$u(0,t)=0 \tag{i}$$

$$u(l,t)=0 \tag{ii}$$

$$u(x,0)=\begin{cases} x, & \text{if } 0\le x\le \dfrac{l}{2} \\ l-x, & \text{if } \dfrac{l}{2}\le x\le l \end{cases} \tag{iii}$$

The solution of Eq. (236) is given by

$$u(x,t)=(c_1\cos\omega x+c_2\sin\omega x)e^{-c^2\omega^2 t} \tag{237}$$

From condition (i), Eq. (237) leads to $c_1 = 0$ and Eq. (237) now reduces to

$$u(x,t)=c_2\sin\omega x e^{-c^2\omega^2 t} \tag{238}$$

which on using condition (ii) gives $u(l,t)=c_2\sin\omega l e^{-c^2\omega^2 t}=0$ and this leads to $\sin\omega l=0$ so that $\omega l=n\pi$, as $c_2\neq 0$. Thus, $\omega=n\pi/l$, where $n$ is an integer. Equation (238) now takes the form

$$u(x,t)=b_n\sin\frac{n\pi x}{l}e^{-c^2n^2\pi^2t/l^2},\quad b_n=c_2$$

Adding all such solutions, the most general solution under conditions (i) and (ii) is given by

$$u(x,t)=\sum_{n=1}^{\infty}b_n\sin\frac{n\pi x}{l}e^{-c^2n^2\pi^2t/l^2} \tag{239}$$

Putting $t = 0$ in this equation, we get

$$u(x,0)=\sum_{n=1}^{\infty}b_n\sin\frac{n\pi x}{l} \tag{240}$$

Now, in order that condition (iii) be satisfied, condition (iii) and Eq. (240) must be the same. This requires the expansion of $u(x, 0)$ [given by condition (iii)] as the Fourier series, where

$$b_n = \frac{2}{l}\int_0^l u(x,0)\sin\frac{n\pi x}{l}\,dx$$

$$= \frac{2}{l}\int_0^{l/2} x\sin\frac{n\pi x}{l}\,dx + \int_{l/2}^{l}(l-x)\sin\frac{n\pi x}{l}\,dx$$

$$= \frac{4l}{n^2\pi^2}\sin\frac{n\pi}{2}$$

(after simplifications).
Therefore, from Eq. (239), the required temperature is

$$u(x,t) = \frac{4l}{\pi^2}\sum_{n=1}^{\infty}\frac{1}{n^2}\sin\frac{n\pi}{2}\sin\frac{n\pi x}{l}e^{-c^2n^2\pi^2t/l^2}$$

**EXAMPLE 9.91** A rod of length 10 cm has its ends $P$ and $Q$ kept at 50 °C and 100 °C, respectively, until steady state conditions prevail. The temperature at $P$ is suddenly raised to 90 °C and that at $Q$ is lowered to 60 °C and these temperatures are maintained. Find the temperature at a distance $x$ from one end at time $t$.

***Solution*** We know that the one-dimensional heat flow is governed by the partial differential equation

$$\frac{\partial u}{\partial t} = c^2\frac{\partial^2 u}{\partial x^2} \tag{241}$$

Given that

$$u(0, t) = 50\text{ °C} \tag{i}$$

$$u(0, t) = 100\text{ °C} \tag{ii}$$

For steady state condition, $\frac{\partial u}{\partial t} = 0$ and Eq. (241) reduces to $\frac{\partial^2 u}{\partial x^2} = 0$ whose solution is

$$u(x) = Ax + B \tag{242}$$

Making use of the given conditions (i) and (ii) in Eq. (242), we get $A$ = 5 and $B$ = 50 so that Eq. (242) becomes

$$u(x) = 5x + 50 \tag{243}$$

Also, given that there is a sudden change in temperature at the end points $P$ and $Q$ so we again have the transient state. If $u_1(x, t)$ denotes the subsequent temperature, then the boundary conditions are

$$u_1(0,t) = 90\text{ °C},\ u_1(10,t) = 60\text{ °C} \tag{iii}$$

and the initial condition is

$$u_1(x,0) = 5x + 50 \tag{iv}$$

If $u_s(x)$ denotes the subsequent steady state function, then $\dfrac{\partial^2 u_s}{\partial x^2} = 0$ which has solution as

$$u_s(x) = A_1 x + B_1$$

As $u_1 \to u_s$ at the end points, using condition (iii), this equation leads to $A_1 = -3, B_1 = 90$ and thus

$$u_s(x) = -3x + 90 \tag{v}$$

Therefore, the total temperature distribution in the rod at time $t$ is

$$u(x,t) = u_s(x) + u_T(x,t)$$

where $u_T(x,t)$ is the temperature in the transient state. From condition (v), this equation becomes

$$u(x,t) = -3x + 90 + u_T(x,t) \tag{244}$$

Moreover, from Eq. (241), we have

$$\frac{\partial u_T}{\partial t} = c^2 \frac{\partial^2 u_T}{\partial x^2}$$

whose solution is

$$u_T(x,t) = (e_1 \cos \omega x + e_2 \sin \omega x) e^{-c^2 \omega^2 t} \tag{245}$$

where $u_T(x,t)$ satisfies the following boundary conditions [from conditions (iii) – (v)]

$$u_T(0,t) = u_1(0,t) - u_s(0) = 0 \tag{vi}$$

$$u_T(l,t) = u_1(10,t) - u_s(10) = 0 \tag{vii}$$

$$u_T(x,0) = u_1(x,0) - u_s(x) = 8x - 40 \tag{viii}$$

Using conditions (vi) and (vii) in Eq. (245), we get $e_1 = 0$ and $\omega = n\pi/10$. Eq. (245) now becomes

$$u_T(x,t) = \sum_{n=1}^{\infty} b_n \sin \frac{n\pi x}{10} e^{-c^2 n^2 \pi^2 t/100} \tag{246}$$

where $b_n = e_2$. Also

$$b_n = \frac{2}{l} \int_0^l u(x,0) \sin \frac{n\pi x}{l} dx = \frac{2}{10} \int_0^{10} (8x - 40) \sin \frac{n\pi x}{10} dx$$

so that after simplification, we get

$$b_n = \begin{cases} 0, & \text{when } n \text{ is odd} \\ -\dfrac{160}{n\pi}, & \text{when } n \text{ is even} \end{cases}$$

Equation (246) thus becomes

$$u_T(x,t) = -\frac{80}{\pi}\sum_{m=1}^{\infty}\frac{1}{m}\sin\frac{m\pi x}{5}e^{-c^2m^2\pi^2t/25}, \quad (n = 2m)$$

Hence, from Eq. (244), the required temperature distribution is

$$u(x,t) = -3x + 90 - \frac{80}{\pi}\sum_{m=1}^{\infty}\frac{1}{m}\sin\frac{m\pi x}{5}e^{-c^2m^2\pi^2t/25}$$

## EXERCISES

Form the partial differential equations (by eliminating the arbitrary constants) from:

**1.** $z = (x^2 + a)(y^2 + b)$.

**2.** $z = xy + y\sqrt{(x^2 - a^2)} + b$.

**3.** Find the differential equations of all planes which are at a constant distance $r$ from the origin.

**4.** Find the differential equation of all spheres whose centre lies on the $z$-axis.

Form the differential equations (by eliminating the arbitrary functions) from:

**5.** $z = y^2 + 2F\left(\frac{1}{x} + \log y\right)$.

**6.** $z = \frac{1}{r}[\psi_1(r - at) + \psi_2(r + at)]$.

**7.** $F(x^2 + y^2, z - xy) = 0$.

Solve the following differential equations:

**8.** $\frac{\partial^2 z}{\partial y \partial x} = 4x \sin(3xy)$.

**9.** $\frac{\partial^3 u}{\partial x^2 \partial y} = \cos(2x + 3y)$.

**10.** $\frac{\partial^2 z}{\partial y^2} = z$ when $y = 0$, $z = e^x$ and $\frac{\partial z}{\partial y} = e^{-x}$.

**11.** $yq - xp = z$.

**12.** $(y - z)p + (x - y)q = z - x$.

**13.** $(x^2 - yz)p + (y^2 - zx)q = z^2 - xy$.

**14.** $x(z^2 - y^2)p + y(x^2 - z^2)q = z(y^2 - x^2)$.

**15.** $x^2(y - z)p + y^2(z - x)q = z^2(x - y)$.

Find the complete solution of the following equations:

**16.** $pq + p + q = 0$.

**17.** $p(1 + q^2) = q(z - a)$.

**18.** $yp + xq + pq = 0$.

**19.** $p + q = \sin x + \sin y$.

**20.** $(p^2 - q^2)z = x - y$.

**21.** $z = px + qy + \sqrt{1 + p^2 + q^2}$.

**22.** $z = px + qy + p^2q^2$.

**23.** Find the singular solution of $z = px + qy + \log pq$.

Solve the following equations:

**24.** $z = p^2x + q^2y$.

**25.** $z^2 = pqxy$.

**26.** $pxy + pq + qy = yz$.

**27.** $z^2(p^2 + q^2) = x^2 + y^2$.

**28.** $\dfrac{\partial^2 z}{\partial x^2} + \dfrac{\partial^2 z}{\partial x \partial y} - 2\dfrac{\partial^2 z}{\partial y^2} = 0$.

**29.** $\dfrac{\partial^2 z}{\partial x^2} - 2\dfrac{\partial^2 z}{\partial x \partial y} + \dfrac{\partial^2 z}{\partial y^2} = \sin x$.

**30.** $\dfrac{\partial^2 z}{\partial x^2} + 4\dfrac{\partial^2 z}{\partial x \partial y} - 5\dfrac{\partial^2 z}{\partial y^2} = y^2$.

**31.** $\dfrac{\partial^2 z}{\partial x^2} - \dfrac{\partial^2 z}{\partial x \partial y} - 6\dfrac{\partial^2 z}{\partial y^2} = xy$.

**32.** $(2D^2 - 5DD' + 2D'^2)z = 5 \sin (2x + y)$.

**33.** $(D^2 + 2DD' + D'^2)z = 2(y - x) + \sin (x - y)$.

**34.** $4r + 12s + 9t = e^{3x-2y}$.

**35.** $r + s - 6t = \cos (2x + y)$.

**36.** $(2DD' + D'^2 - 3D')z = 3 \cos (3x - 2y)$.

**37.** $(D + D' - 1)(D + 2D' - 3)z = 4 + 3x + 6y$.

**38.** $(D^2 - DD' + D' - 1)z = \cos (x + 2y) + e^y$.

**39.** $(D^2 + DD' + 2D'^2)z = 5e^{x+2y}$

**40.** $(D^2 + DD' + D' - 1)z = \sin(x + 2y)$.

**41.** $(D^2 - D'^2 + D + 3D' - 2)z = e^{x-y} - x^2y$.

**42.** $(D^2 - 4DD' + 4D'^2 + D - 2D')z = e^{x+y}$.

**43.** $(D^2 - D'^2)z = \cos(x - 3y)$.

**44.** $\dfrac{\partial^2 \psi}{\partial x^2} = \dfrac{\partial \psi}{\partial t}$.

**45.** $\dfrac{\partial^2 \psi}{\partial x^2} = \dfrac{\partial^2 \psi}{\partial y^2} = n^2\psi$.

**46.** $(2D^2 - D')(D^2 - D')z = 0$.

**47.** $(D - D'^2)z = 0$.

**48.** $(D^2 + D'^2)z = \cos mx \cos ny$.

**49.** $(D^3 - 4D^2D' + 4D'^3)z = \cos(2x + y)$

**50.** $(D^2 + 3DD' + 2D'^2)z = 2x + 3y$.

**51.** $(D^2 + DD' - 6D'^2)z = y \sin x$.

**52.** $(D^2 - DD' + 2D'^2)z = (y - 1)e^x$.

**53.** $yt - q = xy$.

**54.** $x^2\dfrac{\partial^2 z}{\partial x^2} + 2xy\dfrac{\partial^2 z}{\partial x \partial y} + y^2\dfrac{\partial^2 z}{\partial y^2} = (x^2 + y^2)^{n/2}$.

**55.** $x^2 \dfrac{\partial^2 z}{\partial x^2} + 2xy \dfrac{\partial^2 z}{\partial x \partial y} + y^2 \dfrac{\partial^2 z}{\partial y^2} = x^m y^n.$

**56.** $(x^2D^2 - y^2D'^2 - yD' + xD)z = 0.$

**57.** $x^2r + y^2t + xp - yq = \log x.$

Solve the following partial differential equations by Monge's method:

**58.** $r - t \cos^2 x + p \tan x = 0.$

**59.** $t - r \sec^4 y - 2q \tan y = 0$

**60.** $x^2r + 2xys + y^2t = 0$

**61.** $q(1 + q)r - (p + q + 2pq)s + p(1 + p)t = 0.$

**62.** $(x - y)(xr - xs - ys + yt) = (x + y)(p - q).$

**63.** $qr - ps = p^3.$

**64.** $q^2r - 2pqs + p^2t = qr - ps.$

**65.** $(1 + q)^2r - 2(1 + p + q + pq)s + (1 + p)^2t = 0.$

**66.** $y^2r - 2ys + t = p + 6y.$

**67.** $pq - pxs + qxr = 0.$

**68.** $pt - qs - q^3 = 0.$

Use the method of separation of variables to solve the following equations:

**69.** $\dfrac{\partial u}{\partial x} = 2\dfrac{\partial u}{\partial t} + u,$ when $u(x, 0) = 6e^{-3x}.$

**70.** $4\dfrac{\partial z}{\partial x} + \dfrac{\partial z}{\partial y} = 3z,$ when $x = 0,\ z = 3e^{-y} - e^{-5y}.$

**71.** Find a Fourier series of $f(x) = x + x^2$ defined in $(-\pi, \pi)$.

**72.** Expand $f(x) = x \sin x,\ 0 < x < 2\pi$ in a Fourier series.

**73.** Obtain the Fourier series for $f(x) = x^2$ in $(-\pi, \pi)$, and show that $\dfrac{\pi^2}{6} = \sum_{n=1}^{\infty} \dfrac{1}{n^2}$

**74.** Expand $f(x)$ in a Fourier series in $(-2, 2)$ if

$$\begin{aligned} f(x) &= 0, \quad -2 < x < 0 \\ &= 1, \quad 0 < x < 2 \end{aligned}$$

**75.** Find the Fourier series of

$$f(x) = \begin{cases} 0 & \text{when} \quad -2 < x < -1 \\ A & \text{when} \quad -1 < x < 1 \\ 0 & \text{when} \quad 1 < x < 2 \end{cases}$$

**76.** Show that the Fourier series of $f(x) = |x|$ in $-\pi < x < \pi$ is

$$f(x) = \frac{1}{2}\pi - \frac{4}{\pi}\left(\cos x + \cos\frac{3x}{3^2} + \cos\frac{5x}{5^2} + \cdots\right)$$

**77.** A tightly stretched string, whose end points are fixed at $x = 0$ and $x = l$, is initially in a position given by $y = y_0 \sin^3(x\pi/l)$. If it is released from this position, find the displacement $y(x, t)$.

**78.** The ends of a tightly stretched string are fixed at $x = 0$ and $x = l$. At $t = 0$, the shape of the string is defined by $F(x) = kx(l - x)$, where $k$ is a constant. If the string is released from this position, find the displacement of any point $x$ of the string at any time $t > 0$.

**79.** If a string of length $l$ is initially at rest in its equilibrium position and each of its points is given the velocity $(\partial y/\partial t)_{t=0} = k \sin^3(x\pi/l)$. Find $y(x, t)$.

**80.** Solve Eq. (97), when the length of string is $l$, both the ends are fixed and $y(0, t) = 0$, $y(l, t) = 0$, $y(x, 0) = f(x)$ and $\partial y(x, 0)/\partial t = 0$, $0 < x < l$.

**81.** Solve Example 9.55, if the change consists of raising the temperature of $A$ to 20°C and reducing that of $B$ to 80°C.

**82.** A homogeneous rod of conducting material of length 100 cm has its ends kept at zero temperature and the initial temperature is

$$\begin{aligned} u(x, 0) &= x, && 0 \le x \le 50 \\ &= 100 - x, && 50 \le x \le 100 \end{aligned}$$

Find the temperature $u(x, t)$ at any time.

**83.** The ends $P$ and $Q$ of a rod 20 cm long have the temperatures at 30°C and 80°C until the steady state condition prevails. The temperatures of the ends are changed to 40°C and 60°C, respectively. Find the temperature distribution in the rod at time $t$.

**84.** The temperature at one end of a bar, 50-cm long with insulated sides, is kept at 0°C and that the other end is kept at 100°C until steady state conditions prevail. Then two ends are suddenly insulated so that the temperature gradient is zero at each end thereafter. Find the temperature distribution. Also show that the sum of the temperatures at any two points equidistant from the centre is always 100°C.

**85.** Find the solution of Eq. (106) if $u = u_0 \sin nt$, when $x = 0$ for all values of $t$ and $u = 0$ when $x$ is very large.

**86.** A long rectangular plate of width $a$ cm with an insulated surface has its temperature $u$ equal to zero on both the long sides and one of the short sides so that $u(0, y) = 0$, $u(a, y) = 0$, $u(x, \infty) = 0$, $u(x, 0) = \mu x$. Find the steady state temperature within the plate.

**87.** Solve Eq. (120) for $0 < x < \pi$, $0 < y < \pi$ and $u(0, y) = u(\pi, y) = u(x, \pi) = 0$, $u(x, 0) = \sin^2 x$.

**88.** The temperature $u$ is maintained at 0° along three edges of a square plate 100 cm long and the fourth edge is maintained at 100° until steady state conditions prevail. Find $u(x, t)$ at any time $t$. Hence, show that the temperature at the centre of the plate is

$$\frac{200}{\pi}\left[\frac{1}{\cosh(\pi/2)} - \frac{1}{3\cosh(3\pi/2)} + \frac{1}{5\cosh(5\pi/2)} + \cdots\right]$$

**89.** The circumference of a semi-circular plate of radius $a$ kept at a temperature $u(a, \theta) = \mu\theta(\pi - \theta)$, while the temperature of the boundary diameter is kept at 0°C. Find the steady state temperature $u(r, \theta)$, if the lateral surfaces of the plate are to be insulated.

**90.** A semi-circular plate of radius 10 cm has insulated faces and heat flows in plane curves. The bounding diameter is kept at 0°C and the distribution of the temperature on the circumference is maintained by

$$u(10, \theta) = \frac{400}{\pi}(\pi\theta - \theta^2), \quad 0 \le \theta \le \pi$$

Find $u(r, \theta)$.

**91.** The shape of a plate is that of a truncated quadrant of a circle. It is bounded by $r = a$, $r = b$, $\theta = 0$, $\theta = \pi/2$, its faces are insulated and the heat flows in plane curves. The plate is kept at 0°C along three of the edges while along the edge $r = a$, it is kept at temperature $\theta(\pi/2 - \theta)$. Obtain the temperature distributions.

**92.** Find the steady state temperature in a circular plate of radius $a$ which has one-half of its circumference at 0°C and the other half at 60°C.

**93.** In a telephone wire of length $l$, a steady voltage distribution of 20 V at the sending end and 12 V at the receiving end is maintained. The receiving end is grounded at $t = 0$. Neglecting the leakage and inductance, find the voltage and current $t$ seconds later.

**94.** The length of a telephone wire is 3000 miles. It has resistance of 4 Ω/mile and a capacitance $5 \times 10^{-7}$ F/mile. Initially both the ends are grounded so that the line is uncharged. At $t = 0$, a constant emf $E$ is applied to one end, while the other end is left grounded. Show that the steady state current of the grounded end at the end of 1 second is 5.3 per cent (leakage and inductance are being neglected).

**95.** A very long telephone cable is of very high resistance but of negligible inductance and leakage. If a pulsating voltage $E = E_0 \sin \omega t$ is applied at the sending end, show that the current leads the voltage by nearly 45° everywhere and at all times.

**96.** Show that a transmission line with negligible resistance and leakage propagates waves of current and a potential with a velocity equal to $1/\sqrt{LC}$, where $L$ is the self-inductance and $C$ the capacitance.

**97.** Find the solution of Eq. (152) when a periodic emf $v_0 \cos \omega t$ is applied at the end $x = 0$ of the line.

**98.** A telephone line has a resistance of 8.8 Ω/mile and a leakage of 2 μΩ/mile. The line is broken, leaving it with an open circuit transmission. If the resistance of the line measured at the source is 3000 Ω, how far away is the break. [*Hint*: Use Eq. (157)].

**99.** Obtain Eq. (161) for a spherical reactor (use spherical coordinates $x = r \sin \theta \cos \phi$, $y = r \cos \theta \cos \phi$, $z = r \cos \theta$), and solve it by assuming uniform and symmetrical distribution of material, so that $u$ depends only on $r$. If the initial conditions are $u(a) = 0$ and $u(r)$ is bounded, then show that $u(r) = (A/r) \sin (\pi r/a)$ and $(\text{vol})_{\min} = 130/B^3$. Also, obtain the equation analogous to (161) for cylindrical reactors (use cylindrical coordinates $x = r \cos \theta$, $y = r \sin \theta$, $z = z$).

CHAPTER 10

# Calculus of Variations and Its Applications

## 10.1 INTRODUCTION

Calculus of variation is one of the most important branches of theoretical and applied mathematics. Variational principles are of great scientific significance as they provide a unified approach to various physical and mathematical problems and give fundamental exploratory ideas. The first problem in variational calculus was considered by Johann Bernoulli in 1696 which may be stated as follows:

*Let A and B be two points which do not lie on a vertical line such that a particle sliding down the curve joining A and B, under the influence of gravity (neglecting any type of resistance), in the shortest time. Then what is the curve?*

The required curve which turned out to be a cycloid was known as *Brachistochrone* (the curve of quickest descent). The brachistochrone problem played key role in the development of calculus of variations. This problem was also solve independently by James Bernoulli, Leibnitz, Newton and L'Hospital. However, the systematic development of the calculus of variation as an independent mathematical discipline was due to Euler during 1707–1783.

The above problem and a number of other similar mechanical and physical problems led to the development of the calculus of variation, although the topic attracted little attention for about two centuries. In the last fifty years, however, the interest in this subject has been revived due to its applications to the problems of optimization and control, e.g. the path of the rocket trajectory and optimal economic growth. The optimal control theory has also a number of applications in biological and medical sciences. It can be applied to study the spread of a contagious disease, pest control, cancer chemotherapy and immune system, etc.

Some of the classical problems of calculus of variation are described as under:

(i) Let $A(x_1, y_1)$ and $B(x_2, y_2)$ be two points in the $xy$-plane on a plane curve $C$. Let $s$ be the arc length so that

$$s = \int_A^B \left[1 + \left(\frac{dy}{dx}\right)^2\right]^{1/2} dx \quad (1)$$

then the problem is to *find the curve C from A to B so that the arc length is minimum*. That is, we have to find the conditions under which the integral (1) is minimum.

(ii) Given two points $X(x_1, y_1, z_1)$ and $Y(x_2, y_2, z_2)$ on a surface whose equation is $g(x, y, z) = 0$. The arc length $l$ between $X$ and $Y$ is

$$l = \int_{x_1}^{x_2} \left[ 1 + \left(\frac{dy}{dx}\right)^2 + \left(\frac{dz}{dx}\right)^2 \right]^{1/2} dx \tag{2}$$

The problem is then: *Of all possible curves joining X and Y determine that curve which has the shortest possible length.*

(iii) A plane curve $y = y(x)$ passing through two fixed points $(x_1, y_1)$ and $(x_2, y_2)$ is made to revolve about $y$-axis (or $x$-axis). The area of the surface of revolution about $y$-axis is $A = \int 2\pi x \, ds$ so that if $s$ is the arc length along the curve, we have

$$A = 2\pi \int_{x_1}^{x_2} x \left[ 1 + \left(\frac{dy}{dx}\right)^2 \right]^{1/2} dx \tag{3}$$

Here we have to *find the curve* $y = y(x)$ *so that the area of surface of revolution is minimum.* This problem was first solved by Jacob Bernoulli in 1698, but a general method for solving such problems was given by Euler.

In the problems (ii) and (iii) above, we have to find the conditions under which the integrals (2) and (3) are minimum.

(iv) Let $C$ be a curve defined on $[a, b]$ and $l$, the length of $C$ between the points $A(a, 0)$ and $B(0, b)$, then the problem is to *find the curve C such that the region bounded by it and the segment AB (on the x-axis) has the largest area.* That is, if $C$ is a curve given by $r = r(\theta)$ with length

$$l = \int_{0}^{\pi} \left[ r^2 + \left(\frac{dr}{d\theta}\right)^2 \right]^{1/2} d\theta \tag{4}$$

and the area bounded by the line segment $AB$ and the curve $r = r(\theta)$ is

$$A = \frac{1}{2} \int_{0}^{2\pi} r^2 \, d\theta \tag{5}$$

then we have *to find the curve* $r = r(\theta)$ *for which A is maximum subject to the condition* (4).

(v) It is known that the light propagates in an optically non-uniform medium with velocity $c(x, y, z)$. It is required *to find the path of a light ray which joins the points* $A(x_1, y_1, z_1)$ *and* $B(x_2, y_2, z_2)$. From Fermat's principle, we know that the path of the propagation is such as to enable the light to travel from $A$ to $B$ in the shortest possible time. If the equation of the desired path of the light ray is $y = y(x)$ and $z = z(x)$, then the time from $A$ to $B$ is

$$T = \int_{x_1}^{x_2} \frac{ds}{c} = \int_{x_1}^{x_2} \frac{[1 + (dy/dx)^2 + (dz/dx)^2]^{1/2}}{c(x, y, z)} dx \tag{6}$$

and we have to find the conditions under which integral (6) is minimum. This problem is similar to Brachistochrone problem.

(vi) *A problem of gas dynamics.* Find the shape of a solid of revolution moving in a flow of gas with least resistance.

We now mention few elementary examples of extrema from basic geometry.

1. Of all the triangles with given baseline and given perimeter the isosceles triangle has the largest area.
2. Of all the triangles with given baseline and given area the isosceles triangle has the least perimeter.
3. *A problem of Steiner.* Given three points $A$, $B$, $C$ which form an acute triangle $ABC$, a fourth point $P$ can be found inside the triangle so that $PA + PB + PC$ is as small as possible.
4. *Isoperimetric problem for polygon.* Among all polygons which are not self-intersecting and have an even number $2n$ of sides and a given perimeter $2l$, find the one with the greatest area.

In all the above examples (i)–(v) (and the other examples, too) we have seen that the problem of variational calculus involves the determination of the conditions under which a given integral is an extremum, and the problems which involves the investigation of extremum are called *variational problems*. If a certain law of mechanics or theoretical physics is expressed in terms of a line integral and is such that the line integral achieved an extremum subject to some given conditions, then such a physical law is termed as a *variational principle*. Some of the examples of variational principles and their applications include principle of least action, principles of conservation of linear and angular momentums, conservation of energy principle and a number variational principles of classical and relativistic field theories.

It may be noted that in each of the integrals (1)–(6), some quantities are involved which may be obtained by one or several functions. These quantities are known as *functionals*. A functional can thus be defined as a rule which assigns a real number to each function belonging to some class of functions and so the domain of a functional is a set of admissible functions rather than a region of a coordinate space.

In this chapter, we shall study the methods for the optimization (determination of maximum or minimum) of different types of functionals (or integrals). Some of them may have the following forms:

$$\left.\begin{aligned}
I_1 &= \int_{x_1}^{x_2} F(x, y(x)\ y'(x))dx \\
I_2 &= \int_{x_1}^{x_2} F(x, y(x), y'(x), \ldots, y^n(x))dx \\
I_3 &= \int_{x_1}^{x_2} F(x, y_1(x)\ y_2(x), \ldots, y_n(x), y'_1(x), y'_2(x), \ldots, y'_n(x))dx \\
I_4 &= \iint_D F\left(z, y, z(x, y), \frac{\partial z}{\partial x}, \frac{\partial z}{\partial y}\right) dxdy
\end{aligned}\right\} \quad (7)$$

Here the function $F$ is given and the functions $y(x)$, $y_1(x)$, $y_2(x)$, ..., $y_n(x)$, $z(x, y)$ are the arguments of the functionals and a dash denotes the differentiation with respect to $x$.

**Definition 10.1** A variable quantity $I[y(x)]$ is a *functional* [depending upon the function $y = y(x)$] if to each function belonging to a certain class $C$ of functions there is a specific value of $I$. Thus there is a correspondence between a given function $y = y(x)$ and a number $I$.

The right-hand sides of Eqs. (1)–(3), (5) and (6) are the examples of the functionals. Other examples of the functionals, which usually occur in engineering, are as follows:

1. The path $S$ traced out by an automobile in time $t_1$ is a functional of the velocity $v$ of the automobile and is given by

$$S = \int_0^{t_1} v\,dt$$

2. The cost of construction of a railway track between two points is a functional of the distance to be covered. Here one has to evolve a method so that the functional is minimum, and for the flat earth the solution comes out to be a straight line.
3. In aerodynamics or ship dynamics, the time of motion, the fuel consumption and other parameters are functionals of elevator and rudder control laws. In such cases, one is concerned about how to achieve the highest velocity and greatest economy.

The determination of a functional has been illustrated through the following examples.

**EXAMPLE 10.1** Evaluate the functional

$$I = \int_0^1 \left(y +, x\frac{dy}{dx}\right) dx$$

along the paths (i) $y = x$, (ii) $y = 2x^2$ and (iii) $y = e^x$, from (0, 0) to (1, 1).

***Solution*** (i) Here $y = x$ and $\dfrac{dy}{dx} = 1$. Thus

$$I = \int_0^1 (x + x)\,dx = 1$$

(ii) Here $y = 2x^2$ and $\dfrac{dy}{dx} = 4x$. Thus

$$I = \int_0^1 [2x^2 + x(4x)]\,dx = 2$$

(iii) Here $y = e^x$ and $\dfrac{dy}{dx} = e^x$. Thus

$$I = \int_0^1 (e^x + xe^x)dx = e$$

**EXAMPLE 10.2** Evaluate the functional

$$I = \int_0^1 \left[ y^2 + \left(\frac{dy}{dx}\right)^2 \right] dx$$

along the paths (i) $y = x^2$ (ii) $y = (e^x - 1)/(e - 1)$.

***Solution*** (i) Here $y = x^2$ and $\dfrac{dy}{dx} = 2x$. Thus

$$I = \int_0^1 (x^4 + 4x^2)dx = \frac{23}{15}$$

(ii) Here $y = \dfrac{e^x - 1}{e - 1}$ and $\dfrac{dy}{dx} = \dfrac{e^x}{e - 1}$. Thus

$$I = \int_0^1 \left[ \left(\frac{e^x - 1}{e - 1}\right)^2 + \left(\frac{e^x}{e - 1}\right)^2 \right] dx = \frac{e^2 - 2e + 2}{(e - 1)^2}$$

## 10.2 THE VARIATION OF A FUNCTIONAL AND EULER'S EQUATIONS

In this section, the concept of variation of a functional is given along with the related results. These concepts will then be used to find the conditions under which a given functional may attain an extremum. The variation (or differential) of a functional is analogous to the derivative of a function of $n$ variables.

The distance $d(y_1, y_2)$ between two curves $y = y_1(x)$ and $y = y_2(x)$ is defined as

$$d(y_1, y_2) = \max_{x_0 \le x \le x_1} |y_1(x) - y_2(x)|$$

A functional $I[y]$ is said to be linear if

$$I[cy_1(x) + cy_2(x)] = cI[y_1(x)] + cI[y_2(x)]$$

The increment $\Delta I$ is defined as

$$\Delta I = I[y(x) + \delta y(x)] - I[y(x)]$$

which may also be expressed as

$$\Delta I = L[y(x), \delta y] + \beta[y(x), \delta y] \max |\delta y|$$

where $L[y(x), \delta y]$ is a linear functional in $\delta y$ and $\beta[y(x), \delta y] \to 0$ as the maximum value of $\delta y$ approaches 0. $L[y(x), \delta y]$ is the principal part of the increment $\Delta I$ and is known as the *variation* of the functional $I$ and is denoted by $\delta I$.

From the differential calculus, we have

**Definition 10.2** A differentiable function $f(x_1, x_2, ..., x_n)$ of $n$ variables is said to have a (relative) extremum at the points $(\overline{x}_1, \overline{x}_2, ..., \overline{x}_n)$ if

$$\Delta f = f(x_1, x_2, ..., x_n) - f(\overline{x}_1, \overline{x}_2, ..., \overline{x}_n)$$

has same sign for all points $(x_1, x_2, ..., x_n)$ belonging to some neighbourhood of the points $(\overline{x}_1, \overline{x}_2, ..., \overline{x}_n)$. The extremum is minimum if $\Delta f \ge 0$ and maximum if $\Delta f \le 0$.

In an analogous way, we have

**Definition 10.3** The functional $I[y]$ has a (relative) extremum for $y = \overline{y}$ if $I[y] - I[\overline{y}]$ does not change its sign in some neighbourhood of the curve $y = y(x)$.

We now state the following necessary condition (also known as *variational principle*) for the extremum of a functional as

**Theorem 10.1** For a differentiable functional $I[y]$ to have an extremum for $y = \overline{y}(x)$, the necessary condition is that its variation vanishes for $y = \overline{y}(x)$. That is $\delta I = 0$ for $y = \overline{y}(x)$.

It may be noted that for a function $f$ to have a minimum it is not only necessary that $df = 0$ but also that $d^2f$ be non-negative.

We shall use the following lemma in our later considerations.

**Lemma 10.1** If for all arbitrary functions $g(x)$ defined on $[a, b]$ and continuous through the second derivative

$$\int_a^b h(x)g(x)\,dx = 0$$

where $h(x)$ is continuous in $[a, b]$ then $h(x) \equiv 0$ on $[a, b]$.

One of the main problem of calculus of variation is to find the curve for which some given line integral (functional) has an extremum. Starting with the simplest variational problem, we wish to find a function (or curve) $y(x)$ which will cause the line integral

$$I = \int_{x_1}^{x_2} F\left(x, y(x), y'(x)\right) dx \tag{8}$$

to have an extremum subject to the conditions $y(x_1) = y_1$ and $y(x_2) = y_2$, where $y_1$ and $y_2$ are prescribed at the fixed end (or boundary) points $x_1$ and $x_2$.

Here $y(x)$ is a path and there are number of ways to travel from $x_1$ to $x_2$ (Fig. 10.1). Label all these possible travel paths $y(x)$ with an infinitesimal parameter $\alpha$. Such a path may be denoted by $y(x, \alpha)$ with $y(x, 0)$ representing the correct path. For example, if we choose any function $\eta(x)$ which vanishes at $x = x_1$ and $x = x_2$ then a possible set of varied path is

$$y(x, \alpha) = y(x, 0) + \alpha\eta(x) \tag{9}$$

Thus Eq. (8) can be written as

$$I = \int_{x_1}^{x_2} F\left(x, y(x, \alpha), y'(x, \alpha)\right) dx \tag{10}$$

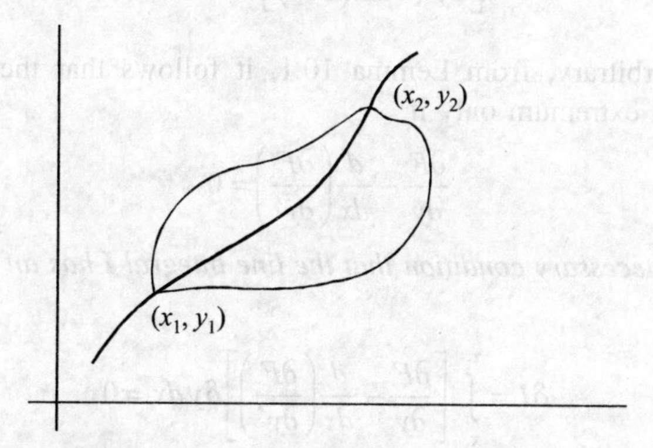


**Fig. 10.1** Varied paths in the extremum problem.

Differentiating Eq. (10), we get

$$\frac{dI}{d\alpha} = \int_{x_1}^{x_2} \left( \frac{\partial F}{\partial x}\frac{\partial x}{\partial \alpha} + \frac{\partial F}{\partial y}\frac{\partial y}{\partial \alpha} + \frac{\partial F}{\partial y'}\frac{\partial y'}{\partial \alpha} \right) dx$$

As $x$ is not a function of $\alpha$, the above equation reduces to

$$\frac{dI}{d\alpha} = \int_{x_1}^{x_2} \left( \frac{\partial F}{\partial y}\frac{\partial y}{\partial \alpha} + \frac{\partial F}{\partial y'}\frac{\partial y'}{\partial \alpha} \right) dx = \int_{x_1}^{x_2} \left( \frac{\partial F}{\partial y}\frac{\partial y}{\partial \alpha} + \frac{\partial F}{\partial y'}\frac{\partial^2 y}{\partial x \partial \alpha} \right) dx \tag{11}$$

Now, integrating the second term of Eq. (11) by parts, remebering that the end points are fixed and same for all curves so that at the end points $\partial y/\partial\alpha = 0$, after simplification, we obtain

$$\frac{dI}{d\alpha} = \int_{x_1}^{x_2} \left[ \frac{\partial F}{\partial y}\frac{\partial y}{\partial \alpha} - \frac{d}{dx}\left(\frac{\partial F}{\partial y'}\right)\frac{\partial y}{\partial \alpha} \right] dx$$

or
$$\frac{dI}{d\alpha} = \int_{x_1}^{x_2} \left[ \frac{\partial F}{\partial y} - \frac{d}{dx}\left(\frac{\partial F}{\partial y'}\right)\right]\frac{\partial y}{\partial \alpha}\, dx \tag{12}$$

The differential quantities

$$\left(\frac{\partial y}{\partial \alpha}\right)_{\alpha=0} d\alpha = \delta y \quad \text{and} \quad \left(\frac{\partial I}{\partial \alpha}\right)_{\alpha=0} d\alpha = \delta I$$

respectively, represent the infinitesimal departure of the varied path from the correct path $y(x)$ at the point $x$ and the infinitesimal variation of $I$ about the correct path. Thus, from Theorem 10.1, Eq. (12) reduces to

$$\int_{x_1}^{x_2} \left[ \frac{\partial F}{\partial y} - \frac{d}{dx}\left(\frac{\partial F}{\partial y'}\right)\right] \delta y\, dx = 0 \tag{13}$$

Since $\delta y$ is arbitrary, from Lemma 10.1, it follows that the integral $I$ in Eq. (8) has an extremum only if

$$\frac{\partial F}{\partial y} - \frac{d}{dx}\left(\frac{\partial F}{\partial y'}\right) = 0 \tag{14}$$

Thus *the necessary condition that the line integral I has an extremum can be written as*

$$\delta I = \int_{x_1}^{x_2} \left[ \frac{\partial F}{\partial y} - \frac{d}{dx}\left(\frac{\partial F}{\partial y'}\right)\right] \delta y\, dx = 0 \tag{15}$$

*requiring that* $y(x)$ *satisfies the differential equation* (14).

Equation (14) is known as *Euler's equation*. This equation plays a key role in the calculus of variation and was found by Euler in 1744. The integral curves of Euler's equation are called *extremals*. Euler's equation is a second-order differential equation and its solution will, in general, consists of two constants which can be obtained from the given boundary conditions.

We shall now deal with some special circumstances under which Euler's equation either reduces to a first-order differential equation or its solution can be obtained by means of quadratures.

*Case (i):* When $I = \int_{x_1}^{x_2} F(x, y)\, dx$

Here $F$ is a function of $x$ and $y$ and Euler's equation (14) reduces to $\partial F/\partial y = 0$, which when solved does not involve arbitrary constants and, in general, it is not possible to find the solution satisfying the boundary conditions; and as such this variational problem does not, in general, admits a solution.

*Case (ii):* When $F$ is linearly dependent of $y'$

Here $F = F(x, y, y') = M(x, y) + N(x, y)y'$. We also have

$$I = \int_{x_1}^{x_2} [M(x, y) + N(x, y)y']dx, \qquad \frac{\partial F}{\partial y} = \frac{\partial M}{\partial y} + \frac{\partial N}{\partial y} y', \qquad \frac{\partial F}{\partial y'} = N(x, y)$$

and
$$\frac{d}{dx} N(x, y) = \frac{\partial N}{\partial x}\frac{dx}{dx} + \frac{\partial N}{\partial y}\frac{dy}{dx} = \frac{\partial N}{\partial x} + \frac{\partial N}{\partial y} y'$$

so that from Eq. (14), we have

$$\frac{\partial M}{\partial y} - \frac{\partial N}{\partial x} = 0 \tag{16}$$

Equation (16) leads to a finite equation which does not satisfy the boundary conditions; and thus the variation problem, in such a case, does not have any solution. But if

$$\frac{\partial M}{\partial y} - \frac{\partial N}{\partial x} \equiv 0 \tag{17}$$

then $Mdx + Ndy = 0$ is an exact differential and

$$I = \int_{x_1}^{x_2} \left[ M(x, y) + N(x, y)\frac{dy}{dx} \right] dx \tag{18}$$

is independent of the path of integration and $I$ has a constant value. In such cases, the variational problem becomes meaningless.

*Case (iii):* When $I = \int_{x_1}^{x_2} F(x, y')\, dx$

Here $F$ does not depend upon $y$, which leads to $\partial F/\partial y = 0$ and thus Eq. (14) reduces to

$$\frac{d}{dx}\left(\frac{\partial F}{\partial y'}\right) = 0 \tag{19}$$

The solution of Eq. (19) is

$$\frac{\partial F}{\partial y'} = c \tag{20}$$

where $c$ is a constant. Equation (20) is a first-order differential equation which does not contain $y$.

*Case (iv):* When $I = \int_{x_1}^{x_2} F(y, y')dx$

Here $F$ does not depend upon $x$. Multiplying Eq. (14) by $y'$, we get

$$y'\frac{\partial F}{\partial y} - y'\frac{d}{dx}\left(\frac{\partial F}{\partial y'}\right) = 0 \tag{21}$$

But

$$\frac{d}{dx}\left(y'\frac{\partial F}{\partial y'}\right) = y''\frac{\partial F}{\partial y'} + y'\frac{d}{dx}\left(\frac{\partial F}{\partial y'}\right)$$

leads to

$$y'\frac{d}{dx}\left(\frac{\partial F}{\partial y'}\right) = \frac{d}{dx}\left(y'\frac{\partial F}{\partial y'}\right) - y''\frac{\partial F}{\partial y'}$$

which when substituted in Eq. (21) yields

$$y'\frac{\partial F}{\partial y} + y''\frac{\partial F}{\partial y'} - \frac{d}{dx}\left(y'\frac{\partial F}{\partial y'}\right) = 0 \tag{22}$$

Since $F = F(y, y')$, we have

$$\frac{dF}{dx} = \frac{\partial F}{\partial y}\frac{dy}{dx} + \frac{\partial F}{\partial y'}\frac{dy'}{dx} = \frac{\partial F}{\partial y}y' + \frac{\partial F}{\partial y'}y''$$

so that Eq. (22) reduces to

$$\frac{dF}{dx} - \frac{d}{dx}\left(y'\frac{\partial F}{\partial y'}\right) = 0$$

Thus, when $F = F(y, y')$, the first integral of the Euler's equation is

$$F - y'\frac{\partial F}{\partial y'} = c \tag{23}$$

where $c$ is a constant.

*Case (v):* When $I = \int_{x_1}^{x_2} F(y')\,dx$

In this case, Euler's equation (14) reduces to

$$\frac{d}{dx}\left(\frac{\partial F}{\partial y'}\right) = 0$$

and, therefore

$$\frac{\partial F(y')}{\partial y'} = \text{constant}$$

which leads to $y' = \text{constant} = a$, whose solution is $y = ax + b$.

Hence, if the functional in Eq. (8) depends only on the derivative of the function $y = y(x)$, then the extremals are always straight lines.

We shall now illustrate applications of Euler's equation and its special cases through the following examples.

**EXAMPLE 10.3** Test for the extremals of the functional

$$I = \int_a^b (x - y)^2 \, dx$$

***Solution*** Here,

$$F = F(x, y) = (x - y)^2, \qquad \frac{\partial F}{\partial y} = -2x + 2y, \qquad \frac{\partial F}{\partial y'} = 0$$

and Eq. (14) leads to a finite equation $x - y = 0$, which is a straight line. However, the integral $I$ vanishes along this line.

**EXAMPLE 10.4** Test for the extremals of the functional

$$I = \int_0^1 (xy + y^2 - 2y^2 y') \, dx$$

under the conditions $y(0) = 1$, $y(1) = 2$.

***Solution*** Here,

$$F = xy + y^2 - 2y^2 y', \quad \frac{\partial F}{\partial y} = x + 2y - 4yy', \quad \frac{\partial F}{\partial y'} = -2y^2$$

and Eq. (14) leads to $y = -x/2$ which is an extremal for the given integral. But this extremal does not satisfy the given boundary conditions, and thus an extremal cannot be achieved for this given functional.

**EXAMPLE 10.5** Investigate the extremals of the functional

$$I = \int_0^1 (y^2 + x^2 y') dx$$

under the conditions $y(0) = 0$, $y(1) = A$.

***Solution*** The given integral is

$$I = \int_0^1 \left( y^2 + x^2 \frac{dy}{dx} \right) dx = \int_0^1 (y^2 dx + x^2 dy)$$

Thus,

$$M = y^2, \qquad N = x^2, \qquad \frac{\partial M}{\partial y} = 2y, \qquad \frac{\partial N}{\partial x} = 2x$$

and Eq. (16) leads to $y = x$ which is a straight line. The first condition $y(0) = 0$ is satisfied, while $y(1) = A$ is satisfied only when $A = 1$; and if $A \neq 1$ then there is no extremal which satisfies the given boundary conditions.

**EXAMPLE 10.6** Test for the extremals of the functional

$$I = \int_{x_1}^{x_2} (y + xy')dx$$

such that $y(x_1) = y_1$, $y(x_2) = y_2$.

***Solution*** The given integral can be expressed as

$$I = \int_{x_1}^{x_2} (ydx + xdy)$$

for which Eq. (17) is identically satisfied. Thus, $M dx + N dy = 0$ is an exact differential and

$$I = \int_{x_1}^{x_2} (ydx + xdy) = \int_{x_1}^{x_2} d(xy) = x_2 y_2 - x_1 y_1$$

is independent of the path of integration. Therefore, the present variational problem is meaning less.

**EXAMPLE 10.7** Find the extremals of the functional

$$I = \int_{x_1}^{x_2} (y^2 + y'^2 - 2y \sin x)\, dx$$

***Solution*** Here,

$$F = y^2 + y'^2 - 2y \sin x, \quad \frac{\partial F}{\partial y} = 2y - 2 \sin x, \quad \frac{\partial F}{\partial y'} = 2y', \quad \frac{d}{dx}\left(\frac{\partial F}{\partial y'}\right) = 2y''$$

Equation (14) thus leads to

$$y'' - y = -\sin x$$

The general solution of this differential equation is

$$y = c_1 e^x + c_2 e^{-x} + \frac{1}{2} \sin x$$

which is the required extremal.

**EXAMPLE 10.8** Investigate the extremals of the functional

$$I = \int_{x_1}^{x_2} \frac{y'^2}{x^3}\, dx$$

***Solution*** Here,

$$F = \frac{y'^2}{x^3}, \qquad \frac{\partial F}{\partial y} = 0, \qquad \frac{\partial F}{\partial y'} = \frac{2y}{x^3}$$

Equation (14) thus leads to

$$\frac{d}{dx}\left(\frac{2y'}{x^3}\right) = 0$$

so that

$$\frac{dy}{dx} = \frac{c_1}{2} x^3$$

The general solution of this equation is

$$y = kx^4 + c$$

($k = c_1/8$) which is the required extremal.

**EXAMPLE 10.9** Find the extremals of the functional

$$I = \int_{0}^{\pi/2} (y'^2 - y^2)\, dx$$

under the boundary conditions $y(0) = 0$, $y(\pi/2) = 1$.

***Solution*** Here,

$$F = y'^2 - y^2, \qquad \frac{\partial F}{\partial y'} = -2y, \qquad \frac{\partial F}{\partial y'} = 2y', \qquad \frac{d}{dx}\left(\frac{\partial F}{\partial y'}\right) = 2y''$$

Equation (14) thus leads to $y'' + y = 0$, which is a second-order differential equation, whose solution is

$$y = c_1 \cos x + c_2 \sin x \tag{24}$$

Using the given boundary conditions, Eq. (24) leads to $c_1 = 0$, $c_2 = 1$. Thus, Eq. (24) becomes $y = \sin x$, which is the extremal of the given functional.

**EXAMPLE 10.10** Obtain the extremals for the functional

$$I = \int_{1}^{2} \frac{(1 + y'^2)^{1/2}}{x}\, dx$$

under the boundary conditions $y(1) = 0$, $y(2) = 2$.

***Solution*** Here,

$$F = \frac{(1+y'^2)^{1/2}}{x}, \quad \frac{\partial F}{\partial y'} = \frac{y'}{x(1+y'^2)^{1/2}}$$

Thus, from Eq. (20), we have

$$\frac{y'}{x(1+y'^2)^{1/2}} = c_1 \quad \text{or} \quad y' = c_1 x(1 + y'^2)^{1/2}$$

Squaring and simplifying this equation, we get

$$y'^2 = \frac{c_1^2 x^2}{1 - c_1^2 x^2} \quad \text{or} \quad \frac{dy}{dx} = \frac{c_1 x}{\sqrt{1 - c_1^2 x^2}}$$

Therefore,

$$dy = \frac{c_1 x}{\sqrt{1 - c_1^2 x^2}} dx$$

which on integration yields

$$y = -\frac{1}{c_1}\sqrt{1 - c_1^2 x^2} + c_2$$

or,

$$x^2 + (y - c_2)^2 = \frac{1}{c_1^2} \tag{25}$$

Equation (25) represents a family of circles whose centre is on $y$-axis. The constants $c_1$ and $c_2$ of integration can be obtained by using the given boundary conditions $y(1) = 0$, $y(2) = 2$ in Eq. (25) and we have $c_1 = 4/\sqrt{65}$, $c_2 = 7/4$, and thus from Eq. (25), the extremal of the given functional is a circle

$$x^2 + y^2 - \frac{7}{2}y - 1 = 0$$

whose centre is (0, 7/4) and radius is $\sqrt{65}/4$ units.

**EXAMPLE 10.11** Investigate the extremals of the functional

$$I = \int_1^2 (y' + x^2 y'^2) dx$$

under the boundary conditions $y(1) = 0$, $y(2) = 1$.

***Solution*** Here,

$$F = y' + x^2 y'^2, \quad \frac{\partial F}{\partial y} = 0, \quad \frac{\partial F}{\partial y'} = 1 + 2x^2 y'$$

Euler's Eq. (14) reduces to

$$\frac{d}{dx}(1 + 2x^2y') = 0 \quad \text{or} \quad 1 + 2x^2y' = c_1$$

which on integration yields

$$y = -\frac{c_1 - 1}{2}\frac{1}{x} + c_3 = c_2\frac{1}{x} + c_3 \tag{26}$$

Using the boundary conditions $y(1) = 0$, $y(2) = 1$ in Eq. (26), we get

$$0 = c_2 + c_3, \qquad 1 = \frac{1}{2}c_2 + c_3$$

which on solving yields $c_2 = -2$, $c_3 = 2$. Thus, from Eq. (26), the extremal of the given functional is

$$y = 2 - \frac{2}{x}$$

**EXAMPLE 10.12** Investigate the extremals of the functional

$$I = \int_0^{\pi/2} (y'^2 + 2y \sin x)dx$$

under the boundary conditions $y(0) = 0$, $y(\pi/2) = 1$.

***Solution*** Here,

$$F = y'^2 + 2y \sin x, \qquad \frac{\partial F}{\partial y} = 2 \sin x, \qquad \frac{\partial F}{\partial y'} = 2y'$$

and Euler's equation (14) leads to $y'' = \sin x$, which has the general solution as

$$y = -\sin x + c_1x + c_2 \tag{27}$$

Now from the boudary conditions and Eq. (27), we have $c_1 = 4/\pi$, $c_2 = 0$; and thus the extremal for the given functional is

$$y = \frac{4}{\pi}x - \sin x$$

**EXAMPLE 10.13** Find the extremals of

$$I = \int_{x_1}^{x_2} \frac{1 + y^2}{y'^2} dx$$

***Solution*** Here,

$$F = \frac{1 + y^2}{y'^2}, \qquad \frac{\partial F}{\partial y'} = \frac{-2(1 + y^2)}{y'^3}$$

and, from Eq. (23), we have

$$y'^2 = \frac{3}{c}(1 + y^2)$$

so that

$$y' = \frac{dy}{dx} = c_1\sqrt{1 + y^2}, \quad c_1 = \sqrt{3/c}$$

Separating the variables and integrating, we get

$$c_1 x = \sinh^{-1} y + c_2$$

Thus, the required extremal is

$$y = \sinh (c_1 x - c_2)$$

**EXAMPLE 10.14** Find the shortest distance between two points in a plane.

***Solution*** Let $s$ be the arc length of a curve in a plane, then an element of the arc length is

$$ds = \sqrt{dx^2 + dy^2} = \sqrt{1 + \left(\frac{dy}{dx}\right)^2}\, dx$$

Thus, the total length of the curve between two points $A(x_1, y_1)$ and $B(x_2, y_2)$ is

$$I = \int_{x_1}^{x_2} ds = \int_{x_1}^{x_2} \sqrt{1 + y'^2}\, dx$$

In this example, we have to find the curve for which $I$ is minimum. Here $F = (1 + y'^2)^{1/2}$ and Euler's Equation (14) reduces to

$$\frac{d}{dx}\left(\frac{y'}{\sqrt{1 + y'^2}}\right) = 0$$

so that

$$\frac{y'}{\sqrt{1 + y'^2}} = c$$

which after simplification leads to

$$y' = \frac{c}{\sqrt{1 - c^2}} = a$$

The solution of this differential equation is

$$y = ax + b \tag{28}$$

which is a straight line and the constants $a$ and $b$ can be obtained from the condition that the curve passes through the points $A$ and $B$. Hence, the shortest path connecting two points in a plane is only a straight line [See also case (v) after Eq. (23)]. If the curve (28) is passing through the points (1, 1) and (3, 3), then from Eq. (28) $a = 1$, $b = 0$ and thus the curve passing through the points (1, 1) and (3, 3) is the straight line $x = y$.

**EXAMPLE 10.15** A curve is lying above $x$-axis and passing through two fixed points. What will be shape of the curve which when rotated about $x$-axis give surface of revolution of minimum surface area?

***Solution*** Let $y = y(x)$ be a curve lying above $x$-axis and passing through the points $P(x_1, y_1)$ and $Q(x_2, y_2)$ (Fig. 10.2). We know that the area of the surface of revolution is

$$A = \int_{x_1}^{x_2} 2\pi y \, ds \tag{29}$$

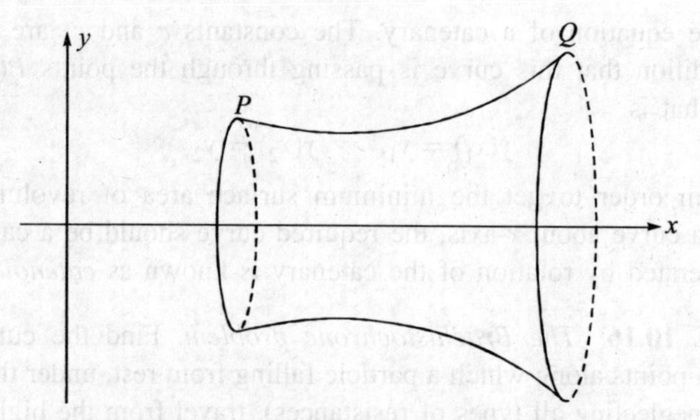


**Fig. 10.2** Minimum surface of revolution.

But $ds = \sqrt{1 + y'^2}\, dx$. Equation (29) thus becomes

$$A = 2\pi \int_{x_1}^{x_2} y\sqrt{1 + y'^2}\, dx \tag{30}$$

We have to find the extremal for the functional in Eq. (30). Here, $F = y\sqrt{1 + y'^2} = F(y, y')$ and the first integral of Euler's equation is given by [see case (iv), Eq. (23)]

$$F - y'\frac{\partial F}{\partial y'} = c$$

That is

$$y(1 + y'^2)^{1/2} - \frac{yy'^2}{(1 + y'^2)^{1/2}} = c$$

which after simplification leads to

$$y' = \frac{dy}{dx} = \frac{(y^2 - c^2)^{1/2}}{c}$$

Separating the variables, we get

$$\frac{dy}{(y^2 - c^2)^{1/2}} = \frac{1}{c}\,dx$$

which on integration yields

$$\cosh^{-1}\frac{y}{c} = \frac{x + c_1}{c}$$

so that

$$y = c\cosh\frac{x + c_1}{c} \tag{31}$$

which is the equation of a catenary. The constants $c$ and $c_1$ are determined by the condition that this curve is passing through the points $P(x_1, y_1)$ and $Q(x_2, y_2)$. That is

$$y(x_1) = y_1, \qquad y(x_2) = y_2$$

Therefore, in order to get the minimum surface area of revolution by the rotation of a curve about $x$-axis, the required curve should be a catenary. The surface generated by rotation of the catenary is known as *catenoid*.

**EXAMPLE 10.16** *The Brachistochrone problem.* Find the curve passing through two points along which a particle falling from rest, under the influence of gravity (neglecting all types of resistances), travel from the higher position to the lower position in least possible time.

***Solution*** Let $A$ and $B$ be two points, which do not lie in the same vertical line (Fig. 10.3), such that a particle sliding down the curve joining $A$ and $B$ in the shortest possible time (under constant gravity and in the absence of all types of resistances), then we have to find the shape of the curve.

Let $v$ be the speed of the particle (of mass $m$) along the curve $C$, then the time taken in travelling a small arc $ds$ is $ds/v$, and thus the total time taken by the particle in moving from $A$ to $B$ is

$$I = \int_A^B \frac{ds}{v} = \int_A^B \frac{\sqrt{1 + y'^2}}{v}\,dx \tag{32}$$

If $y$ is the vertical distance of fall up to point $B$, then from the conservation theorem for energy (loss in kinetic energy is equal to the gain in potential energy) we have

$$\frac{1}{2}mv^2 = mgy$$

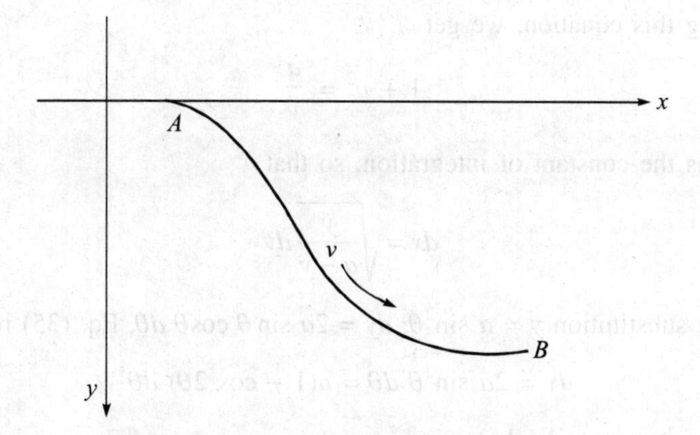


**Fig. 10.3** The Brachistochrone problem.

which yields $v = \sqrt{2gy}$ and Eq. (32) takes the form

$$I = \frac{1}{\sqrt{2g}} \int_A^B \frac{\sqrt{1+y'^2}}{\sqrt{y}} dx \tag{33}$$

Here $F = \sqrt{(1+y'^2)/y}$, and with this $F$, Eq. (14) reduces to

$$1 + y'^2 + 2yy'' = 0 \tag{34}$$

which is a second-order differential equation and its solution is obtained as follows:

Since there is no $x$ in Eq. (34), we take

$$p = \frac{dy}{dx} = y'$$

so that

$$\frac{d^2y}{dx^2} = y'' = \frac{dp}{dx} = \frac{dp}{dy}\frac{dy}{dx} = y'\frac{dp}{dy} = p\frac{dp}{dy}$$

Equation (34) now becomes

$$1 + p^2 + 2yp\frac{dp}{dy} = 0$$

Separating the variables, we have

$$\frac{2p\,dp}{1+p^2} + \frac{dy}{y} = 0$$

Integrating this equation, we get

$$1 + p^2 = \frac{a}{y}$$

where $a$ is the constant of integration, so that

$$dx = \sqrt{\frac{y}{a-y}}\,dy \tag{35}$$

Using the substitution $y = a \sin^2\theta$, $dy = 2a \sin\theta \cos\theta \, d\theta$, Eq. (35) reduces to

$$dx = 2a \sin^2\theta \, d\theta = a(1 - \cos 2\theta) \, d\theta$$

which on integration leads to

$$x = a\left(\theta - \frac{1}{2}\sin 2\theta\right) + b$$

or,

$$x = \frac{1}{2} a(2\theta - \sin 2\theta) + b \tag{36a}$$

where $b$ is the constant of integration. But

$$y = a \sin^2\theta = \frac{1}{2} a(1 - \cos 2\theta) \tag{36b}$$

Equations (36) are the parametric representation of the required curve.

For simplicity, we assume that the higher point $A$ coincides with the origin and since the curve (36) passes through the point $A$, we have $b = 0$. If we take $\phi = 2\theta$ and $c = a/2$, then Eq. (36) becomes

$$x = c(\phi - \sin \phi), \qquad y = c(1 - \cos \phi) \tag{37}$$

which are the parametric equations of a *cycloid* (Fig. 10.4).

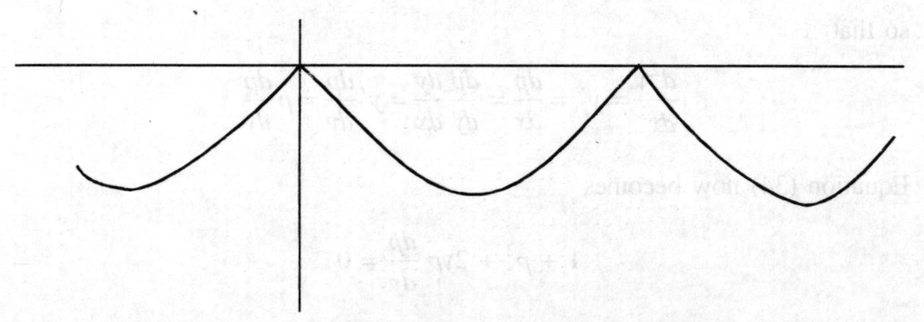

**Fig. 10.4** A cycloid.

Hence, in order that a particle travel along a curve from the higher position to the lower position in minimum time (under the influence of gravity), the curve must be a cycloid. (*see* also Example 8.30).

The term brachistochrone is derived from the Greek *brachistos* and *chronos* which means shortest and time, respectively. The techniques which we have used here were not available to Johann Bernoulli.

## 10.3 FUNCTIONALS DEPENDING ON *n* UNKNOWN FUNCTIONS

The fundamental problem of calculus of variation which was discussed in Section 10.2 (cf. Euler's equation) can easily be generalized when $F$ is a function of a number of independent variables $y_j$ and their derivatives $y'_j$.

Consider the functional

$$I = \int_{x_1}^{x_2} F(x, y_1(x), y_2(x), \ldots, y_n(x), y'_1(x), y'_2(x), \ldots, y'_n(x))\, dx \tag{38}$$

which depends on $n$ functions $y_j(x)$ and their derivatives $y'_j(x)$ satisfying the boundary conditions

$$y_j(x_1) = A_j, \quad y_j(x_2) = B_j \tag{39}$$

Then the necessary condition for the existence of extremals for (38) is given by

**Theorem 10.2** A necessary condition for the curve $y_j(x) = y_j$ to be an extremal of the functional (38) is that the function $y_j(x)$ satisfy the equation

$$\frac{\partial F}{\partial y_j} - \frac{d}{dx}\left(\frac{\partial F}{\partial y'_j}\right) = 0 \tag{40}$$

This theorem can be proved by varying only one of the functions $y_j(x)$, keeping others fixed. Then functional (38) reduces to a functional which depends only on one of the functions $y_j(x)$. Therefore, using the techniques of Section 10.2 [cf. derivation of Eq. (14)], the function $y_j(x)$ having a continuous derivative must satisfy Eq. (40) along with the boundary conditions (39) (for details, see [8]). Equation (40) are known as *Euler–Lagrange's equations* and is a system of $n$ second-order differential equations. The general solution of Eq. (40) represents curves for which the variation of (38) vanishes, and contains $2n$ arbitrary constants which have to be determined by using the boundary condition (39).

The problem of finding the extremals of the functional of the type (38) usually occur in obtaining the equation of geodesic (which is the shortest curve joining two points of some manifold). Similar type of problems occur in geometric optics where one has to obtain the path of light rays in an inhomogeneous medium [cf. Eq. (6) and Example 10.43].

**EXAMPLE 10.17** Find the extremals of the functional

$$I = \int_{x_1}^{x_2} (1 + y'^2 + z'^2)^{1/2} dx$$

***Solution*** Since $F = (1 + y'^2 + z'^2)^{1/2}$ is a function of two variables and the Euler–Lagrange's equations

$$\frac{\partial F}{\partial y} - \frac{d}{dx}\left(\frac{\partial F}{\partial y'}\right) = 0$$

$$\frac{\partial F}{\partial z} - \frac{d}{dx}\left(\frac{\partial F}{\partial z'}\right) = 0$$

for this $F$ lead to the following differential equations

$$\frac{y'}{(1 + y'^2 + z'^2)^{1/2}} = c_1, \qquad \frac{z'}{(1 + y'^2 + z'^2)^{1/2}} = c_2$$

The solutions of these differential equations are

$$y = ax + b, \qquad z = dx + e$$

Thus the extremals of the given integral are the straight lines in space.

**EXAMPLE 10.18** Find the extremals of the functional

$$I = \int_{x_1}^{x_2} (2yz - 2y^2 + y'^2 - z'^2)\, dx$$

***Solution*** Here

$$F = 2yz - 2y^2 + y'^2 - z'^2, \quad \frac{\partial F}{\partial y} = 2z - 4y, \quad \frac{\partial F}{\partial y'} = 2y', \quad \frac{\partial F}{\partial z} = 2y, \quad \frac{\partial F}{\partial z'} = -2z'$$

and Eq. (40) leads to

$$z = y'' + 2y, \qquad y + z'' = 0$$

Eliminating $z$ from these equations, we get

$$y^{iv} + 2y'' + y = 0$$

The general solution of this equation is

$$y = (c_1 + c_2 x) \cos x + (c_3 + c_4 x) \sin x$$

which along with $z = y'' + 2y$ is the required extremal.

## 10.4 FUNCTIONALS DEPENDING ON HIGHER-ORDER DERIVATIVES

Quite often the problems of calculus of variation (for example, the problems of elasticity theory) involve functionals which not only depend upon the first derivative but also on the higher-order derivatives of the given functions. Such a functional is of the form

$$I[y(x)] = \int_{x_1}^{x_2} F(x, y(x), y'(x), y''(x), y'''(x), \ldots, y^{(n)}(x))\, dx \tag{41}$$

The necessary condition for the integral (41) to have extremum is

$$\frac{\partial F}{\partial y} - \frac{d}{dx}\left(\frac{\partial F}{\partial y'}\right) + \frac{d^2}{dx^2}\left(\frac{\partial F}{\partial y''}\right) + \cdots + (-1)^n \frac{d^n}{dx^n}\left(\frac{\partial F}{\partial y^{(n)}}\right) = 0 \tag{42}$$

along with the boundary conditions

$$y(x_1) = A_0, \quad y'(x_1) = A_1, \quad \ldots, \quad y^{(n-1)}(x_1) = A_{n-1} \tag{43a}$$

$$y(x_2) = B_0, \quad y'(x_2) = B_1, \quad \ldots, \quad y^{(n-1)}(x_2) = B_{n-1} \tag{43b}$$

where $F$ is assumed to be $(n + 2)$ times differentiable with respect to all its arguments. Equation (42) is known as *Euler–Poisson's equation*. This is a differential equation of order $2n$ and its general solution has $2n$ arbitrary constants which can be obtained from the boundary conditions (43). If

$$I[y(x), z(x)] = \int_{x_1}^{x_2} F(x, y(x), y'(x), y''(x), \ldots, y^{(n)}(x), z(x), z'(x), z''(x), \ldots, z^{n}(x))\, dx \tag{44}$$

then the corresponding Euler–Poisson's equations [the necessary condition for the existence of the extremum for (44)] are

$$\frac{\partial F}{\partial y} - \frac{d}{dx}\left(\frac{\partial F}{\partial y'}\right) + \frac{d^2}{dx^2}\left(\frac{\partial F}{\partial y''}\right) + \cdots + (-1)^n \frac{d^n}{dx^n}\left(\frac{\partial F}{\partial y^{(n)}}\right) = 0 \tag{45a}$$

$$\frac{\partial F}{\partial z} - \frac{d}{dx}\left(\frac{\partial F}{\partial z'}\right) + \frac{d^2}{dx^2}\left(\frac{\partial F}{\partial z''}\right) + \cdots + (-1)^m \frac{d^m}{dx^m}\left(\frac{\partial F}{\partial z^{(m)}}\right) = 0 \tag{45b}$$

Thus, the necessary condition for the integral

$$I[y_1(x), y_2(x), \ldots, y_n(x)] = \int_{x_1}^{x_2} F(x, y_1(x), y'_1(x), \ldots, y_1^{(n)}(x), y_2(x), y_2'(x), \ldots, y_2^{(n)}(x), \ldots, y_m(x), y'_m(x), \ldots, y_m^{(n_m)}(x))dx$$

to have an extremum is

$$\frac{\partial F}{\partial y_i} - \frac{d}{dx}\left(\frac{\partial F}{\partial y_i'}\right) + \frac{d^2}{dx^2}\left(\frac{\partial F}{\partial y_i''}\right) + \cdots + (-1)^{n_i} \frac{d^{n_i}}{dx^{n_i}}\left(\frac{\partial F}{\partial y_i^{(n_i)}}\right) = 0 \tag{46}$$

In the following example, we shall illustrate the proof of the Euler-Poisson's equation for a functional which depends upon first and second derivatives of the desired function.

**EXAMPLE 10.19** If $F = F(y, y', y'', x)$ and the variational principle holds with zero variation of both $y$ and $y'$ at the end points, prove that the corresponding Euler–Poisson's equations are

$$\frac{\partial F}{\partial y} - \frac{d}{dx}\left(\frac{\partial F}{\partial y'}\right) + \frac{d^2}{dx^2}\left(\frac{\partial F}{\partial y''}\right) = 0 \tag{47}$$

***Solution*** Consider the integral

$$I[y(x)] = \int_{x_1}^{x_2} F(y(x), y'(x), y''(x), x)dx \tag{48}$$

where the variation of $y$ and $y'$ is zero at the end points. Our aim is to find the extremum condition for this functional.

Taking into consideration the derivation of Euler's equation (14), and differentiating equation (48) under the integral sign, we have

$$\frac{dI}{d\alpha} = \int_{x_1}^{x_2} \left(\frac{\partial F}{\partial y}\frac{\partial y}{\partial \alpha} + \frac{\partial F}{\partial y'}\frac{\partial y'}{\partial \alpha} + \frac{\partial F}{\partial y''}\frac{\partial y''}{\partial \alpha} + \frac{\partial F}{\partial x}\frac{\partial x}{\partial \alpha}\right)dx$$

As the end points are fixed and $x$ is not a function of $\alpha$, we have

$$\frac{dI}{d\alpha} = \int_{x_1}^{x_2} \left(\frac{\partial F}{\partial y}\frac{\partial y}{\partial \alpha} + \frac{\partial F}{\partial y'}\frac{\partial y'}{\partial \alpha} + \frac{\partial F}{\partial y''}\frac{\partial y''}{\partial \alpha}\right)dx \tag{49}$$

Integrating by parts the second and the third terms of Eq. (49), we get after simplification

$$\left(\frac{dI}{d\alpha}\right)_{\alpha=0} d\alpha = \int_{x_1}^{x_2} \left[\frac{\partial F}{\partial y} - \frac{d}{dx}\left(\frac{\partial F}{\partial y'}\right) + \frac{d^2}{dx^2}\left(\frac{\partial F}{\partial y''}\right)\right]\left(\frac{\partial y}{\partial \alpha}\right)_{\alpha=0} d\alpha\, dx$$

or,
$$\delta I = \int_{x_1}^{x_2} \left[\frac{\partial F}{\partial y} - \frac{d}{dx}\left(\frac{\partial F}{\partial y'}\right) + \frac{d^2}{dx^2}\left(\frac{\partial F}{\partial y''}\right)\right]\delta y\, dx \tag{50}$$

Since the variational principle demands that $\delta I = 0$ at the end points, therefore, from Lemma 10.1, Eq. (50) leads to

$$\frac{\partial F}{\partial y} - \frac{d}{dx}\left(\frac{\partial F}{\partial y'}\right) + \frac{d^2}{dx^2}\left(\frac{\partial F}{\partial y''}\right) = 0$$

This is a fourth-order differential equation and its solution represents the extremals of the functional (48). The constants of integration appearing in the solution can be determined from the boundary conditions.

**EXAMPLE 10.20** Find the extremals of

$$I = \int_{x_1}^{x_2} \left( \frac{m}{2} y y'' + \frac{k}{2} y^2 \right) dx$$

and interpret.

***Solution*** Here

$$\left.\begin{aligned} F &= \frac{m}{2} y y'' + \frac{k}{2} y^2, \\ \frac{\partial F}{\partial y} &= \frac{m}{2} y'' + ky, \\ \frac{\partial F}{\partial y'} &= 0, \\ \frac{\partial F}{\partial y''} &= \frac{m}{2} y, \\ \frac{d^2}{dx^2}\left(\frac{\partial F}{\partial y''}\right) &= \frac{m}{2} y'' \end{aligned}\right\} \tag{51}$$

Since the extremal of the given integral is the solution of Eq. (47), therefore, from Eq. (51), Eq. (47) leads to

$$y'' + \frac{k}{m} y = 0$$

which is a second-order linear differential equation whose general solution is

$$y = c_1 \cos \omega x + c_2 \sin \omega x \tag{52}$$

where $\omega = \sqrt{k/m}$ and $c_1$, $c_2$ are constants of integration.

Equation (52) represents the motion of a particle of mass $m$ along a curve $y = y(x)$; and in fact, the particle goes under simple harmonic motion (cf. Section 6.1).

**EXAMPLE 10.21** Find the extremals of the functional

$$I = \int_{x_1}^{x_2} (2xy + y'''^2) dx$$

***Solution*** Here $F = 2xy + y'''^2$ and from Eq. (42), the corresponding Euler–Poisson's equation is

$$\frac{\partial F}{\partial y} - \frac{d}{dx}\left(\frac{\partial F}{\partial y'}\right) + \frac{d^2}{dx^2}\left(\frac{\partial F}{\partial y''}\right) - \frac{d^3}{dx^3}\left(\frac{\partial F}{\partial y'''}\right) = 0 \tag{53}$$

Now

$$\frac{\partial F}{\partial y} = 2x, \qquad \frac{\partial F}{\partial y'} = \frac{\partial F}{\partial y''} = 0, \qquad \frac{\partial F}{\partial y'''} = 2y'''$$

so that Eq. (53) reduces to

$$\frac{d^3}{dx^3} y''' = x$$

Integrating this equation, we get

$$y = \frac{x^7}{7!} + c_1 \frac{x^5}{5!} + c_2 \frac{x^4}{4!} + c_3 \frac{x^3}{3!} + c_4 \frac{x^2}{2!} + c_5 \frac{x}{1!} + c_6$$

as the required extremal.

## 10.5 VARIATIONAL PROBLEMS IN PARAMETRIC FORM

Let $x = x(t)$, $y = y(t)$, $t_0 \le t \le t_1$ be the parametric relationships between the variables $x$ and $y$. Consider the integral

$$I[y(x)] = \int_{x_1}^{x_2} F(x, y(x), y'(x))\, dx \tag{54}$$

where

$$y' = \frac{dy}{dx} = \frac{dy/dt}{dx/dt} = \frac{\dot{y}}{\dot{x}}, \qquad \dot{x}dt = dx \tag{55}$$

From Eq. (55), Eq. (54) takes the form

$$I = \int_{t_1}^{t_2} F\left(x, y, \frac{\dot{y}}{\dot{x}}\right) \dot{x}\, dt \tag{56}$$

We write

$$G(x, y, \dot{x}, \dot{y}) = F\left(x, y, \frac{\dot{y}}{\dot{x}}\right) \dot{x} \tag{57}$$

Then the corresponding Euler's equations for the extremum of Eq. (56) are

$$\frac{\partial G}{\partial x} - \frac{d}{dt}\left(\frac{\partial G}{\partial \dot{x}}\right) = 0 \tag{58a}$$

$$\frac{\partial G}{\partial y} - \frac{d}{dt}\left(\frac{\partial G}{\partial \dot{y}}\right) = 0 \tag{58b}$$

If $x = x(t)$, $y = y(t)$, ..., $z = z(t)$, then the extremal for the functional

$$I = \int_{t_1}^{t_2} F(x, y, ..., z, \dot{x}, \dot{y}, ..., \dot{z}, t)dt \tag{59}$$

are given by the Euler's equations

$$\begin{aligned} \frac{\partial F}{\partial x} - \frac{d}{dt}\left(\frac{\partial F}{\partial \dot{x}}\right) &= 0 \\ \frac{\partial F}{\partial y} - \frac{d}{dt}\left(\frac{\partial F}{\partial \dot{y}}\right) &= 0 \\ &\vdots \\ \frac{\partial F}{\partial z} - \frac{d}{dt}\left(\frac{\partial F}{\partial \dot{z}}\right) &= 0 \end{aligned} \tag{60}$$

**EXAMPLE 10.22** Find the extremal of the functional

$$I = \int_{t_1}^{t_2} (x\dot{y} - y\dot{x})\, dt$$

***Solution*** Here

$$G = x\dot{y} - y\dot{x}, \qquad \frac{\partial G}{\partial x} = \dot{y}, \qquad \frac{\partial G}{\partial y} = -\dot{x}, \qquad \frac{\partial G}{\partial \dot{x}} = -y, \qquad \frac{\partial G}{\partial \dot{y}} = x$$

and Eqs. (58) reduce to

$$\dot{y} - \frac{d}{dt}(-y) = 0, \qquad -\dot{x} - \frac{d}{dt}(x) = 0$$

so that $dy/dt = 0$ and $dx/dt = 0$. The solutions of these equations are $x = c_1$ and $y = c_2$ which on squaring and adding leads to

$$x^2 + y^2 = a^2$$

where $a^2 = c_1^2 + c_2^2$. Hence the extremal of the given functional is a circle.

## 10.6 ISOPERIMETRIC PROBLEM

The determination of the extremum conditions for the given functional

$$I[y(x)] = \int_{x_1}^{x_2} F(x, y(x), y'(x))\, dx \tag{61}$$

subject to the subsidiary conditions (isoperimetric constraints)

$$J[y(x)] = \int_{x_1}^{x_2} G(x, y(x), y'(x))\, dx \tag{62}$$

which possesses a given value, is called an *isoperimetric problem.* Here, both $y(x_1) = y_1$, $y(x_2) = y_2$ are given and the given functions $F$ and $G$ are both twice differentiable.

The variational problem *among all closed curves of a givem perimeter (length) find the one which encloses maximum area,* is the isoperimetric problem and hence the name, isoperimetric, i.e. with the same perimeter. This problem is sometimes referred to as *Dido's problem.*

The Euler's equation satisfied by the function $y = y(x)$ which extremize functional (61) under condition (62) is

$$\frac{\partial f^*}{\partial y} - \frac{d}{dx}\left(\frac{\partial f^*}{\partial y'}\right) = 0 \tag{63}$$

where $f^* = F + \lambda G$ and the constant $\lambda$ is the undetermined multiplier whose value has to be found by the condition of each individual problem. This isoperimetric problem may be generalised as follows:

$$I = \int_{x_1}^{x_2} F(x, y_1(x), y_2(x), \ldots, y_n(x), y_1'(x), y_2'(x), \ldots, y_n'(x))dx \tag{64}$$

subject to the conditions

$$y_i(x_1) = a_i, \qquad y_i(x_2) = b_i, \qquad (i = 1, 2, \ldots, n)$$

and

$$J[y(x)] = \int_{x_1}^{x_2} G_j(x, y_1(x), y_2(x), \ldots, y_n(x), y'_1(x), y'_2(x), \ldots, y'_n(x))dx = l_j \tag{65}$$

where $j = 1, 2, \ldots, k$ and $l_j$ are constants. Here $k$ may be greater than, equal to or less than $n$. Then the necessary conditions (Euler's equations) for extremum are

$$\frac{\partial}{\partial y_i}\left(F + \sum_{j=1}^{k} \lambda_j G_j\right) - \frac{d}{dx}\left[\frac{\partial}{\partial y_i'}\left(F + \sum_{j=1}^{k} \lambda_j G_j\right)\right] = 0 \tag{66}$$

The $2n$ constants in the solution of Eq. (66) and the values of the $k$ parameters $\lambda_1, \lambda_2, \ldots, \lambda_k$ are obtained from Eq. (65). The parameters $\lambda_1, \lambda_2, \ldots, \lambda_k$ are called the *Lagrange's multiplier.*

REMARKS. 1. It may be noted that the extremals for the functional

$$I = \int_{x_1}^{x_2} F(x, y(x), y'(x))\, dx$$

subject to the conditions (constraints)

$$J = \int_{x_1}^{x_2} G(x, y(x), y'(x))\, dx = \text{constant}$$

coincide with the extremals of $J$ under the constraint $I$ = constant. In such cases, the Euler's equations are same. This property is known as *Reciprocity principle*.

Thus, the problem of finding the curve enclosing the maximum area bounded by a closed curve of given perimeter and the problem of minimum length of a closed curve enclosing a given area are reciprocal. These two problems admit a common extremum—a circle.

2. The isoperimetric problem when the functions are defined in terms of the parametric relations takes the following form:

Consider the functional

$$I = \int_{t_1}^{t_2} F(x, y, ..., z, \dot{x}, \dot{y}, ..., \dot{z}, t)\, dt \tag{67}$$

subject to some given boundary conditions and the conditions

$$y_i(t_1) = A_i, \qquad y_i(t_2) = B_i, \qquad G_j(x, y, \ldots, z, \dot{x}, \dot{y}, \ldots, \dot{z}, t) = 0 \tag{68}$$

where $i = 1, 2, ..., n$ and $j = 1, 2, \ldots, k$. Then for the extremals of $I$ the necessary conditions are

$$\left.\begin{aligned} \frac{\partial f^*}{\partial x} - \frac{d}{dt}\left(\frac{\partial f^*}{\partial \dot{x}}\right) &= 0 \\ \frac{\partial f^*}{\partial y} - \frac{d}{dt}\left(\frac{\partial f^*}{\partial \dot{y}}\right) &= 0 \\ &\vdots \\ \frac{\partial f^*}{\partial z} - \frac{d}{dt}\left(\frac{\partial f^*}{\partial \dot{z}}\right) &= 0 \end{aligned}\right\} \tag{69}$$

where $f^* = F + \Sigma_{j=1}^{k} \lambda_i G_j$. The $2n$ constants $\lambda_j$ may be obtained by using the boundary conditions and the constraint conditions (68).

## Particular cases

(i) Consider the integral

$$I = \int_{t_1}^{t_2} F(y, \dot{y}, \ddot{y}, t)\, dt$$

and write $\dot{y} = z$ so that $z - \dot{y} = 0 \equiv g$ and

$$I = \int_{t_1}^{t_2} F(y, z, \dot{z}, t)\, dt$$

and the extremals satisfy the Euler's equation

$$\frac{\partial F}{\partial y} - \frac{d}{dt}\left(\frac{\partial F}{\partial \dot{y}}\right) + \frac{d^2}{dt^2}\left(\frac{\partial F}{\partial \ddot{y}}\right) = 0$$

and thus the necessary condition for the extremum of

$$I = \int_{t_1}^{t_2} F(y, \dot{y}, \ddot{y}, \ldots, y^{(n)}\, t)\, dt \tag{70}$$

is

$$\frac{\partial F}{\partial y} - \frac{d}{dt}\left(\frac{\partial F}{\partial \dot{y}}\right) + \frac{d^2}{dt^2}\left(\frac{\partial F}{\partial \ddot{y}}\right) + \cdots + (-1)^n \frac{d^n}{dt^n}\left(\frac{\partial F}{\partial y^{(n)}}\right) = 0 \tag{71}$$

(ii) Consider the functional

$$I = \int_{x_1}^{x_2} F(x, y, y', z, z')\, dx \tag{72}$$

subject to the conditions

$$G(x, y, z) = 0, \quad y(x_1) = a_1, \; y(x_2) = b_1, \quad z(x_1) = a_2, \quad z(x_2) = b_2 \tag{73}$$

Then $y = y(x)$, $z = z(x)$ are the extremals if

$$\left.\begin{aligned} \frac{\partial f^*}{\partial y} - \frac{d}{dx}\left(\frac{\partial f^*}{\partial y'}\right) = 0 \\ \frac{\partial f^*}{\partial z} - \frac{d}{dx}\left(\frac{\partial f^*}{\partial z'}\right) = 0 \end{aligned}\right\} \tag{74}$$

where $f^* = F + \lambda G$. The constants $\lambda$ may be obtained by using the boundary conditions.

**EXAMPLE 10.23** Of all closed non-intersecting plane curves whose length (perimeter) is given, find the one for which the enclosed area is maximum.

***Solution*** Consider the parametric representation

$$x = x(t), \quad y = y(t) \tag{75}$$

for any closed curve such that the functions (75) are continuous differentiable with respect to $t$. Since the curve is closed, we have

$$x(t_1) = x(t_2) = x_0, \quad y(t_1) = y(t_2) = y_0 \tag{76}$$

Without loss of generality, we can suppose that Eq. (75) describe any given curve in the anticlockwise sense as $t$ increases from $t_1$ to $t_2$.

It is known [16] that the area enclosed by a simple closed curve $C$ in $xy$-plane is

$$A = \frac{1}{2}\int_C (x\,dy - y\,dx) = \frac{1}{2}\int_{t_1}^{t_2}\left(x\frac{dy}{dt} - y\frac{dx}{dt}\right)dt = \frac{1}{2}\int_{t_1}^{t_2}(x\dot{y} - y\dot{x})\,dt \tag{77}$$

where $\dot{x} = dx/dt$, $\dot{y} = dy/dt$.

The total length of the curve $C$ is

$$B = \int_{t_1}^{t_2} ds = \int_{t_1}^{t_2}\sqrt{\dot{x}^2 + \dot{y}^2}\,dt \tag{78}$$

Now, we have to find the curve $C$ so that area $A$ is maximum (i.e. integral (77) has an extremum) along with the given length (perimeter) given by Eq. (78) such that the conditions (76) are satisfied.

Since we are considering the parametric representation, the Euler's equations are

$$\left.\begin{aligned} \frac{\partial f^*}{\partial x} - \frac{d}{dt}\left(\frac{\partial f^*}{\partial \dot{x}}\right) = 0 \\ \frac{\partial f^*}{\partial y} - \frac{d}{dt}\left(\frac{\partial f^*}{\partial \dot{y}}\right) = 0 \end{aligned}\right\} \tag{79}$$

and $$f^* = F + \sum_{j=1}^{k}\lambda_j G_j = \frac{1}{2}(x\dot{y} - y\dot{x}) + \lambda\sqrt{\dot{x}^2 + \dot{y}^2}$$

Thus

$$\left.\begin{aligned} &\frac{\partial f^*}{\partial x} = \frac{1}{2}\dot{y}, && \frac{\partial f^*}{\partial y} = -\frac{1}{2}\dot{x} \\ &\frac{\partial f^*}{\partial \dot{x}} = -\frac{1}{2}y + \frac{\lambda\dot{x}}{\sqrt{\dot{x}^2 + \dot{y}^2}}, && \frac{\partial f^*}{\partial \dot{y}} = \frac{1}{2}x + \frac{\lambda\dot{y}}{\sqrt{\dot{x}^2 + \dot{y}^2}} \end{aligned}\right\} \tag{80}$$

Using Eq. (80) in Eq. (79), we get after simplification

$$\frac{d}{dt}\left(y - \frac{\lambda \dot{x}}{\sqrt{\dot{x}^2 + \dot{y}^2}}\right) = 0, \quad \frac{d}{dt}\left(-x - \frac{\lambda \dot{y}}{\sqrt{\dot{x}^2 + \dot{y}^2}}\right) = 0 \tag{81}$$

Now integrating Eq. (81), we get

$$y - \frac{\lambda \dot{x}}{\sqrt{\dot{x}^2 + \dot{y}^2}} = k, \quad x + \frac{\lambda \dot{y}}{\sqrt{\dot{x}^2 + \dot{y}^2}} = h$$

or,

$$y - k = \frac{\lambda \dot{x}}{\sqrt{\dot{x}^2 + \dot{y}^2}}, \quad x - h = -\frac{\lambda \dot{y}}{\sqrt{\dot{x}^2 + \dot{y}^2}} \tag{82}$$

Squaring and adding Eq. (82), we get

$$(x - h)^2 + (y - k)^2 = \lambda^2 \tag{83}$$

which represents a circle with centre as $(h, k)$ and radius as

$$\lambda = \left(\frac{\text{Perimeter}}{2\pi}\right)^{1/2}$$

Hence, the closed curve of given perimeter which encloses maximum area is a circle.

## 10.7 CANONICAL FORM OF THE EULER'S EQUATION

In a number of problems of theoretical physics, it is often convenient to express the Euler's equations and also the variational principles (i.e. the extremal properties of certain functionals) in terms of some other variables which may have some specific physical meanings.

Consider the Euler's equation

$$\frac{\partial F}{\partial y} - \frac{d}{dx}\left(\frac{\partial F}{\partial y'}\right) = 0 \tag{14}$$

Here, $F = F(x, y, y')$. This equation can be expressed in another form by means of a proper transformation of variables. Let one such transformation of variable, instead of $x$ and $y$, be

$$p = \frac{\partial F}{\partial y'} \tag{84}$$

and

$$H = -F + y'\frac{\partial F}{\partial y'} = -F + y'p \tag{85}$$

From Eq. (84) Euler's equation (14) leads to

$$\frac{\partial F}{\partial y} = \frac{dp}{dx} \tag{86}$$

while from Eqs. (85) and (86), we have

$$\frac{\partial H}{\partial y} = -\frac{dp}{dx}, \quad \frac{\partial H}{\partial p} = \frac{dy}{dx} \tag{87}$$

The system of Eqs. (87) is equivalent to the second-order Euler's equation (14). This system is known as *Hamiltonian* or *canonical* form of the Euler's equation (14). The variables $H$ and $p$ are called canonical variables and the quantity $H$ defined by Eq. (85) is known as the *Hamiltonian.*

Equation (87) may be generalized to the functionals which depends upon $n$ unknown functions. We have seen earlier that the Euler's equation corresponding to the functional

$$I = \int_{x_1}^{x_2} F(x, y_1(x), y_2(x), \ldots, y_n(x), y_1'(x), y_2'(x), \ldots, y_n'(x))\, dx \tag{38}$$

which depends on $n$ functions $y_j(x)$ and their derivatives $y'_j(x)$ satisfying the boundary conditions

$$y_j(x_1) = A_j,\ y_j(x_2) = B_j \tag{39}$$

is

$$\frac{\partial F}{\partial y_j} - \frac{d}{dx}\left(\frac{\partial F}{\partial y_j'}\right) = 0 \tag{40}$$

which is a system of $n$ second-order differential equations.

This system can be reduced to a system of first-order differential equation if we replace $x$, $y_j$, $y'_j$ by $x$, $y_j$, $p_j$, where $j = 1, 2, \ldots, n$ and $p_j$ is defined as

$$p_j = \frac{\partial F}{\partial y_j'} \tag{88}$$

Corresponding to the functional (38), we define Hamiltonian as

$$H = -F(y_j, y_j', x) + \sum_{i=1}^{n} y_j' p_j \tag{89}$$

and the canonical form of the Euler's equation is

$$\frac{\partial H}{\partial y_j} = -\frac{dp_j}{dx}, \quad \frac{\partial H}{\partial p_j} = \frac{dy_j}{dx} \tag{90}$$

Equation (90) is a system of $2n$ first-order differential equations equivalent to Eq. (40). The variables $x$, $y_j$, $p_j$, $H$ are the canonical variable corresponding to the functional (38).

*Alternative derivation of equation (90)*: Consider $H$ to be a function of $y_j$, $p_j$ and $x$, i.e.

$$H = H(y_j, p_j, x) \tag{91}$$

Taking the differential of Eq. (91), we get

$$dH = \frac{\partial H}{\partial y_j} dy_j + \frac{\partial H}{\partial p_j} dp_j + \frac{\partial H}{\partial x} dx \tag{92}$$

Now taking the differential of Eq. (89), we have

$$dH = y_j' \, dp_j + p_j dy_j' - \frac{\partial F}{\partial y_j} dy_j - \frac{\partial F}{\partial y_j'} dy_j' - \frac{\partial F}{\partial x} dx \tag{93}$$

But $\quad p_j = \dfrac{\partial F}{\partial y_j'}, \quad p_j' = \dfrac{\partial F}{\partial y_j}$

and Eq. (93) thus reduces to

$$dH = y_j' dp_j - p_j' dy_j - \frac{\partial F}{\partial x} dx \tag{94}$$

Hence, from Eqs. (92) and (94), we have

$$-p_j' = \frac{\partial H}{\partial y_j}, \quad y_j' = \frac{\partial H}{\partial p_j}, \quad \frac{\partial H}{\partial x} = -\frac{\partial F}{\partial x} \tag{95}$$

which are the required canonical form of the Euler's equation (40).

## 10.8 FUNCTIONALS DEPENDING ON FUNCTIONS OF SEVERAL INDEPENDENT VARIABLES

So far we have dealt with various forms of Euler's equations for determining the extremals of various kinds of functionals. These various forms of Euler's equations are ordinary differential equations. We shall now consider the problem of finding the extremal of the functional which involve multiple integrals.

Consider the functional of the form

$$I\,[z = f(x, y)] = \iint_D F\left(x, y\, z, \frac{\partial z}{\partial x}, \frac{\partial z}{\partial y}\right) dx\, dy \tag{96}$$

defined on some surface $z = f(x, y)$. Let $z = f(x, y)$ be given on the boundary of the domain $D$ (i.e. the edges of the surface are fixed), then the necessary condition for the extremal of the functional (96) is

$$\frac{\partial F}{\partial z} - \frac{\partial}{\partial x}\left(\frac{\partial F}{\partial p}\right) - \frac{\partial}{\partial y}\left(\frac{\partial F}{\partial q}\right) = 0 \tag{97}$$

where

$$p = \frac{\partial z}{\partial x}, \quad q = \frac{\partial z}{\partial y}$$

In fact, here, we are solving the problem: Find the surface with fixed edges $z = f(x, y)$ on which the functional (96) takes the extremum value {for the proof of Eq. (97), see [7]}.

Equation (97) is a second-order partial differential equation, in contrast to a second-order ordinary differential equation (14), known as *Ostrogradsky equation*; named after Russian mathematician M.V. Ostrogradsky who obtained the general formula for the variation of multiple integrals in 1834. The solution of Eq. (97) give rise to the extremal of functional (96). Hence, the solution of the variational problem reduces to the solution of partial differential equation (97).

For the extremal of the functional

$$I[z=z(x_1, x_2, \ldots, x_n)] = \int_D \cdots \int F(x_1, x_2, \ldots, x_n, z, p_1, p_2, \ldots, p_n)\, dx_1 dx_2, \ldots, dx_n \tag{98}$$

the corresponding Ostrogradsky equation (or, the necessary condition) is

$$\frac{\partial F}{\partial z} - \sum_{i=1}^{n} \frac{\partial}{\partial x_i}\left(\frac{\partial F}{\partial p_i}\right) = 0 \tag{99}$$

where

$$p_i = \frac{\partial z}{\partial x_i}$$

If the integrand $F$ of the functional $I$ depends on higher derivatives, then for the extremal of the functional we have an equation similar to that of Euler–Poisson's equation (42). For example, for the extremal of the functional

$$I[z(x, y)] = \iint_D F\left(x, y, z, \frac{\partial z}{\partial x}, \frac{\partial z}{\partial y}, \frac{\partial^2 z}{\partial x^2}, \frac{\partial^2 z}{\partial x \partial y}, \frac{\partial^2 z}{\partial y^2}\right) dx\, dy \tag{100}$$

the necessary condition (or the Ostrogradsky equation) is

$$\frac{\partial F}{\partial z} - \frac{\partial}{\partial x}\left(\frac{\partial F}{\partial p}\right) - \frac{\partial}{\partial y}\left(\frac{\partial F}{\partial q}\right) + \frac{\partial^2}{\partial x^2}\left(\frac{\partial F}{\partial r}\right) + \frac{\partial^2}{\partial x \partial y}\left(\frac{\partial F}{\partial s}\right) + \frac{\partial^2}{\partial y^2}\left(\frac{\partial F}{\partial t}\right) = 0 \tag{101}$$

where

$$p = \frac{\partial z}{\partial x}, \quad q = \frac{\partial z}{\partial y}, \quad r = \frac{\partial^2 z}{\partial x^2}, \quad s = \frac{\partial^2 z}{\partial x \partial y}, \quad t = \frac{\partial^2 z}{\partial y^2} \tag{102}$$

**EXAMPLE 10.24** Obtain Ostrogradsky equation for the functional

$$I[z(x, y)] = \iint_D \left[\left(\frac{\partial z}{\partial x}\right)^2 + \left(\frac{\partial z}{\partial y}\right)^2\right] dx\, dy$$

***Solution*** The given functional $I$ can be written as

$$I = \iint_D (p^2 + q^2)\, dx\, dy$$

Here $F = p^2 + q^2$, $\partial F/\partial p = 2p$, $\partial F/\partial q = 2q$ and Eq. (97) thus leads to

$$\frac{\partial^2 z}{\partial x^2} + \frac{\partial^2 z}{\partial y^2} = 0 \tag{103}$$

Equation (103) is the well-known Laplace's equation in two dimensions [cf. Eq. (120), Chapter 9)]. This equation is often expressed as

$$\nabla^2 z = 0 \tag{104}$$

where

$$\nabla^2 = \frac{\partial^2}{\partial x^2} + \frac{\partial^2}{\partial y^2}$$

is the two-dimensional Laplacian.

**EXAMPLE 10.25** Obtain Ostrogradsky equation for the functional

$$I = \iint_D \left[\left(\frac{\partial z}{\partial x}\right)^2 + \left(\frac{\partial z}{\partial y}\right)^2 + 2zf(x,y)\right] dx\,dy$$

***Solution*** The given functional $I$ can be expressed as

$$I = \iint_D \left[p^2 + q^2 + 2z\,f(x,y)\right] dx\,dy$$

Here, $F = 2\,zf(x, y) + p^2 + q^2$, $\partial F/\partial z = 2f(x, y)$, $\partial F/\partial p = 2p$, $\partial F/\partial q = 2q$. With these values, Eq. (97) thus reduces to

$$\frac{\partial^2 z}{\partial x^2} + \frac{\partial^2 z}{\partial y^2} = f(x,y)$$

which may also be expressed as

$$\nabla^2 z = f(x,y) \tag{105}$$

Equation (105) is known as *Poisson's equation* and reduces to Laplace's equation when $f(x, y) = 0$.

**EXAMPLE 10.26** Find Ostrogradsky equation for the functional

$$I = \iint_D \left[\left(\frac{\partial^2 z}{\partial x^2}\right)^2 + \left(\frac{\partial^2 z}{\partial y^2}\right)^2 + 2\left(\frac{\partial^2 z}{\partial x \partial y}\right)^2\right] dx\,dy$$

***Solution*** Using the notation of Eq. (102), the given integral can be expressed as

$$I = \iint_D (r^2 + t^2 + 2s^2)\, dx\,dy$$

Here, $F = r^2 + t^2 + 2s^2$. Then

$$\frac{\partial F}{\partial z} = 0, \quad \frac{\partial F}{\partial p} = 0, \quad \frac{\partial F}{\partial q} = 0, \quad \frac{\partial F}{\partial r} = 2r, \quad \frac{\partial F}{\partial s} = 4s, \quad \frac{\partial F}{\partial t} = 2t$$

Thus Ostrogradsky equation (101) leads to

$$\frac{\partial^2 r}{\partial x^2} + 2\frac{\partial^2 s}{\partial x \partial y} + \frac{\partial^2 t}{\partial y^2} = 0$$

or
$$\frac{\partial^4 z}{\partial x^4} + 2\frac{\partial^4 z}{\partial x^2 \partial y^2} + \frac{\partial^4 z}{\partial y^4} = 0$$

which may also be expressed as

$$\nabla^2 \nabla^2 z = 0 \qquad (106)$$

Equation (106) is known as *biharmonic equation.*

## 10.9 LAGRANGE'S EQUATIONS OF MOTION

In mechanics there is an important application of Eq. (40). It is known that the Newton's law F = *ma* is a fundamental equation in elementary physics. To describe the motion of the mechanical system for which all the forces (except for the forces due to constraints) are derivable from generalised scalar potential that may be function of coordinates, velocities and time, we need an integral principle. This integral principle, given by W. R. Hamilton (1805 – 1865) is one of the most and important principle of mechanics and mathematical physics. This principle (a variational principle) is called *Hamilton's principle.* It states that the motion of a particle or system of particles is such that $\int_{t_1}^{t_2} L dt$ is stationary, where $L = T - V$ is the Lagrangian; $T$ is the kinetic energy and $V$ is the potential energy of the particle or system.

Hamilton's principle may also be explained as follows:

Out of all possible paths by which the particle or system of particles could travel from its position at time $t_1$ to its position at time $t_2$, it actually travelled along the path for which the line integral

$$I = \int_{t_1}^{t_2} L dt \qquad (107)$$

is an extremum. The integral $I$ in Eq. (107) is often known as *action.*

If we make the following identification

$$x \to t,\ y_j \to q_j,\ F(y_j, y_j', x) \to L(q_j, \dot{q}_j, t) \qquad (108)$$

in Eq. (38), then Eq. (40) leads to

$$\frac{\partial L}{\partial q_j} - \frac{d}{dt}\left(\frac{\partial L}{\partial \dot{q}_j}\right) = 0 \tag{109}$$

where $q_j, j = 1, 2, \ldots, n$ are the generalised coordinates.

Equation (109), which are the extremum condition for the line integral (107) are known as *Lagrange's equations of motion.*

We shall now discuss the applications of Lagrange's equations (109) as follows:

**EXAMPLE 10.27** A particle of mass $m$ is projected with an initial velocity $u$ making and angle $\alpha$ with the horizontal. Neglecting the resistance due to air, discuss the motion of the projectile using Lagrange's equations (see also Section 7.5).

***Solution*** Let $P(x, y)$ be the position of the particle at time $t$, then $T = \frac{1}{2}m(\dot{x}^2 + \dot{y}^2), V = mgy$ and

$$L = T - V = \frac{1}{2}m(\dot{x}^2 + \dot{y}^2) - mgy \tag{110}$$

Equation (109) in this case are

$$\frac{d}{dt}\left(\frac{\partial L}{\partial \dot{x}}\right) - \frac{\partial L}{\partial x} = 0 \tag{111a}$$

$$\frac{d}{dt}\left(\frac{\partial L}{\partial \dot{y}}\right) - \frac{\partial L}{\partial y} = 0 \tag{111b}$$

Solving these equations using Eq. (110), we get

$$\ddot{x} = 0 \quad \text{and} \quad \ddot{y} = -g \tag{112}$$

The initial conditions are

$$x = 0, y = 0, \dot{x} = u\cos\alpha, \dot{y} = u\sin\alpha \text{ when } t = 0 \tag{113}$$

Solving the pair of equations in Eq. (112) using Eq. (113), we get

$$\frac{dx}{dt} = u\cos\alpha, \frac{dy}{dt} = u\sin\alpha - gt \tag{114}$$

$$x = u\cos\alpha t, \; y = u\sin\alpha t - \frac{1}{2}gt^2 \tag{115}$$

Eliminating $t$ from Eq. (115), we get

$$y = x\tan\alpha - \frac{g}{2u}x^2\sec^2\alpha \tag{116}$$

which is the path of the projectile.

Equations (114)–(116) provide all information about the motion of the projectile.

**EXAMPLE 10.28** Obtain the equation of motion for a particle of mass $m$ moving, under gravity, near the earth.

***Solution*** Here

$$L = T - V = \frac{1}{2}m(\dot{x}^2 + \dot{y}^2 + \dot{x}^2) - mgz \tag{117}$$

and the Lagrange's Eq. (109) are given by

$$\begin{aligned} \frac{d}{dt}\left(\frac{\partial L}{\partial \dot{x}}\right) - \frac{\partial L}{\partial x} &= 0 \\ \frac{d}{dt}\left(\frac{\partial L}{\partial \dot{y}}\right) - \frac{\partial L}{\partial y} &= 0 \\ \frac{d}{dt}\left(\frac{\partial L}{\partial \dot{z}}\right) - \frac{\partial L}{\partial z} &= 0 \end{aligned} \tag{118}$$

From Eqs. (117) and (118), we have

$$\ddot{x} = 0,\ \ddot{y} = 0,\ \ddot{z} + g = 0 \tag{119}$$

so that

$$\dot{x} = \text{constant},\ \dot{y} = \text{constant},\ \ddot{z} = -g \tag{120}$$

which shows that the horizontal velocity is constant and the vertical acceleration is $-g$ in the gravitational field near the surface of the earth.

**EXAMPLE 10.29** Two mass particles of masses $m_1$ and $m_2$ are connected by a string passing through a hole in a smooth table so that $m_1$ rest on the table surface and $m_2$ hangs suspended. Assumimg mass $m_2$ moves up and down in a verticle line, find the Lagrange's equations of motion and interpret.

***Solution*** Let $P(r, \theta)$ be the coordinates of the mass $m_1$ lying on the table (Fig. 10.5). Let $PQ = 2l$, the length of the string and $OP = r$ so that $OQ = PQ - OP = 2l - r$.

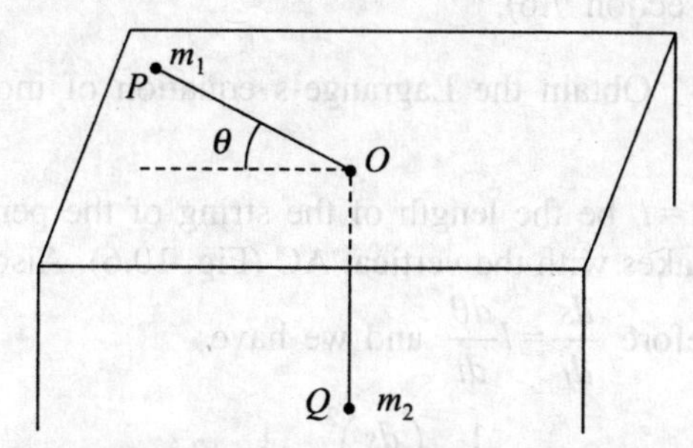


**Fig. 10.5** Diagram for Example 10.29.

The mass $m_1$ has radial and transverse components of velocity as $\dot{r} = \dfrac{dr}{dt}$

and $r\dot{\theta} = r\dfrac{d\theta}{dt}$, respectively, while the mass $m_2$ has downward velocity as $\dfrac{d}{dt}(OQ) = \dfrac{d}{dt}(2l - r) = -\dot{r}$.

Thus, for $m_1$ and $m_2$, the kinetic energies, respectively, are

$$T_1 = \frac{1}{2}m_1(\dot{r}^2 + r^2\dot{\theta}^2) \text{ and } T_2 = \frac{1}{2}m_2\dot{r}^2$$

while the potential energies are

$$V_1 = 0 \text{ and } V_2 = -m_2 g(2l - r)$$

Therefore, the Lagrangian is

$$L = T - V = \frac{1}{2}m_1(\dot{r}^2 + r^2\dot{\theta}^2) + \frac{1}{2}m_2\dot{r}^2 + m_2 g(2l - r) \tag{121}$$

where $T = T_1 + T_2$ and $V = V_1 + V_2$.

The Lagrange's equation (109) in term of polar coordinates $(r, \theta)$ are

$$\frac{d}{dt}\left(\frac{\partial L}{\partial \dot{r}}\right) - \frac{\partial L}{\partial r} = 0 \tag{122}$$

$$\frac{d}{dt}\left(\frac{\partial L}{\partial \dot{\theta}}\right) - \frac{\partial L}{\partial \theta} = 0 \tag{123}$$

From Eqs. (121) and (122), we have

$$(m_1 + m_2)\ddot{r} - m_1 r\dot{\theta}^2 = -m_2 g \tag{124}$$

while from Eqs. (121) and (123), we have

$$m_1 r^2\dot{\theta} = \text{constant} \tag{125}$$

Equations (124) and (125) are required Lagrange's equations of motion. Equation (125) shows that the areal velocity is constant and thus the mass $m_1$ is moving under the influence of a central force attracted towards the fixed point O (see also section 7.6).

**EXAMPLE 10.30** Obtain the Lagrange's equation of motion for a simple pendulum.

***Solution*** Let $AB = l$ be the length of the string of the pendulum and $\theta$, the angle which AB makes with the vertical AC (Fig. 10.6). Also, let $\text{arc}(BD) = s$.

Since $s = l\theta$, therefore $\dfrac{ds}{dt} = l\dfrac{d\theta}{dt}$ and we have

$$T = \frac{1}{2}m\left(\frac{ds}{dt}\right)^2 = \frac{1}{2}ml^2\dot{\theta}^2$$

and $$V = mg(CD) = mg(AD - AC) = mg(AB - AC) = mg(l - l\cos\theta)$$

so that
$$L = T - V = \frac{1}{2}ml^2\dot{\theta}^2 - mg(l - l\cos\theta) \quad (126)$$

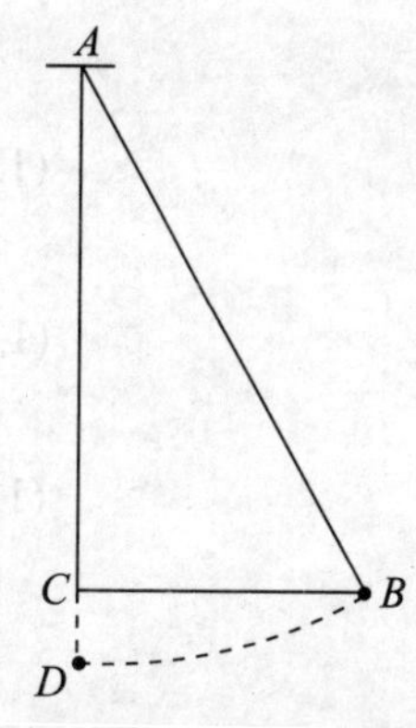


Fig. 10.6 Simple pendulum.

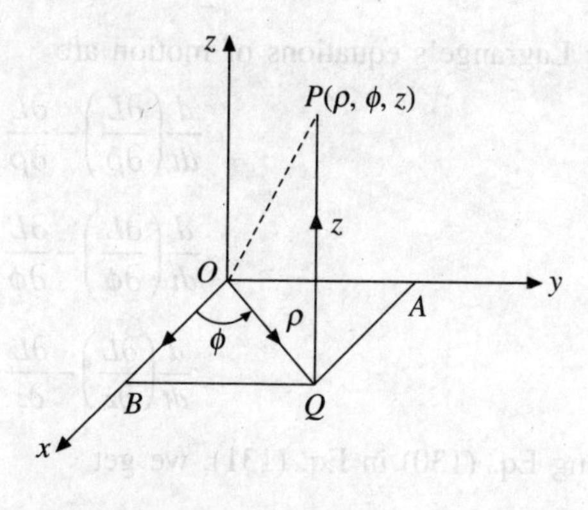


Fig. 10.7 Cylindrical coordinates.

Here, the Lagrange's equation is

$$\frac{d}{dt}\left(\frac{\partial L}{\partial \dot{\theta}}\right) - \frac{\partial L}{\partial \theta} = 0 \quad (127)$$

which using Eq. (126) leads to

$$\ddot{\theta} + \frac{g}{l}\sin\theta = 0 \quad (128)$$

Equation (128) is the Lagrange's equation of motion for a simple pendulum. For a complete solution of Eq. (128), the reader is referred to Section 6.2.

**EXAMPLE 10.31** A particle of mass $m$ is moving in a conservative force field. Find the Lagrangian and Lagrange's equations of motion in cylindrical coordinates $(\rho, \phi, z)$.

***Solution*** Let $\angle BOQ = \phi$, $\mathbf{OP} = \mathbf{r}$, $OQ = \rho$, $QP = z$, then from the Fig. 10.7, we have

$$\mathbf{OP} = \mathbf{OQ} + \mathbf{QP} = (\mathbf{OB} + \mathbf{BQ}) + \mathbf{QP} \quad (129)$$

Also, from $\Delta BOQ$, $\mathbf{OB} = \rho\cos\phi\hat{i}$, $\mathbf{BQ} = \rho\sin\phi\hat{j}$ and Eq. (129) now becomes

$$OP = \mathbf{r} = \rho\cos\phi\hat{i} + \rho\sin\phi\hat{j} + z\hat{k}$$

so that
$$\mathbf{v} = \frac{d\mathbf{r}}{dt} = (\dot{\rho}\cos\phi - \rho\sin\phi\dot{\phi})\hat{i} + (\dot{\rho}\sin\phi + \rho\cos\phi\dot{\phi})\hat{j} + \dot{z}\hat{k}$$

and
$$T = \frac{1}{2}mv^2 = \frac{1}{2}m(\dot{\rho}^2 + \rho^2\dot{\phi}^2 + \dot{z}^2)$$

Also, since the force field is conservative, we have $V = V(\rho, \phi, z)$. Thus, the Lagrangian is given by

$$L = T - V = \frac{1}{2}m(\dot{\rho}^2 + \rho^2\dot{\phi}^2 + \dot{z}^2) - V(\rho,\phi,z) \tag{130}$$

The Lagrange's equations of motion are

$$\frac{d}{dt}\left(\frac{\partial L}{\partial \dot{\rho}}\right) - \frac{\partial L}{\partial \rho} = 0 \tag{131a}$$

$$\frac{d}{dt}\left(\frac{\partial L}{\partial \dot{\phi}}\right) - \frac{\partial L}{\partial \phi} = 0 \tag{131b}$$

$$\frac{d}{dt}\left(\frac{\partial L}{\partial \dot{z}}\right) - \frac{\partial L}{\partial z} = 0 \tag{131c}$$

Using Eq. (130) in Eq. (131), we get

$$m[\ddot{\rho} - \rho\dot{\phi}^2] = -\frac{\partial V}{\partial \rho} \tag{132a}$$

$$m\rho^2\ddot{\phi} = -\frac{\partial V}{\partial \phi} \tag{132b}$$

$$m\ddot{z} = -\frac{\partial V}{\partial z} \tag{132c}$$

Equation (132) are the required Lagrange's equations of motion.

**EXAMPLE 10.32** A particle is constrained to move on the plane curve $xy = c$ where $c$ is a constant. Obtain Lagrange's equation of motion.

***Solution*** Given that the particle is constrained to move on the plane curve

$$y = \frac{c}{x} \tag{133}$$

Also

$$T = \frac{1}{2}mv^2 = \frac{1}{2}m(\dot{x}^2 + \dot{y}^2) \tag{134}$$

$$V = mgy \tag{135}$$

Here $x$ and $y$ are not linearly independent as they are related through the equation of contraints given by Eq. (133). Eliminating $y$ from Eqs. (134) and (135) by using Eq. (133), we get

$$T = \frac{1}{2}m\dot{x}^2(1 + \frac{c^2}{x^4}),\ V = mgc(\frac{1}{x})$$

so that

$$L = T - V = \frac{1}{2}m\dot{x}^2(1 + \frac{c^2}{x^4}) - mgc(\frac{1}{x}) \tag{136}$$

and the Lagrange's equation of motion

$$\frac{d}{dt}\left(\frac{\partial L}{\partial \dot{x}}\right) - \frac{\partial L}{\partial x} = 0$$

using Eq. (136) reduces to

$$m\ddot{x}\left(1 + \frac{c^2}{x^4}\right) - \frac{2c^2 m}{x^5}\dot{x}^2 - \frac{mcg}{x^2} = 0$$

which is the required Lagrange's equation of motion.

**EXAMPLE 10.33** Two masses $m_1$ and $m_2$ are attached to the ends of a string of constant length $l$. Mass $m_1$ moves without friction on the surface of a cone, while the mass $m_2$ can move vertically up and down. For this system, obtain the Lagrange's equations of motion.

***Solution*** Consider the spherical coordinates $(r, \theta, \phi)$ for the mass $m_1$ and $z$ for the mass $m_2$, Since

$$ds^2 = dr^2 + r^2 d\theta^2 + r^2 \sin^2 \theta d\phi^2$$

therefore, for mass $m_1$

$$v^2 = \left(\frac{ds}{dt}\right)^2 = \dot{r}^2 + r^2\dot{\theta}^2 + r^2 \sin^2 \theta\dot{\phi}^2$$

while for $m_2$

$$v^2 = \left(\frac{dz}{dt}\right)^2 = \dot{z}^2$$

Also, the total kinetic energy of the system is equal to the kinetic energy of mass $m_1$ plus the kinetic energy for the mass $m_2$. Thus

$$T = \frac{1}{2}m_1(\dot{r}^2 + r^2\dot{\theta}^2 + r^2 \sin^2\theta\dot{\phi}^2) + \frac{1}{2}m_2\dot{z}^2$$

while, the total potential energy of the system is equal to the potential energy of mass $m_1$ plus the potential energy for the mass $m_2$. Thus

$$V = m_1 gr \cos\theta - m_2 gz$$

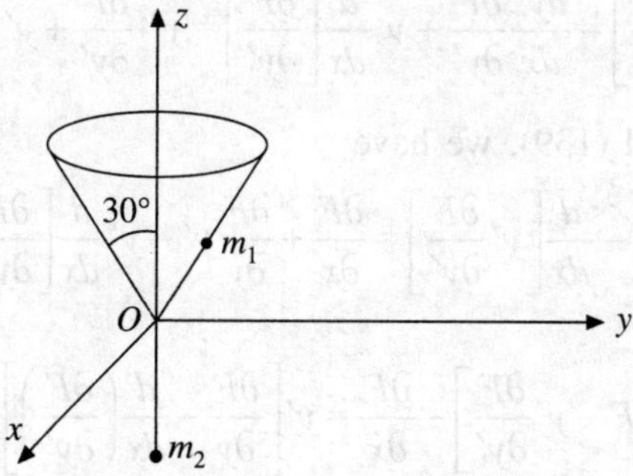


**Figure 10.8** Surface of a cone.

In this problem we are using four variables: $r$, $\theta$, $\phi$ and $z$ and we may have four Lagrange's equations of motion; but this is not so, as the motion is constrained. Thus, we have to eliminate $\theta$ and either $r$ or $z$. For a cone (see Fig. 10.8) $\theta$ = 30°, sin $\theta$ = 1/2, $\cos\theta = \sqrt{3}/2$, $\dot{\theta} = 0$ and $z = -|z| = -(l - r)$ so that $\dot{z} = \dot{r}$ and thus the Lagrangian can be expressed as

$$L = T - V = \frac{1}{2}m_1\left(\dot{r}^2 + \frac{r^2\dot{\phi}^2}{4}\right) + \frac{1}{2}m_2\dot{r}^2 - \frac{1}{2}m_1 gr\sqrt{3} + m_2 g(l - r)$$

Hence, we have two Lagrange's equations of motion in terms of $r$ and $\phi$ which are given by

$$\frac{d}{dt}[(m_1 + m_2)\dot{r}] - \frac{m_1 r\dot{\phi}^2}{4} + \left(\frac{\sqrt{3}}{2}m_1 + m_2\right)g = 0$$

and

$$\frac{d}{dt}\left(\frac{m_1}{4}\rho^2\dot{\phi}\right) = 0 \Rightarrow \rho^2\dot{\phi} = \text{constant}$$

## 10.10 MISCELLANEOUS SOLVED EXAMPLES

In this section, we shall solve a number of problems which are based on different sections of this chapter and we have

**EXAMPLE 10.34** Show that Eq. (14) can also be expressed as

(i) $$\frac{d}{dx}\left[F - y'\frac{\partial F}{\partial y'}\right] - \frac{\partial F}{\partial x} = 0 \text{ and}$$

(ii) $$\frac{\partial F}{\partial y} - \frac{\partial^2 F}{\partial x \partial y} - y'\frac{\partial^2 F}{\partial y \partial y'} - y''\frac{\partial^2 F}{\partial y'^2} = 0 \tag{137}$$

***Solution*** (i) Here $F = F(x, y, y')$ and thus from the chain rule for partial differentiation, we have

$$\frac{dF}{dx} = \frac{\partial F}{\partial x}\frac{dx}{dx} + \frac{\partial F}{\partial y}\frac{dy}{dx} + \frac{\partial F}{\partial y'}\frac{dy'}{dx} = \frac{\partial F}{\partial x} + \frac{\partial F}{\partial y}y' + \frac{\partial F}{\partial y'}y'' \tag{138}$$

Also

$$\frac{d}{dx}\left[y'\frac{\partial F}{\partial y'}\right] = \frac{dy'}{dx}\frac{\partial F}{\partial y'} + y'\frac{d}{dx}\left[\frac{\partial F}{\partial y'}\right] = y''\frac{\partial F}{\partial y'} + y'\frac{d}{dx}\left[\frac{\partial F}{\partial y'}\right] \tag{139}$$

From Eqs. (138) and (139), we have

$$\frac{dF}{dx} - \frac{d}{dx}\left[y'\frac{\partial F}{\partial y'}\right] = \frac{\partial F}{\partial x} + \frac{\partial F}{\partial y}y' - y'\frac{d}{dx}\left[\frac{\partial F}{\partial y'}\right]$$

which leads to

$$\frac{d}{dx}\left[F - y'\frac{\partial F}{\partial y'}\right] - \frac{\partial F}{\partial x} = y'\left[\frac{\partial F}{\partial y} - \frac{d}{dx}\left(\frac{\partial F}{\partial y'}\right)\right] = 0$$

[from Eq. (14)].

(ii) Since $\frac{\partial F}{\partial y'} = F(x, y, y')$, we have

$$\frac{d}{dx}\left[\frac{\partial F}{\partial y'}\right] = \frac{\partial}{\partial x}\left[\frac{\partial F}{\partial y'}\right]\frac{dx}{dx} + \frac{\partial}{\partial y}\left[\frac{\partial F}{\partial y'}\right]\frac{dy}{dx} + \frac{\partial}{\partial y'}\left[\frac{\partial F}{\partial y'}\right]\frac{dy'}{dx} = \frac{\partial^2 F}{\partial x \partial y'} + \frac{\partial^2 F}{\partial y \partial y'}y' + \frac{\partial^2 F}{\partial y'^2}y''$$

Using this value in Eq. (14), we get the required result.

**EXAMPLE 10.35** Find the extremals of the following functionals

(i) $I = \int_{x_1}^{x_2} (y + \frac{y^3}{3})\, dx$

(ii) $I = \int_1^3 (3xy - y^2)\, dx$ such that $y = 1$ when $x = 1$ and $y = 9/2$ when $x = 3$.

(iii) $I = \int_0^1 (e^y + xy')\, dx$ such that $y = 0$ when $x = 0$ and $y = 0$ when $x = 1$.

***Solution***

(i) Here $F = y + \frac{y^3}{3}$ and Eq. (14) leads to $y^2 + 1 = 0$ which on solving yields $y = \pm i$. This is not possible and therefore the given functional has no extremal.

(ii) Here $F = 3xy - y^2$ and Eq. (14) in this case leads to $3x - 2y = 0$ which is the only extremal for the given functional. But this extremal does not satisfy the given condition that $y = 1$ when $x = 1$. Thus, this problem has no solution.

(iii) In this case, Euler's Eq. (14) leads to $e^y = 1$ or $y = 0$ which is the only extremal. Moreover, the given conditions are also satisfied. Hence, $y = 0$ is the required extremal.

**EXAMPLE 10.36** Find the curve passing through the points $(0,2)$ and $(2,6)$ such that its length between these points is minimum.

***Solution*** From Example 10.14, the equation of the curve is $y = ax + b$ where $a$ and $b$ are constants. Since the curve passes through the points $(0,2)$ and $(2,6)$, we obtain $a = 2$ and $b = 2$ and therefore the required curve is $y = 2x + 2$ (a straight line).

**EXAMPLE 10.37** Find the extremal of the functional $I = \int_0^1 (y'^2 + 12xy)\, dx$ such that $y(0) = 0, y(1) = 1$.

***Solution*** From Euler's Eq. (14), we have

$$y'' = 6x$$

which on integration leads to

$$y = x^3 + c_1 x + c_2 \tag{140}$$

Using the given conditions in Eq. (140), we get $c_1 = c_2 = 0$ and thus the required extremal is $y = x^3$.

**EXAMPLE 10.38** Find the extremal of the functional $I = \int_0^{\pi/2} (y'^2 - y^2 + 2xy)\, dx$ such that $y(0) = 0, y(\pi/2) = 0$.

***Solution*** Here, $F = y'^2 - y^2 + 2xy$ and Euler's Eq. (14) leads to

$$y'' + y = x$$

which is second order differential equation whose solution is

$$y = y_c + y_p = c_1 \cos x + c_2 \sin x + x \tag{141}$$

Using the given conditions in Eq. (141), we get $c_1 = 0$ and $c_2 = N\pi/2$. Thus, the required extremal is $y = x - \frac{\pi}{2} \sin x$.

**EXAMPLE 10.39** If $F$ in Euler's Eq. (14) is depending only on $y'$ then show that the extremals are straight lines.

***Solution*** Here we shall use Eq. (137). Since $F = F(y')$, therefore $\frac{\partial F}{\partial y} = 0, \frac{\partial F}{\partial x} = 0, \frac{\partial^2 F}{\partial x \partial y'} = 0$ and $\frac{\partial^2 F}{\partial y \partial y'} = 0$. Eq. (137) now reduces to

$$y'' \frac{\partial^2 F}{\partial y'^2} = 0$$

This equation leads to

*Case (i):* If $y'' = 0$ then $y = ax + b$ which represents a two-parameter family of straight lines.

*Case (ii):* If $\frac{\partial^2 F}{\partial y'^2} = 0$. This equation leads to one or several real roots $y' = c_n$ so that $y = c_n x + d$ which is again a straight line. Thus, if $F = F(y')$ then the extremals are always straight lines.

**EXAMPLE 10.40** In polar coordinates, the arc length of a curve joining two points $(r_1, \theta_1)$ and $(r_2, \theta_2)$ is given by

$$I = \int_{\theta_1}^{\theta_2} [r^2 + (\frac{dr}{d\theta})^2]^{1/2}\, d\theta$$

Find the extremal of this functional.

***Solution*** Here

$$F = F(r, \theta, r') = (r^2 + r'^2)^{1/2} \tag{142}$$

where $r' = \dfrac{dr}{d\theta}$ and Euler's equation, in terms of $(r, \theta)$

$$\frac{\partial F}{\partial r} - \frac{d}{d\theta}\left[\frac{\partial F}{\partial r'}\right] = 0 \tag{144}$$

leads to

$$rr'' - 2r' - r^2 = 0 \tag{143}$$

To obtain the solution of Eq. (143), put $p = r' = \dfrac{dr}{d\theta}$ then $r'' = r'\dfrac{dp}{dr}$ so that Eq. (143) reduces to

$$\frac{dp}{dr} = \frac{2p}{r} + \frac{r}{p} \tag{144}$$

Now take $t = p/r$ so that $p = rt$ and Eq, (144) becomes

$$t + r\frac{dt}{dr} = 2t + \frac{1}{t}$$

Separating the variables, we get

$$\frac{2dr}{r} = \frac{2tdt}{t^2 + 1}$$

which on integration yields

$$2\log r = \log a^2 + \log(t^2 + 1)$$

This equation may also be expressed as

$$\log(t^2 + 1) = \log\frac{r^2}{a^2}$$

which after simplification leads to

$$d\theta = a\frac{dr}{r(r^2 - a^2)^{1/2}}$$

Integrating this equation, we get

$$\theta = a\left[\frac{1}{a}\sec^{-1}\frac{r}{a}\right] + b$$

which can be expressed as

$$r\cos(\theta - b) = a$$

This is the equation of a straight line in polar form and is the required extremal. Here $a$ and $b$ are constants of integration.

**EXAMPLE 10.41** Find the equation of the geodesic on the cone $z^2 = 8(x^2 + y^2)$.

***Solution*** It is known that the curve along a surface which represents the shortest distance between two neighbouring points is a geodesic. Here we shall use cylindrical coordinates ($x = r\cos\theta, y = r\sin\theta, z = z$) so that $x^2 + y^2 = r^2$ and the given equation of the cone becomes $z^2 = 8r^2$ which gives $dz = \sqrt{8}dr$. Now

$$ds^2 = dr^2 + r^2 d\theta^2 + dz^2 = 9dr^2 + r^2 d\theta^2$$

so that

$$I = \int_{r_1}^{r_2} ds = \int_{r_1}^{r_2} \sqrt{9dr^2 + r^2 d\theta^2} = \int_{r_1}^{r_2} \sqrt{9 + r^2\theta'^2}\, dr$$

We have to minimize this integral. It may be noted that we are using $r$ as the integration variable since the integrand contains $r$ but not $\theta$. Then $\dfrac{\partial F}{\partial \theta} = 0$ and the Euler's equation

$$\frac{d}{dr}\left[\frac{\partial F}{\partial \theta'}\right] - \frac{\partial F}{\partial \theta} = 0$$

leads to

$$\frac{r^2\theta'}{\sqrt{9 + r^2\theta'^2}} = a$$

where $a$ is a constant. Simplifying this equation, we get

$$\frac{d\theta}{dr} = \theta' = \frac{3a}{r(r^2 - a^2)^{1/2}}$$

which on integration leads to

$$\theta + b = 3\cos^{-1}\left(\frac{a}{r}\right)$$

Simplifying this equation, we get

$$a = r\cos\left(\frac{\theta + b}{3}\right)$$

which is the equation of geodesic on the given cone.

**EXAMPLE 10.42** Find the geodesic of a right circular cylinder of radius $a$.

***Solution*** The element of $ds$ on the surface of a cylinder of radius $a$ (in terms of cylindrical coordinates) is

$$ds^2 = d\rho^2 + \rho^2 d\theta^2 + dz^2 \tag{145}$$

But given that $\rho = a$ so that $d\rho = 0$ and Eq. (145) reduces to

$$ds^2 = a^2 d\theta^2 + dz^2$$

which can be expressed as

$$ds = (a^2 + z'^2)^{1/2}\, d\theta$$

where $z' = dz/d\theta$. Therefore, the arc length $s$ between two points A and B is given by

$$s = \int_A^B (a^2 + z'^2)^{1/2}\, d\theta$$

Here, $F = F(\theta, z, z') = (a^2 + z'^2)^{1/2}$ and Euler's equation

$$\frac{\partial F}{\partial z} - \frac{d}{d\theta}\left[\frac{\partial F}{\partial z'}\right] = 0$$

leads to

$$\frac{z'}{(a^2 + z'^2)^{1/2}} = \text{constant} = c$$

Simplifying this equation, we get

$$\frac{dz}{d\theta} = b$$

which on integration leads to

$$\theta = bz + d \tag{146}$$

(where $b = \sqrt{1 - c^2} / ca$). Equation (146) represents circular helix. Thus, the geodesics on a right circular cylinder are circular helix.

**EXAMPLE 10.43** If the index of refraction (in polar coordinates) is proportional to $(1/r^2)$, find the path followed by a light ray.

***Solution*** From Fermat's principle, it is known that the path of propagation of a light ray is such that the light travels from a point A to another point B in shortest possible time. If the speed of light in a medium of index of refraction $n$ is $v = \dfrac{ds}{dt} = \dfrac{c}{n}$ then $dt = \dfrac{n}{c} ds$ and the time taken for the light ray to travel from the point A to the point B is

$$I = \int_A^B \frac{n}{c}\, ds$$

which, in polar coordinates, reduces to (given that $n \propto \dfrac{1}{r^2}$)

$$I = \int_{r_1}^{r_2} \frac{kr^{-2}}{c} (dr^2 + r^2 d\theta^2)^{1/2} = \frac{k}{c} \int_{r_1}^{r_2} r^{-2} (1 + r^2 \theta'^2)^{1/2}\, dr$$

Here $F = r^{-2}(1 + r^2\theta'^2)^{1/2}$, $\theta' = \dfrac{d\theta}{dr}$ and Euler's equation leads to

$$\frac{d}{dr}\left[\frac{\theta'}{(1+r^2\theta'^2)^{1/2}}\right]=0$$

which give rise to

$$\frac{\theta'}{(1+r^2\theta'^2)^{1/2}}=\text{constant}=a$$

Simplifying this equation, we get

$$\frac{d\theta}{dr}=\theta'=\frac{a}{(1-a^2r^2)^{1/2}}$$

so that after integration, we have

$$\theta=\sin^{-1}ar+b$$

which is the required path. If the coordinates of the points A and B are known then the constants of integration in the above equation can be determined.

**EXAMPLE 10.44** Find the extremals of the functional $I=\int_0^{\pi/2}(y'^2+z'^2+2yz)dx$ such that $y(0)=0, y(\pi/2)=1, z(0)=0, z(\pi/2)=-1$.

***Solution*** Here $F=y'^2+z'^2+2yz$ and Euler's Eq. (40) leads to

$$z-y''=0 \tag{147}$$

$$y-z''=0 \tag{148}$$

Eliminating $z$ from these equations, we get

$$y^{iv}-y=0$$

The general solution of this equation is

$$y=c_1e^x+c_2e^{-x}+c_3\cos x+c_4\sin x \tag{149}$$

Differentiating Eq. (149) twice and substituting the value in Eq. (147), we get

$$z=c_1e^x+c_2e^{-x}-c_3\cos x-c_4\sin x \tag{150}$$

Using the given conditions in Eqs. (149) and (150), after simplification, we get $c_1=c_2=c_3=0, c_4=1$. Thus, from Eqs. (149) and (150), the required extremals are

$$y=\sin x,\quad z=-\sin x$$

**EXAMPLE 10.45** Find the extremals of the functional $I=\int_0^1(1+y''^2)dx$ such that $y(0)=0, y'(0)=1, y(1)=1, y'(1)=1$.

***Solution*** Here $F=1+y''^2$ and Euler–Poisson Eq. (42) leads to

$$y^{iv}=0$$

which has the general solution as

$$y = c_1 + c_2 x + c_3 x^2 + c_4 x^3 \tag{151}$$

Also, from Eq. (151) we have

$$y' = c_2 + 2c_3 x + 3c_4 x^2 \tag{152}$$

Now using the given conditions in Eqs. (151) and (152), after simplification we get $c_1 = c_3 = c_4 = 0, c_2 = 1$. Thus, from Eq. (151), the required curve on which the given functional can attain extremal is

$$y = x$$

which is a straight line passing through origin.

**EXAMPLE 10.46** Show that $I = \int_{x_1}^{x_2} (y'^2 + yy'')dx$ does not possess an extremal.

***Solution*** Here $F = y'^2 + yy''$ and Euler–Poisson's Eq. (42) leads to $y'' - 2y'' + y'' = 0$ which is not a differential equation. Hence, there is no extremal for the given functional.

**EXAMPLE 10.47** Find the extremals of the functional $I = \int_0^1 (2x + \dot{x}^2 + \dot{y}^2)dt$ such that $x(0) = 1, y(0) = 1, x(1) = 1.5, y(1) = 1$, where $\dot{x} = \dfrac{dx}{dt}$ and $\dot{y} = \dfrac{dy}{dt}$.

***Solution*** Here $G = 2x + \dot{x}^2 + \dot{y}^2$ and Euler's Eq. (58) yield

$$\ddot{x} = 1 \quad \text{and} \quad \ddot{y} = 0$$

Integrating these equations, we get

$$x = \frac{t^2}{2} + c_1 t + c_2 \tag{153}$$

$$y = c_3 t + c_4 \tag{154}$$

Using the given conditions in Eqs. (153) and (154) after simplification, we get $c_1 = c_3 = 0, c_2 = c_4 = 1$. Thus the required extremals, from Eqs. (153) and (154), are

$$x = \frac{1}{2}(t^2 + 2) \quad \text{and} \quad y = 1.$$

**EXAMPLE 10.48** Find the extremals of the functional $I = \int_{t_1}^{t_2} (\dot{x}^2 + \dot{y}^2)^{1/2} dt$.

***Solution*** Here

$$G = (\dot{x}^2 + \dot{y}^2)^{1/2} \tag{155}$$

Euler's Eq. (58 a) now leads to

$$\frac{\dot{x}}{(\dot{x}^2+\dot{y}^2)^{1/2}}=c_1 \tag{156}$$

which after simplification yields

$$\frac{dy}{dx}=\frac{(1-c_1^2)^{1/2}}{c_1}=c_2 \text{ (say)}$$

Integrating this equation we get the required extremal as

$$y=c_2x+c_3 \tag{157}$$

which is a straight line.

***Note:***

(i) If the constant $c_1$ in Eq. (156) is zero, then Eq. (156) leads to the extremal $x=c_4$.

(ii) It can easily be seen that Eq. (58 b) with Eq. (155) leads to the same extremals.

**EXAMPLE 10.49** Find the extremals of the functional $I=\int_{t_1}^{t_2} y(\dot{x}^2+\dot{y}^2)^{1/2}\,dt$.

***Solution*** Here $G=y(\dot{x}^2+\dot{y}^2)^{1/2}$ and Euler's Eqs. (58 a) and (58 b) lead to the same equation as

$$\frac{y\dot{x}}{(\dot{x}^2+\dot{y}^2)^{1/2}}=c \tag{158}$$

which may be expressed as

$$\left(\frac{dy}{dx}\right)^2=\frac{y^2-c^2}{c^2}$$

After simplification this equation yields

$$\frac{c\,dy}{(y^2-c^2)^{1/2}}=dx$$

Integrating this equation, we get

$$c\cosh^{-1}\left(\frac{y}{c}\right)=x+c_1$$

or

$$y=c\cosh\left(\frac{x+c_1}{c}\right)$$

which are the required extremals, where $c_1$ is an arbitrary constant.

***Note:*** If the constant $c$ in Eq. (158) is zero then Eq. (158) leads to $y\dot{x}=0$ which means $y = 0$ and $\dot{x}=0$. The extremals in this case are $y=0$ and $x=c_2$.

**EXAMPLE 10.50** Find the extremals of the functional $I=\int_{x_1}^{x_2}(y'^2-y^2)dx$ subject to the constraints $\int_{x_1}^{x_2} ydx=1$.

***Solution*** Here $I=\int_{x_1}^{x_2}(y'^2-y^2)dx,\ J=\int_{x_1}^{x_2} ydx=1$ and

$$f^* = F+\lambda G=(y'^2-y^2)+\lambda y \tag{159}$$

From Eq. (66), the Euler's equation in our case is

$$\frac{\partial f^*}{\partial y}-\frac{d}{dx}\left[\frac{\partial f^*}{\partial y'}\right]=0$$

which from Eq. (159), after simplification, reduces to

$$y''+y=\frac{\lambda}{2}$$

The solution of this second-order differential equation is

$$y=c_1\cos x+c_2\sin x+\frac{\lambda}{2}$$

which is required extremal.

**EXAMPLE 10.51** Find the extremals of the functional $\int_{x_1}^{x_2}(x^2+y'^2)dx$ subject to the constraint condition $\int_{x_1}^{x_2} y^2 dx=2$.

***Solution*** Here

$$f^* = F+\lambda G=x^2+y'^2+\lambda y^2 \tag{160}$$

and Euler's Eq. (66), using Eq. (160), reduces to

$$y''-\lambda y=0 \tag{161}$$

For convenience put $\lambda=\omega^2$, then the general solution of Eq. (161) is

$$y=c_1e^{\omega x}+c_2e^{-\omega x}$$

which is the required extremal. It may however be noted that if $\lambda$ in Eq. (161) is zero, then $y=c_3x+c_4$ are the required extremals.

**EXAMPLE 10.52** Let $l$ be the length of a curve $C$ passing through the points $P(x_1, 0)$ and $Q(x_2, 0)$. Find the curve $C$ such that the region bounded by it and the segment $PQ$ (on the $x$-axis) has largest area.

***Solution*** Let $A$ denotes the area enclosed by a plane curve $C$ and the $x$-axis. Given that the length of the curve $C$ is $l$, then

$$A = \int_{x_1}^{x_2} y dx \quad \text{and} \quad l = \int_{x_1}^{x_2} (1 + y'^2)^{1/2}\, dx$$

Here, $f^* = F + \lambda G = y + \lambda(1 + y'^2)^{1/2}$. Substituting this value if $f^*$ in Eq. (66), after simplification, we get

$$1 - \frac{d}{dx}\left[\frac{\lambda y'}{(1 + y'^2)^{1/2}}\right] = 0$$

which on integration leads to

$$x - \left[\frac{\lambda y'}{(1 + y'^2)^{1/2}}\right] = c_1$$

Simplifying this equation, we get

$$y'^2 = \frac{(x - c_1)^2}{\lambda^2 - (x - c_1)^2}$$

so that

$$dy = \frac{x - c_1}{\{\lambda^2 - (x - c_1)^2\}^{1/2}}\, dx$$

Integrating this equation, we get

$$y = \{\lambda^2 - (x - c_1)^2\}^{1/2} + c_2$$

which leads to

$$(x - c_1)^2 + (y - c_2)^2 = \lambda^2$$

Thus, the required curve which encloses largest area is a circle.

**EXAMPLE 10.53** For a given surface, show that the sphere is the solid figure of revolution which has maximum volume.

***Solution*** If $S$ and $V$ denote the surface and volume generated by rotating the arc $OAB$ of the curve about $x$-axis, where $O$ is the origin, point $A$ is lying in $xy$-plane and point $B$ is on the positive $x$-axis such that $OB = a$. Then

$$S = \int_0^a 2\pi y(1 + y'^2)^{1/2}\, dx = \text{constant} \tag{162}$$

and
$$V = \int_0^a \pi y^2 \, dx \tag{163}$$

Here we have to find the condition under which the functional given by Eq. (163) attains maximum subject to the constraint condition given by Eq. (162). We have

$$f^* = F + \lambda G = \pi y^2 + \lambda[2\pi y(1 + y'^2)^{1/2}] \tag{164}$$

Since $f^* = f^*(y, y')$, therefore Euler's Eq. (66) reduces to

$$f^* - y'\frac{\partial f^*}{\partial y'} = \text{constant} = c$$

which from Eq. (164) leads to

$$\frac{2\lambda\pi y}{(1 + y'^2)^{1/2}} + \pi y^2 = c \tag{165}$$

Since the curve *OAB* passes through $O(0,0)$ and $B(a,0)$, this equation leads to $c = 0$ and Eq. (165) now reduces to

$$\frac{2\lambda\pi y}{(1 + y'^2)^{1/2}} + \pi y^2 = 0$$

which after simplification leads to

$$\frac{dy}{dx} = y' = \frac{(4\lambda^2 - y^2)^{1/2}}{y}$$

Separating the variables in this equation, we get

$$dx = \frac{ydy}{(4\lambda^2 - y^2)^{1/2}}$$

so that on integration, we have

$$x = -(4\lambda^2 - y^2)^{1/2} + c_1$$

Since the curve is passing through $O$, this equation yields $c_1 = 2\lambda$ and thus we have

$$x = 2\lambda - (4\lambda^2 - y^2)^{1/2}$$

Simplifying this equation, we get

$$(x - 2\lambda)^2 + y^2 = (2\lambda)^2$$

which is the equation of a circle with centre at $(2\lambda, 0)$ and radius $2\lambda$. Therefore, the solid figure formed by the revolution of an arc of a curve (a circle) is a sphere.

## EXERCISES

1. Evaluate the functional

$$I = \int_0^1 \left[ y^2 + \left( \frac{dy}{dx} \right)^2 \right] dx$$

along the path $y = x$ from (0, 0) to (1, 1).

2. Find the extremal of the functional

$$I = \int_0^1 (y'^2 + 12xy)\, dx, \quad y(0) = 0, \quad y(1) = 1.$$

3. Test for the extremum of the functional

$$I = \int_{x_1}^{x_2} (y^2 + 2xyy')\, dx$$

along with the conditions $y(x_1) = y_1$, $y(x_2) = y_2$.

4. Find the extremal of the functional

$$I = \int_{x_1}^{x_2} (y'^2 + 2yy' - 16y^2)\, dx$$

5. Find the extremal of the functional

$$I = \int_{x_1}^{x_2} \frac{\sqrt{1 + y'^2}}{y}\, dx$$

6. Obtain the extremal of the functional

$$I = \int_{x_1}^{x_2} (y^2 + y'^2 + 2ye^x)\, dx$$

7. Obtain the extremal of the functional

$$I = \int_{x_1}^{x_2} (y^2 - y'^2 - 2y \sin x)\, dx$$

8. Find the extremal of the functional

$$I = \int_{x_1}^{x_2} (x^2 y'^2 + 2y^2 + 2xy)\, dx$$

**9.** Find the extremal of the functional

$$I = \int_0^{\pi} (4y\cos x + y'^2 - y^2)dx, \quad y(0) = y(\pi) = 0.$$

**10.** Obtain the extremal for the functional

$$I = \int_0^1 (y^2 + y'^2)dx, \quad y(0) = 2, \quad y(1) = 1.$$

**11.** Find the extremal of

$$I = \int_1^2 y'(1 + x^2 y')dx, \quad y(1) = 3, \quad y(2) = 5.$$

**12.** Find the curve $x = x(t)$ which minimizes the functional

$$I = \int_0^1 (\dot{x}^2 + 1)\, dt, \quad x(0) = 1,\ x(1) = 2$$

**13.** Prove that the shortest distance between two points in a space is a straight line.

**14.** Show that the geodesics of a spherical surface are great circles.

**15.** Find the equation of the geodesic on a right circular cone of semi-vertical angle $\alpha$.

**16.** Two rings, each of radius $a$, are placed parallel with their centres $2b$ apart and on a common normal. An axially symmetric soap film is formed between them but does not cover the ends of the rings. Find the shape assumed by the film.

**17.** Determine the shape of a solid of revolution moving in a flow of gas with least resistance.

**18.** Obtain the extremal for the functional

$$I = \int_{-l}^{l} \left( -\frac{1}{2}\mu y''^2 + \rho y \right) dx$$

with boundary conditions $y(-l) = y(l) = 0$, $y'(-l) = y'(l) = 0$.

**19.** Investigate the extremal of the functional

$$I = \int_{x_1}^{x_2} (16y^2 - y''^2 + x^2)\, dx$$

**20.** Find the extremal of the functional

$$I = \int_{x_1}^{x_2} (y''^2 - 2y'^2 + y^2 - 2y \sin x)\, dx$$

**21.** Investigate the extremal of the functional

$$I = \int_{x_1}^{x_2} F(y', z')\, dx$$

**22.** Obtain the extremal of the functional

$$I = \int_0^{\pi/2} (y'^2 + z'^2 + 2yz)\, dx$$

along with the boundary conditions

$$y(0) = 0,\ y\left(\frac{\pi}{2}\right) = 1, \quad z(0) = 0,\ z\left(\frac{\pi}{2}\right) = -1.$$

**23.** Find the extremal of the functional

$$I = \int_0^1 (1 + y'^2 + z'^2),\ y(0) = 0,\ y(1) = 2;\ z(0) = 0,\ z(1) = 4.$$

**24.** Find the curve $x = x(t)$ which minimizes the functional

$$I = \int_0^1 (\dot{x}^2 + 1)\, dt$$

where $x(0) = 1$ but $x$ can take any value at $t = 1$.

**25.** Show that an isosceles triangle has the smallest perimeter for a given area and given base.

**26.** Among all curves of length $l$ in the upper half-plane passing through the points $(-a, 0)$ and $(a, 0)$ find the one which together with the interval $[-a, a]$ encloses the largest area.

**27.** Minimize the cost functional

$$I = \frac{1}{\sqrt{2g}} \int_0^a \frac{(1 + \dot{y}\dot{z})^{1/2}}{x^{1/2}}\, dx$$

where the variables $y = y(x)$, $z = z(x)$ are subject to the constraints $y = z + 1$ and where $y(0) = 0$, $y(a) = b$.

**28.** Minimize the cost functional

$$I = \frac{1}{2} \int_0^2 (\ddot{x})^2\, dt$$

where $x = x(t)$ satisfies $x(0) = 1$, $\dot{x}(0) = 1$, $x(2) = 1$, $\dot{x}(2) = 0$.

**29.** Find the extremal of the functional

$$I = \int_0^1 (\dot{x}_1 \dot{x}_2 + x_1 x_2)\, dt$$

under the boundary conditions $x_1(0) = 4$, $x_2(0) = 2$, $\partial F / \partial \dot{x}_i = 0$ at $t = 1$ $(i = 1, 2)$.

**30.** Find the stationary path $x = x(t)$ for the functional

$$I = \int_0^1 \left(1 + \frac{d^2 x}{dt^2}\right) dt$$

subject to the boundary conditions

$$x(0) = 0, \qquad \left(\frac{dx}{dt}\right)_0 = 1, \qquad x(1) = 1, \qquad \left(\frac{dx}{dt}\right)_1 = 1.$$

**31.** A mechanical system is described by the differential equation

$$\frac{d^2 x}{dt^2} = u$$

where $x$ is the state variable and $u$ the control variable. Find the optimal control $u$ which takes the system from $x = 1$, $dx/dt = 1$ at $t = 0$ to $x = 0$, $dx/dt = 0$ at $t = 1$ and which minimizes the functional

$$J = \int_0^1 u^2 dt$$

**32.** A perfectly flexible rope of uniform density hangs at rest with its end points fixed. Determine its shape.

**33.** Among all curves lying on the sphere $x^2 + y^2 + z^2 = a^2$ and passing through two fixed points $(x_1, y_1, z_1)$ and $(x_2, y_2, z_2)$. Find the one which has the least length.

**34.** Find the extremal of $I = \int_0^2 y'^2 dx$ under the constraint $J = \int_0^2 y\, dx = 1$, such that $y(0) = 0$ and $y(2) = 1$.

**35.** Find the extremal of $I = \int_0^1 y'^2 dx$ under the constraint $J = \int_0^1 y\, dx = 2$, such that $y(0) = 0$ and $y(1) = 1$.

**36.** Find the extremal of the functional

$$I = \int_0^1 (y'^2 + z'^2 - 4xz' - 4z)\, dx$$

subject to the conditions $y(0) = 0$, $z(0) = 0$, $y(1) = 1$, $z(1) = 1$, and

$$\int_0^1 (y'^2 - xy' - z'^2)\, dx = 2.$$

**37.** Find Ostrogradsky equation for the functional

(i) $$I = \iint_D \left[1 + \left(\frac{\partial z}{\partial x}\right)^2 + \left(\frac{\partial z}{\partial y}\right)^2\right]^{1/2} dx\, dy$$

(ii) $$I = \iint_D \left[\left(\frac{\partial z}{\partial x}\right)^2 - \left(\frac{\partial z}{\partial y}\right)^2\right] dx\, dy$$

(iii) $$I[f(x, y, z)] = \iiint_D \left[\left(\frac{\partial f}{\partial x}\right)^2 + \left(\frac{\partial f}{\partial y}\right)^2 + \left(\frac{\partial f}{\partial z}\right)^2 + 2fg(x,y,z)\right] dx\, dy\, dz$$

(iv) $$I = \iint_D \left[\left(\frac{\partial^2 z}{\partial x^2}\right)^2 + \left(\frac{\partial^2 z}{\partial y^2}\right) + 2\left(\frac{\partial^2 z}{\partial x \partial y}\right) - 2z f(x,y)\right] dx\, dy$$

# Answers to Exercises

## Chapter 1

**1.** $xy' = y - 2$.

**2.** $y' + y = 0$.

**3.** $2(x + 1)y' = y$.

**4.** $(x^2 - y)y' = xy$.

**5.** $y'' - y' = 0$.

**6.** $y'' - 2y' + 2y = 0$.

**7.** $y'' + \omega^2 y = 0$.

**8.** $y'' - k^2 y = 0$.

**9.** $y'' - 8y' + 16y = 0$.

**10.** $xy'' + y' = 0$.

**11.** $y''' - 6y'' + 11y' - 6y = 0$.

**12.** $(1 + \cos\theta)\dfrac{dr}{d\theta} + r\sin\theta = 0$.

**13.** $xy'' + 2y' = xy - x^2 + 2$.

**14.** $y''' = 7y' - 6y$.

**15.** $y[1 - (y')^2] = 2xy'$.

**16.** $y'' + \dfrac{2}{x}y' = 0$.

**17.** $xy' - y = 0$.

**18.** $2xyy' = y^2 - x^2$.

**19.** $(x^2 - y^2)y' = 2xy$.

**20.** $2xy' = y$.

**21.** $y''' = 0$.

**22.** $yy'' + (y')^2 = 0$.

**23.** $L\dfrac{di(t)}{dt} + Ri(t) = E(t)$.

**24.** $\dfrac{dv}{dt} + k\dfrac{v^2}{m} = g$.

**25.** $\dfrac{dx}{dt} = r - kx, \quad k > 0$.

**26.** Using $\tan\phi = \dfrac{x}{y}$, $\quad \tan\left(\dfrac{\pi}{2} - \theta\right) = \dfrac{dy}{dx}$, $\quad \tan\theta = \dfrac{dy}{dx}$

$\tan\phi = \tan 2\theta = \dfrac{2\tan\theta}{1 - \tan^2\theta}$ we get $x\left(\dfrac{dx}{dy}\right)^2 + 2y\dfrac{dx}{dy} = x$.

**27.** Water lost in time $\Delta t$ = change in volume of water $\times \frac{1}{4}\sqrt{2gh}\,\Delta t = -\pi r^2 \Delta h$, where $\Delta h$ is the change in height for a small change in time $\Delta t$, and $r$ is the radius of $A$. Division by $\Delta t$ yields $\Delta h/\Delta t = \sqrt{2gh}/4\pi r^2$. Now, $r = \frac{2h}{5}$ at any time. Thus, as $t \to 0$,

$$\frac{dh}{dt} = \frac{25\sqrt{2g}}{16} h^{-1/2}$$

**28.** $\dfrac{dy}{dx} = \dfrac{-y}{\sqrt{s^2 - y^2}}$.

## Chapter 2

**1.** $(x^2 + 1)(y^2 + 1) = c$.

**2.** $(\sin x)(e^y + 1) = c$.

**3.** $c(y + 1) = e^{(x^2/2) + x}$.

**4.** $(4x + y + 1) = 2 \tan(2x + c)$.

**5.** $\log x - \log\left[1 + \sqrt{1 + x^2}\right] + \sqrt{1 + x^2} + \sqrt{1 + y^2} = c$.

**6.** $x(x + 2a)^3 = e^{2(y+c)/a}$

**7.** $2xy(x + y)^{-2} + \log(x + y) = c$.

**8.** $y = ce^{x^3/(3y^3)}$.

**9.** $\dfrac{x}{y} + \log(xy) = c$.

**10.** $x + ye^{x/y} = c$.

**11.** $x = ce^{\cos(y/x)}$.

**12.** $x = c \sin \dfrac{y}{x}$.

**13.** $\tan^{-1} \dfrac{2y + 1}{2x + 1} = \log\left[c\sqrt{x^2 + y^2 + x + y + \dfrac{1}{2}}\right]$.

**14.** $\log(x - y - 1) = x - 2y - c$.

**15.** $3x + 2y + c + 2 \log(1 - x - y) = 0$.

**16.** $(x + 2y - 5) = c(2x - y)^2$.

**17.** $x + y + \dfrac{4}{3} = ce^{3(x-2y)}$.

**18.** $c = (y + x - 1)^5(y - x + 1)^2$.

**19.** $(x - 1)y = x^2(x^2 - x + c)$.

**20.** $1 + y = \tan x + ce^{-\tan x}$.

**21.** $16x^2y = 4x^4 \log x - x^4 + c$.

**22.** $y(1 + x^3) = \dfrac{1}{2}x - \dfrac{1}{4} \sin 2x + c$.

**23.** $\dfrac{y(1 + \sqrt{x})}{1 - \sqrt{x}} = x + \dfrac{2}{3}x^{3/2} + c$.

**24.** $xe^{\tan^{-1} y} y = \dfrac{1}{2}e^{2 \tan^{-1} y} + c$

**25.** $x\sqrt{\cot y} = c + \sqrt{\tan y}$.

**26.** $y^{-2}e^{x^2} = 2x + c$.

**27.** $x = ce^y - (y + 2)$.

**28.** $xy^2 = 2y^5 + c$.

**29.** $(2 + cx)xy^2 = 1.$

**30.** $\frac{1}{xy} = c - \int x^{-1} \sin x \, dx.$

**31.** $(x+1)^2 y^3 = \frac{1}{6}x^6 + \frac{2}{5}x^5 + \frac{1}{4}x^4 + c.$

**32.** $y^2 = x^2 + cx - 1.$

**33.** $y^{-1} \sec^2 x = c - \frac{1}{3}\tan^3 x.$

**34.** $y^{-1} = c \cos x + \sin x.$

**35.** $x \log y = e^x(x - 1) + c.$

**36.** $\frac{1}{y^2} = ce^{x^2} + x^2 + 1.$

**37.** $\tan y = c(1 - e^x)^3.$

**38.** $\frac{1}{x} = 2 - y^2 - ce^{-y^2/2}.$

**39.** $x^3 - y^3 + 3xy(x - y) = c.$

**40.** $y \cos 2x + 2y + \frac{2}{3}y^3 = c.$

**41.** $xy + y \log x + x \cos y = c.$

**42.** $(e^y + 1) \sin x = c.$

**43.** $-y - 1 + x^2 - x \cos y = cx.$

**44.** $e^x + (c + x^2)y = 0.$

**45.** $cx + y + \log x + 1 = 0.$

**46.** $\frac{2}{3}x^3 - \frac{1}{2}y^2 + \frac{e^x}{y} = c.$

**47.** $\frac{y}{x} = \log cy.$

**48.** $xy - \frac{1}{xy} - \log y^2 = c.$

**49.** $\log \frac{x}{y} - xy = c.$

**50.** $\frac{1}{2}x^2y^2 + \log \frac{x}{y} - \frac{1}{xy} = c.$

**51.** $cy \cos xy = x.$

**52.** $x + \frac{y^2}{x^2} = c.$

**53.** $e^x(x^2 + y^2) = c.$

**54.** $6x^3y^3 - 27xy^4 - 3y^3 \log y - y^3 = c.$

**55.** $x^3y^3 + x^2 = cy.$

**56.** $\left(4x^5 + 2x^4y + \frac{4}{3}x^3y^2 + x^3y^3\right)y = c.$

**57.** $x^2y^4(x + y^2) = c.$

**58.** $x^3y^2(x + y^3) = c.$

**59.** $x^2y^2(x + 4y^4) = c.$

**60.** $2x^{-36/13}y^{24/13} - 12x^{-10/13}y^{-15/13} = c.$

**61.** $\sqrt{x^2 + y^2} = a \sin\left(\tan^{-1}\frac{y}{x} + c\right).$

**62.** $xy + yz + zx = c^2.$

**63.** $2xz = y^2z^3 + cy^2.$

**64.** $y(x + z) = c(x + y + z)$.

**65.** $\frac{1}{x} = \frac{1}{y} + c_1, \quad z = c_2 - \frac{nxy}{y - x} \log \frac{x}{y}$.

**66.** $lx + my + nz = c_1, \quad x^2 + y^2 + z^2 = c_2$.

**67.** $x^2 - y^2 - 2xy = c_1, \quad x^2 - y^2 - z^2 = c_2$.

**68.** $xyz = c_1, \quad x^2 + y^2 + z^2 = c_2$.

**69.** $y = c_1 z, \quad x^2 + y^2 + z^2 = c_2 z$.

**70.** $\tan^{-1} \frac{y^2 + x}{x} - \frac{1}{2} \log [x^2 + (y^2 + x)^2] = \log c$.

## Chapter 3

**1.** $(y - 4x - c)(y - 3x - c) = 0$.

**2.** $(y - x - c)(x^2 + y^2 - c^2) = 0$.

**3.** $(y - c)(y + x^2 - c)(xy + cy + 1) = 0$.

**4.** $(x^2 + 2y^2 - c)(x^3 + y^2 - c) = 0$.

**5.** $(y - c)(x + y - c)(xy + x^2 + y^2 - c) = 0$.

**6.** $\sin^{-1} \frac{y}{x} = \pm \log cx$.

**7.** $2y = cx^2 + \frac{a}{c}$.

**8.** $y = 2c\sqrt{x} + \tan^{-1} c^2$.

**9.** $y = 3x + \log \frac{3}{1 - ce^{3x}}$.

**10.** $x = c\sqrt{p} - \frac{1}{3}p^2, y = \frac{c\sqrt{p} - (p^2/3)}{p} + p\left(c\sqrt{p} - \frac{1}{3}p^2\right)$

**11.** $(c - y)(1 + p^2) = 1$, with the given relation.

**12.** $y = c + \frac{a}{2}\left[p\sqrt{1 + p^2} - \log\left(p + \sqrt{1 + p^2}\right)\right]$, with the given relation.

**13.** $x = c + 2p + 2 \log (p - 1), \quad y = c + p^2 + 2p + 2 \log (p - 1)$.

**14.** $y\sqrt{1 - p^2} + (1 - p^2)^{3/2} = c$, with the given relation.

**15.** $y^2 = 2cx + c^n$.

**16.** $x = \log p^2 + 6p + c$.

**17.** $y - c = \sqrt{x - x^2} - \tan^{-1} \sqrt{\frac{1 - x}{x}}$.

**18.** $y^2 + cyx^{2/\sqrt{5}-1} = c^2 x^{4/\sqrt{5}-1}$.

**19.** $y^2 = 2cx + c^2$.

**20.** $y = cx - \frac{ac^2}{c + 1}$.

**21.** $y = cx - \sin^{-1} c$.

**22.** $y^2 = cx^2 + (1 + c)$.

**23.** $x^2 + y^2 + 2c(x + y) + c^2 = 0$.

**24.** $xy = cy + c$.

**25.** $y^2 = cx^2 - \dfrac{bc}{ac+1}$.

## Chapter 4

**1.** $2^{48} \approx 2.815 \times 10^{14}$.

**2.** 283%.

**3.** 22,400 yr.

**4.** 6541.48 yr.

**5.** 15,512 yr.

**6.** 59 mg, 1315.8 yr.

**7.** Given in text.

**8.** $\dfrac{\log 2}{0.07} \sim 9.9$ yr.

**9.** $r = 10.986\%$.

**10.** 0.206.

**11.** 23 min. (approx.).

**12.** 34.1°C, 37.4°C, 8.5 min.

**13.** $\dfrac{dV}{dt} = \dfrac{dV}{dr}\dfrac{dr}{dt} = 4\pi r^2 \dfrac{dr}{dt} = kA = 4k\pi r^2$, 5.7 months.

**14.** $N(60) \sim 23.0$ million in 1850, $N(110) \sim 76.5$ million in 1900, and $N(160) \sim 148.4$ million in 1950 (the census figures for these years were 23.2, 76.2 and 151.3 million), $a/b \sim 197$ million.

**15.** 31.1 yr.

**16.** $P(t) = P_0 e^{(k_1-k_2)t}$; $k_1 > k_2$, births surpass deaths, and so population increases, $k_1 = k_2$, a constant population since the number of both births and deaths are equal, and $k_1 < k_2$, deaths surpass births, and thus population decreases.

**17.** 1,000,000, 52.9 months.

**18.** $P(t) = e^{a/b} e^{-ce^{-bt}}$, $c = \dfrac{c_2}{b}$, $c_2 = -e^{bc_1}$; $c = \dfrac{a}{b} - \log P_0$.

**19.** $P(45) = 8.99$ billion.

**20.** (i) 61.3 cm, (ii) $N = \dfrac{61.27}{1 + 1.22e^{-1.118t}}$, where $t = 0, 1, 2$, correspond to 16, 32 and 48, respectively, (iii) 19.1 cm, 36.6 cm, 58.8 cm.

**21.** (i) $y = \dfrac{38.47}{1 + 31.06(0.0303605)^t}$

(ii) $y = 1.20, 3.59, 9.52, 19.7, 29.7, 35.2, 37.4$ corresponding to $t = 0, 1, \ldots, 6$, respectively.

(iii) 38.5 cm$^2$.

**22.** $-c \log y + dy = a \log x - bx + k$.

**23.** $x(t)=\dfrac{k_1ac_1e^{ak_1t}}{1+k_1c_1e^{ak_1t}}, y(t)=c_2(1+k_1c_1e^{ak_1t})^{k_2/k_1}.$

**24.** 37.2%, 66.5%, 86.9%, 95.7%. **25.** (b) 3, 6, 14, 23, 24, 16, 8, 3, 1.

**26.** max $x = x_0e^c$, min $x = x_0e^{-c}$. **27.** 917 companies.

**28.** 1834; 2000.

**29.** (a) $G=\dfrac{A}{100\,aV}+\left(G_0-\dfrac{A}{100\,aV}\right)e^{-at}$, (b) $\dfrac{A}{100\,aV}$.

**30.** $A(t) = 200 - 170\, e^{-t/50}$.

**31.** The DE is $\dfrac{dA}{dt}+\dfrac{4A}{100+2t}=3$ and $A(t) = A(30) = 64.38$ lb.

**32.** (a) $0.08\,(1-e^{-0.6})$, $0.08(1-e^{-2.4})$, (b) 0.08, (c) 50 log 2, 100 log 2.

**33.** $Q(4) \sim 7.39$ g, $t \sim 2.63$ hr. **34.** 63.3 g, 80 min.

**35.** (a) 51.1 lb, 6.6 lb, 7.6 lb, (b) 3 hr. 53 min.

**36.** –2.94 cal/s.

**37.** $T = 450r^{-1} - 25$; $T(7) = 39\frac{2}{7}$, $T(8) = 31\frac{1}{9}$, $T(9) = 25$.

**38.** $i = 5 - 5e^{-5t}$, $i(0.2) \simeq 3.16$ A.

**39.** $i = 2e^{-2t}\sin 25t + 20e^{-2t}$, $i(0.5) \sim 7.516836$ A.

**40.** $Q = 0.05e^{-10t}$, $i = -0.005e^{-10t}$

**41.** (i) $I=I_0e^{(2a/3)t^{3/2}}$, (ii) $I=-k_1^{-1}\log\left(-ak_1t+e^{-k_1I_0}\right)$.

**44.** $r=e^{\theta/\sqrt{3}}$, $r=2e^{(\theta-\pi)/\sqrt{3}}$.

**45.** (a) 38.2 min., (b) 11.2 min., (c) 2.04 ft.

**46.** $B(h) = \pi r^2$, $r^2 = h^2\tan^2 30° = h^2/3$, $dh/dt = -12.7\,h^{-3/2}$, $0.4\,h^{5/2} = c - 12.7t$, $c = 4\times 10^4$, $t = c/12.7 \sim 53$ min.

**47.** $2(y-2)^2 + (x-1)^2 = c_2^2$.

**48.** $y = 2 - x + 3e^{-x}$.

**49.** (a) $2\log|y| = x^2 + y^2 + c_2$. (b) $y^2 = 2x + c_2$.

(c) $x^3 + y^3 = c_2$. (d) $y^2 - x^2 = c_2x^3$.

(e) $y=\frac{1}{4}-\frac{1}{6}x^2+c_2x^{-4}$. (f) $2y^3 = 3x^2 + c_2$.

(g) $2\log(\cos hy) + x^2 = c_2$. (h) $y^{5/3} = x^{5/3} + c_2$.

**50.** About 2 miles.

**51.** $\Delta A = -kA\Delta x$, $A$ = amount of incident light, $A$ = absorbed light, $\Delta x$ = thickness, $-k$ = constant of proportionality. Let $\Delta x \to 0$, then $A' = -kA$. Hence, $A(x) = A_0 e^{-kx}$ is the amount of light in a thick layer at depth $x$ from the surface of incidence.

**52.** $pV$ = constant = $c$.

**53.** $I\dfrac{d\omega}{dt} = -k\omega^{1/2}, \quad k > 0, \quad \omega(t) = \left(\sqrt{\omega_0} - \dfrac{k}{2I}t\right)^2,$

$$t_1 = \frac{2I\sqrt{\omega_0}\,(1-2^{-1/2})}{k}, \qquad t_2 = \frac{2I\sqrt{\omega_0}}{k}.$$

**54.** $p(h) = p_0 e^{-0.000116h}$.

**55.** $v(t) = \dfrac{W-B}{k}(1-e^{-kt/m}), \qquad m = \dfrac{W}{g} = \dfrac{2254}{9.8} = 230 \text{ kg},$

$$y(t) = \frac{W-B}{k}\left[t - \frac{m}{k}(1-e^{-kt/m})\right], \qquad t_{cr} = 17.3 \text{ s}, \qquad y_{cr} = 106 \text{ m}.$$

## Chapter 5

**1.** $y = c_1 e^{-ax} + c_2 e^{-bx}$.

**2.** $y = c_1 e^{2x} + c_2 \cos 2x + c_3 \sin 2x$.

**3.** $y = c_1 e^{2x} + e^{-1/2x}\left(c_2 \cos \dfrac{\sqrt{3}}{2}x + c_2 \sin \dfrac{\sqrt{3}}{2}x\right)$.

**4.** $y = (c_1 x + c_2)e^x + c_3 e^{-x}$.

**5.** $y = (c_1 x + c_2)e^x + c_3 e^{2x}$.

**6.** $y = (c_1 + c_2 x)e^x + c_2 \cos x + c_3 \sin x$.

**7.** $y = 2e^x \sin 2x$.

**8.** $y = e^x[(c_1 + c_2 x)\cos x + (c_3 + c_4 x)\sin x]$.

**9.** $y = c_1 e^{-x} + c_2 e^{2x} - \dfrac{1}{2}e^x$.

**10.** $y = 2e^x + c_1 e^{-2x} + c_2 e^{-x}$.

**11.** $y_p = 2$.

**12.** $y_p = x^2 - 2x$.

**13.** $y_p = \dfrac{x}{2} + \dfrac{3}{4}$.

**14.** $y_p = \dfrac{x^3}{3}$.

**15.** $y_p = -2\left(\dfrac{x^5}{20} + \dfrac{x^4}{4} + x^3 + 3x^2\right)$.

**16.** $y_p = 3x^2$.

**17.** $y_p = 2e^x$.

**18.** $y_p = \dfrac{3}{5}e^{-2x}$.

**19.** $y_p = -\dfrac{1}{2}\sin x$.

**20.** $y_p = -\dfrac{1}{2}\cos x$.

**21.** $y_p = \frac{3}{10}(\sin x + 3\cos x)$. **22.** $y_p = -e^x(x^2 + 6x + 26)$.

**23.** $y_p = \frac{1}{3}e^x(3x^2 - 6x + 4)$. **24.** $y_p = \frac{x^3 e^{-x}}{60}(20 - x^2)$.

**25.** $y_p = xe^x$. **26.** $y_p = -xe^{-2x}$.

**27.** $y_p = \frac{x^3 e^{2x}}{30}$. **28.** $y_p = \frac{7x^2 e^x}{2}$.

**29.** $y_p = 4 + e^x + \frac{1}{2}(\sin x - 3\cos x)$.

**30.** $y_p = x^2 - 3x + \frac{7}{2} - 2xe^{-2x}$. **31.** $y_p = x^2 + \frac{x^2}{8}(\cos x - \sin x)$.

**32.** $y_p = -\frac{1}{27}(9x^3 + 9x^2 + 33x - 34)$.

**33.** $y = c_1 \cos ax + c_2 \sin ax - \frac{1}{a^2}\cos ax \log\tan\left(\frac{\pi}{4} + \frac{1}{2}ax\right)$.

**34.** $y = c_1 e^{mx} + c_2 e^{-mx} + c_3 \cos mx + c_4 \sin mx - \frac{x}{4m^3}\sin mx + \frac{x}{4m^3}\sinh mx$.

**35.** $y = (c_1 + c_2 x)e^x + c_3 e^{-x} + \frac{1}{8}e^{-x}x^2 + \frac{1}{8}xe^{-x} + x + 1$.

**36.** $y = c_1 \cos x + c_2 \sin x + \frac{1}{25}e^{2x}(5x - 4)$.

**37.** $y = c_1 \cos x + c_2 \sin x + \frac{1}{2}e^{-x} + \frac{1}{2}x \sin x + x^3 - 6x + \frac{1}{5}e^x(2\sin x + \cos x)$.

**38.** $y = e^{-x}\left(c_1 + c_2 x + c_3 x^2 + \frac{1}{6}x^3\right)$.

**39.** $y = c_1 e^{-\sqrt{7/2}x}\cos\left(\frac{3}{2}x + c_2\right) + c_3 e^{\sqrt{7/2}\,x}\sin\left(\frac{3}{2}x + c_4\right) + x^2 + \frac{127}{8}$.

**40.** $y = c_1 \cos x + c_2 \sin x + c_3 \cos 2x + c_4 \sin 2x + \frac{1}{12}x\left(\sin x - \frac{1}{2}\sin 2x\right)$.

**41.** $y = c_1 + c_2 e^x + c_3 e^{-x} + c_4 \cos x + c_5 \sin x + 3xe^x + 2x \sin x + x^2$.

**42.** $y = (c_1 + c_2 x)e^{2x} + \frac{1}{4}\left(x^2 + 2x + \frac{3}{2}\right) + e^x + \frac{1}{8}\cos 2x$.

**43.** $y = c_1 \cos x + c_2 \sin x + \frac{1}{2}x \sin x + \frac{1}{25}e^{2x}(5x - 4) - \frac{1}{5}e^x(2\cos x - \sin x)$.

**44.** $y = c_1e^{-2x} + c_2e^{-x} + 2e^x$.

**45.** $y = c_1e^{-2x} + c_2e^{-x} + \frac{1}{10}(\sin x - 3\cos x)$.

**46.** $y = e^{-x/2}\left(c_1 \cos\frac{\sqrt{3}}{2}x + c_2 \sin\frac{\sqrt{3}}{2}x\right) + x^2 - 2x$.

**47.** $y = c_1e^{4x} + c_2e^{-2x} - xe^x - 2e^{-x}$.

**48.** $y = c_1 + c_2e^{3x} - \frac{e^{2x}}{5}(3 \sin x + \cos x)$.

**49.** $y = (c_1 + c_2x)e^x + (c_3 + c_4x)e^{-x} + x - \frac{\sin x}{4}$.

**50.** $y = c_1 + c_2e^{-x} + \frac{x^3}{3}$.

**51.** $y = c_1e^{-2x} + c_2e^{-x} - 2e^x$.

**52.** $y = c_1e^{-x} + c_2xe^{-x} + \frac{x^4e^{-x}}{12}$.

**53.** $y = c_1 \cos x + c_2 \sin x - x^2 \cos x + x \sin x$.

**54.** $y = (c_1 + c_2x)\, e^{-x} + \frac{1}{4}x^2e^{-x}(2 \log x - 3)$.

**55.** $y = (c_1 + c_2x)e^x + x^2e^x\left(\frac{1}{2}\log x - \frac{3}{4}\right)$.

**56.** $y = c_1 + c_2 \cos 2x + c_3 \sin 2x - \frac{1}{4}x \cos 2x + \frac{1}{8}(\sin 2x \log|\cos 2x|)$
$+ \frac{1}{8}(\log|\sec 2x + \tan 2x|)$.

**57.** $y = (c_1 + c_2x + c_3x^2)e^{2x} + \frac{1}{2}(x^2\log|x|e^{2x})$.

**58.** $y = c_1x + c_2x^2$.

**59.** $y = c_1x + c_2(2x^2 - 1)$.

**60.** $y = c_1x + c_2x^2 + \frac{1}{2}x^3 \log x - \frac{3}{4}x^3$.

**61.** $y = c_1x + c_2x^{-1} + e^{-x}(1 + x^{-1})$.

**62.** $y = (c_1 + c_2 \log x)x + 2 \log x + 4$.

**63.** $y = c_1x^{-1} + c_2x + c_3x \log x + \frac{1}{4}x^{-1} \log x$.

**64.** $y = x[c_1 \cos(\log x) + c_2 \sin(\log x)] + x \log x$.

**65.** $y = c_1x^4 + c_2x^{-5} - \frac{1}{14}x^2 - \frac{1}{9}x - \frac{1}{20}$.

**66.** $y = c_1x^{-1} + c_2x^{-2} + x^{-2}e^x$.

**67.** $y = c_1(2x+3)^{-1} + c_2(2x+3)^3 - \frac{3}{4}(2x+3) + 3.$

**68.** $y = c_1(x+3)^2 + c_2(x+3)^3 + \frac{1}{6}\log(x+3) + \frac{5}{36}.$

**69.** $y = -x^2 \sin x - 4x \cos x + 6 \sin x + c_1 x + c_2.$

**70.** $y = \frac{x^4}{24} + \frac{x^3}{6}\log x - \frac{11}{36}x^3 + c_1 x^2 + c_2 x + c_3.$

**71.** $2(y^{1/4} - 1) = x.$

**72.** $\sqrt{y^2 - 8y} + 8\sinh^{-1}\left(\sqrt{\frac{y}{8} - 1}\right) = 3x.$

**73.** $y_1 = c_0\left(1 + \frac{1}{3}x + \frac{1\cdot 4}{3\cdot 6}x^2 + \cdots\right), y_2 = c_0 x^{7/3}\left(1 + \frac{8}{10}x + \frac{8\cdot 11}{10\cdot 13}x^2 + \cdots\right)$

**74.** $y_1 = c_0\left(1 + 3x^2 + \frac{4}{5}x^2 + \cdots\right),\quad y_2 = c_0 x^{3/2}\left(1 + \frac{3}{8}x^2 - \frac{1\cdot 3}{8\cdot 16}x^4 + \cdots\right).$

**75.** $y_1 = c_0\left(1 - 3x + \frac{3x^2}{1\cdot 3} - \cdots\right), y_2 = x^{1/2}(1 - x).$

**76.** $y_1 = c_0 x^{-1}\left(1 + \frac{x^2}{2\cdot 1} + \frac{x^4}{2\cdot 4\cdot 1\cdot 5} + \cdots\right),$

$y_2 = c_0 x^{1/2}\left(1 + \frac{x^2}{2\cdot 7} + \frac{x^4}{2\cdot 4\cdot 7\cdot 11} + \cdots\right).$

**77.** $y = c_0\left(1 - \frac{x^2}{2} + \frac{x^4}{8} + \cdots\right) + c_1\left(x - \frac{x^3}{2} + \frac{x^5}{40} + \cdots\right).$

**78.** $y = c_0\left(1 - \frac{x^2}{4} - \frac{x^3}{12} - \frac{5x^4}{96} + \cdots\right) + c_1\left(x - \frac{x^3}{6} - \frac{x^4}{24} + \cdots\right).$

**79.** $y = c_0(1 - x^2) + c_1\left(x - \frac{x^3}{3} - \frac{x^5}{15} + \cdots\right).$

**80.** $y = c_0\left(1 - x^2 - \frac{x^4}{3} + \cdots\right) + c_1 x.$

**81.** $y_1 = -\frac{Ax^2}{16}\left(1 - \frac{x^2}{2\cdot 6} + \frac{x^4}{2\cdot 4\cdot 8\cdot 6} + \cdots\right),$

$y_2 = B\left[y_1 \log x + x^{-2}\left(1 + \frac{x^2}{2^2} + \frac{x^4}{2^2\cdot 4^2} + \cdots\right)\right].$

**82.** $y_1 = a(1\cdot 2x^2 + 2\cdot 3x^3 + 3\cdot 4x^4 + \cdots),$

$y_2 = b[y_1 \log x + (-1 + x + 3x^2 + 5x^3 + \cdots)]$

**83.** $y_1 = a(2x + 2x^2 + x^3 - x^4 + \cdots),$

$y_2 = b[y_1 \log x + (1 - x - 5x^2 - x^3 + \cdots)].$

**84.** $y = c_0 x - c_1 \log x + c_1\left(1 + x - \sum_{n=2}^{\infty} \frac{x^n}{n-1}\right)$

**85.** $y_1 = (y)_{k=0} = c_0[1 + 2^2 x + 3^2 x^2 + 4^2 x^3 + \cdots),$

$$y_2 = \left(\frac{\partial y}{\partial k}\right)_{k=0} = c_0[y_1 \log x - 2(1\cdot 2x + 2\cdot 3x^2 + \cdots)].$$

**86.** $y_1 = c_0\left(1 - 2x - \frac{3x^2}{2!} - \frac{4x^3}{3!} - \cdots\right),$

$$y_2 = c_0\left[y_1 \log x + 2\left(2 - \frac{1}{2}\right)x - \frac{3}{2!}\left(2 + \frac{1}{2} - \frac{1}{3}\right)x^2 + \cdots\right].$$

**87.** $y = (c_0 + c_1 \log x)(1 + 2x + x^2) + c_1\left[-3x - 3x^2 + \sum_{n=3}^{\infty} \frac{(-1)^n 2x^n}{n(n-1)(n-2)}\right].$

**88.** $y_1 = c_0 x^{1/2}\left(1 + \frac{1\cdot 3}{4^2}x^2 + \frac{1\cdot 3\cdot 5\cdot 7}{4^2\cdot 8^2}x^4 + \cdots\right)$

$$y_2 = c_0\left\{y_1 \log x + 2x^{1/2}\left[\frac{1\cdot 3}{4^2}\left(1 + \frac{1}{3} - \frac{1}{2}\right)x^2 + \frac{1\cdot 3\cdot 5\cdot 7}{4^2\cdot 8^2}\left(1 + \frac{1}{3} - \frac{1}{2} + \frac{1}{5} + \frac{1}{7} - \frac{1}{4}\right)x^4 + \cdots\right]\right\}$$

**89.** $y = c_0 \sum_{n=0}^{\infty} \frac{(4x)^n}{2n!} + c_0 x^{1/2} \sum_{n=0}^{\infty} \frac{(4x)^n}{(2n+1)!}$

**90.** $y = A(1 + 2x^2 + 3x^4 + 4x^6 + \cdots) + Bx^{-1}(1 + 3x^2 + 5x^4 + 7x^6 + \cdots).$

**91.** $y = A\left(\frac{1}{2} + \frac{1\cdot 4}{5!}x^3 + \frac{1\cdot 4\cdot 7}{8!}x^6 + \cdots\right) + Bx^{-2}\left(1 + \frac{2}{3!}x^3 + \frac{2\cdot 5}{6!}x^6 + \cdots\right).$

**92.** $y_1 = x^{1/2}\left[1 + \sum_{n=1}^{\infty} \frac{(-1)^n}{2^n n!}x^n\right], y_2 = x\left[1 + \sum_{n=1}^{\infty} \frac{(-1)^n}{3\cdot 5\cdots(2n+1)}x^n\right], c_0 = 1$

**93.** $y_1 = x^{1/2} + \sum_{n=1}^{\infty} \dfrac{(-1)^n \, 3x^{(n+1)/2}}{2^n n!(2n-3)(2n-1)(2n+1)},$

$y_2 = 1 + 2x + \dfrac{1}{3}x^2, \ c_0 = 1.$

**94.** $y_1 = 1 + \sum_{n=1}^{\infty} \dfrac{(-1)^{n+1} x^n}{4n^2 - 1}, \ y_2 = x^{-1/2} + x^{1/2}, \ c_0 = 1.$

**95.** $y_1 = x + \dfrac{1}{15} \sum_{n=1}^{\infty} (2n+3)(2n+5)x^{n+1},$

$y_2 = x^{1/2} + \dfrac{1}{2} \sum_{n=1}^{\infty} (n+1)(n+2)x^{(n+1)/2}, c_0 = 1.$

**96.** $y_1 = 1 + 3 \sum_{n=1}^{\infty} \dfrac{5^n x^n}{n!(2n+1)(2n+3)}, \ y_2 = x^{-3/2} + 10x^{-1/2}, c_0 = 1.$

**97.** $y_1 = x^{3/2} + \sum_{n=1}^{\infty} \dfrac{(-1)^{n+1} \, x^{(n+3)/2}}{3^{n-1}(2n-1)(2n+1)(2n+3)},$

$y_2 = 1 + \dfrac{2}{3}x + \dfrac{1}{9}x^2, c_0 = 1$

## Chapter 6

**1.** $x = (1/4)\cos 16t$, $v = -4 \sin 16t$, amplitude = (1/4) ft, period = ($\pi/8$) s, frequency = ($8/\pi$) cycles/s, $x = (\sqrt{2}/8)$ ft, $v = -2\sqrt{2}$ ft/s, $a = -32\sqrt{2}$ ft/s$^2$.

**2.** (a) $x = \dfrac{1}{4}\sin(8t)$ ft, $v = 2\cos(8t)$ ft.

(b) Amplitude = $\dfrac{1}{4}$ ft, $T = \dfrac{\pi}{4}$ s, $f = \dfrac{4}{\pi}$ cycles/s.

(c) 1.89 ft/s, 5.3 ft/s$^2$.

**3.** $\dfrac{\sqrt{2}\pi}{8}$.

**4.** $x(t) = -\dfrac{1}{4}\cos 4\sqrt{6}\,t$.

**5.** $x(t) = \dfrac{1}{2}\cos 2t + \dfrac{3}{4}\sin 2t = \dfrac{\sqrt{13}}{4}\sin(2t + 0.5880)$.

**6.** (b) $x = \dfrac{e^{-8t}}{2}(\sin 8t + \cos 8t)$, taking downward as positive.

(c) $x = \frac{\sqrt{2}}{2} e^{-8t} \sin\left(8t + \frac{\pi}{4}\right)$, $A(t) = \frac{\sqrt{2}}{2} e^{-8t}$, $\omega = 8, \phi = \frac{\pi}{4}$, quasi period $= \frac{2\pi}{8} = \frac{\pi}{4}$ s.

**7.** $x = 0.125\, e^{-3t} (3 \sin 4t + 4 \cos 4t) = 0.625\, e^{-3t} (\sin (4t + \phi))$ where $\cos \phi = 3/5, \sin \phi = 4/5$, or $\phi = 0.927$ radians $= 53°$.

**8.** $x = e^{-16t} \sin (8t + 0.5)$, critically damped.

**9.** $s/4$, $s/2$, $x/2 = e^{-2}$, i.e. the weight is approximately 0.14 ft below the equilibrium position.

**10.** (a) $x(t) = \frac{4}{3} e^{-2t} - \frac{1}{3} e^{-8t}$. (b) $x(t) = -\frac{2}{3} e^{-2t} + \frac{5}{3} e^{-8t}$.

**11.** (a) $x(t) = e^{-2t}\left(-\cos 4t - \frac{1}{2} \sin 4t\right)$, (b) $x(t) = \frac{\sqrt{5}}{2} e^{-2t} \sin (4t + 4.249)$.

(c) $t = 1.294$ s.

**12.** (a) $\beta > \frac{5}{2}$; (b) $\beta = \frac{5}{2}$; (c) $0 < \beta < \frac{5}{2}$. **13.** $\beta = 16$.

**14.** Use Archimedes' principle to obtain $\frac{W}{g} \frac{d^2x}{dt^2} = -6250x$ and period $T = \sqrt{6250\, g} = 200\sqrt{5}\, W = 1270$ lb (approx.).

**15.** $C = 5.1$ lb ft/s. **16.** $T = 1.418$ hr., $|v| = 17{,}725$ miles/hr.

**17.** $x = -\frac{1}{3} e^{-12t} + \frac{2}{3} e^{-6t}$, $v = 4e^{-12t} - 4e^{-6t}$.

**18.** $x = e^{-t}\left[\frac{1}{2} \cos \sqrt{15}\, t - \frac{1}{6}(19\sqrt{15}) \sin \sqrt{15}\, t\right] + 12 \sin 4t$,

$v = e^{-t}\left[-48 \cos \sqrt{15}\, t + \frac{1}{3}(8\sqrt{15}) \sin \sqrt{15}\, t\right] + 48 \cos 4t$.

**19.** (a) $x = 0.960e^{-1.56t} - 0.695e^{-6.45t} - 0.298 \sin 10t - 0.265 \cos 10t$.

(b) Steady state part $= -0.298 \sin 10t - 0.265 \cos 10t = 0.397 \sin(10t + 3.87)$

(c) Amplitude $= 0.397$, period $= \frac{2\pi}{10} = \frac{\pi}{5}$, frequency $= \frac{5}{\pi}$ cycles/s.

**20.** $x(t) = e^{-t/2}\left(-\frac{4}{3} \cos \frac{\sqrt{47}}{2} t - \frac{64}{3\sqrt{47}} \sin \frac{\sqrt{47}}{2} t\right) + \frac{10}{3}(\cos 3t + \sin 3t)$.

**21.** $x(t) = -\frac{1}{2} \cos 4t + \frac{9}{4} \sin 4t + \frac{1}{2} e^{-2t} \cos 4t - 2e^{-2t} \sin 4t$.

**22.** The solution may be written as $x = 2A(t) \sin 9t$, where $A(t) = \sin t$, the time varying amplitude, and is a slowly varying function (period $= 2\pi$) in comparison with the wave sin $9t$ (period = $2\pi/9$). The wave sin $9t$ is said to be *amplitude modulated.* In the theory of acoustics, these fluctuations of amplitude are called *beats,* the loud sound corresponding to the large amplitudes. Beats may occur when two tuning forks having nearly equal frequencies are set into vibrations simultaneously. A practical use of this is in tuning of pianos (or other instruments) where successful tuning is marked by adjusting the frequency of a note to that of a standard note until beats are eliminated. The phenomenon is also important in the theories of electricity and optics.

**24.** $\theta = \theta_0 \cos \omega t + (v_0/\omega) \sin \omega t$, where $\omega = \sqrt{2k/mr^2}$.

**25.** $q(t) = -\dfrac{1}{2} e^{-10t}(\cos 10t + \sin 10t) + \dfrac{3}{2}$, $q(t) = \dfrac{3}{2}$ coulombs.

**26.** (a) $q = 4 - 2e^{-5t/2}(2 \cos 5t + \sin 5t)$, $I = 25e^{-5t/2} \sin 5t$.

(b) Transient terms for $q$ and $I$ are $-2e^{-5t/2}$ $(2 \cos 5t + \sin 5t)$ and $25e^{-5t/2}$ $\sin 5t$, respectively. The steady state term for $q$ is 4.

(c) $q = 4$, $I = 0$.

**27.** $q = 2(\cos 40t - \cos 50t)$, $I = 20(5 \sin 50t - 4 \sin 40t)$.

**28.** $q = 1 + \cos 10t$, $I = -10 \sin 10t$.

**29.** (a) $\dfrac{d^2q}{dt^2} + 12\dfrac{dq}{dt} + 100q = 48 \sin 10t$, $q = I = 0$ at $t = 0$.

(b) $q = \dfrac{1}{10} e^{-6t}(4 \cos 8t + 3 \sin 8t) - \dfrac{2}{5} \cos 10t$.

**30.** $y = \dfrac{x^2}{625}$, $-250 \le x \le 250$, absolute value of slope = 0.8.

**31.** (a) 15.6 lb (b) $y = 31.2\left[\cosh\left(\dfrac{x}{31.2}\right) - 1\right]$, choosing the minimum point at (0, 0).

**32.** (a) $y = \dfrac{S}{6EI}(3Lx^2 - x^3)$, $0 \le x \le L$. (b) $\dfrac{SL^3}{3EI}$

**33.** (a) 2.4 in.

**34.** (a) 0.1152 in. (b) 0.1617 in.

**35.** $y = \dfrac{-W}{48El}(2x^4 - 5lx^3 + 3l^2x^2)$.

**37.** $c_i - c_e = \dfrac{Qr_0}{3h}$.

**40.** (a) $\bar{P} = 8$, $P = -13e^{-t} + 7e^{-2t} + 8$, stable.

(b) $\bar{P} = 1$, $P = e^t\left(-\dfrac{3}{2} \sin 2t + 2 \cos 2t\right) + 1$, unstable.

## Chapter 7

**1.** $\frac{1}{x} = t + c_1, \quad y = c_2 e^{-t}.$

**2.** $x = -3e^{-t} + c_1, \quad y = \frac{3}{2} e^{-t} + c_1 te^t + c_2 e^t.$

**3.** $x = t^2 + c_1, \; y = t^3 + t^2 + 3c_1 t + c_2.$

$z = t^4 + \frac{5}{3}t^3 + \left(6c_1 + \frac{1}{2}\right)t^2 + (c_1 + 4c_2)t + c_3.$

**4.** $x = c_1 e^t - \frac{1}{2}(\sin t + \cos t), \; y = t - 1 + c_2 e^{-t}.$

**5.** $x = (2t + c_1)\, e^{3t}, \qquad y = -\frac{3 \sin 2t + 2 \cos 2t}{15} - e^{3t}(3t^2 + 3c_1 t + 2t + c_2).$

**6.** $x = c_1 e^{t/3} + c_2 e^{-t/3} - 6t, \qquad y = -2c_1 e^{t/3} - c_2 e^{-t/3} + \frac{1}{2} e^t + 9t + 9.$

**7.** $x = \sin t + c_1 \sin 2t + c_2 \cos 2t,$

$y = \frac{1}{2} \cos t - \frac{2}{5} c_1 \cos 2t + \frac{2}{5} c_2 \sin 2t + c_3 e^t + c_4 e^{-t}.$

**8.** $x = c_1 e^t + c_2 \sin t + c_3 \cos t - \frac{1}{15}(3 \cos 2t + 4 \sin 2t),$

$y = c_1 e^t + (c_2 - c_3) \sin t + (c_2 + c_3) \cos t - \frac{4}{15}(2 \cos 2t + \sin 2t).$

**9.** $x = c_1 e^{-t} + c_2 e^{t/2} + 2e^t + 2t, \quad y = -c_1 e^{-t} + \frac{c_2}{2} e^{t/2} + 2e^t + \frac{t^2}{2} - t + c_3.$

**10.** $x = t^2 + t - 1, \quad y = -t.$

**11.** $x = \frac{Em}{eH^2}\left(1 - \cos \frac{eH}{m} t\right), \quad y = \frac{E}{H} t - \frac{Em}{eH^2} \sin \frac{eH}{m} t$. These are the parametric equations of a cycloid.

**12.** No solutions.

**13.** Infinitely many. Define $x(t)$ arbitrarily and solve for $y(t)$.

**14.** Infinitely many; $x(t) = -y(t)$.

**15.** (a) $x = c_1 e^t, \quad y = c_2 e^{3t} - \frac{c_1}{2} e^t, \qquad z = c_3 e^{2t} - \frac{3c_1}{2} e^t - c_2 e^{3t}.$

(b) $x = c_1 e^t, \quad y = -\frac{3c_1}{4} e^t + c_2 e^{-t}, \quad z = -\frac{3c_1}{2} e^t + \frac{2c_2}{5} e^{-t} + c_3 e^{3t/2}$

**16.** (b) $x = 80t, \quad y = 80\sqrt{3}\, t - 16t^2.$

(c) Range = $400\sqrt{3}$ ft, max-height = 300 ft, time of flight = 8.66 s.

(d) Position after 2 s = (160, 213),
Position after 4 s = (320, 298),
Velocity after 2 s = 109 ft/s,
Velocity after 4 s = 80.7 ft/s.

**17.** Max. range = 165 miles, height = 41.5 miles, time of flight = 3 min. 5 s.

**18.** (a) 4 s (b) 200 ft.

**19.** 283 ft/s; 15.8 s.

**21.** $f(r) = 2mh^2r^{-3}$.

**22.** $e = 0.278 < 1$, $\beta(\pi) \cong 7010$ miles, $T = 138$ min.

**23.** $r_0 \cong 4289$ miles, altitude = 329 miles.

**24.** Orbital speed $\approx$ 16,860 miles/h, Escape speed $\approx$ 23,844 miles/h.

**26.** $x = \cos \omega_1 t$, $y = \sqrt{3} \cos \omega_1 t$.

**28.** $i_1 = i_2 + i_3$ and any two of $4i_1 + 20 \dfrac{di_3}{dt} - \sin 2t = 0$,

$$4i_2 + 4\frac{di_2}{dt} + 20\frac{di_3}{dt} - \sin 2t = 0, \quad 20\frac{di_3}{dt} - 2\frac{di_2}{dt} = 0.$$

**29.** $i_3 = i_1 + i_2$, and any two of $30i_1 + 20q - 10i_2 - 2\dfrac{di_2}{dt} = 0$,

$$2\frac{di_2}{dt} + 10i_2 - 120 = 0, \quad 30i_1 + 20q - 120 = 0, \text{ where } i_1 = dq/dt.$$

**30.** $I_1 = 3 - 2e^{-5t} - e^{-20t}$, $I_2 = 4e^{-5t} - e^{-20t} - 3$. Steady state currents = 3 and $-3$.

**31.** $I_1 = 2e^{-5t} + e^{-20t} + 3 \sin 10t - 3 \cos 10t$,
$I_2 = 3 \cos 10t - 4e^{-5t} + e^{-20t}$.

**32.** $y_{\max} = 4.5$ lb, $t = \dfrac{25}{3} \log 2 \cong 5.78$ min.

**33.** $\dfrac{dx}{dt} = (A - B)x + Cy$, $\dfrac{dy}{dt} = Bx - (D + C)y + Ez$, $\dfrac{dz}{dt} = Dy - (E + F)z$.

**34.** When $t = \dfrac{100}{2\sqrt{3}} \log \dfrac{3 + \sqrt{3}}{3 - \sqrt{3}} \approx 38$ min., $\approx 39$ lb salt.

**35.** (a) $x' = -0.134x + 0.02y$, $y' = 0.036x - 0.02\,y$.

(b) $x = 10c_1e^{-0.14t} + c_2e^{-0.014t}$, $y = -3c_1e^{-0.14t} + 6c_2e^{-0.014t}$.

**37.** (a) $x_1 = \frac{10}{17} + \frac{24}{17} e^{-0.85t}$, $x_2 = \frac{24}{17}(1 - e^{-0.85t})$, (b) 10/17 mg, 24/17 mg.

**38.** $x(t) = 4e^{-t} + 3e^{-7t} + 1$, $y(t) = 8e^{-t} - 3e^{-7t} + 2$. Stable arms race.

**39.** $x(t) = e^{-6t} + 6e^{2t} - 2$, $y(t) = -e^{-6t} + 6e^{2t} - 3$. Run away arms race.

**41.** (a) 500 predators, 4000 prey.

(b) 100 $\pi$ or 314 days.

(c) $\frac{(x-4000)^2}{16c^2} + \frac{(y-500)^2}{c^2} = 1.$.

**42.** (a) $\frac{x''}{x} - \left(\frac{x'}{x}\right)^2 + b_1 x' = \left(a_1 - b_1 x - \frac{x'}{x}\right)\left[-a_2 + c_2 x + \frac{b_2}{c_1}\left(a_1 - b_1 x - \frac{x'}{x}\right)\right]$.

(b) $\left(0, \frac{a_2}{b_2}\right), \left(\frac{a_1}{b_1}, 0\right), \left(\frac{a_1 b_2 - a_2 c_1}{b_1 b_2 - c_1 c_2}, \frac{a_2 b_1 - a_1 c_2}{b_1 b_2 - c_1 c_2}\right)$.

(c) $y = cx$.

(d) $\frac{dy}{dx} = \frac{y}{x}\left[\frac{b_2(y/x) + c_2}{b_1 + c_1(y/x)}\right]$; Starvation of both species.

## Chapter 8

**1.** $\frac{1}{s-2} + \frac{24}{s^4} + \frac{3(s-2)}{s^2+9}$.

**2.** $\frac{s \cos b - a \sin b}{s^2 + a^4}$.

**3.** $\frac{2\,abs}{[s^2 + (a+b)^2][s^2 + (a-b)^2]}$.

**4.** $\frac{18}{s(s^2+36)}$.

**5.** $\frac{6}{(s+3)^4}$.

**6.** $\frac{4}{s^2 + 4s + 20}$.

**7.** $\frac{a(s^2 + 2a^2)}{s^4 + 4a^4}$.

**8.** $\frac{6(s+4)}{(s^2 + 8s + 25)^2}$.

**9.** $\frac{1 + e^{-\pi s}}{s^2 + 1}$.

**10.** $\sqrt{\frac{\pi}{s}} e^{-1/4s}$.

**11.** $\frac{1}{s+a}, \frac{\omega}{s^2 + \omega^2}$.

**13.** $\frac{3}{2} - 3x^2 + \frac{1}{4}x^4$.

**14.** $e^{ax} \cos bx$.

**15.** $-4e^{-2x} \cos 2x$.

**16.** $4e^{3x} - e^{-x}$.

**17.** $\frac{1}{3}x(e^x - e^{-2x})$.

**18.** $\cos 2x$.

**19.** $\frac{1}{2}(\sin x - xe^{-x})$.

**20.** $e^{-x}(1 - \cos x)$.

**21.** $\frac{2}{\sqrt{3}} \sinh\left(\frac{1}{2}x\right) \sin\left(\frac{1}{2}\sqrt{3}\, x\right)$.

**22.** $\cos ax \sinh ax$.

**23.** $-\frac{1}{6}e^{-x} - \frac{4}{3}e^{2x} + \frac{7}{2}e^{3x}$.

**24.** $-\frac{1}{3}e^{-x} - \frac{7}{2}x^2 e^{2x} + 4xe^{2x} + \frac{1}{3}e^{2x}$.

**25.** $2e^x - 2\cos x + \sin x$.

**26.** $\frac{1}{3}e^{-x}(\sin x + \sin 2x)$ (**Hint:** Express the given fraction as equal to $\frac{As+B}{s^2+2s+2} + \frac{Cs+D}{s^2+2s+5}$).

**28.** $\frac{1}{6}x^3 e^{-5x}$.

**29.** $\frac{2as}{(s^2-a^2)^2}$.

**30.** $\frac{ax - \sin ax}{a^3}$.

**31.** $(1 - ax)e^{-ax}$.

**32.** $\frac{s}{s^2+a^2}, \frac{a}{s^2+a^2}$.

**33.** $\frac{1}{s-a}, \frac{1}{(s-a)^2}$.

**34.** $\frac{54(s^2+12)}{s^2(s^2+36)^2}$.

**35.** $\frac{2s^3 - 6a^2 s}{(s^2+a^2)^3}$.

**36.** $\log\left(\frac{s+b}{s+a}\right)$.

**37.** $\cos^{-1}\frac{s}{2}$.

**38.** $\frac{1-e^{-x}}{x}$.

**39.** $x = e^t \cos t$.

**40.** $y = \frac{1}{3}(5e^t + e^{-2t}) - e^{-t}$.

**41.** $x = 1 + 2t + 2\cos t - \sin t$.

**42.** $x = \frac{1}{2} + \frac{1}{2}e^{2t} - te^{3t}$.

**43.** $x = \frac{at}{2}\sinh t$.

**44.** $y = e^t + \frac{1}{4}e^{-2t} - \frac{t}{2} - \frac{1}{4}$.

**45.** $y = 2t + 3 + \frac{1}{2}(e^{3t} - e^t) - 2e^{2t}$.

**46.** $y = \frac{t}{2\omega}\sin \omega t$.

**47.** $x = \frac{1}{5}(4\cos 3t + 4\sin 3t + \cos 2t)$.

**48.** $y = \frac{e^{-2x}}{20}(39\cos 2x + 47\sin 2x) + \frac{1}{20}(2\sin 2x + \cos 2x)$.

**49.** $x = e^t(\cos t + \sin t) - e^{-t}\sinh 2t$.

**50.** $y = x$.

**51.** $y = 5e^{-x}$.

**52.** $y = 1 - (4 - c)xe^{-2x}, y'(0) = c - 4$.

**53.** $x = \frac{1}{2}(e^t + \cos t + 2 \sin t - t \cos t), \quad y = \frac{1}{2}(t \sin t - e^t + \cos t - \sin t)$.

**54.** $x = 2(1 - e^{-t} - te^{-t}), \quad y = -t(1 + 2e^{-t}) + 2(1 - e^{-t})$.

**55.** $4x^2 + 4xy + 5y^2 = 4$.

**56.** $i_1 = \frac{a}{p+w}(\sin wt + \sin pt), \quad i_2 = \frac{a}{p+w}(\cos wt - \cos pt)$.

**57.** $x = \frac{1}{3}e^{4t} - 2e^{-2t} + \frac{5}{3}e^t, \; y = \frac{1}{2}e^{4t} + \frac{3}{2}e^{-2t}, \; z = 3e^{4t}$.

**58.** Write $\frac{s}{(s^2 + a^2)^2}$ as $\frac{s}{s^2 + a^2}\frac{1}{s^2 + a^2}, \frac{x \sin ax}{2a}$.

**59.** $\frac{1}{4}\sin 2x + \frac{x}{2}\cos 2x$.

**60.** $\frac{a \sin ax - b \sin bx}{a^2 - b^2}$.

**61.** $\frac{1}{4}(e^x - e^{-3x})$.

**62.** $\frac{1}{16}(e^{2x} - e^{-2x} - 4xe^{-2x})$.

**63.** $\frac{1}{2}(\sin x - \cos x + e^{-x})$.

**64.** $\frac{1}{64}x(\sin 2x - 2x \cos 2x)$.

**65.** $\frac{1}{s}\tanh\left(\frac{sa}{4}\right)$.

**66.** $\frac{\frac{1}{s^2 T} - e^{-sT}}{s(1 - e^{-sT})}$.

**67.** $\frac{1}{s^2}\tanh\left(\frac{1}{2}as\right)$.

**68.** $\frac{1}{(1 - e^{-\pi s})(s^2 + 1)}$.

**69.** $\frac{E\omega}{s^2 + \omega^2}\coth\left(\frac{\pi s}{2\omega}\right)$.

**70.** $y = e^{2x}$.

**71.** $x = 1 + 2t$.

**72.** $V(t) = 8 - 8e^{-4t}$.

**73.** $x(t) = \frac{1}{3}\cos 8t + \frac{1}{4}\sin 8t$, amplitude $= \frac{5}{2}$, period $= \frac{\pi}{2}$, frequency $= \frac{2}{\pi}$.

**74.** $x(t) = -\frac{3}{2}e^{-7t/2}\cos\frac{\sqrt{15}}{2}t - \frac{7\sqrt{15}}{10}e^{-7t/2}\sin\frac{\sqrt{15}}{2}t$.

**75.** $(n \sin at - a \sin nt)\frac{F}{mn(n^2 - a^2)}, n^2 = \frac{k}{m}$.

**76.** $\dfrac{dx}{dt}=\dfrac{I}{m}e^{-\mu t/(2m)}\left(\cos nt-\dfrac{\mu}{2mn}\sin nt\right), n^2=\dfrac{k}{m}-\dfrac{2}{4m^2}, x=\dfrac{I}{mn}e^{-\mu t/(2m)}\sin nt.$

**77.** $x=\begin{cases}0, & t<t_0\\ \dfrac{P_0(t-t_0)}{m}, & t>t_0\end{cases}$

**80.** $\dfrac{Wl^2}{16EI}x^2-\dfrac{Wl}{12\,EI}x^3+\dfrac{W}{24EI}x^4-\dfrac{W}{24EI}\left(x-\dfrac{l}{2}\right)^4 u\left(x-\dfrac{l}{2}\right).$

**82.** $i(t) = 20e^{-2t} + 2e^{-2t}\sin 25t.$

**83.** $i_1=5\dfrac{5e^{-2t}}{\sqrt{6}}\sinh\dfrac{\sqrt{6}}{2}t-5e^{-2t}\cosh\dfrac{\sqrt{6}}{2}t,$

$i_2=5-\dfrac{10}{\sqrt{6}}e^{-2t}\sinh\dfrac{\sqrt{6}}{2}t-5e^{-2t}\cosh\dfrac{\sqrt{6}}{2}t,$

$i_3=\dfrac{5}{\sqrt{6}}e^{-2t}\sinh\dfrac{\sqrt{6}}{2}t.$

**84.** $i(t)=\dfrac{e^{-10t}}{101}-\dfrac{\cos t}{101}+\dfrac{10}{101}\sin t-\dfrac{10}{101}e^{-10(t-3\pi/2)}.$

$u\left(t-\dfrac{3\pi}{2}\right)+\dfrac{10}{101}\cos\left(t-\dfrac{3\pi}{2}\right)u\left(t-\dfrac{3\pi}{2}\right)+\dfrac{1}{101}\sin\left(t-\dfrac{3\pi}{2}\right)u\left(t-\dfrac{3\pi}{2}\right).$

**85.** $q=\dfrac{3}{5}e^{-10t}+6t\,e^{-10t}-\dfrac{3}{5}\cos 10t,\quad i=-60\,te^{-10t}+6\sin 10t.$

Steady state current = $6\sin 10t$.

## Chapter 9

**1.** $pq = 4xyz.$

**2.** $px + qy = pq.$

**3.** $z = px + qy + r\sqrt{1+p^2+q^2}.$

**4.** $py - qx = 0.$

**5.** $px^3 + qx = 2y^2.$

**6.** $\dfrac{\partial^2 v}{\partial t^2}=\dfrac{a^2}{r^2}\dfrac{\partial}{\partial r}\left(r^2\dfrac{\partial v}{\partial r}\right).$

**7.** $py - qx = y^2 - x^2.$

**8.** $z = 24x\cos 3xy - 36x^2\sin 3xy + f(x) + \phi(y).$

**9.** $u = f(x) + x\phi(y) + \psi(y) - \dfrac{1}{12}\sin(2x + 3y).$

**10.** $z = e^y\cosh x + e^{-y}\sinh x.$

**11.** $\phi(xy, xz) = 0.$

**12.** $\dfrac{x^2}{2} + yz = f(x + y + z).$

**13.** $f\left(\dfrac{x-y}{y-z},\dfrac{y-z}{z-x}\right)=0.$

**14.** $f(x^2 + y^2 + z^2, xyz) = 0.$ **15.** $xy + yz + zx = \phi(xyz).$

**16.** $z = ax - \dfrac{ay}{1+a} + b.$ **17.** $\pm(x + by) + c = 2(bz - ab - 1).$

**18.** $2z = ay^2 - \dfrac{a}{a+1}x^2 + b.$ **19.** $z = a(x - y) - (\cos x + \cos y) + b$

**20.** $z^{3/2} = (x + a)^{3/2} + (y + a)^{3/2} + b.$

**21.** $z = ax + by + \sqrt{1 + a^2 + b^2}.$ **22.** $z = ax + by + a^2b^2.$

**23.** $z = -2 - \log(xy).$

**24.** $z = \left(\sqrt{a} + \sqrt{x}\right)^2 + \left(\sqrt{b} + \sqrt{y}\right)^2.$ **25.** $z = ax^b y^{1/b}.$

**26.** $\log(z - ax) = y - a\log(a + y) + b.$

**27.** Put $z^2 = Z$, $z = b + x\sqrt{(x^2 + a^2)} + a\log\left[x + \sqrt{x^2 + a^2}\right]$

$$+ y\sqrt{y^2 - a^2} + a\log\left[y + \sqrt{y^2 - a^2}\right].$$

**28.** $z = f_1(y + x) + f_2(y - 2x).$ **29.** $z = f_1(y + x) + xf_2(y + x) - \sin x.$

**30.** $z = f_1(y + x) + f_2(y - 5x) + \dfrac{x^2y^2}{2} - \dfrac{4x^2y}{3} + \dfrac{7x^4}{4}.$

**31.** $z = f_1(y - 2x) + f_2(y + 3x) + \dfrac{x^3y}{6} + \dfrac{x^4}{24}.$

**32.** $z = f_1(y + 2x) + f_2(2y + x) - \dfrac{5}{3}x\cos(2x + y).$

**33.** $z = f_1(y - x) + xf_2(y - x) + x^2y - x^3 + \dfrac{1}{2}x^2\sin(x - y).$

**34.** $z = f_1(2y - 3x) + xf_2(2y - 3x) + \dfrac{1}{2}x^2e^{3x-2y}.$

**35.** $z = f_1(y + 2x) + f_2(y - 3x) + \dfrac{1}{5}\sin(2x + y).$

**36.** $z = f_1(x) + e^{3y}f_2(2y - x) + \dfrac{3}{50}[4\cos(3x - 2y) + 3\sin(3x - 2y)].$

**37.** $z = e^x f_1(y - x) + e^{3x}f_2(y - 2x) + x + 2y + 6.$

**38.** $z = e^x f_1(y) + e^{-x}f_2(x + y) + \dfrac{1}{2}\sin(x + 2y) - xe^y.$

**39.** $z = f_1(y + x) + f_2(y - 2x) - e^{x+2y}$

**40.** $z = e^{-x}f_1(y) + e^x f_2(y - x) - \dfrac{1}{10}[\cos(x + 2y) + 2\sin(x + 2y)].$

**41.** $z = e^{-2x} f_1(y+x) + e^x f_2(y-x) - \frac{1}{4}e^{x-y} + \frac{1}{2}\left(x^2y + xy + \frac{3}{2}x^2 + \frac{3}{2}y + 3x + \frac{21}{4}\right).$

**42.** $z = f_1(y+2x) + e^{-x} f_2(y+2x) - xe^{x+y}.$

**43.** $z = \sum ce^{k^2x+ky} + \frac{1}{82}[\sin(x-3y) + 9\cos(x-3y)].$

**44.** $\psi = \sum ce^{hx+h^2t}.$ **45.** $\psi = \sum ce^{n(x\cos\alpha + y\sin\alpha)}.$

**46.** $z = \sum c_1 e^{h_1x+h_1^2y} + \sum c_2 e^{h_2x+2h_2^2y}.$

**47.** $z = \sum ce^{np}, p = x\cos\alpha + y\sin\alpha.$

**48.** $z = f_1(y+ix) + f_2(y-ix) - (m^2+n^2)^{-1}\cos mx \cos ny.$

**49.** $z = f_1(y) + f_2(y+2x) + xf_3(y+2x) + \frac{1}{4}x^2\sin(x+2y).$

**50.** $z = f_1(y-x) + f_2(y-2x) - \frac{7}{6}x^2 + \frac{3}{2}x^2y.$

**51.** $z = f_1(y-3x) + f_2(y+2x) - y\sin x - \cos x.$

**52.** $z = f_1(y+2x) + f_2(y-x) + ye^x.$

**53.** $z = f_1(x) + y^2 f_2(x) + \frac{1}{2}xy^2\log y.$

**54.** $z = f_1\left(\frac{y}{x}\right) + x f_2\left(\frac{y}{x}\right) + \frac{1}{n(n-1)}(x^2+y^2)^{n/2}.$

**55.** $z = f_1\left(\frac{y}{x}\right) + x f_2\left(\frac{y}{x}\right) + \frac{1}{(m+n)(m+n-1)}x^m y^n$

**56.** $z = f_1(xy) + f_2\left(\frac{y}{x}\right).$

**57.** $z = f_1(xy) + f_2\left(\frac{y}{x}\right) + \frac{1}{6}(\log x)^3.$

**58.** $z = \phi(y - \sin x) + \psi(y + \sin x).$

**59.** $2z = \psi(x + \tan y + \phi(x - \tan y)).$

**60.** $z = \psi\left(\frac{y}{x}\right) + x\phi\left(\frac{y}{x}\right).$

**61.** $x + \phi(z) = \psi(x+y+z).$

**62.** $\phi(xy) + z = \psi(x+y).$

**63.** $x - yz + \phi(z) = \psi(y)$.

**64.** $x + \phi(z) = \psi(y - z)$.

**65.** $y + x\phi(x + y + z) = \psi(x + y + z)$.

**66.** $z + y\phi(y^2 + 2x) - y^3 = \psi(y^2 + 2x)$.

**67.** $\psi(z) = \log x + \phi(y)$.

**68.** $y = xz - \psi(z) - \phi(x)$.

**69.** $u = 6e^{-3x-2t}$.  **70.** $z = 4e^{x-y} - e^{2x-5y}$.

**71.** $\dfrac{\pi^2}{3} + 4\sum_{n=1}^{\infty} (-1)^n \cos\dfrac{nx}{n^2} - 2\sum_{n=1}^{\infty} (-1)^n \sin\dfrac{nx}{n}$.

**72.** $-1 + \pi \sin x - \dfrac{1}{2}\cos x + 2\sum_{n=2}^{\infty} \dfrac{\cos nx}{n^2 - 1}$

**73.** $\dfrac{\pi^2}{3} + 4\sum_{n=1}^{\infty} (-1)^n \dfrac{\cos nx}{n^2}$.

**74.** $\dfrac{1}{2} + \dfrac{2}{\pi}\left[\sin\dfrac{\pi x}{2} + \dfrac{\sin(3\pi x/2)}{3} + \dfrac{\sin(5\pi x/2)}{5} + \dots\right]$.

**75.** $\dfrac{A}{2} + \dfrac{2A}{n\pi}\sin\dfrac{n\pi x}{2}\cos\dfrac{n\pi x}{2}$.

**77.** $y = \dfrac{y_0}{l}\left(3\sin\dfrac{\pi x}{l}\cos\dfrac{c\pi t}{l} - \sin\dfrac{3\pi x}{l}\cos 3\dfrac{c\pi t}{l}\right)$.

**78.** $y = \dfrac{8kl^2}{\pi^3}\sum_{n=1}^{\infty} \dfrac{1}{(2n-1)^3}\sin\dfrac{2n-1}{l}\cos\dfrac{(2n-1)\pi ct}{l}$.

**79.** $y = \dfrac{3kl}{4\pi a}\sin\dfrac{\pi x}{l}\sin\dfrac{\pi at}{l} - \dfrac{kl}{12\pi a}\sin\dfrac{3\pi x}{l}\sin\dfrac{3\pi at}{l}$.

**80.** $y = \sum_{n=1}^{\infty} b_n \sin\dfrac{n\pi x}{l}\cos\dfrac{n\pi ct}{l}$, where $b_n = \dfrac{2}{l}\int_0^l f(x)\sin\dfrac{n\pi x}{l}\,dx$

**81.** $u = \dfrac{40x}{l} + 20 - \dfrac{40}{\pi}\sum_{m=1}^{\infty} \dfrac{1}{m}\sin\dfrac{2m\pi x}{l}\, e^{-4c^2m^2\pi^2t/l^2}$

**82.** $u(x, t) = \dfrac{400}{\pi^2}\sum_{n=1}^{\infty} \dfrac{(-1)^n}{(2n+1)^2} e^{-[(2n+1)c\pi/100]^2 t}\sin\dfrac{(2n+1)\pi x}{100}$.

**83.** $u = x + 40 - \dfrac{20}{\pi}\sum_{n=1}^{\infty} \dfrac{1}{n}(2\cos n\pi + 1)\sin\dfrac{n\pi x}{20}\, e^{-\alpha^2n^2\pi^2t/400}$.

**84.** $u(x,t) = 50 - \dfrac{400}{\pi^2}\sum_1^{\infty}\dfrac{1}{(2n-1)^2}\cos\dfrac{(2n-1)x\pi}{50}\, e^{-c^2\pi^2(2n-1)^2 t/2500}$

**85.** $u = u_0 e^{-\sqrt{(n/2k)}\,x}\sin[nt - \sqrt{n/(2k)}\,x]$.

**86.** $u(x, y) = \dfrac{2a\mu}{\pi}\dfrac{(-1)^{n+1}}{n}e^{-n\pi y/a}\sin\dfrac{n\pi x}{a}$.

**87.** $u(x, y) = \dfrac{2}{\pi}\sum_{n=1,3,5,\ldots}^{\infty}\dfrac{2n^2+\pi^2}{n(n^2+\pi^2)}\dfrac{\sin nx \sinh x(n-y)}{\sinh n\pi}$.

**89.** $u(r, \theta) = \dfrac{8\mu}{\pi}\sum_{n=1}^{\infty}\left(\dfrac{r}{a}\right)^{2n-1}\dfrac{\sin(2n-1)\theta}{(2n-1)^2}$

**90.** $u(r, \theta) = \dfrac{3200}{\pi^2}\sum_1^{\infty}\left(\dfrac{r}{10}\right)^{2n-1}\dfrac{\sin(2n-1)\theta}{(2n-1)^2}$.

**91.** $u(r, \theta) = \dfrac{2}{\pi}\sum_{n=1,3,5,\ldots}^{\infty}\left(\dfrac{a}{r}\right)^{2n}\dfrac{r^{4n}-b^{4n}}{a^{4n}-b^{4n}}\dfrac{\sin(2n\theta)}{n^3}$

**92.** $u(r, \theta) = 50 - \dfrac{200}{\pi}\sum_{n=1}^{\infty}\dfrac{1}{2n-1}\left(\dfrac{r}{a}\right)^{2n-1}\sin(2n-1)\theta$.

**93.** $v = \dfrac{20(l-x)}{l} + \dfrac{24}{\pi}\sum_1^{\infty}\dfrac{(-1)^{n+1}}{n}\sin\dfrac{n\pi x}{l}\, e^{-n^2\pi^2 t/(RCl^2)}$.

$i = \dfrac{20}{lR} + \dfrac{243}{lR}\sum(-1)^n \sin\dfrac{n\pi x}{l}\, e^{-n^2\pi^2 t/(RCl^2)}$.

**97.** $v = v_0\cos(\omega t - \omega x\sqrt{LC})$.

**98.** 200 miles (approx.).

**99.** $\dfrac{\partial^2 u}{\partial r^2} + \dfrac{2}{r}\dfrac{\partial u}{\partial r} + \dfrac{1}{r^2\sin^2\theta}\dfrac{\partial}{\partial\theta}\left(\sin\theta\dfrac{\partial u}{\partial\theta}\right) + \dfrac{1}{r^2\sin^2\theta}\dfrac{\partial^2 u}{\partial\theta^2} + B^2 u = 0.$

$u = \dfrac{c_1}{r}\sin Br + \dfrac{c_2}{r}\cos Br,\quad \dfrac{\partial^2 u}{\partial r^2} + \dfrac{1}{r}\dfrac{\partial u}{\partial r} + \dfrac{1}{r^2}\dfrac{\partial^2 u}{\partial\theta^2} + \dfrac{\partial^2 u}{\partial z^2} + B^2 u = 0.$

## Chapter 10

**1.** 4/3.

**2.** $y = x^3$.

**3.** Meaningless variational problem.

**4.** $y = c_1\sin(4x - c_2)$.

**5.** $(x - c_1)^2 + y^2 = c_2^2$.

**6.** $y = c_1 e^x + c_2 e^{-x} + \frac{1}{2} x e^x$.

**7.** $y = c_1 \cos x + c_2 \sin x - \frac{1}{2} x \cos x$.

**8.** $y = c_1 x + c_2 x^{-2} + \frac{1}{3} x \log x$.

**9.** $y = (c_2 + x) \sin x$, a family of extremals, where $c_2$ is an arbitrary constant.

**10.** $y = \dfrac{\sin hx}{\sin hl}$.

**11.** $y = 7 - \dfrac{4}{x}$.

**12.** The curve is $x = t + 1$ and the minimum value is 2.

**15.** $r = a \sec(\phi \sin \alpha + b)$.

**16.** The curve is a catenary $\rho/k = \cosh (z/k)$ with minimum distance from the axis equal to $k$, and the radius of the soap film at height $z$ is $\rho(z)$ with $\rho\,(\pm b) = a$.

**17.** $y(x) = R(x/l)^{3/4}$ where $y(0) = 0$, $y(l) = R$.

**18.** $y = -\dfrac{\rho}{24\mu}(x^4 - 2l^2 x^2 + l^4)$.

**19.** $y = c_1 e^{2x} + c_2 e^{-2x} + c_3 \cos 2x + c_4 \sin 2x$.

**20.** $y = (c_1 + c_2 x) \cos x + (c_3 + c_4 x) \sin x - \frac{1}{4} x^2 \sin x$.

**21.** $y = c_1 x + c_2$, $z = c_3 x + c_4$.

**22.** $y = \sin x$, $z = -\sin x$.

**23.** $y = 2x$, $z = 4x$.

**24.** The curve is $x(t) = 1$, $0 \le t \le 1$ and the minimum value is 1.

**26.** $(x - c_1)^2 + (y - c_2)^2 = \lambda^2$ – a family of circles.

**27.** $\dfrac{\dot{y}}{[x(1 + \dot{y})]^{1/2}} = \dfrac{A}{2}$.

**28.** $x_1 = \frac{1}{2} t^3 - \frac{7}{4} t^2 + t + 1$, $x_2 = \frac{3}{2} t^2 - \frac{7}{2} t + 1$, and the minimum value is 13/4.

**29.** $x_1 = 4(e^t + e^{-t+1})/(1 + e^2)$, $x_2 = 2(e^t + e^{-t+1})/(1 + e^2)$.

**30.** The curve is $x = t$, $0 \le t \le 1$, and the minimum value is 1.

**31.** $u = 2(9t - 5)$, $0 \le t \le 1$.

**32.** A catenary.

**33.** $c_1 x + c_2 y - z = 0$, this equation of the plane together with the given equation of the sphere gives the arc of the great circle – a geodesic.

**34.** $y = \dfrac{x}{2}$.

**35.** $y = 1 - 6x - 6x^2$.

**36.** $y = -\frac{5}{2}x^2 + \frac{7}{2}x,\ z = x.$

**37.** (i) $\frac{\partial^2 z}{\partial x^2}\left[1 + \left(\frac{\partial z}{\partial y}\right)^2\right] - 2\frac{\partial z}{\partial x}\frac{\partial z}{\partial y}\frac{\partial^2 z}{\partial x \partial y} + \frac{\partial^2 z}{\partial y^2}\left[1 + \left(\frac{\partial z}{\partial x}\right)^2\right] = 0.$

(ii) $\frac{\partial^2 z}{\partial x^2} - \frac{\partial^2 z}{\partial y^2} = 0$ (iii) $\frac{\partial^2 f}{\partial x^2} + \frac{\partial^2 f}{\partial y^2} + \frac{\partial^2 f}{\partial z^2} = g(x, y, z)$

(iv) $\nabla^2\nabla^2 z = f(x, y)$.

# Bibliography


[1] Andronow, A.A. and C.E. Chaikin, *Theory of Oscillations,* Princeton University Press, Princeton (1953).

[2] Arfken, G., *Mathematical Methods for Physists*, 2nd ed., Academic Press, New York (1970).

[3] Braun, M., *Differential Equations and Their Applications*, Springer-Verlag, Berlin (1975).

[4] Chiang, A.C., *Fundamental Methods of Mathematical Economics*, McGraw-Hill, New York (1967).

[5] Christie, D.E., *Vector Mechanics*, McGraw-Hill, New York (1964).

[6] Churchil, R.V., *Operational Mathematics*, 2nd ed., McGraw-Hill, New York (1958).

[7] Elsgolts, L., *Differential Equations and the Calculus of Variations,* Mir Publishers, Moscow (1970).

[8] Goldstein, H., *Classical Mechanics,* 2nd ed., Narosa Publishing House, New Delhi (1993).

[9] Griffith, J.S., *Mathematical Neurobiology: An Introduction to the Mathematics of Nervous System,* Academic Press, New York (1971).

[10] Kaplan, W., *Ordinary Differential Equations*, Addison-Wesley, Reading Mass. (1962).

[11] Keish, B., 'Dating works of arts through their natural radioactivity—improvements and applications', *Science*, **160,** 413–415 (1963).

[12] Kelly, L.M., *Elementary Differential Equations*, 6th ed., McGraw-Hill, New York (1965).

[13] Krishnamurthi, V., V.P. Mainra, and J.L. Arora, *An Introduction to Linear Algebra,* East-West Press, New Delhi (1956).

[14] Lotka, A.J., *Elements of Mathematical Biology*, Dover, New York (1956).

[15] Margenau, H. and G.M. Murphy, *The Mathematics of Physics and Chemistry*, Van Nostrand, Reinhold, New York (1956).

[16] Pathan, M.A. and Zafar Ahsan, *Vector Analysis*, Pragati Prakashan, Meerut (2003).

[17] Rapport, A., *Flight, Games and Debates*, University of Michigan Press, Michigan (1960).

[18] Rashevsky, N., *Mathematical Biophysics*, University of Chicago Press, Chicago (1960).

[19] Redheffer, R.M., *Mathematics of Physics and Modern Engineering*, McGraw-Hill, New York (1960).

[20] Saaty, T.L., *Mathematical Model of Arms Control and Disarmament*, John Wiley & Sons, New York (1968).

[21] Shames, I.H., *Engineering Mechanics*, Prentice Hall, Englewood Cliffs, New Jersey (1966).

[22] Sokolnikoff, I.S., *Mathematical Theory of Elasticity*, 2nd ed., McGraw-Hill, New York (1966).

[23] Thrall, R.M., J.A. Mortioner, K.R. Rebman, and R.F. Braum, *Some Mathematical Models in Biology*, University of Michigan Press, Michigan (1967).

[24] Tenenbaum, M. and A. Pollard, *Ordinary Differential Equations*, Harper & Row, New York (1963).

[25] Tierney, J.A., *Calculus and Analytical Geometry*, 4th ed., Allyn & Bacon, Boston (1979).

[26] Yates, R.C., *Curves and Their Properties*, National Council for Teachers of Mathematics, Reston, Virginia (1974).

[27] Yeh, H. and J.I. Abrams, *Principle of Mechanics of Solids and Fluids*, Vol. 1, Mapel Press, New York (1960).

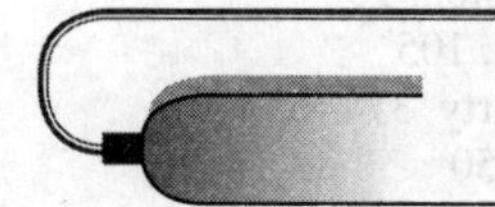

# Index